W9-ACT-561

OPTIMAL CONTROL OF DISTRIBUTED PARAMETER SYSTEMS

OPTIMAL CONTROL OF DISTRIBUTED PARAMETER SYSTEMS

N. U. AHMED
Department of Electrical Engineering
University of Ottawa
Canada

K. L. TEO
Department of Applied Mathematics
The University of New South Wales
Australia

NORTH HOLLAND
New York • Oxford

Elsevier North Holland, Inc.
52 Vanderbilt Avenue, New York, New York 10017

Sole Distributors outside the United States and Canada:
Elsevier Science Publishers B. V.
P.O. Box 211, 1000 AE Amsterdam, The Netherlands

Library of Congress Cataloging in Publication Data

Ahmed, N. U. (Nasir Uddin)
Optimal control of distributed parameter systems.

Bibliography: p.
Includes index.
1. Distributed parameter systems. 2. Control theory.
I. Teo, K. L. II. Title.
QA402.A35 003 81-3923
0-444-00559-5 AACR2

Desk Editor Michael Cantwell
Design Edmée Froment
Design Editor Glen Burris
Production Manager Joanne Jay
Compositor Science Typographers, Inc.
Printer Haddon Craftsmen

Manufactured in the United States of America

In Memory of

Feroza

My grandmother,
Tey Tan

Dedicated to

My uncle and my parents
Masuda
Lisa
Shockley
Mona
Rebeka
Pamela

My parents
Lyanne
James

CONTENTS

PREFACE

A recent, most productive development is the theory of optimal control of distributed parameter systems. There are already two celebrated books in the area. The first one, *Distributed Control Systems*, was written by A. G. Butkovskiy and the second one, by J. L. Lions, is *Optimal Control of Systems Governed by Partial Differential Equations*. Both books were written in the late sixties. Since their publication, there have been numerous developments in the area. Among these are developments in optimal control of distributed parameter systems with controls appearing in the system coefficients, linear and nonlinear hyperbolic systems, general linear evolution equations amenable to semigroup approach, and nonlinear evolution equations on Banach space. We feel that there is a need of an additional book covering some of these recent developments. To keep the size of the book reasonable we have chosen sampled topics of current interest without trying to be exhaustive. Therefore, we have been forced to exclude many other important contributions in the subject.

The book contains five chapters under the following headings:

Chapter One. Mathematical Background

Chapter Two. Partial Differential Equations, Integral Equations, and Evolution Equations

Chapter Three. Optimal Control of Parabolic Partial Differential Systems with Controls in the Coeffeicients

Chapter Four. Optimal Control of Hyperbolic Partial Differential Systems

Chapter Five. Optimal Control of Evolution Equations on Banach Spaces

In Chapter 1 we present necessary background materials from Mathematics to follow the rest of the book. Chapter 2 deals with the background materials on partial differential equations, integral equations, and abstract evolution equations on Banach spaces that are required in the later chapters of the book. Some of the materials of this chapter have not appeared in any previous book. In Chapter 3 we study the problems of optimal control of systems described by parabolic partial differential equations with first boundary conditions. The controls appear in the system coefficients. This class of problems has not been treated in any previous book. Chapter 4 is devoted to the study of optimal control of linear and nonlinear hyperbolic systems with controls appearing either on the boundary or in the coefficients of the system equations. Chapter 5 concerns the problems of optimal control of abstract differential equations on Banach spaces. In this chapter we have considered both linear and nonlinear problems. In the first three sections we have studied linear control problems involving both coercive and noncoercive operators. The results of the first section are included to preserve the continuity of development. In the last section new results on control problems of nonlinear evolution equations have been added.

The prerequisites for reading the book are elements of measure theory and functional analysis usually covered in the senior undergraduate Mathematics courses.

We address ourselves to three categories of readers:

i. graduate students of Mathematics and Physical and Engineering Sciences,
ii. research scientists interested in the applications of optimization theory involving distributed parameter systems, and
iii. control theorists.

Since all the results (in Chapters 3–5) appear with detailed proof, the readers of the first category will find the book extremely useful in their research work. We believe that familiarity with the materials of the book will help the reader to follow current literature without difficulty. Thus, the book can serve as a graduate text for a two-semester course in Mathematics, System Sciences, and Engineering. Since necessary conditions of optimality and several computational methods are discussed, the reader from the second category will find the book useful. The book contains many original results from recent publications and we have also added research notes at the end of each chapter. Thus, the reader of the third category will also find the book interesting.

The authors have consulted the books of Professors A. V. Balakrishnan, J. L. Lions, and A. G. Butkovskiy, to whom they are deeply indebted.

The contents of our book have been influenced directly or indirectly by many outstanding contributors in the field. It is impossible to list all their names. We are greatly indebted to all of them.

We express our sincere thanks to Professor M. N. Oguztöreli for encouragement and many valuable suggestions. The senior author is very grateful to the late Professor George Glinski, who introduced him to the field of system theory. The junior author wishes to acknowledge Dr. D. J. Clements and Mr. Z. S. Wu for their valuable discussion and kind assistance.

We wish to express our appreciation to our colleagues, and acknowledge financial assistance form the University of Ottawa and the University of New South Wales. We would also like to thank Mrs. H. Langley, Mrs. R. Metzker, and Mrs. L. Dooley for typing the manuscript. Finally, we wish to express our appreciation to the editors and staff of Elsevier North Holland for their expert cooperation.

N. U. AHMED
K. L. TEO

OPTIMAL CONTROL OF DISTRIBUTED PARAMETER SYSTEMS

CHAPTER ONE

MATHEMATICAL BACKGROUND

Functional analysis plays a central role in modern control theory. For the convenience of the reader, we summarize, in this chapter, some of the very essential results from this broad area. These results are used in the later chapters, with appropriate references given wherever necessary.

In Section 1.1, the basic references for the results are Dunford and Schwartz [DS.1], Hermes and LaSalle [HL.1], Hille and Phillips [HP.1], Larsen [Lar.1], and Vainberg [V.1]. Most of the results of Sections 1.1.1–1.1.7 are available in the books of Dunford and Schwartz [DS.1] and Larsen [Lar.1]. For the vector-valued functions given in Section 1.1.8, the best reference is Hille and Phillips [HP.1]. For Section 1.1.9, on extremality conditions, a good reference is Vainberg [V.1].

In Section 1.2, most of the results can be found in the book of Ladyžhenskaja, Solonikov, and Ural'ceva [LSU.1]. For distribution theory, used in this section, the reader may consult any standard text on the subject, e.g., Barros-Neto [Bn.1] and Adams [A.1].

In Section 1.3, the best references for semigroup theory are Hille and Phillips [HP.1], Butzer and Berens [BB.1] and Dunford and Schwartz [DS.1]. Most of the results presented in this section can be found in Butzer and Berens [BB.1].

In Section 1.4, most of the results have been collected from recent literature [HJV.1, J.1, KR.1, Pl.1].

1.1. Elements of Functional Analysis

1.1.1. Topological Notions

Let X be a nonempty set. A *topology* $\mathfrak{T}$ for X is a family of subsets of X satisfying the following axioms:

i. X and the empty set $\varnothing \in \mathfrak{T}$.
ii. For arbitrary $F(\neq\varnothing)\subset\mathfrak{T}$, $\bigcup_{A\in F} A\in\mathfrak{T}$.
iii. For any finite set $F\subset\mathfrak{T}$, $\bigcap_{A\in F} A\in\mathfrak{T}$.

The members of $\mathfrak{T}$ are called *open sets* and the pair $(X,\mathfrak{T})$ is called a *topological space*. A subset K of X is a *closed set* if and only if its complement K' is an open set, that is, $K'\in\mathfrak{T}$. An element $G\in\mathfrak{T}$ containing a point x of X is called a *neighborhood* of x.

Let $\mathfrak{T}_1$ and $\mathfrak{T}_2$ be two topologies on X and suppose that every open set of $\mathfrak{T}_1$ is an open set of $\mathfrak{T}_2$. Then the topology $\mathfrak{T}_1$ is said to be *weaker* than the topology $\mathfrak{T}_2$, or, equivalently, the topology $\mathfrak{T}_2$ is said to be *stronger* than the topology $\mathfrak{T}_1$. The stronger topology contains more open sets. If every open set of $\mathfrak{T}_1$ is an open set of $\mathfrak{T}_2$ and, conversely, every open set of $\mathfrak{T}_2$ is an open set of $\mathfrak{T}_1$, then the two topologies are *equivalent*. The topology $\mathfrak{T}$ of a topological space $(X,\mathfrak{T})$ is said to be a *Hausdorff topology* if it satisfies the *separation axiom*: For each pair of distinct points $x_1, x_2\in X$ there exist disjoint open sets $O_1, O_2\in\mathfrak{T}$ such that $x_1\in O_1$ and $x_2\in O_2$.

Let $(X,\mathfrak{T})$ and $(Y,\mathfrak{U})$ be two topological spaces. A function f: $(X,\mathfrak{T})\to(Y,\mathfrak{U})$ is said to be *continuous at the point* $x\in X$ if the inverse image of every $\mathfrak{U}$-neighborhood of the point $f(x)\in Y$ is a $\mathfrak{T}$-neighborhood of the point $x\in X$. The function f: $(X,\mathfrak{T})\to(Y,\mathfrak{U})$ is said to be (*simply*) *continuous* if for every $A\in\mathfrak{U}$, $f^{-1}(A)\in\mathfrak{T}$. We omit $\mathfrak{T}$ if the topology is understood and write X for $(X,\mathfrak{T})$. A sequence $\{x_n\}$ of points of the topological space $(X,\mathfrak{T})$ is said to *converge to a point* $x\in X$, denoted by

$$\lim_{n\to\infty} x_n = x, \quad \text{or} \quad x_n \xrightarrow{\mathfrak{T}} x,$$

if, for each open set $G\in\mathfrak{T}$ containing x, there exists an integer $n_0\geq 0$ such that $x_n\in G$ for all $n>n_0$, that is, G contains all but a finite number of points of the sequence.

Let $(X,\mathfrak{T})$ be a topological space and let A be a nonempty subset of X. The class $\mathfrak{T}_A\equiv\{A\cap B, B\in\mathfrak{T}\}$ of all intersections of A with elements of $\mathfrak{T}$ is a topology for A. This is called the *relative topology of A induced by the topology* $\mathfrak{T}$ *of X*.

Consider a topological space $(X,\mathfrak{T})$, $A\in\mathfrak{T}$, and a family $\mathfrak{G}=\{G_i, G_i\in\mathfrak{T}\}$ of subsets of $\mathfrak{T}$ such that $A\subset\bigcup G_i$. Then $\mathfrak{G}$ is called an *open covering* of A. A is *compact* if every open covering of A contains a finite subfamily that covers A. In other words, if A is compact and $A\subset\cup G_i$, where $G_i\in\mathfrak{T}$, then one can select a finite number of elements from the sequence $\{G_i\}$, say $G_{i_1}, G_{i_2},\ldots, G_{i_m}$, so that $A\subset\bigcup_{k=1}^{m} G_{i_k}$. A is *sequentially compact* if every sequence in A contains a subsequence that converges to a point of A. A is *conditionally sequentially compact* if every sequence in A contains a subsequence that converges to a point of X. A closed subset of a compact set is compact. A compact subset of a Hausdorff topological space is closed. Let $(X,\mathfrak{T})$ be a topological space. A family $\mathfrak{B}$ of open sets of $\mathfrak{T}$ is called a *base* for the topology $\mathfrak{T}$ if the following is satisfied:

i. to each $x \in X$ there corresponds some $B \in \mathfrak{B}$ such that $x \in B$.
ii. if $B_1, B_2 \in \mathfrak{B}$ and $x \in B_1 \cap B_2$, there exists some $B_3 \in \mathfrak{B}$ such that $x \in B_3 \subset B_1 \cap B_2$.

Let $\mathfrak{S}$ be a nonempty collection of elements of $\mathfrak{T}$. Then $\mathfrak{S}$ is a *subbase* for the topology $\mathfrak{T}$ if the collection of all finite intersections of elements of $\mathfrak{S}$ is a base for $\mathfrak{T}$.

If $(X, \mathfrak{T})$ and $(Y, \mathfrak{U})$ are two topological spaces, we define a *topology of the product space* $X \times Y$ by taking as a base the collection of all sets of the form $F \times G$, where F, G are open sets of $\mathfrak{T}$ and $\mathfrak{U}$, respectively.

If h is a one-to-one mapping of X onto Y and if both h and h^{-1} are continuous, then h is called a *homeomorphic mapping*, or a *homeomorphism*. The two topological spaces $(X, \mathfrak{T})$ and $(Y, \mathfrak{U})$ are *equivalent* if there exists a homeomorphism between X and Y.

A topological space is said to be *separable* if it contains a countable dense subset. Let X be a topological space. A subset K of X is said to be *nowhere dense* if its closure $\bar{K}$ has an empty interior; K is called a set of *first category* in X if K is the union of a countable set of nowhere dense sets. If K is not of first category, then K is of *second category*.

1.1.2. Metric Spaces

A *metric* d for a nonempty set X is a real-valued function defined on $X \times X$ such that for all $x, y, z \in X$, (i) $d(x, y) \geq 0$ and $d(x, y) = 0$ if and only if $x = y$, (ii) $d(x, y) = d(y, x)$ (symmetry), and (iii) $d(x, z) \leq d(x, y) + d(y, z)$ (triangle inequality). $d(x, y)$ is called the *distance* between x and y. The collection of open spheres $\{B(x, \varepsilon), x \in X, \varepsilon \geq 0\}$, where $B(x, \varepsilon) = \{y \in X : d(y, x) < \varepsilon\}$, constitutes a base for a topology $\mathfrak{T}_d$ of X. This topology is called the *metric topology* and the set X together with the metric topology $\mathfrak{T}_d$ is called a *metric space* and is written as (X, d) instead of $(X, \mathfrak{T}_d)$. This topology separates points in X, that is, (X, d) is a Hausdorff space and, moreover, it separates disjoint closed sets in X (*normal space*). A metric space is separable if and only if it has a countable base. In a metric space a sequence $\{x_n\}$ converges to a limit point x if and only if $\lim_n d(x_n, x) = 0$; any convergent sequence satisfies the *Cauchy condition*: $d(x_n, x_m) \to 0$ as $n, m \to \infty$. A metric space is said to be *complete* if every Cauchy sequence has a limit point in it.

A homeomorphism between two metric spaces that leaves distance invariant is called an *isometry*. Each incomplete metric space X can be uniquely completed (up to an isometry) into a complete metric space $\tilde{X}$ in which X is isometrically imbedded. A complete metric space is of second category.

A metric space X is said to be *totally bounded* if, for every $\varepsilon > 0$, there is a finite collection of open spheres of radius ε that cover X. A metric space is compact if and only if it is both complete and totally bounded. A compact metric space is separable. In a metric space compactness and sequential compactness are equivalent.

1.1.3. Linear Topological Vector Spaces

Let X be a real or complex linear vector space. A topology $\mathfrak{T}$ on X is said to be *compatible* with the algebraic structure of X if the mappings

i. $(x, y) \to x + y$ of $X \times X$ into X and
ii. $(\alpha, x) \to \alpha x$ of $R \times X$ into X

are continuous. A linear vector space, with a compatible topology that is Hausdorff, is called a *linear topological vector space*. Since the translate of X by a fixed element of X is a homeomorphism of X onto itself, the topology is completely determined by a neighborhood base of the zero element. A topological linear space is called a *locally convex linear topological space* if every neighborhood of the zero element of X contains a convex neighborhood of zero.

1.1.4. Banach Spaces

Let X be a real or complex linear vector space. A real-valued function $x \to \|x\|$ defined on X is called a *norm* if for any $x, y \in X$ and $\lambda \in R/C$ (R and C are the fields of real and complex numbers, respectively),

i. $\|x\| \geq 0$ and $\|x\| = 0$ if and only if $x = 0 \in X$, and
ii. $\|x+y\| \leq \|x\| + \|y\|$ and $\|\lambda x\| = |\lambda| \, \|x\|$.

A linear vector space X for which a norm is defined becomes a metric space if we define $d(x, y) = \|x - y\|$. With the metric, X becomes a locally convex linear topological space and the topology is called the *strong topology* for X. A linear vector space that carries the topology generated by the above metric is called a *normed linear space*, and if with respect to this topology it is complete, then it is called a *Banach space*. A sequence $\{x_n\} \in X$ is said to be *strongly convergent to an element x of X* if $\|x_n - x\| \to 0$ as $n \to \infty$. A closed linear subspace of X is again a Banach space. Let X be a linear vector space and suppose two norms $\|\cdot\|_1$ and $\|\cdot\|_2$ are defined on X. These norms are said to be *equivalent* (or alternatively they define the same topology) on X if there exist two positive numbers c_1 and c_2 such that, for every $x \in X$, $c_1 \|x\|_1 \leq \|x\|_2 \leq c_2 \|x\|_1$.

Let X and Y be two real or complex Banach spaces. A linear transformation T with domain $D(T) \subset X$ (D assumed to be a linear manifold) and range $R(T) \subseteq Y$ is *bounded on* $D(T)$ if there is a constant c such that

$\|Tx\|_Y \le c\|x\|_X$ for all $x \in D$. We call the smallest such constant the *norm of* T and denote it by $\|T\|$. Thus $\|T\| = \sup\{\|Tx\| : x \in D, \|x\| \le 1\}$. A bounded linear operator T on D always has a bounded linear extension $\bar{T}$ on $\bar{D}$ that preserves norm $\|\bar{T}\| = \|T\|$. For a linear operator T on $D \subset X$ to Y the following three statements are equivalent: (i) T is bounded on D; (ii) T is continuous at a point of D; (iii) T is uniformly continuous on D in the sense that for any $\varepsilon > 0$ there exists a $\delta = \delta(\varepsilon) > 0$ such that $\|Tx_1 - Tx_2\| < \varepsilon$ whenever $\|x_1 - x_2\| < \delta$, $x_1, x_2 \in D$. The linear (vector) space of all bounded linear operators from X into Y (X, Y Banach), denoted by $\mathfrak{L}(X, Y)$, forms a Banach space. If T is a linear operator from $D \subset X$ *onto* $R \subset Y$, then the inverse T^{-1} exists and is bounded if and only if there exists a constant $c > 0$ such that $\|Tx\| \ge c\|x\|$ for all $x \in D$.

A linear transformation T on $D(T) \subset X$ to Y is said to be *closed* if its graph $\gamma(T) \equiv \{(x, Tx) : x \in D\}$ is a closed subspace of $X \times Y$: in other words, whenever $x_n \xrightarrow{s} x$ (strongly) in X, $x_n \in D$, and $Tx_n \xrightarrow{s} y$ in Y, it follows that $x \in D$ and $y = Tx$. A linear operator T from $D \subset X$ *into* Y has a closed extension if and only if no element of the form $(0, y)$, $y \ne 0$, belongs to the closure of the graph $\gamma(T)$ in $X \times Y$, where $X \times Y$ is given any suitable norm, for example, $\|(x, y)\|_{X \times Y} = \|x\|_X + \|y\|_Y$. In this case the closure of the graph $\gamma(T)$ defines the smallest closed linear extension of T. A bounded linear transformation on a closed domain $D \subset X$ to Y is closed. The inverse of a closed linear operator, if it exists, is closed.

Theorem 1.1.1 (Open Mapping Theorem). *A continuous linear operator T from X* onto *Y maps open sets into open sets and, in particular, if T is one-to-one, it is an isomorphism.*

Theorem 1.1.2 (Closed Graph Theorem). *A closed linear transformation on X to Y is continuous. A continuous linear transformation from X to Y is closed.*

Theorem 1.1.3 (Uniform Boundedness Principle). *Let $\{T_\alpha, \alpha \in \Lambda\}$ be a family of operators from $\mathfrak{L}(X, Y)$. If for each $x \in X$ there is a constant c_x such that* $\sup\{\|T_\alpha x\|, \alpha \in \Lambda\} \le c_x$, *then the operators $\{T_\alpha\}$ are uniformly bounded.*

Theorem 1.1.4 (Banach–Steinhaus Theorem). *Let $\{T_n\} \in \mathfrak{L}(X, Y)$ such that (i) $\|T_n\| \le C$ for all n and (ii) $\lim_n T_n x$ exists for a dense subset of X. Then $\lim_n T_n x$ exists for all $x \in X$ and the limit defines a linear operator T with $\|T\| \le \underline{\lim}_n \|T_n\|$.*

By a real or complex *Hilbert space H* we mean a real or complex Banach space for which there is defined on $H \times H$ a real- or complex-valued function (x, y), called the *scalar product of x and y*, with the following properties: (i) $(\alpha_1 x_1+\alpha_2 x_2, y)=\alpha_1(x_1, y)+\alpha_2(x_2, y)$; (ii) $(x, y)=(y, x)$ [or $\overline{(y, x)}$, the complex conjugate of (x, y)], and (iii) $(x, x)=\|x\|^2$.

1.1.5. Linear Functionals and Dual (or Conjugate) Spaces

The fundamental result on linear functionals is the *Hahn–Banach extension theorem*, which states that given a bounded linear functional f defined on a linear manifold M of a Banach space X, there exists a bounded linear functional f^* defined on all of X such that $f^*(x)=f(x)$ for all $x\in M$ and the norm of f^* on X equals the norm of f on M. That is,

$$\|f^*\|\equiv\sup\{|f^*(x)|, x\in X, \|x\|\leq 1\}=\|f\|\equiv\sup\{|f(x)|, x\in M, \|x\|\leq 1\}.$$

The Hahn–Banach theorem has many important consequences in functional analysis and is, in fact, regarded as the cornerstone of functional analysis. One of the many consequences of the Hahn–Banach theorem is the existence and abundance of bounded linear functionals on Banach spaces. For each point $x_0\in X$, $x_0\neq 0$, there exists a bounded linear functional f on X such that $f(x_0)=\|x_0\|$ and $\|f\|=1$. Given a linear manifold M of X and a point $x_0\in X$ at a positive distance $d=\inf\{\|x_0-x\|: x\in M\}$ from M, there exists a bounded linear functional f on X such that (i) $f(M)=0$, (ii) $f(x_0)=1$, and (iii) $\|f\|=1/d$. A bounded linear functional that vanishes on a dense linear subset of X vanishes on all of X. If x_1, x_2 are any two distinct points of X, then there exists a bounded linear functional f on X such that $f(x_1)\neq f(x_2)$. In other words, there are enough bounded linear functionals on X to distinguish points of X: that is, if $f(x)=0$ for all bounded linear functionals in X, then $x=0$.

The space X^* of all bounded linear functionals on X is called the *dual* (*conjugate* or *adjoint*) *space of* X. Since $\mathfrak{L}(X, Y)$ is a Banach space whenever Y is a Banach space and X a normed space, for $Y=R/C$, we have $\mathfrak{L}(X, Y)=\mathfrak{L}(X, R/C)=X^*$, and hence X^* is a Banach space. The values of the functional $f\in X^*$ at the points $x\in X$ are denoted by either $f(x)$ or $\langle f, x\rangle$. It is clear that for $f\in X^*$, $x_1, x_2\in X$, and $\alpha, \beta\in R$, $\langle f, \alpha x_1+\beta x_2\rangle=\alpha\langle f, x_1\rangle+\beta\langle f, x_2\rangle$; and similarly for $f_1, f_2\in X^*$ and $x\in X$ we have $\langle\alpha f_1+\beta f_2, x\rangle=\alpha\langle f_1, x\rangle+\beta\langle f_2, x\rangle$ and $|\langle f, x\rangle|\leq\|f\|_{X^*}\|x\|$. We may consider $\langle\cdot,\cdot\rangle$ to be a bilinear form on $X^*\times X$. The set of all linear functions on X (not necessarily bounded), denoted by X', is called the *algebraic dual* of X; and X^*, which consists of all continuous linear functionals on X, is called the (*topological*) *dual*. Clearly $X^*\subset X'$.

The dual of X^*, also known as the *second dual of* X, is denoted by X^{**}. It is convenient to denote elements of X by $\{x\}$ and those of X^* and X^{**} by $\{x^*\}$ and $\{x^{**}\}$, respectively. We have seen that an element x^* of X^* defines a continuous linear functional with values $\langle x^*, x\rangle=x^*(x)$ at $x\in X$. For a fixed $x\in X$, it is clear that the bilinear form also defines a continuous linear functional on X^* and we can write it as $x^*(x)\equiv J_x(x^*)$. The correspondence $x \to J_x$ from X into X^{**} is called the *canonical map*. Define $X_0^{**} \equiv \{x^{**} \in X^{**} : x^{**} = J_x$ for some $x \in X\}$. It is clear that X_0^{**} is a linear manifold of X^{**}. Since $\|J_x\| = \|x\|$ and X^* separates points of X, the canonical map $x \to J_x$ from X into X^{**} is one-to-one, norm preserving, and thus an isometric isomorphism of X *onto* X_0^{**}. In this sense we may regard X as a subset of X^{**}. If, under the canonical embedding, $X = X^{**}$, then X is called *reflexive*. If X is reflexive, then so is X^*, and a closed linear manifold of a reflexive space is also reflexive. If X^* is separable, so is X. If X is a Hilbert space, then, for each bounded linear functional x^* on X, there exists a $y \in X$ such that $x^*(x) = (y, x)$ for all $x \in X$. Further, $\|x^*\| = \|y\|$, and thus the dual X^* is isometrically isomorphic to X. Thus every Hilbert space is a reflexive Banach space.

1.1.6. Weak and Weak Star Topologies and Operator Topologies

Let X be a Banach space and X^* its dual. Elements of X^* can be used to generate a new topology for X called the *weak topology*. Note that the norm topology on X was called the strong topology. The weak topology is obtained by taking as a *base* all sets (neighborhoods) of the form $N(x_0, F^*, \varepsilon) \equiv \{y \in X : |x^*(y - x_0)| < \varepsilon, x^* \in F^*\}$, where $x_0 \in X$, F^* is any finite subset of X^*, and $\varepsilon > 0$. Let $\mathfrak{T}_w$ denote the weak topology. Endowed with the topology $\mathfrak{T}_w$, X becomes a locally convex linear topological vector space, denoted by $(X, \mathfrak{T}_w)$ or X. The topology $\mathfrak{T}_w$ is weaker (coarser) than the strong (norm) topology. Particularly, the linear functionals on X that are continuous in the weak topology are precisely the functionals in X^*.

The concepts of open (closed) sets, compactness, convergence, etc., are topological, hence they must be qualified by referring to the topology involved. In the case of normed linear spaces, when one speaks of open (closed) sets, compactness, convergence, etc., one refers to the strong (norm) topology, while, with reference to its weak topology, they are called weakly open (weakly closed) sets, weak compactness, weak convergence, etc. Thus a sequence $\{x_n\} \in X$ is said to converge weakly to an element $x \in X$, written $x_n \xrightarrow{w} x$ or $x_n \xrightarrow{\mathfrak{T}_w} x$, if, for every $x^* \in X^*$, $x^*(x_n) \to x^*(x)$. Every weakly convergent sequence is bounded. Every strongly convergent sequence is weakly convergent, but the converse is not true. Every weakly closed set is strongly closed, but the converse need not be true unless some additional conditions are satisfied.

Theorem 1.1.5 (Mazur). *A* convex *subset of a normed linear space X is weakly closed if and only if it is strongly closed.*

Corollary 1.1.1 (Banach–Saks–Mazur). *Let X be a normed linear space and $\{x_n\}$ a sequence in X converging weakly to x_0. Then there exists a finite convex combination of $\{x_n\}$ that converges strongly to x_0. In other words,* *for every integer* n_0 *and* $\varepsilon>0$, *there exists a finite set of real numbers* $\{\alpha_i\}$, $\alpha_i \geq 0$, $\Sigma\alpha_i = 1$, *such that*

$$\left\|\sum \alpha_i x_{n_0+i} - x_0\right\| < \varepsilon.$$

Theorem 1.1.6 (Eberlein–Šmulian) [DS.1, Theorem 1, p. 430]. *A subset of a Banach space X is weakly compact if and only if it is weakly sequentially compact.*

A Banach space X is said to be weakly complete if every weak Cauchy sequence has a weak limit point in X. Every reflexive Banach space is weakly complete.

Theorem 1.1.7. *A subset of a reflexive Banach space is conditionally weakly compact if and only if it is (strongly) bounded; and it is weakly compact if and only if it is bounded and weakly closed.*

Theorem 1.1.8 (Banach–Dunford) [DS.1, Theorem 15, p. 422]. *Let T be a linear operator from a Banach space X into a Banach space Y. Then T is continuous with respect to the strong topologies in X and Y if and only if it is continuous with respect to the weak topologies.*

Any dual space X^*, of a Banach space X, has at least three topologies: (i) the strong (norm) topology; (ii) the weak topology, that is, the topology generated by X^{**} on X^*; and (iii) the topology induced by X on X^*. Since under the canonical mapping $X \subset X^{**}$, the topology induced by X on X^* is weaker than the weak topology. This topology is called the *weak** (w*-) *topology* and is sometimes denoted by $\mathfrak{T}_{w^*}$. The system of neighborhoods of the form $N(x_0^*, F, \varepsilon) \equiv \{x^* \in X^* : |x^*(x) - x_0^*(x)| < \varepsilon, x \in F\}$, where $x_0^* \in X^*$, F is a finite subset of X, and $\varepsilon > 0$, constitutes a base for the topology $\mathfrak{T}_{w^*}$. It is easy to see that $\mathfrak{T}_{w^*}$ is a Hausdorff topology and, equipped with this topology, X^* becomes a locally convex linear topological vector space $(X^*, \mathfrak{T}_{w^*})$.

The set of all continuous linear functionals on $(X^*, \mathfrak{T}_{w^*})$ consists precisely of the elements of X. A sequence $\{x_n^*\} \subset X^*$ is said to converge in the w*-topology to a point $x^* \in X^*$, denoted by $x_n^* \xrightarrow{w^*} x^*$ or $x_n^* \xrightarrow{\mathfrak{T}_{w^*}} x^*$, if, for every $x \in X$, $x_n^*(x) \to x^*(x)$. A w*-convergent sequence is bounded in the norm topology of X^*, that is, $\sup_n \|x_n^*\| < \infty$. A linear manifold $M^* \subset X^*$ is w*-closed if and only if for each $x_0^* \notin M^*$, there exists an $x_0 \in X$ such that $x_0^*(x_0) = 1$ and $x^*(x_0) = 0$ for all $x^* \in M^*$.

Theorem 1.1.9 (Alaoglu). *The closed unit sphere of* X^* *is compact in the w*-topology.*

Theorem 1.1.10. *A subset of* X^* *is* w*-*compact if and only if it is (strongly) bounded and* w*-*closed.*

Theorem 1.1.11 (Geometric Hahn–Banach theorem). *Let $(X, \mathfrak{T})$ be a real or complex linear topological vector space and let $C_1, C_2 \subset X$ be nonempty disjoint convex sets.*

i. *If the interior of C_1, written $\operatorname{int} C_1$, is nonempty, then there exists a closed real hyperplane $L = \{x \in X : x'(x) = a\}$, for some $x' \in X^*$ and $a \in R$, that separates C_1 and C_2.*
ii. *If C_1 and C_2 are open, then there exists a closed real hyperplane that strictly separates C_1 and C_2.*

Corollary 1.1.2. *Let $(X, \mathfrak{T})$ be a locally convex real or complex linear topological vector space, C a nonempty closed convex subset of X, and $x_0 \in X$. If $x_0 \notin C$, then there exists a closed real hyperplane L that strictly separates C and $\{x_0\}$.*

Corollary 1.1.3. *Let B be a normed linear space, C a closed convex subset of B, and $x_0 \notin C$. Then there exists an $f^* \in B^*$, a constant α, and an $\varepsilon > 0$ such that*

$$f^*(x_0) \geq \alpha > \alpha - \varepsilon \geq f^*(y) \quad \text{for all } y \in C.$$

Theorem 1.1.12. *Let C_1, C_2 be convex subsets of the real linear topological vector space $(X, \mathfrak{T})$ and suppose $\operatorname{int} C_1 \neq \varnothing$. Then C_1 and C_2 can be separated by a closed real hyperplane if and only if $\operatorname{int}(C_1) \cap C_2 = \varnothing$.*

Corollary 1.1.4. *A closed convex subset with nonempty interior of a real linear topological vector space is supported at every boundary point by a closed real hyperplane.*

Theorem 1.1.13. *The* w*-*topology of the closed unit sphere of the dual space X^* for a Banach space X is a metric topology if and only if X is separable.*

Theorem 1.1.14 (Krein–Šmulian) [DS.1, Theorem 9, p. 939]. *The closed convex hull of a weakly compact subset of a Banach space is weakly compact.*

Let X, Y be normed linear spaces, and let $\mathfrak{L}(X, Y)$ denote the collection of all bounded linear operators from X into Y. $\mathfrak{L}(X, Y)$ is a normed linear space. The space $\mathfrak{L}(X, Y)$ can be given many different topologies out of which those that are most interesting for applications are the *uniform operator topology*, *the strong operator topology*, and the *weak operator topology*.

The uniform operator topology, denoted by $\mathfrak{T}_{uo}$, is the topology determined by the operator norm $\|T\| = \sup\{\|Tx\|_Y, x \in X, \|x\|_X \leq 1\}$. If Y is a Banach space, then $\mathfrak{L}(X, Y)$ is also a Banach space with respect to the uniform operator topology.

For the strong operator topology define the system of neighborhoods of the zero element of $\mathfrak{L}(X, Y)$ as $N(0, F, \varepsilon) = \{T \in \mathfrak{L}(X, Y) : \|Tx\| < \varepsilon, x \in F\}$, where F is any finite subset of X and $\varepsilon > 0$. The strong operator topology, denoted by $\mathfrak{T}_{so}$, is determined by the system of neighborhoods $\{N(0, F, \varepsilon)$, F finite subset of X, $\varepsilon > 0\}$ that constitutes a base for the topology. Endowed with this topology, $(\mathfrak{L}(X, Y), \mathfrak{T}_{so})$ is a locally convex linear topological vector space. A sequence $\{T_n\} \subset \mathfrak{L}(X, Y)$ is said to converge in the strong operator topology to an element T, denoted by s: $\lim_n T_n = T$, or $T_n \overset{\mathfrak{T}_{so}}{\to} T$, if, for every $x \in X$, $T_n x \overset{s}{\to} Tx$ in Y. Every sequence $\{T_n\}$ that converges in the strong operator topology, is bounded (a consequence of uniform boundedness principle). If Y is a Banach space, then $\mathfrak{L}(X, Y)$, equipped with the topology $\mathfrak{T}_{so}$, is sequentially complete.

For the weak operator topology define the system of neighborhoods of the zero element of $\mathfrak{L}(X, Y)$ as

$$N(0, F, F^*, \varepsilon) \equiv \{T \in \mathfrak{L}(X, Y) : |\langle y^*, Tx \rangle| < \varepsilon, x \in F, y \in F^*\},$$

where F and F^* are finite subsets of X and Y^*, respectively, and $\varepsilon > 0$. The weak operator topology, denoted by $\mathfrak{T}_{wo}$, is determined by the system of neighborhoods $\{N(0, F, F^*, \varepsilon)$, F, F^* finite subsets of X and Y^*, and $\varepsilon > 0\}$ that forms a base for the topology. A sequence $\{T_n\} \in \mathfrak{L}(X, Y)$ is said to converge in the weak operator topology to T if, for every $x \in X$ and $y^* \in Y^*$, $\langle y^*, T_n x \rangle \to \langle y^*, Tx \rangle$. If Y is weakly sequentially complete, then $(\mathfrak{L}(X, Y), \mathfrak{T}_{wo})$ is a sequentially complete locally convex linear topological space.

It is clear from the above discussion that the uniform operator topology is stronger than the strong operator topology, which, in turn, is stronger than the weak operator topology.

1.1.7. Extremal Sets, Extremal Points and the Krein–Milman Theorem

Let K be a subset of a real or complex vector space X. A nonempty subset E of K is said to be an *extremal subset of* K if a proper convex combination $\alpha x_1 + (1-\alpha)x_2$, $0 < \alpha < 1$, of two points x_1, x_2 of K lies in E only if both x_1 and x_2 are in E. In other words, a subset E of K is an extremal subset if, whenever it contains an interior point of a line segment in K, the entire line segment lies in E.

An extremal subset of K consisting of just one point is called an *extreme* or *extremal point of* K. In other words, an extremal point of K is any point in K that is not an interior point of a line segment in K. The set of all extremal points of K will be denoted by $\text{ext}(K)$.

The sides and corners of a rectangle in the plane are extremal subsets of the rectangle, but only the corners are extremal points. The surface of the closed unit ball in R^n is an extremal subset of the ball, and every point on it is an extreme point. The vertices of an n-simplex are the extremal points of the simplex and the faces are the extremal subsets.

Theorem 1.1.15 (Krein–Milman) [DS.1, Theorem 4, p. 440]. *A nonempty compact convex subset K of a locally convex linear topological space X has at least one extremal point and* $\mathrm{cl}\,\mathrm{co}(\mathrm{ext}\,K)=K$, *that is, the closed convex hull of the extremal points of K equals K.*

Let B_1^p denote the closed unit ball of an L_p-space and S_1^p the surface of the ball B_1^p. For $1<p<\infty$, $\mathrm{ext}(B_1^p)\neq\varnothing$ and $\mathrm{ext}(B_1^p)=S_1^p$; for $p=1$, $\mathrm{ext}(B_1^1)=\varnothing$; but for $p=\infty$, $\mathrm{ext}(B_1^\infty)\neq\varnothing$ and $\mathrm{ext}(B_1^\infty)=\{$the set of characteristic functions$\}$. Note that, for $1<p<\infty$, B_1^p is compact (in the weak topology) but B_1^1 is not. Similarly, B_1^∞ is (w*-) compact. Thus compactness is an essential condition for the Krein–Milman theorem. By Alaoglu's theorem we know that the closed unit ball of X^* is always (w*-) compact and hence $\mathrm{ext}(B_1^\infty)\neq\varnothing$. This also shows that L_1-space is not the dual of any locally convex linear topological space.

1.1.8. Vector-Valued Functions

Let X be a Banach space with dual X^* and suppose I is an open interval in R. A mapping x from I into X is called a *vector-valued function*. The vector-valued function x is said to be *strongly continuous at a point* $t_0\in I$ if

$$\operatorname*{s\,lim}_{t\to t_0} x(t)=x(t_0), \quad \text{or equivalently} \quad \lim_{t\to t_0}|x(t)-x(t_0)|_X=0;$$

and it is said to be *weakly continuous at* $t_0\in I$ if

$$\operatorname*{w\,lim}_{t\to t_0} x(t)=x(t_0), \quad \text{or equivalently} \quad \lim_{t\to t_0}\langle x^*, x(t)\rangle=\langle x^*, x(t_0)\rangle$$

for each $x^*\in X^*$. The vector-valued function x is strongly (weakly) continuous on I if it is strongly (weakly) continuous at each point in I. It is clear that a vector-valued function is weakly continuous on I if and only if the scalar-valued function $t\to\langle x^*, x(t)\rangle$ is continuous on I for each $x^*\in X^*$. If the function $t\to\langle x^*, x(t)\rangle$ is continuous on I, then it is bounded on any compact subset of I and hence, due to uniform boundedness principle (Theorem 1.1.3), $|x(t)|_X$ is bounded on any compact subset of I.

Let Y be another Banach space such that $X\subseteq Y$. A vector-valued function x defined on I with values in X is said to be *strongly differentiable on I* if, for each $t_0\in I$, there exists an element $y(t_0)\in Y$ such that

$$\operatorname*{s\,lim}_{h\to 0}\left[x(t_0+h)-x(t_0)/h\right]=y(t_0),$$

or equivalently

$$\lim_{h\to 0}\left|\frac{x(t_0+h)-x(t_0)}{h}-y(t_0)\right|_Y=0;$$

and it is said to be *weakly differentiable on* I if, for each $t_0 \in I$, there exists an element $y(t_0) \in Y$ such that

$$\lim_{h \to 0} \left\langle y^*, \quad \frac{x(t_0+h)-x(t_0)}{h} - y(t_0) \right\rangle = 0 \quad \text{for each } y^* \in Y^*.$$

If x is weakly differentiable on I, then the scalar-valued function $t \to \langle y^*, x(t) \rangle$ is differentiable on I for each $y^* \in Y^*$, but the converse is not generally true.

If x is weakly differentiable on I and the weak derivative equals zero on I, then $x(t)$ is a constant vector-valued function.

Let X, Y be any two arbitrary Banach spaces and $\mathcal{L}(X, Y)$ the space of bounded linear operators from X into Y. An operator-valued function, $t \to T(t)$, defined on I and taking values in $\mathcal{L}(X, Y)$ is said to be *continuous in the uniform operator topology* if, for each $t_0 \in I$, $\lim_{t \to t_0} \|T(t) - T(t_0)\| = 0$. It is said to be *continuous in the strong operator topology* if for each $t_0 \in I$ and $x \in X$, $\lim_{t \to t_0} |T(t)x - T(t_0)x|_Y = 0$. It is said to be *continuous in the weak operator topology if* for each $t_0 \in I$, $x \in X$, and $y^* \in Y^*$, $\lim_{t \to t_0} \langle y^*, T(t)x \rangle = \langle y^*, T(t_0)x \rangle$.

For the differentials of operator-valued functions, we again have three possibilities depending on whether the incremental ratio converges to the derivative in the uniform, strong, or weak operator topology.

Let $(X, \mathfrak{T})$ be a linear topological vector space and x a function on $I = [a, b]$ with values in X. Let $C(I, X_{\mathfrak{T}})$ denote the space of functions on I with values in X and continuous in the topology $\mathfrak{T}$. Let π denote the partition of the interval $I \equiv [a, b]$, $a = t_0 < t_1 \leq \cdots \leq t_n = b$, together with points $\{s_\kappa \in [t_{\kappa-1}, t_\kappa], \kappa = 1, 2, \ldots, n\}$, and let $|\pi| \equiv \max_{1 \leq \kappa \leq n}(t_\kappa - t_{\kappa-1})$ denote the diameter of the partition π. Define $s_\pi(x) = \sum_{\kappa=1}^{n} x(s_\kappa) \cdot (t_\kappa - t_{\kappa-1})$. If $s_\pi(x)$ converges in the topology $\mathfrak{T}$ to an element $\mathcal{G}(a, b; x)$, then we define the limit $\mathcal{G}(a, b; x)$ to be the *Riemann integral*

$$\mathcal{G}(a, b; x) \equiv \int_a^b x(t)\, dt$$

of x *on* $I = [a, b]$ *with respect to the topology* $\mathfrak{T}$. Thus if X is a Banach space and $\mathfrak{T}$ the strong (norm) topology, then we obtain the definition for *strong Riemann integral*. If $\mathfrak{T}$ is the weak topology, we obtain the *weak Riemann integral*, and if X is the dual of some other Banach space, so that $\mathfrak{T}$ is the w*-topology, then we have w*-*Riemann integral*.

The strong Riemann integral satisfies all the basic properties that integrals of scalar-valued functions satisfy, namely,

i. $\mathcal{G}(a, b; \alpha_1 x_1 + \alpha_2 x_2) = \alpha_1 \mathcal{G}(a, b; x_1) + \alpha_2 \mathcal{G}(a, b; x_2)$,
ii. $\mathcal{G}(a, b; x) = \mathcal{G}(a, c; x) + \mathcal{G}(c, b; x)$ for $c \in [a, b]$,
iii. $|\mathcal{G}(a, b; x)|_X \leq (b-a) \cdot \max\{|x(t)|_X, t \in [a, b]\}$,
iv. $t \to \mathcal{G}(a, t; x)$ is strongly differentiable and the derivative equals $x(t)$,
v. if $x_n(t) \overset{s}{\to} x(t)$ in X uniformly on I, then $\mathcal{G}(a, b; x_n) \overset{s}{\to} \mathcal{G}(a, b; x)$.

If A is a closed operator with domain $D(A)\subset X$ and range in Y (a Banach space), x is any vector-valued function with values $x(t)\in D(A)$ for all $t\in I$, and $Ax(t)$ is strongly continuous on I to Y, then $A\mathcal{G}(a, b; x)=\mathcal{G}(a, b; Ax)$.

We are also interested in measurable vector-valued functions. Let X be a real or complex Banach space and I a finite Lebesgue measurable subset of R. Let x be a vector-valued function defined on I with values $x(t)\in X$. x is said to be a *simple function* if there exists a partition of I into a finite number of disjoint measurable subsets $\{I_j\}$ of I such that x is a constant (vector) function on each of the sets I_j. x is *countably valued* if it assumes at most a countable set of values in X, assuming each value different from zero on a measurable subset. x is said to be *separably valued* if its range $x(I)$ is separable, and it is *almost separably valued* if there exists a null set I_0 such that $x(I\setminus I_0)$ is separable. x is said to be a *strongly measurable* function if there exists a sequence $\{x_n\}$ of simple functions converging strongly almost everywhere (a.e.) on I to x, and it is said to be *weakly measurable* if the scalar-valued function $t\to\langle x^*, x(t)\rangle$ is measurable for each $x^*\in X^*$.

Theorem 1.1.16. *A vector-valued function is strongly measurable if and only if it is weakly measurable and almost separably valued.*

As a consequence, if x is strongly measurable, then the scalar-valued function $t\to|x(t)|_X$ is measurable.

Corollary 1.1.5. *If X is separable, then strong and weak measurability are equivalent.*

Strongly measurable vector-valued functions have properties analogous to those of measurable scalar-valued functions. If x is the strong limit a.e. of a sequence $\{x_n\}$ of strongly measurable functions, then x is strongly measurable. If x is the weak limit a.e. of a sequence of strongly measurable functions, then x is also strongly measurable. These results are also valid if *limit* a.e. is replaced by *limit in measure*.

If x is weakly measurable and for each $x^*\in X^*$, $t\to x^*(x(t))$ is integrable, then there exists an $x^{**}\in X^{**}$ such that

$$x^{**}(x^*)=\int_I x^*(x(t))\,dt \quad \text{for all } x^*\in X^*.$$

As a consequence we can set

$$x^{**}=\int_I x(t)\,dt.$$

In general, x^{**} can not be replaced by an element of X; if such a replacement can be made, then the integral is called the *Pettis integral*. A function x from I into X is said to be *Pettis integrable* if and only if there exists an element $e \in X$ such that

$$x^*(e) = \int_I x^*(x(t))\,dt \quad \text{for all } x^* \in X^*,$$

where the integral on the right-hand side is assumed to exist in the Lebesgue sense. Then e is the Pettis integral of x and it is denoted by

$$e = P\int_I x(t)\,dt.$$

A stronger version of the integral is known as the *Bochner integral*. If x is a simple function from I into X, that is, if $x(t) = \Sigma x_\kappa C_{I_\kappa}(t)$, C_{I_κ} being the characteristic function of the set I_κ, we define the Bochner integral of x over a measurable set J by

$$B\int_J x(t)\,dt = \sum x_\kappa l(J \cap I_\kappa),$$

where l denotes the Lebesgue measure on I. In general, a function x on I to X is said to be *Bochner integrable* if there exists a sequence of simple functions $\{x_n\}$ on I to X that converges strongly to x a.e. on I in such a way that

$$\lim_n \int_I |x(t) - x_n(t)|_X\,dt = 0.$$

Thus, by definition, the Bochner integral of x over any measurable set $J \subset I$ is

$$B\int_J x(t)\,dt = \mathrm{s}\lim_n B\int_J x_n(t)\,dt.$$

A necessary and sufficient condition that x be Bochner integrable is that x is strongly measurable and

$$\int_I |x(t)|_X\,dt < \infty.$$

We denote the set of all Bochner integrable functions on I to X by $B(I, X)$. If X is the field of scalars, the Bochner integral reduces to the usual Lebesgue integral. $B(I, X)$ becomes a linear vector space under the natural definition of addition and scalar multiplication.

The integral $B\int_J x(t)\,dt$ for J any measurable set defines a linear transformation from $B(I, X)$ into X. Moreover, the following properties hold for the Bochner integral:

i. If $x \in B(I, X)$, then

$$\left| B\int_J x(t)\,dt \right|_X \leq \int_J |x(t)|\,dt, \qquad J \text{ measurable.}$$

ii. If

$$\{x_n\} \in B(I, X) \quad \text{and} \quad \lim_{n, m \to \infty} \int_I |x_n(t) - x_m(t)|_X \, dt = 0,$$

then there exists an $x \in B(I, X)$, uniquely determined except on a set of measure zero, such that

$$\lim_{n \to \infty} \int_I |x_n(t) - x(t)|_X \, dt = 0.$$

iii. If $\{x_n\} \in B(I, X)$ and $x_n(t) \xrightarrow{s} x(t)$ in X a.e. on I, and if there exists a scalar function $g \in L_1$ such that $|x_n(t)| \leq g(t)$ a.e. for all integers $n \geq 1$, then $x \in B(I, X)$ and

$$B \int_1 x(t) \, dt = \operatorname{s} \lim_n B \int_I x_n(t) \, dt.$$

This is the dominated convergence theorem.

iv. If $\{I_\kappa\}$ is any sequence of disjoint measurable subsets of I, then for any $x \in B(I, X)$

$$B \int_I x(t) \, dt \equiv B \int_{\cup I_j} x(t) \, dt = \sum_j \left(B \int_{I_j} x(t) \, dt \right),$$

where the sum on the right is absolutely convergent.

v. If $x \in B(I, X)$, then the integral $B \int_J x(t) \, dt$ is absolutely continuous with respect to the Lebesgue measure, that is,

$$\lim_{l(J) \to 0} B \int_J x(t) \, dt = 0.$$

If we identify strongly measurable functions of I to X that differ only on sets of measure zero, the linear space $B(I, X)$ becomes a Banach space, and we denote it by $L_1(I, X)$.

In general, the above results are valid for any complete σ-finite measure space (Ω, β, μ). The space $L_1(\Omega, \beta, \mu; X)$, denoted by $L_1(\Omega, X)$, is a Banach space. Similarly, the equivalence classes of strongly measurable X-valued functions on Ω such that

$$\int_\Omega |x(t)|_X^p \, dt < \infty \quad \text{for } 1 \leq p < \infty$$

and

$$\operatorname{ess\,sup} \{ |x(t)|_X, \, t \in \Omega \} < \infty, \quad \text{for } p = \infty$$

form Banach spaces with respect to the norms

$$\|x\| \equiv \left(\int_\Omega |x(t)|_X^p \, dt \right)^{1/p} \quad \text{and} \quad \|x\| \equiv \operatorname{ess\,sup} \{ |x(t)| : t \in \Omega \},$$

respectively. They are denoted by $L_p(\Omega, X)$, $1 \leq p \leq \infty$.

Theorem 1.1.17. *If X is a reflexive Banach space with dual X^* and $1<p<\infty$, then $L_p(\Omega, X)$ is also a reflexive Banach space, and its dual is given by $L_q(\Omega, X^*)$, where $q=p/(p-1)$.*

Theorem 1.1.18. *Let Ω be any measurable subset of R^n and suppose $x\in L_p(\Omega, X)$. Then*

$$\lim_{|h|\to 0}\int_\Omega |x(t+h)-x(t)|_X^p\,dt=0.$$

Theorem 1.1.19. *Let Ω be a measurable subset of R^n with l denoting the Lebesgue measure on Ω. Then for each $x\in L_p(\Omega, X)$ and for almost all $t\in\Omega$,*

$$\lim_{\kappa\to\infty}\frac{1}{l(\Omega_\kappa)}\int_{\Omega_\kappa}|x(\theta)-x(t)|_X^p\,d\theta=0$$

whenever $\{\Omega_\kappa\}$ is a decreasing sequence of measurable subsets of Ω such that $t\in\Omega_\kappa$ for all κ and $\lim_{\kappa\to\infty}l(\Omega_\kappa)=0$. In particular,

$$\lim_{\kappa\to\infty}\frac{1}{l(\Omega_\kappa)}B\int_{\Omega_\kappa}x(\theta)\,d\theta=x(t)\quad\text{for almost all } t\in I.$$

Since, in the rest of the book, integrals of vector-valued functions are understood in the sense of Bochner, unless specifically mentioned, we shall write

$$\int_I x(t)\,dt\quad\text{instead of } B\int_I x(t)\,dt.$$

1.1.9. Conditions for Extremality

We close this section by stating a few classical results from the theory of extremals. A real-valued function f defined on a topological vector space X is said to be *convex* (strictly convex) if, for $x, y\in X$,

$$f(\alpha x+(1-\alpha)y)\le(<)\alpha f(x)+(1-\alpha)f(y)\quad\text{for } \alpha\in[0,1];$$

and it is said to be *concave* (strictly concave) if, for $x, y\in X$,

$$f(\alpha x+(1-\alpha)y)\ge(>)\alpha f(x)+(1-\alpha)f(y)\quad\text{for } \alpha\in[0,1].$$

A set $C\subset X$ is said to be convex if, for $\alpha\in[0,1]$, $\alpha x+(1-\alpha)y\in C$ whenever $x, y\in C$.

Theorem 1.1.20. *Let C be a weakly compact subset of a Banach space X and f a weakly lower semicontinuous functional on C, that is, $f(x_0)\le\underline{\lim}_n f(x_n)$ whenever $x_n \xrightarrow{w} x_0$. Then f attains its minimum on C. Further, if C is also convex and f strictly convex, then it has a unique minimum in C.*

Theorem 1.1.21. *Let C be a weakly compact subset of a Banach space X and f a weakly upper semicontinuous functional on C, that is, $f(x_0)\geq \overline{\lim}^n f(x_n)$ whenever $x_n \xrightarrow{w} x_0$. Then f attains its maximum on C. Further, if C is also convex and f strictly concave, then it has a unique maximum in C.*

Theorem 1.1.22. *Let C be a weakly compact subset of a Banach space X and f a real-valued continuous functional on C. Then f attains both its maximum and minimum on C.*

A functional f defined on a Banach space X is said to be *Gateaux differentiable at the point* $x_0 \in X$ if there exists a homogeneous functional $f'(x_0,\cdot)$ defined on X such that

$$\lim_{\alpha\to 0}\left\{\frac{f(x_0+\alpha h)-f(x_0)}{\alpha}\right\}=f'(x_0,h) \quad \text{for all } h\in X.$$

Theorem 1.1.23. *Let C be a closed convex subset of a Banach space X and f a real-valued Gateaux differentiable functional on C. Suppose further that f is convex and the set $C_0\equiv\{x\in C: f(x)\leq f(y) \text{ for all } y\in C\}$ is nonempty. Then a necessary and sufficient condition for an element x_0 of C to be an element of C_0 is that $f'(x_0, y-x_0)\geq 0$ for all $y\in C$.*

1.2. Some Special Function Spaces

Let Ω be an open connected set (i.e., a domain) in R^n and $\partial\Omega$ its boundary, while $\bar{\Omega}$ is its closure. Suppose that I is any open interval in R and let $Q\equiv\Omega\times I$.

1.2.1. Spaces for Continuous Functions

Let N denote the set of nonnegative integers, and N^n the n-copies of N. If $\beta\in N^n$, the notation D^β represents the partial derivative

$$\frac{\partial^{\beta_1+\cdots+\beta_n}}{\partial x_1^{\beta_1}\cdots\partial x_n^{\beta_n}}.$$

For functions ϕ defined on $Q\equiv\Omega\times I$, $D_t^\alpha D_x^\beta\phi$ represents the partial derivatives of ϕ in which D_x^β, $\beta\in N^n$, stands for D^β with respect to x in Ω, while D_t^α, $\alpha\in N$, stands for D^α with respect to t in I. Let $|\beta|\equiv\sum_{i=1}^n\beta_i$.

For a given nonnegative integer m, let $C^m(\Omega)\equiv\{\phi: D_x^\beta\phi, |\beta|\leq m, \text{ are continuous functions on } \Omega\}$. $C^\infty(\Omega)\equiv\bigcap_{j=0}^\infty C^j(\Omega)$ and $C^0(\Omega)\equiv C(\Omega)$. Let $C_0^m(\Omega)$ $[C_0^\infty(\Omega)]$ denote all those elements from $C^m(\Omega)$ $[C^\infty(\Omega)]$ that have compact supports in Ω.

Since Ω is open, functions from $C^m(\Omega)$ are not necessarily bounded on Ω. However, if $\phi\in C^m(\Omega)$ is bounded and uniformly continuous on Ω, then it has a unique, bounded continuous extension to $\bar{\Omega}$. Let $C^m(\bar{\Omega}) \equiv \{\phi \in C^m(\Omega): D_x^\beta \phi, |\beta| \leq m$, are bounded and uniformly continuous on $\Omega\}$.

It is clear that $C^m(\bar{\Omega})$ is a Banach space with the norm $\|\cdot\|_{C^m(\bar{\Omega})}$ defined by

$$\|\phi\|_{C^m(\bar{\Omega})} \equiv \max\left\{ \sup_{x\in\Omega} |D_x^\beta \phi(x)| : |\beta| \leq m \right\}. \tag{1.2.1}$$

For a given nonnegative integer κ, let $\sum_{|\beta|=\kappa} D^\beta \phi$ denote the sum of partials of ϕ of order exactly κ.

For j an integer define

$$\langle \phi \rangle_{\bar{\Omega}}^{(j)} \equiv \sum_{|\beta|=j} \left\{ \sup_{x\in\Omega} |D_x^\beta \phi(x)| \right\},$$

and for $\alpha \in (0,1)$ let

$$\langle \phi \rangle_{\bar{\Omega}}^{(\alpha)} \equiv \sup_{x, x' \in \Omega} \left\{ \frac{|\phi(x) - \phi(x')|}{|x - x'|^\alpha} \right\}.$$

In general, for λ a nonintegral positive number define

$$|\phi|_{\bar{\Omega}}^{(\lambda)} \equiv \sum_{|\beta|=[\lambda]} \langle D_x^\beta \phi \rangle_{\bar{\Omega}}^{(\alpha)} + \sum_{j=0}^{[\lambda]} \langle \phi \rangle_{\bar{\Omega}}^{(j)}, \tag{1.2.2}$$

where $\alpha = \lambda - [\lambda]$ with $[\lambda]$ being the largest integer less than λ.

For a given nonintegral positive number λ, let $\mathcal{H}^\lambda(\bar{\Omega})$ be the linear vector space of all elements $\{\phi\}$ from $C^{[\lambda]}(\bar{\Omega})$ such that $|\phi|_{\bar{\Omega}}^{(\lambda)} < \infty$. It is known that $\mathcal{H}^\lambda(\bar{\Omega})$ is a Banach space with respect to the norm $|\cdot|_{\bar{\Omega}}^{(\lambda)}$, and it is called the *Hölder space*.

For functions $\{\phi\}$ defined on $Q \equiv \Omega \times I$, we can define a similar Hölder space with the norm

$$\begin{aligned} |\phi|_{\bar{Q}}^{(\lambda, \lambda/2)} \equiv & \sum_{2\alpha + |\beta| = [\lambda]} \| D_t^\alpha D_x^\beta \phi \|_{x, \bar{Q}}^{(\gamma_1)} \\ & + \sum_{0 < \lambda - 2\alpha - |\beta| < 2} \| D_t^\alpha D_x^\beta \phi \|_{t, \bar{Q}}^{(\gamma_2)} + \sum_{j=0}^{[\lambda]} \|\phi\|_{\bar{Q}}^{(j)}, \end{aligned} \tag{1.2.3}$$

where $\gamma_1 = \lambda - [\lambda]$, $\gamma_2 = (\lambda - 2\alpha - |\beta|)/2$, and

$$\|\psi\|_{x, \bar{Q}}^{(\gamma_1)} \equiv \sup_{(x,t), (x',t) \in Q} \left\{ \frac{|\psi(x,t) - \psi(x',t)|}{|x - x'|^{\gamma_1}} \right\},$$

$$\|\psi\|_{t, \bar{Q}}^{(\gamma_2)} \equiv \sup_{(x,t), (x,t') \in Q} \left\{ \frac{|\psi(x,t) - \psi(x,t')|}{|t - t'|^{\gamma_2}} \right\},$$

and

$$\|\psi\|_{\bar{Q}}^{(j)} \equiv \sum_{2\alpha + |\beta| = j} \sup_{(x,t) \in Q} \left\{ |D_t^\alpha D_x^\beta \psi(x,t)| \right\}.$$

Again, $\mathcal{H}^{\lambda, \lambda/2}(\bar{Q})$ is a Banach space with respect to the norm $|\cdot|_{\bar{Q}}^{(\lambda, \lambda/2)}$.

1.2.2. Some Concepts of Distributions

In what follows, we shall present a brief description of certain concepts from the theory of distributions [A.1, Bn.1] with special emphasis on the notion of *generalized derivatives of integrable functions*. A generalized derivative is also called a *weak* or *distributional derivative*. One of the standard definitions of Sobolev spaces is phrased in terms of such derivatives.

Let Ω be an open subset of R^n. A distribution on Ω is a continuous linear functional on $C_0^\infty(\Omega) \equiv \mathfrak{D}(\Omega)$. The space of distributions is denoted by $\mathfrak{D}'(\Omega)$. For example, every $f \in L_1^{\text{loc}}(\Omega)$ defines a distribution in the sense that

$$T_f(\phi) = \int_\Omega \{f(x)\phi(x)\}\,dx \quad \text{for all } \phi \in \mathfrak{D}(\Omega).$$

In fact, T_f is a *Radon measure* on Ω that is merely a continuous linear functional on $C_0(\Omega)$. Thus, $T_f \in M(\Omega) \equiv$ the space of Radon measures and hence $M(\Omega) \subset \mathfrak{D}'(\Omega)$. In particular, the elements of $C^m(\Omega)$, $0 \leq m \leq \infty$, $L_p(\Omega)$, $1 \leq p \leq \infty$, define distributions on Ω. In other words, under the mapping $f \to T_f$, $C^m(\Omega)$, $L_p(\Omega) \subset \mathfrak{D}'(\Omega)$ for all $0 \leq m \leq \infty$, $1 \leq p \leq \infty$. The *Dirac measure* δ on R^n defined by $\delta(\phi) = \phi(0)$ for all $\phi \in \mathfrak{D}(\Omega)$ is a distribution. In fact, for any $\beta \in N^n$ such that $|\beta| < \infty$,

$$(D^\beta \delta)(\phi) = (-1)^{|\beta|}(D^\beta \phi)(0) \quad \text{for all } \phi \in \mathfrak{D}(\Omega)$$

also defines a distribution on Ω.

A linear functional T is a distribution on Ω if and only if for every compact set $K \subset \Omega$, there exists a constant c and an integer $m \geq 0$ such that

$$|T(\phi)| \leq c \sup\{|(D^\beta \phi)(x)| : x \in \Omega, |\beta| \leq m\}$$

for all $\phi \in \mathfrak{D}(\Omega)$ with $\operatorname{supp} \phi \subset K$.

By identifying $f \in L_1^{\text{loc}}(\Omega)$ with $T_f \in \mathfrak{D}'(\Omega)$, we can define distributional derivatives of f in the sense that

$$(D^\beta f)(\phi) = (-1)^{|\beta|} f(D^\beta \phi)$$

for all $|\beta| < \infty$ and all $\phi \in \mathfrak{D}(\Omega)$. In particular, if there exists a $g \in L_1^{\text{loc}}(\Omega)$ such that for a given $f \in L_1^{\text{loc}}(\Omega)$, $(D^\beta f)(\phi) = g(\phi)$ for all $\phi \in \mathfrak{D}(\Omega)$, then g is the distributional (or generalized) derivative of f of order $|\beta|$.

Let ρ be a real-valued function belonging to $C_0^\infty(R^n)$ having these properties:

i. $\rho(x) \geq 0$ for all $x \in R^n$;
ii. $\rho(x) = 0$ for all $|x| \geq 1$; and
iii. $\int_{R^n} \{\rho(x)\}\,dx = 1$.

As an example, we may take

$$\rho(x) = \begin{cases} K \exp\{-1/(1-|x|^2)\} & \text{if } |x| < 1 \\ 0 & \text{if } |x| \geq 1, \end{cases}$$

where $K > 0$ is chosen so that condition (iii) is satisfied.

For $\varepsilon>0$, the function $\rho(\cdot;\,\varepsilon)$ defined by $\rho(x;\,\varepsilon)\equiv\varepsilon^{-n}\rho(x/\varepsilon)$ belongs to $\mathfrak{D}(\Omega)$ and satisfies these properties:

i. $\rho(x;\,\varepsilon)\geq 0$ for all $x\in R^n$;
ii. $\rho(x;\,\varepsilon)=0$ for all $|x|\geq\varepsilon$; and
iii. $\int_{R^n}\{\rho(x;\,\varepsilon)\}\,dx=1$.

For $\phi\in L_1^{\mathrm{loc}}(R^n)$ and $\varepsilon>0$, define

$$\phi_\varepsilon(x)\equiv\int_{R^n}\{\rho(x-y;\,\varepsilon)\phi(y)\}\,dy. \tag{1.2.4}$$

This function belongs to $C^\infty(R^n)$ and is called the *integral average* of ϕ. Clearly,

$$\phi_\varepsilon(x)=\frac{1}{\varepsilon^n}\int_{|y-x|\leq\varepsilon}\left\{\rho\left(\frac{x-y}{\varepsilon}\right)\phi(y)\right\}dy.$$

Theorem 1.2.1. *Let ϕ be a function which is defined on R^n and equal to zero identically outside the domain Ω. Then the following statements are valid.*

i. *If $\phi\in L_1^{\mathrm{loc}}(\Omega)$, then $\phi_\varepsilon\in C^\infty(R^n)$, where ϕ_ε is as defined by (2.4).*
ii. *Suppose that $\phi\in L_1^{\mathrm{loc}}(\Omega)$ and that there exists a compact subset K of Ω such that $\phi(x)\equiv 0$ for $x\notin K$. Then, for $\varepsilon>0$ but smaller than the distance between K and $\partial\Omega$, $\phi_\varepsilon\in C_0^\infty(\Omega)\equiv\mathfrak{D}(\Omega)$.*
iii. *If $\phi\in L_p(\Omega)$, where $1\leq p<\infty$, then $\phi_\varepsilon\in L_p(\Omega)$, $\|\phi_\varepsilon\|_{p,\Omega}\leq\|\phi\|_{p,\Omega}$, and $\lim_{\varepsilon\downarrow 0}\|\phi_\varepsilon-\phi\|_{p,\Omega}=0$, where $\|\psi\|_{p,\Omega}\equiv\|\psi\|_{L_p(\Omega)}$.*
iv. *Suppose that $\phi\in C(\Omega)$ and that K is a subdomain strictly contained in Ω. Then, $\lim_{\varepsilon\downarrow 0}\phi_\varepsilon(x)=\phi(x)$ uniformly on K.*
v. *If $\phi\in C(\bar{\Omega})$, then $\lim_{\varepsilon\downarrow 0}\phi_\varepsilon(x)=\phi(x)$ uniformly on Ω.*

Note that $C_0(\Omega)$ is dense in $L_p(\Omega)$ for $p\in[1,\infty)$. In fact, $C_0^\infty(\Omega)$ is also dense in $L_p(\Omega)$ for $p\in[1,\infty)$

Theorem 1.2.2. *Let ϕ and ϕ_ε be as in Theorem 1.2.1 and suppose that ϕ, $D^\beta\phi\in L_1^{loc}(\Omega)$. Then the following two statements are valid.*

i. *$D^\beta\phi_\varepsilon=[D^\beta\phi]_\varepsilon$, where $[\psi]_\varepsilon$ denotes the integral average of ψ and $D^\beta\phi$ the distributional derivative of ϕ of order $|\beta|$.*
ii. *If $D^\beta\phi\in L_p(\Omega)$, then $D^\beta\phi_\varepsilon\overset{s}{\rightarrow}D^\beta\phi$ in $L_p(\Omega)$.*

Consider the Banach space $L_p(\Omega)$, $1\leq p\leq\infty$, and let

$$W_p^m(\Omega)\equiv\left\{\phi\in L_p(\Omega):\,D^\beta\phi\in L_p(\Omega),\,|\beta|\leq m\right\}.$$

$W_p^m(\Omega)$ is a (linear) vector space and is a Banach space with respect to the norm

$$\|\phi\|^{(m)}_{p,\Omega} \equiv \sum_{|\beta| \le m} \|D^{\beta}\phi\|_{p,\Omega}. \tag{1.2.5}$$

Note that $W_p^m(\Omega)$ is separable for $p \in [1,\infty)$ and is also reflexive if $p \in (1,\infty)$. In particular, $W_2^m(\Omega)$ is a Hilbert space with the natural inner product. It is clear that $W_p^0(\Omega) = L_p(\Omega)$.

Let $\mathring{W}_p^m(\Omega)$ denote the closure of $\mathfrak{D}(\Omega) \equiv C_0^\infty(\Omega)$ in the norm topology of $W_p^m(\Omega)$, and $H_p^m(\Omega)$ the closure of $C^m(\Omega)$ in the same topology. Clearly $\mathring{W}_p^0(\Omega) = L_p(\Omega)$ for $p \in [1,\infty)$. Further, it is known that $W_p^m(\Omega) = H_p^m(\Omega)$ for $p \in [1,\infty]$.

To proceed further, we introduce certain definitions and notation relating to domains, their boundaries, and functions defined on the boundaries.

Definition 1.2.1. A bounded domain Ω is said to have *piecewise smooth boundary* if its closure can be represented in the form

$$\bar{\Omega} = \bigcup_{k=1}^{N} \bar{\Omega}_k,$$

where $\Omega_k \cap \Omega_l = \emptyset$ for $k \ne l$, while each part $\bar{\Omega}_k$ can be homeomorphically mapped onto the unit ball or cube by means of functions $z_i^k \in C^1(\bar{\Omega}_k)$ $(i = 1, \ldots, n,\ k = 1, \ldots, N)$ having Jacobians $|\text{grad}\, z^k|$ greater than some positive number δ.

In the next definition, some smoothness conditions are introduced for the boundary $\partial\Omega$.

Definition 1.2.2. Let Ω be a bounded domain with boundary $\partial\Omega$. We say that $\partial\Omega$ *is of class* C^m if, for each point $x^0 \in \partial\Omega$, there exists a ball N_0 with center x^0 such that $\partial\Omega \cap N_0$ can be represented in the form

$$x_i = h(x_1, \ldots, x_{i-1}, x_{i+1}, \ldots, x_n),$$

for some $i \in \{1, \ldots, n\}$, where $h \in C^m$.

Note that, in the above definition, $x_1, \ldots, x_{i-1}, x_{i+1}, \ldots, x_n$ are called *local parameters* of $\partial\Omega$.

For piecewise smooth boundary $\partial\Omega$,

$$\mathring{W}_p^1(\Omega) = \left\{ \phi \in W_p^1(\Omega) : \phi(x) = 0 \text{ for all } x \in \partial\Omega \right\}.$$

Theorem 1.2.3. *Let Ω be a bounded domain with piecewise smooth boundary. Then, for every $\phi \in W_p^m(\Omega)$,*

i. $\|\phi\|_{q,\partial\Omega} \le K \|\phi\|^{(m)}_{p,\Omega}$ *for* $q \in [1, p(n-1)/n-pm)]$ *if* $n > pm$ *and it is true for* $q \in [1,\infty)$ *if* $n = pm$; *and*

ii. $|\phi|^{(\alpha)}_{\Omega} \le K \|\phi\|^{(m)}_{p,\Omega}$ *for* $n < pm$ *with* $\alpha \in [0,1)$ *and* $\alpha \le (pm-n)/p$.

The constant K depends only on n, p, m, q, Ω, and $\partial\Omega$.

The proof of the above theorem is based on a theorem given in Ladyzhénskaya and Ural'tseva [LU.1, p. 43].

The next theorem gives an important inequality for functions from $\mathring{W}_p^1(\Omega)$.

Theorem 1.2.4 [LSU.1, Theorem 2.2, pp. 62–63]. *Let* Ω *be a bounded domain with piecewise smooth boundary,* $p \geq 1$, $\phi \in \mathring{W}_p^1(\Omega)$ *and* $m \geq 1$. *Define*

$$\|\phi_x\|_{p,\Omega} \equiv \left(\int_\Omega \left\{ \sum_{i=1}^n \left[\phi_{x_i}(x)\right]^2 \right\}^{p/2} dx \right)^{1/p}$$

and

$$\alpha = \left(\frac{1}{m} - \frac{1}{q} \right) \left(\frac{1}{n} - \frac{1}{p} + \frac{1}{m} \right)^{-1}.$$

Then

$$\|\phi\|_{q,\Omega} \leq \beta (\|\phi_x\|_{p,\Omega})^\alpha (\|\phi\|_{m,\Omega})^{1-\alpha}$$

for

i. $$q \in [m, \infty] \quad \textit{and} \quad \beta = \left(1 + \frac{(p-1)m}{p}\right)^\alpha \quad \textit{if } p \geq n = 1,$$

ii. $$q \in \left[m, \frac{np}{n-p} \right] \quad \textit{and} \quad \beta = \left(\frac{(n-1)p}{n-p} \right)^\alpha \quad \textit{if} \quad n > 1, p < n,$$

and $m \leq np/(n-p)$,

iii. $$q \in \left[\frac{np}{n-p}, m \right] \quad \textit{and} \quad \beta = \left(\frac{(n-1)p}{n-p} \right)^\alpha \quad \textit{if } n > 1,$$

$p < n$, *and* $m \geq np/(n-p)$, *and*

iv. $$q \in [m, \infty) \quad \textit{and} \quad \beta = \max\left\{ \left(\frac{q(n-1)}{n} \right)^\alpha, \left(1 + \frac{(p-1)m}{p}\right)^\alpha \right\}$$

if $p \geq n > 1$.

The next theorem presents a Sobolev inequality.

Theorem 1.2.5. *Let* Ω *be a bounded domain in* R^n *with boundary* $\partial\Omega$, *let* $\partial\Omega$ *be piecewise smooth and let* ϕ *be an element of* $W_p^m(\Omega)$, $1 \leq p \leq \infty$. *Then*

$$\|\phi\|_{q,\Omega} \leq K \|\phi\|_{p,\Omega}^{(m)},$$

where $q > 1$ *if* $n = 1$, *while* $1 < q < np/(n-mp)$ *if* $n \geq 2$. *The constant* K *depends only on* Ω, m *and* p.

The proof of the above theorem follows easily from Theorems 1, 2 (SO. 1, pp. 56, 57) and a Lemma (SO, 1, p. 66) (see also Fr.3, Theorem 10.2, p. 28).

1.2.3. *Certain Sobolev Spaces on Q*

In what follows, we shall consider Sobolev spaces $W_p^{2,1}(Q)$, $W_2^{1,0}(Q)$, $W_2^{1,1}(Q)$, $V_2(Q)$, and $V_2^{1,0}(Q)$ and state some of their important properties.

To start with, let $L_{p,r}(Q)$, $1 \le p, r \le \infty$, denote the Banach space of all measurable real-valued functions $\{f\}$ defined on Q with norm $\| f \|_{p,r,Q}$ defined by

$$\| f \|_{p,r,Q} \equiv \left\{ \int_I \left[\int_\Omega (|f(x,t)|^p)\, dx \right]^{r/p} dt \right\}^{1/r} \quad \text{for } 1 \le p, r < \infty,$$

$$\| f \|_{p,\infty,Q} \equiv \operatorname*{ess\,sup}_{t \in I} \{ \| f(\cdot, t) \|_{p,\Omega} \} \quad \text{for } 1 \le p < \infty, \quad r = \infty,$$

$$\| f \|_{\infty,r,Q} \equiv \left\{ \int_I (\| f(\cdot, t) \|_{\infty,\Omega})^r\, dt \right\}^{1/r} \quad \text{for } p = \infty, \quad 1 \le r < \infty,$$

and

$$\| f \|_{\infty,\infty,Q} \equiv \operatorname*{ess\,sup}_{(x,t) \in Q} |f(x,t)| \quad \text{for } p = \infty, \quad r = \infty.$$

Recall that $\| \cdot \|_{p,\Omega}$ and $\| \cdot \|_{\infty,\Omega}$ are, respectively, the norms of the Banach spaces $L_p(\Omega)$ and $L_\infty(\Omega)$ defined before. For simplicity, we denote $L_{p,p}(Q)$ by $L_p(Q)$ and the norm $\| \cdot \|_{p,p,Q}$ by $\| \cdot \|_{p,Q}$.

Let p be a real number such that $p \ge 1$, and let $W_p^{2,1}(Q)$ denote the Banach space of all those functions $\{\phi\}$ from $L_p(Q)$ with the finite norm

$$\| \phi \|_{p,Q}^{(2,1)} \equiv \| \phi \|_{p,Q} + \| \phi_t \|_{p,Q} + \sum_{i=1}^{n} \| \phi_{x_i} \|_{p,Q} + \sum_{i,j=1}^{n} \| \phi_{x_i x_j} \|_{p,Q}, \tag{1.2.6}$$

where

$$\phi_t \equiv \frac{\partial \phi}{\partial t}, \qquad \phi_{x_i} \equiv \frac{\partial \phi}{\partial x_i}, \quad \text{and} \quad \phi_{x_i x_j} \equiv \frac{\partial^2 \phi}{\partial x_i \partial x_j}.$$

$W_2^{1,0}(Q)$ is the *Hilbert space* of all real-valued measurable functions defined on Q with finite scalar product

$$\langle z, y \rangle_{W_2^{1,0}(Q)} \equiv \iint_Q \left\{ z(x,t) y(x,t) + \sum_{i=1}^{n} z_{x_i}(x,t) y_{x_i}(x,t) \right\} dx\, dt.$$

$W_2^{1,1}(Q)$ is the *Hilbert space* of all real-valued measurable functions defined on Q with finite scalar product

$$\langle z, y \rangle_{W_2^{1,1}(Q)} \equiv \iint_Q \left\{ z(x,t) y(x,t) + \sum_{i=1}^{n} z_{x_i}(x,t) y_{x_i}(x,t) + z_t(x,t) y_t(x,t) \right\} dx\, dt.$$

$V_2(Q)$ is the *Banach space* of all those $\{\phi\}$ from $W_2^{1,0}(Q)$ with respect to the norm

$$|\phi|_Q \equiv \|\phi\|_{2,\infty,Q} + \|\phi_x\|_{2,Q}, \tag{1.2.7}$$

where

$$\|\phi_x\|_{2,Q} \equiv \left(\iint_Q \left\{ \sum_{i=1}^{n} \left(\phi_{x_i}(x,t) \right)^2 \right\} dx\, dt \right)^{1/2}$$

$V_2^{1,0}(Q)$ is the *Banach space* of all those $\{\phi\}$ from $V_2(Q)$ that are continuous in t [in the norm of $L_2(\Omega)$] and equipped with the norm

$$\|\phi\|_Q \equiv \sup_{t\in I} \{\|\phi(\cdot,t)\|_{2,\Omega}\} + \|\phi_x\|_{2,Q}. \tag{1.2.8}$$

Note that a function ϕ is said to be continuous in t in the norm of $L_2(\Omega)$ if

$$\|\phi(\cdot,t+\Delta t) - \phi(\cdot,t)\|_{2,\Omega} \to 0$$

as $\Delta t \to 0$.

Also note that $V_2^{1,0}(Q)$ is the completion of $W_2^{1,1}(Q)$ in the norm $\|\cdot\|_Q$. The symbol $\circ$ over $W_p^{2,1}(Q)$, $W_2^{1,0}(Q)$, $W_2^{1,1}(Q)$, $V_2(Q)$, or $V_2^{1,0}(Q)$ indicates restriction to those elements of the respective spaces that vanish on $\partial\Omega \times I$. All these spaces are complete.

Theorem 1.2.6 [LSU.1 pp. 74–75]. *Let Ω be a bounded domain with piecewise smooth boundary. If $\phi \in \overset{\circ}{V}_2(Q)$, then the following two inequalities are satisfied:*

i. $\|\phi\|_{q,r,Q} \le \beta(\|\phi\|_{2,\infty,Q})^{1-(2/r)}(\|\phi_x\|_{2,Q})^{2/r}$ *and*
ii. $\|\phi\|_{q,r,Q} \le \beta|\phi|_Q$,

where

$$\frac{1}{r} + \frac{n}{2q} = \frac{n}{4},$$

with

$$\begin{aligned} &r \in [2,\infty], && q \in \left[2, \frac{2n}{n-2}\right] && \textit{for } n>2,\\ &r \in (2,\infty], && q \in [2,\infty) && \textit{for } n=2,\\ &r \in [4,\infty], && q \in [2,\infty] && \textit{for } n=1, \end{aligned} \tag{1.2.9}$$

while the constant β depends only on n and q as indicated in Theorem 1.2.4.

The proof of the above theorem is based on Theorem 1.2.4, Hölder's inequality, and Young's inequality, given by

$$ab \le \frac{1}{m}(\varepsilon)^m(a)^m + \frac{m-1}{m}(\varepsilon)^{-m/(m-1)}(b)^{m/(m-1)},$$

where $a>0$, $b>0$, $\varepsilon>0$, and $m>1$.

REMARK 1.2.1. Let $\phi \in L_{q,r}(Q)$ with $q, r, \geq 1$ and define $\phi_\varepsilon(x,t) = (1/\varepsilon)\int_t^{t+\varepsilon}[\phi(x,\tau)]\,d\tau$, $\varepsilon > 0$. Then it follows from Theorem 1.1.19 that $\phi_\varepsilon \to \phi$ in the norm of $L_{q,r}(\Omega \times (0, T-\delta))$ where $\delta \in (0, T)$.

Theorem 1.2.7. *Let $\phi \in V_2^{1,0}(Q)$ with ϕ_ε as defined in Remark 1.2.1 and $\delta \in (0,T)$. Then the following conclusions are valid:*

i. $\phi_\varepsilon \in W_2^{1,1}(\Omega \times (0, T-\delta))$;
ii. $\lim_{\varepsilon \downarrow 0} \| \phi_\varepsilon - \phi \|_{\Omega \times (0, T-\delta)} = 0$;
iii. $\lim_{\varepsilon \downarrow 0} \| \phi_\varepsilon - \phi \|_{q, r, \Omega \times (0, T-\delta)} = 0$ *with q and r subject to condition* (1.2.9).

The first conclusion of the theorem is obvious, while a proof of the second conclusion can be found in Ladyžhenskaja, Solonikov, and Ural'ceva [LSU.1, pp. 86–87]. The last conclusion follows from (ii) and Theorem 1.2.6 (ii).

Theorem 1.2.8 [LSU.1, Lemma 4.12, p. 89]. *Let Ω be a bounded domain with piecewise smooth boundary. Suppose that $\{\psi^k\}$ is a fundamental system of basis vectors for $\mathring{W}_2^1(\Omega)$ and that $\{d_k\}$ are arbitrary smooth functions equal to zero for $t = T$. Then the class of functions of the form $\sum_{k=1}^N d_k(t)\psi^k(x)$ is dense in $\tilde{W}_2^{1,1}(Q) \equiv \{z \in \mathring{W}_2^{1,1}(Q) : z(x,T) = 0$ for all $x \in \Omega\}$.*

1.2.4. Convergence in Various Topologies

In what follows, we shall present several convergence results that will be found useful in the later chapters.

Theorem 1.2.9 [NT.2, Lemma 4.2, p. 52]. *Let Ω be a bounded measurable subset of R^n. Suppose that $\{f^n\}$ and $\{g^n\}$ are two sequences of real-valued measurable functions defined on Ω such that the following assumptions are satisfied:*

i. *$f^n \to f$ a.e. on Ω and there exists a constant $M_1 > 0$ such that*

$$\sup_n |f^n(x)| \leq M_1 \quad a.e. \text{ on } \Omega$$

ii. *$g^n \to g$ weakly in $L_2(\Omega)$ and there exists a constant $M_2 > 0$ such that*

$$\sup_n \|g^n\|_{2,\Omega} \equiv \sup_n \|g^n\|_{L_2(\Omega)} \leq M_2.$$

Then $f^n g^n \to fg$ weakly in $L_2(\Omega)$.

Theorem 1.2.10 [NT.2. Lemma 4.3, p. 53]. *Let $Q \equiv \Omega \times I$ be a bounded measurable subset of R^{n+1} and $\{f^n\}$ a bounded sequence from $L_{2,r}(Q)$, where $r \in (1,2]$. If $f^n \to f$ a.e. on Q, then there exists a subsequence $\{f^{n(k)}\}$ of $\{f^n\}$ such that $f^{n(k)} \to f$ weakly in $L_{2,r}(Q)$.*

Theorem 1.2.11 [NT.2, Lemma 4.4 p. 53]. *Let $Q \equiv \Omega \times I$ be a bounded measurable subset of R^{n+1} and $\{f^n\}$ a sequence of measurable functions from $V_2^{1,0}(Q)$. If $f^n \to f$ strongly in $V_2^{1,0}(Q)$, then $f^n \to f$ weakly in $L_{2,p}(Q)$, where $p \geq 1$.*

Theorem 1.2.12 [Ar.1, Lemma 3, p. 633]. *Let $Q \equiv \Omega \times I$ be a measurable subset of R^{n+1} and $\{f^n\}$ a sequence of functions from $L_2(Q)$ that converges to a limit function f weakly in $L_2(Q)$. If $\| f^n \|_{2,\infty,Q} \leq K$ independently of n, then $\| f \|_{2,\infty,Q} \leq K$.*

The proof of the above theorem is based on the Banach–Saks theorem, Minkowski's inequality, and Fatou's lemma.

1.2.5. Conditions for Continuity and Semicontinuity

In what follows, we shall present two theorems concerning the continuity and lower semicontinuity of certain functions.

Theorem 1.2.13 [Kr.1, p. 20]. *Let $p, q \in [1, \infty)$, r, s be positive integers, and g be a Carathéodory function defined on $Q \times R^r$ [i.e., $g(x, y)$ is measurable in x on Q for each $y \in R^r$, and continuous in y on R^r for almost all $x \in Q$]. Define $G\phi = g(\cdot, \phi(\cdot))$ for $\phi \in L_p(Q, R^r)$. If the operator G maps $L_p(Q, R^r)$ into $L_q(Q, R^s)$, then G is continuous.*

The following theorem contains a result on the lower semicontinuity of certain functionals.

Theorem 1.2.14. *Let $\Omega_1 \subset R^r$ and $\Omega_2 \subset R^s$ be such that Ω_1 is open with compact closure and Ω_2 is compact and convex. Let f be a continuous function defined on $\Omega_1 \times \Omega_2$, $f(x, \cdot)$ convex on Ω_2 for each $x \in \Omega_1$, and $\{y^k\}$ a sequence of measurable functions defined on Ω_1 with values in Ω_2. If $y^k \to y^0$ in the weak* topology of $L_\infty(\Omega_1, R^s)$, then*

$$\int_{\Omega_i} \{f(x, y^0(x))\}\, dx \leq \liminf_{k \to \infty} \int_{\Omega_1} \{f(x, y^k(x))\}\, dx.$$

1.2.6. An Abstract Function Space

When we consider evolution equations on a Banach space, it is often required to verify that their solutions have certain regularity properties such as continuity and differentiability. In this regard, distribution theory for vector-valued functions is extremely useful. We present here only a brief description of the material that is directly useful for our purpose—specifically, for the proof of Theorem 1.2.15.

Let I be an open bounded subset of the real line and X a real or complex Banach space with dual X^*. Denote by $\mathfrak{D}(I, X)\equiv C_0^\infty(I, X)$ the space of X-valued C^∞-functions with compact support in I. If X is the field of scalars, then we simply write $\mathfrak{D}(I)$ for $\mathfrak{D}(I, X)$. Let $\mathfrak{D}'(I, X)$ denote the space of X-valued distributions on I and suppose that it is equipped with its strong topology, that is, the topology of uniform convergence on bounded sets in $\mathfrak{D}(I)$. Note that $\mathfrak{D}'(I, X)=\mathcal{L}(\mathfrak{D}(I), X)$, the space of continuous linear transformations from $\mathfrak{D}(I)$ into X.

We can show that for any p, $1\leq p\leq\infty$,

$$\mathfrak{D}(I, X)\subsetneq L_p(I, X)\subsetneq\mathfrak{D}'(I, X)$$

and that the imbeddings are continuous. Further, $\mathfrak{D}(I, X)$ is dense in $L_p(I, X)$ in the strong topology. Therefore, $\mathfrak{D}(I, X)$ and $L_p(I, X)$ are normal spaces of distributions on I with values in X. Consequently, $(L_p(I, X))^*=L_q(I, X^*)\subset\mathfrak{D}'(I, X)$, where $q=p/(p-1)$. Thus, if $x\in L_p(I, X)$ and $\dot{x}(\dot{x}(t)\equiv(d/dt)x(t), t\in I,)\in L_q(I, X^*)$, we can consider $x, \dot{x}\in\mathfrak{D}'(I, X)$, that is, $\dot{x}$ is the distributional derivative of x. In general, identifying $x\in L_p(I, X)$ with the corresponding distribution T_x, we can define distributional derivative of x of any order:

$$(D^m x)(\phi)=(-1)^m x(D^m\phi)\quad\text{for all } \phi\in\mathfrak{D}(I).$$

Theorem 1.2.15. *Let H be a Hilbert space, which is identified with its dual H^*, and suppose that E, a subset of H, has the structure of a reflexive Banach space and that E^* is its dual. Let $E\subset H\subset E^*$ both algebraically and topologically, and let E be dense in H. Let $I\equiv(0, T)$, $T<\infty$ and $x\in L_p(I, E)$, with $\dot{x}\in L_q(I, E^*)$, where the differentiation is understood in the sense of distribution $[\mathfrak{D}'(I, E)]$ and $p^{-1}+q^{-1}=1$, $1<p, q<\infty$. Then $x\in C(\bar{I}, H)$.*

PROOF. Define $W_{p,q}(I)\equiv\{x: x\in L_p(I, E)$ and $\dot{x}\in L_q(I, E^*)\}$ for $1<p, q<\infty$. $W_{p,q}$ is a vector space, and it becomes a Banach space with respect to the norm

$$\|x\|_{W_{p,q}}\equiv\left(\|x\|^2_{L_p(I, E)}+\|\dot{x}\|^2_{L_q(I, E^*)}\right)^{1/2}. \tag{1.2.10}$$

Let $a>0$. Consider the interval $I_a\equiv(-a, T+a)$ and let $\theta\in C^\infty$ equal 1 on $\bar{I}=[0, T]$ and 0 in the neighborhood of the points $\{-a, T+a\}$. We extend every element x of $W_{p,q}(I)$ to the interval $[-a, 0]$ and $[T, T+a]$ by setting, respectively, $x(-t)=x(t)$ and $x(T+t)=x(T-t)$. Define $y=\theta x$; clearly $y=x$ for $t\in I$, $y\in L_p(I_a, E)$, and $\dot{y}\in L_q(I_a, E^*)$. Now by regularization in t we can approximate $y\in W_{p,q}(I_a)$ by a sequence of C^∞ functions $\{y_m\}$ that vanish near $\{-a\}$ and $\{T+a\}$. Indeed, consider $\phi\in\mathfrak{D}(R)\equiv C_0^\infty(R)$ with

$\operatorname{supp}\phi\subset[-1,+1]$, $\phi\geq 0$ and $\int_R\phi(t)\,dt=1$. Define $\phi_m(t)=m\phi(mt)$ and

$$y_m(t)\equiv\int_R d\theta\,\phi_m(t-\theta)y(\theta). \tag{1.2.11}$$

It is clear that y_m vanishes in the neighborhood of the points $\{-a, T+a\}$ for m sufficiently large and $y_m\in C^\infty(I_a, E)$. We show that $y_m\to y$ in the strong topology of $W_{p,q}(I_a)$, or, equivalently,

$$y_m\overset{s}{\to}y \quad \text{in } L_p(I_a, E),$$

$$\dot{y}_m\overset{s}{\to}\dot{y} \quad \text{in } L_q(I_a, E^*). \tag{1.2.12}$$

Since $\int_R d\theta\,\phi_m(t-\theta)=1$ for all m, we can write

$$[y_m(t)-y(t)]=\int_R d\theta\,\phi_m(t-\theta)[y(\theta)-y(t)]. \tag{1.2.13}$$

Clearly

$$|y_m(t)-y(t)|_E\leq\int_R d\theta\,(\phi_m(t-\theta))^{(1/p)+(1/q)}|y(\theta)-y(t)|_E$$

$$\leq\left(\int_R d\theta\,\phi_m(t-\theta)\right)^{1/q}\left(\int_R d\theta\,\phi_m(t-\theta)|y(\theta)-y(t)|_E^p\right)^{1/p}.$$

Therefore

$$|y_m(t)-y(t)|_E^p\leq\left(\int_R d\theta\,\phi_m(t-\theta)|y(\theta)-y(t)|_E^p\right),$$

and hence

$$\int_{I_a} dt\,|y_m(t)-y(t)|_E^p\leq\int_{-1}^{+1}d\xi\,\phi(\xi)\int_{I_a}dt\,|y(t-\xi/m)-y(t)|_E^p. \tag{1.2.14}$$

Since, for any $y\in L_p(I_a, E)$, $\lim_{h\to 0}\int_{I_a}dt\,|y(t+h)-y(t)|_E^p=0$, it follows from (1.2.14) that

$$\int_{I_a}dt\,|y_m(t)-y(t)|_E^p\to 0 \quad \text{as } m\to\infty. \tag{1.2.15}$$

Similarly, we can show that

$$\int_{I_a}dt\,|\dot{y}_m(t)-\dot{y}(t)|_{E^*}^q\to 0 \quad \text{as } m\to\infty. \tag{1.2.16}$$

Thus, from (1.2.15) and (1.2.16) it follows that $y_m\to y$ in the strong topology of $W_{p,q}(I_a)$. Since y_m, as defined by (1.2.11), is an element of $C^\infty(I_a, E)$, it is clear that $y_m(t), \dot{y}_m(t) \in E$ for all $t \in I_a$. Therefore,

$$\langle \dot{y}_m(t), y_m(t) \rangle_{E^*-E} = (\dot{y}_m(t), y_m(t))_H = \frac{1}{2} \frac{d}{dt} (|y_m(t)|_H^2), \quad (1.2.17)$$

where we have used $\langle \cdot,\cdot \rangle_{E^*-E}$ to denote the duality pairing between elements of E^* and E and $(\cdot,\cdot)_H$ to denote the scalar product in the Hilbert space H. Since y_m vanishes near $\{-a\}$, integrating (1.2.17) on $[-a, t]$, $t \in I_a$, we have,

$$\begin{aligned} |y_m(t)|_H^2 &= 2 \int_{-a}^{t} d\theta \langle \dot{y}_m(\theta), y_m(\theta) \rangle \\ &\leq 2 \left(\int_{-a}^{t} d\theta |\dot{y}_m(\theta)|_{E^*}^q \right)^{1/q} \left(\int_{-a}^{t} d\theta |y_m(\theta)|_E^p \right)^{1/p}. \end{aligned} \quad (1.2.18)$$

By using Cauchy's inequality in (1.2.18), we obtain

$$\begin{aligned} |y_m(t)|_H^2 &\leq \left\{ \left(\int_{-a}^{t} d\theta |\dot{y}_m(\theta)|_{E^*}^q \right)^{2/q} + \left(\int_{-a}^{t} d\theta |y_m(\theta)|_E^p \right)^{2/p} \right\} \\ &\leq \| y_m \|_{W_{p,q}(I_a)}^2 \quad \text{for all } t \in I_a. \end{aligned} \quad (1.2.19)$$

Thus, it follows from (1.2.19) that, for any integer $n \geq 0$ and $t \in I_a$,

$$\sup_{t \in I_a} |y_{m+n}(t) - y_m(t)|_H \leq \| y_{m+n} - y_m \|_{W_{p,q}(I_a)}. \quad (1.2.20)$$

Since $y_m \to y$ in the strong topology of $W_{p,q}(I_a)$, $\{y_m\}$ is a Cauchy sequence in $W_{p,q}(I_a)$; hence it follows from (1.2.20) that $\{y_m\}$ is also a Cauchy sequence from $C(I_a, H)$. Since $C(I_a, H)$ is a Banach space with respect to the norm $\|x\| = \sup_{t \in I_a} \|x(t)\|_H$, there exists a $\tilde{y} \in C(I_a, H)$ such that $\sup_{t \in I_a} |y_m(t) - \tilde{y}(t)|_H \to 0$ as $m \to \infty$. We show that $y(t) = \tilde{y}(t)$ a.e. on I_a. Since $y_m \xrightarrow{s} y$ in $L_p(I_a, E)$, $y_m \xrightarrow{s} \tilde{y}$ in $C(I_a, H)$, and $E \subset H$, both algebraically and topologically, it is clear that both y and $\tilde{y} \in L_p(I_a, H)$. Thus

$$\| y - \tilde{y} \|_{L_p(I_a, H)} \leq \| y - y_m \|_{L_p(I_a, H)} + \| y_m - \tilde{y} \|_{L_p(I_a, H)}$$

and, since $L_p(I_a, E) \subset L_p(I_a, H)$, there exists a number $k > 0$ such that

$$\| y - \tilde{y} \|_{L_p(I_a, H)} \leq k \| y - y_m \|_{L_p(I_a, E)} + \| y_m - \tilde{y} \|_{L_p(I_a, H)}. \quad (1.2.21)$$

Since $y_m \to y$ in $L_p(I_a, E)$ and also $y_m \to \tilde{y}$ in $C(I_a, H)$ and hence in $L_p(I_a, H)$, we have $\| y - \tilde{y} \|_{L_p(I_a, H)} = 0$; consequently, $y(t) = \tilde{y}(t)$ for almost all $t \in \bar{I}_a$. Since $a > 0$ is arbitrary, we have $y(t) = \tilde{y}(t)$ a.e. on $\bar{I}$ and hence, neglecting a set of measure zero, we conclude that $y \in C(\bar{I}, H)$. □

1.3. Some Elements of Semigroup Theory

Semigroup theory plays a significant role in control theory on Banach spaces [B.1, B.3, Fa.2., Fa.6, Fa.8, Fa.9]. In this section we wish to present some fundamental results from semigroup theory that are applied in control theory in Chapter 5. The proof of these results can be found in Butzer and Berens [BB.1].

1.3.1. Introduction to Semigroups

Let X be a real or complex Banach space and let $\mathcal{L}(X) \equiv \mathcal{L}(X, X)$ denote the Banach algebra of endomorphisms of X. For $T \in \mathcal{L}(X)$, $\|T\|$ denotes the norm of T. Let $R_0 \equiv [0, \infty)$.

Definition 1.3.1. Let $\{T(t), t \in R_0\}$ be an operator-valued function defined on R_0 with values in $\mathcal{L}(X)$. If $T(t)$, $t \in R_0$, satisfies the following conditions:

i. $T(t_1 + t_2) = T(t_1)T(t_2)$, $t_1, t_2 \in R_0$;
ii. $T(0) = I$ (identity operator);

then $\{T(t), t \in R_0\}$ is called a *one-parameter semigroup* (*s.g.*) of operators in $\mathcal{L}(X)$.

The s.g. $\{T(t), t \in R_0\}$ is said to be *of class* c_0 if it satisfies this additional property:

iii. $\operatorname{slim}_{t \to 0^+} T(t)x = x$, $x \in X$,

referred to as the *strong continuity of* $T(t)$ *at the origin* $t = 0$.

In the rest of this book we shall generally assume, unless otherwise stated, that the family of linear operators $\{T(t), t \in R_0\}$ mapping X into itself is an s.g. of class c_0. Thus, all the three conditions of the above definition are satisfied.

Theorem 1.3.1. *Let* $\{T(t), t \in R_0\}$ *be an s.g. of class* c_0 *in* $\mathcal{L}(X)$. *Then*

i. $\|T(t)\|$ *is bounded on every finite interval of* R_0.
ii. *For each* $x \in X$, $t \to T(t)x$ *is a strongly continuous vector-valued function (with values in* X*) on* R_0.
iii. $w_0 \equiv \inf\{(1/t)\log\|T(t)\|, t \in R_0\} = \lim_{t \to \infty}(1/t)\log\|T(t)\| < \infty$.
iv. *For each* $w > w_0$, *there exists a constant* M_w *such that for all* $t \in R_0$, $\|T(t)\| \le M_w e^{wt}$.

The proof of this theorem follows essentially from Definition 1.3.1.

Due to property (ii), as given in the above theorem, and the definition for continuity of operator-valued functions in the strong operator topology $\mathcal{T}_{so}$ [see Section 1.1.8], we see that $t \to T(t)$ is continuous in the strong operator topology. Thus, the family of operators $\{T(t), t \in R_0\}$ is often called a *strongly continuous semigroup in* $\mathcal{L}(X)$. If, in addition, the map $t \to T(t)$ is continuous in the uniform operator topology ($\mathcal{T}_{uo}$), then $\{T(t), t \in R_0\}$ is said to be a uniformly continuous semigroup in $\mathcal{L}(X)$. In case $\|T(t)\| \leq \beta < \infty$ $(\beta > 1)$ for all $t \geq 0$, then $\{T(t), t \in R_0\}$ is called an *equibounded s.g. of class* c_0, and if $\beta = 1$, it is called a *contraction s.g. of class* c_0.

Definition 1.3.2. The *infinitesimal generator* A of the s.g. $\{T(t), t \in R_0\}$ is defined by

$$Ax = \mathrm{s}\lim_{t \to 0^+} A_t x, \qquad A_t x \equiv \left(\frac{T(t)x - x}{t} \right), \tag{1.3.1}$$

whenever the limit exists. The domain of the operator A, denoted by $D(A)$, is defined by

$$D(A) = \left\{ x \in X : \text{s-}\lim_{t \to 0^+} A_t x \quad \text{exists} \right\}.$$

Theorem 1.3.2

i. *$D(A)$ is a linear manifold in X and A is a linear operator.*
ii. *If $x \in D(A)$, then $T(t)x \in D(A)$ for each $t \geq 0$.*
iii. *For $x \in D(A)$, $(d/dt)(T(t)x) = AT(t)x = T(t)Ax$ for all $t \geq 0$ and*

$$T(t)x - x = \int_0^t T(\theta) Ax \, d\theta.$$

iv. *$D(A)$ is dense in X [that is, $\overline{D(A)}^s = X$] and A is a closed operator [that is, $x_n \in D(A)$, $x_n \xrightarrow{s} x_0$, $Ax_n \xrightarrow{s} y_0$ imply that $x_0 \in D(A)$ and $y_0 = Ax_0$].*

A useful consequence of the fact that the infinitesimal generator A is closed is that its domain $D(A)$ is a Banach space with respect to its graph norm:

$$\|x\|_{D(A)} = \|x\|_X + \|Ax\|_X.$$

Let $\gamma(A) \equiv \{(x, Ax) : x \in D(A)\} \subset X \times X$ denote the graph of the operator A. Clearly, if A is a closed operator, then its graph $\gamma(A)$ is a closed linear manifold of $X \times X$, and conversely. Since a closed linear manifold in a Banach space is also weakly closed, $\gamma(A)$ is also a weakly closed subspace of the product space $X \times X$.

Definition 1.3.3. For nonnegative integers $n \geq 0$ the operator A^n is defined inductively by the relations $A^0 \equiv I$, $A^1 = A$, and $D(A^n) \equiv \{x \in X : x \in D(A^{n-1})$ and $A^{n-1}x \in D(A)\}$, and by

$$A^n x = \text{s-}\lim_{t \to 0^+} A_t(A^{n-1}x) \quad \text{for } x \in D(A^n),$$

where

$$A_t \equiv \left(\frac{T(t) - I}{t} \right).$$

For the operator A^n and its domain $D(A^n)$ our results are similar to those of Theorem 1.3.2.

Theorem 1.3.3

i. *$D(A^n)$ is a linear subspace of X, and A^n is a linear operator.*
ii. *If $x \in D(A^n)$, then $T(t)x \in D(A^n)$ for $t \geq 0$.*
iii. *For $x \in D(A^n)$*

$$\frac{d^n}{dt^n}(T(t)x) = A^n T(t)x = T(t)A^n x.$$

iv. *$D(A^n)$ is dense in X for each $n = 1, 2, 3, \ldots$; moreover $\bigcap_{n=1}^{\infty} D(A^n)$ is also dense in X, and A^n is a closed operator.*

Due to this last property $D(A^n)$ is again a Banach space with respect to its graph norm defined by

$$\|x\|_{D(A^n)} \equiv \left(\sum_{s=0}^{n} \|A^s x\|_X \right).$$

Definition 1.3.4 (Holomorphic s.g.). An s.g. $T(t)$, $t \in R_0$ of class c_0 in $\mathfrak{L}(X)$ is said to be *holomorphic* if $T(t)X \subset D(A)$ for each $t > 0$.

Theorem 1.3.4. *Let $\{T(t), t \in R_0\}$ be a holomorphic s.g. of class c_0. Then*

i. *$T(t)$ is continuous for $t > 0$ in the uniform operator topology, and the same is true for $T^{(n)}(t) \equiv (d^n/dt^n)T(t)$ for all integers $n \geq 0$.*
ii. *$T(t)X \subset D(A^n)$, $t > 0$, for all integers $n \geq 0$.*
iii. *$A^n T(t)$, $n \geq 0$, are bounded linear operators (in X) for each $t > 0$.*
iv. *$(d^n/dt^n)(T(t)x) = A^n T(t)x$ for $t > 0$, $n \geq 0$, $x \in X$.*

Even though, according to the above result, the s.g. $\{T(t), t \in R_0\}$ is continuously differentiable an arbitrary number of times, on the open set $(0, \infty)$, in the uniform operator topology, it does not imply that $t \to T(t)$ is analytic. However, an additional condition guarantees the analyticity, and $T(t)$ admits an holomorphic extension $T(\xi)$ in a sector of the half plane $\operatorname{Re} \xi > 0$.

Theorem 1.3.5. *If $\{T(t),\ t\in R_0\}$ is a holomorphic s.g. of class c_0 and if there is constant $a>0$ such that $\|AT(t)\|<(a/t)$, $0<t\leq 1$, then this s.g. has a holomorphic extension $\{T(\xi),\ \xi\in\Delta\}$, where $\Delta\equiv\{\xi\in C: \operatorname{Re}\xi>0, |\arg\xi|<(1/ea)\}$; thus*

i. *$T(\xi)$ is a holomorphic operator-valued function on Δ with values in $\mathcal{L}(X)$.*
ii. *$T(\xi_1+\xi_2)=T(\xi_1)T(\xi_2)$, $\xi_1, \xi_2\in\Delta$.*
iii. s-$\lim_{\xi\to 0}T(\xi)x=x$ *for all $x\in X$, where $|\arg\xi|\leq(\varepsilon/ea)$ $(0<\varepsilon<1)$, and conversely.*

Due to c_0-property of the s.g. $\{T(t),\ t\in R_0\}$, $X_0=\bigcup_{0<t<\infty}T(t)X$ is dense in X.

If $\{T(t),\ t\in R_0\}$ is a holomorphic s.g. of class c_0 with unbounded infinitesimal generator A, then

$$\limsup_{t\to 0^+}(t\|AT(t)\|)>(1/e), \tag{1.3.2}$$

that is, near $t=0$, $\|AT(t)\|>(1/te)$.

1.3.2. Relation of the Semigroup to the Resolvent of Its Generator

Let A be a closed (not necessarily bounded) operator with domain $D(A)\subset X$ and range $R(A)\subset X$. Define

$$\rho(A)\equiv\left\{\begin{array}{l}\lambda\in C\ (\textit{complex plane}): (\lambda I-A)^{-1}\equiv R(\lambda, A)\\ \text{exists with } D(R(\lambda, A)) \text{ dense in } X\end{array}\right\}.$$

The set $\rho(A)$ is called the *resolvent set of the operator A* and $\sigma(A)\equiv C\backslash\rho(A)$ is called the *spectral set*. If $\lambda\in\rho(A)$, then $R(\lambda, A)$, called the *resolvent of A corresponding to* λ, is closed and bounded with domain dense in X. This implies that the range of the operator $A(\lambda)\equiv(\lambda I-A)$ is all of X and thus that $R(\lambda, A)$ transforms X one-to-one onto $D(A)$. That is,

i. $A(\lambda)R(\lambda, A)x=x$ for all $x\in X$ and
ii. $R(\lambda, A)A(\lambda)x=x$ for all $x\in D(A)$, and, if $\lambda, \mu\in\rho(A)$, then
iii. $R(\lambda, A)-R(\mu, A)=(\mu-\lambda)R(\lambda, A)R(\mu, A)$.

Let $\{T(t),\ t\in R_0\}$ be an s.g. of class c_0 in $\mathcal{L}(X)$. If $\{T(t),\ t\in R_0\}$ is continuous in the uniform operator topology, then the corresponding generator A is a bounded linear operator in X, $T(t)=e^{tA}$, and for $\operatorname{Re}\lambda>\|A\|$

$$R(\lambda, A)=\int_0^\infty e^{-\lambda t}e^{tA}\,dt=(\lambda I-A)^{-1},$$

that is, the resolvent is the Laplace transform of the s.g. $T(t)$, $t\geq 0$. Similar result holds for the s.g. $\{T(t),\ t\geq 0\}$ of class c_0.

Theorem 1.3.6. *Let $\{T(t), t \in R_0\}$ be an s.g. of class c_0 in $\mathcal{L}(X)$ with A as its (infinitesimal) generator. If* $w_0 = \lim_{t\to\infty}(\log\{\|T(t)\|\}/t)$ *and* $\mathrm{Re}\,\lambda > w_0$ *then*

i. $\lambda \in \rho(A)$,
ii. $R(\lambda, A)x = \int_0^\infty e^{-\lambda t} T(t)x\,dt$ *for all* $x \in X$, *and*
iii. $\text{s-lim}_{|\lambda|\to\infty} \lambda R(\lambda, A)x = x$, $(|\arg\lambda| \leq \alpha_0 < \pi/2)$.

We have seen in Theorem 1.3.2 that the generator of a c_0-s.g. $\{T(t), t \in R_0\}$ in $\mathcal{L}(X)$ is a *closed* linear operator with *domain* $D(A)$ *dense in* X. The famous Hille–Yosida theorem goes far beyond this result. It gives a necessary and sufficient condition for an operator $A : D(A) \to X$ to be the (infinitesimal) generator of a (strongly continuous) c_0-s.g. in X.

Theorem 1.3.7 (Hille–Yosida). *A necessary and sufficient condition for a closed linear operator A, with domain $D(A)$ dense in X and range $R(A) \subset X$, to generate an s.g. $\{T(t), t \in R_0\}$ of class c_0 is that there exist numbers M_0 and w_0 such that for every real $\lambda > w_0$*

i. $\lambda \in \rho(A)$ *and*
ii. $\|(R(\lambda, A))^n\| \leq M_0/(\lambda - w_0)^n$, $n = 1, 2, \ldots$. *In this case* $\|T(t)\| \leq M_0 e^{tw_0}$ *for all* $t \geq 0$.

In general an s.g. satisfying the above inequality with arbitrary (M_0, w_0) is called an *exponential s.g.* As corollaries to the above theorem we have the following results.

Corollary 1.3.1. *If A is a closed densely defined linear operator from $D(A)$ into X and $R(\lambda, A)$ exists for all λ larger than some number w_0 and satisfies the inequality $\|R(\lambda, A)\| \leq (\lambda - w_0)^{-1}$, then A is the generator of a c_0-s.g. $\{T(t), t \in R_0\}$ satisfying $\|T(t)\| \leq e^{tw_0}$ for all $t \geq 0$.*

Corollary 1.3.2. *A necessary and sufficient condition for a closed densely defined linear operator A from $D(A)$ into X to generate a contraction s.g. of class c_0 in $\mathcal{L}(X)$ is that $\{\lambda : \lambda > 0\} \subset \rho(A)$ and that $\lambda R(\lambda, A)$ be a contraction (nonexpansive) operator for each $\lambda > 0$. In this case $\|T(t)\| \leq 1$ for all $t \geq 0$.*

Similarly, if the conditions of Theorem 1.3.6 are satisfied with $w_0 = 0$, then we have $\|T(t)\| \leq M_0$ for $t \in R_0$, and the s.g. is said to be bounded. The s.g. is said to be *dissipative* if $w_0 < 0$.

Given a closed densely defined linear operator A on $D(A)$ to X, there can be at most one s.g. whose generator is the given operator A. That is, A generates a unique s.g.

1.3.3. Dual Semigroup

Let X be a real or complex Banach space with dual X^* and $\{T(t),\ t\in R_0\}$ an s.g. of class c_0.

Definition 1.3.5. Let A be a linear operator from $D(A)$ $(\subset X)$ into X with $D(A)$ dense in X. The dual operator A^* of A is a transformation whose domain consists of all those $x^*\in X^*$ for which there exists a $y^*\in X^*$ (dependent on x^*) such that

$$\langle y^*, x\rangle = \langle x^*, Ax\rangle \quad \text{for all } x\in D(A).$$

In this case we set $A^*x^*=y^*$. Since $D(A)$ is dense in X, A^* is uniquely defined.

Theorem 1.3.8. *Let A be a linear operator with domain $D(A)$ dense in X. Then*

i. *A^* is a* w*-*closed linear operator.*
ii. *If A is also closed, then $D(A^*)$ is* w*-*dense in X^*, and if X is reflexive, then $D(A^*)$ is strongly dense in X^*.*

In a reflexive Banach space, weak and weak topologies being equivalent,*

$$\overline{D(A^*)}^{\,\mathrm{s}} = \overline{D(A^*)}^{\,\mathrm{w}} = \overline{D(A^*)}^{\,\mathrm{w}^*} = X^*.$$

If $\{T(t),\ t\in R_0\}$ is an s.g. of class c_0 in $\mathcal{L}(X)$, then $\{T^*(t),\ t\in R_0\}$ is a family of operators belonging to the Banach algebra $\mathcal{L}(X^*)$ with $\|T^*(t)\| = \|T(t)\|$ for $t\in R_0$. Further,

i. $T^*(t_1+t_2)=T^*(t_1)T^*(t_2)$, $t_1, t_2\in R_0$,
ii. $T^*(0)=I^*$ (identity operator in X^*),
iii. $\text{w}^*\text{-}\lim_{t\to 0^+} T^*(t)x^*=x^*$ for all $x^*\in X^*$.

The following result for the dual s.g. T^* is similar in content to the result given in Theorem 1.3.2. for the s.g. T.

Theorem 1.3.9. *Suppose $\{T(t),\ t\in R_0\}$ is a semigroup of class c_0 and A its generator. Then*

i. *The dual A^* is a* w*-*closed linear operator, and its domain $D(A^*)$ is* w*-*dense in X^*.*
ii. *If $x^*\in D(A^*)$, then $T^*(t)x^*\in D(A^*)$ also, and*

$$A^*T^*(t)x^* = T^*(t)A^*x^*,$$

$$\langle (T^*(t)-I^*)x^*, x\rangle = \int_0^t \langle T^*(\theta)A^*x^*, x\rangle\, d\theta$$

for all $x\in X$.

iii. *An element* x^* *belongs to* $D(A^*)$ *if and only if*

$$A_t^* x^* \equiv \left(\frac{T^*(t)x^* - x^*}{t} \right)$$

has a w*-*limit as* $t \to 0^+$, *and*

$$A_t^* x^* \xrightarrow{\text{w}^*} A^* x^* \quad \text{as } t \to 0^+.$$

iv. A^* *is the* w*-*infinitesimal generator of the dual semigroup* $\{T^*(t), t \in R_0\}$.

In general the s.g. $\{T^*(t), t \in R_0\}$ is not a c_0-s.g. If, however, X is a reflexive Banach space, then $\{T^*(t), t \in R_0\}$ is a c_0-s.g. in X^*. Define

$$X_0^* \equiv \left\{ x^* \in X^* : \lim_{t \to 0^+} \|T^*(t)x^* - x^*\| = 0 \right\}; \tag{1.3.3}$$

it is clear that X_0^* is nonempty since it contains at least the zero element. In fact, it is larger.

Theorem 1.3.10

i. X_0^* *is a (strongly) closed invariant linear manifold in* X^*, *that is,* $T^*(t)X_0^* \subset X_0^*$ *for* $t \geq 0$.
ii. $D(A^*) \subset X_0^*$ *and*

$$\|T^*(t)x^* - x^*\| \leq (\sup\{\|T(\theta)\|, 0 \leq \theta \leq t\} \cdot \|A^* x^*\|)t \tag{1.3.4}$$

for all $x^* \in D(A^*)$.

As a consequence of the above theorem, X_0^* is a Banach space with respect to the strong topology of X^*.

Let $\{T_0^*(t), t \in R_0\}$ denote the restriction to X_0^* of the dual s.g. $\{T^*(t), t \in R_0\}$. Since $\{T_0^*(t), t \in R_0\}$ is an s.g. of class c_0 in the Banach algebra $\mathcal{L}(X_0^*)$, by Theorem 1.3.2 its infinitesimal generator, denoted by A_0^*, is a closed densely defined linear operator from $D(A_0^*)$ into X_0^*.

Theorem 1.3.11

i. $D(A_0^*) \subset D(A^*) \subset X_0^* \subset X^*$, X_0^* *is equal to the strong closure of* $D(A^*)$ *in* X^*, *and* X^* *is equal to the* w*-*closure of* $D(A_0^*)$.
ii. A_0^* *is equal to the largest restriction of* A^* *with both domain and range in* X_0^*, *and the graph of* A^* *is equal to the* w*-*closure of the graph of* A_0^* [*i.e.*, $\gamma(A^*) = \overline{\gamma(A_0^*)}^{\text{w}^*}$].
iii. *If* X *is reflexive, then* X^* *is equal to the strong closure of* $D(A^*)$ *in* X^* *and hence* [*due to* (*i*) *and* (*ii*)] $X_0^* = X^*$ *and* $A_0^* = A^*$.

As a consequence of the above result, if X is a reflexive Banach space, then the s.g. $\{T^*(t), t \in R_0\}$ is of class c_0 in X^*.

We close this section by stating the following approximation theorem for semigroups.

Theorem 1.3.12. *Let* X_0^* *be as defined by* (1.3.3). *Then*

i. *If* $x^* \in X_0^*$ *satisfies* $\|T^*(t)x^* - x^*\| = o(t)$, *then* x^* *is invariant under the dual* s.g. [*i.e.*, $T^*(t)x^* = x^*$, $t \geq 0$].
ii. *For an element* $x^* \in X_0^*$ *the following are equivalent*:
    a. $x^* \in D(A^*)$,
    b. $w^*\text{-}\lim_{t \to 0^+} \{ \frac{[T^*(t)x^* - x^*]}{t} \} = y^*$ *with* $y^* \in X^*$,
    c. $\|T^*(t)x^* - x^*\| = O(t)$ *as* $t \to 0^+$

Here $o(t)$ and $O(t)$ are the standard notations for small order of approximation near $t = 0^+$.

REMARK 1.3.1. Theorem 1.3.12 is an easy consequence of Theorems 1.3.9 and 1.3.10.

1.4. Measurable Set-Valued Functions

1.4.1. Introduction

Set-valued mappings, also called multifunctions, play an important role in control theory. They are used to prove the existence of measurable controls. They also are used to show the equivalence of the differential equation $\dot{x}(t) = f(t, x(t), u(t))$, $u(t) \in U(t)$, and the differential inclusion $\dot{x}(t) \in F(t, x(t))$, where $F(t, \xi) \equiv f(t, \xi, U(t))$ and $t \to U(t)$ is a suitable multifunction. This transition from differential equation to differential inclusion is very useful in control theory.

For convenience of reference, we present in this section some of the fundamental results from the theory of measurable multifunctions directly useful for our purpose. For proofs of these results and for additional information we shall refer the reader to the original literature [HJV.1, J.1, KR.1, Pl.1].

Let T be a locally compact Hausdorff space with a given positive Radon measure and X a topological vector space. Let $\mathcal{P}(X)$ denote the set of all nonempty subsets of X. A function $\Gamma: T \to \mathcal{P}(X)$ is said to be a *set-valued function* or a *multifunction* if, for each $t \in T$, $\Gamma(t) \in \mathcal{P}(X)$. If $X = (X, \rho)$ is a metric space, then Γ has closed complete, or compact values if and only if, for any $t \in T$, $\Gamma(t)$ is closed, complete, or compact, respectively. If X is only a topological space, then $\Gamma: T \to \mathcal{P}(X)$ is (weakly) measurable if and only if $\Gamma^{-1}(C) = \{t \in T: \Gamma(t) \cap C \neq \emptyset\}$ is measurable for every closed (open) subset C of X. If X is a perfectly normal topological space, then measurability implies weak measurability. For functions, measurability and weak measurability are equivalent, and this is also true for set-valued functions under certain assumptions.

Let Γ_J denote the restriction of Γ to the set $J \subset T$. Since $\Gamma_J^{-1}(B) = J \cap \Gamma^{-1}(B)$ for all $J \subset T$ and $B \subset X$, it is clear that Γ is (weakly) measurable if and only if Γ_J is (weakly) measurable for every compact J.

It is known from Lusin's theorem that if Γ is any function, then Γ is a measurable function if and only if, for every compact $J \subset T$ and $\varepsilon > 0$, there exists a compact subset $J_\varepsilon \subset J$ such that $\mu(J \backslash J_\varepsilon) < \varepsilon$ and Γ_{J_ε} is continuous. Similar results relating measurability with continuity for set-valued mappings hold. We shall state some of these results later in this section.

1.4.2. Semicontinuity of Set-Valued Maps

In the rest of this section we shall assume, unless otherwise stated, that T is a locally compact metric space. This is sufficient for the application we have in mind. Let $X = (X, \rho)$ be a metric space, $K(X)$ the class of nonempty closed subsets of X, and $\Gamma: T \to K(X)$ a multifunction.

The mapping Γ is said to be pseudo-upper semicontinuous (p.u.s.c.) at $t_0 \in T$ if

$$\bigcap_{\varepsilon > 0} \operatorname{cl} \cup \{\Gamma(t), t \in N_\varepsilon(t_0)\} \equiv \bigcap_{\varepsilon > 0} \operatorname{cl} \Gamma(N_\varepsilon(t_0)) \subseteq \Gamma(t_0), \tag{1.4.1}$$

where $N_\varepsilon(t_0)$ denotes the ε-neighborhood of $t_0 \in T$. Similarly, it is said to be pseudo-lower semicontinuous (p.l.s.c.) at $t_0 \in T$ if

$$\Gamma(t_0) \subset \bigcap_{\varepsilon > 0} \operatorname{cl} \Gamma(N_\varepsilon(t_0)). \tag{1.4.2}$$

The multifunction Γ is said to be p.u.s.c. (p.l.s.c.) on T if the property (1.4.1) [(1.4.2)] holds for all $t_0 \in T$. Γ is said to be pseudocontinuous (p.c.) at $t_0 \in T$ if it is both p.u.s.c. and p.l.s.c. at t_0. Γ is upper semicontinuous (u.s.c.) at $t_0 \in T$ if for each open set $M \subset X$ containing $\Gamma(t_0)$ there exists an $\varepsilon > 0$ such that

$$\Gamma(t) \subset M \quad \text{for all } t \in N_\varepsilon(t_0). \tag{1.4.3}$$

Γ is lower semicontinuous (l.s.c.) at $t_0 \in T$ if for every open set $M \subset X$ with $M \cap \Gamma(t_0) \neq \varnothing$, there exists an $\varepsilon > 0$ such that $\Gamma(t) \cap M \neq \varnothing$ for all $t \in N_\varepsilon(t_0)$. Γ is said to be continuous at $t_0 \in T$ if it is both u.s.c. and l.s.c. at t_0.

Let $\gamma(\Gamma) \equiv \{(t, x) \in T \times X : x \in \Gamma(t)\}$ denote the graph of the set-valued map Γ. The graph $\gamma(\Gamma)$ is said to be closed if, for every sequence $\{(t_n, x_n)\} \in \gamma(\Gamma)$ that converges to $(t, x) \in T \times X$, the limit $(t, x) \in \gamma(\Gamma)$.

1.4.3. Measurability and Continuity

An earlier result relating measurability and continuity is due to Plis [Pl.1].

Theorem 1.4.1. *Let X be a compact metric space. A multifunction $\Gamma: T \to K(X)$ is measurable if and only if, for every $\varepsilon > 0$, there exists a closed set $T_\varepsilon \subset T$ with $\mu(T \setminus T_\varepsilon) < \varepsilon$ such that Γ_ε, the restriction of Γ to T_ε, is continuous.*

This result has been generalized by Jacobs [J2]. For $x \in X$ and $A \subset X$ define

$$\rho(x, A) = \inf\{\rho(x, y), y \in A\}.$$

We assume throughout that $K(X)$ is equipped with the uniformity determined by the metric ρ.

Lemma 1.4.1. *If Γ is a measurable multifunction with values $\Gamma(t) \in \mathcal{P}(X)$ for $t \in T$ and X is separable, then for every $\varepsilon > 0$ there exists a closed set $T_\varepsilon \subset T$ such that $\mu(T \setminus T_\varepsilon) < \varepsilon$ and such that the restriction of the mapping $(t, x) \to \rho(x, \Gamma(t))$ to $T_\varepsilon \times X$ is continuous.*

Theorem 1.4.2. *Let X be a polish space (complete separable metric space) and $\Gamma: T \to K(X)$ a multifunction. If, for every $\varepsilon > 0$, there exists a closed set $T_\varepsilon \subset T$ such that $\mu(T \setminus T_\varepsilon) < \varepsilon$ and Γ_ε($\equiv$restriction of Γ to T_ε) is continuous, then Γ is measurable.*

Lemma 1.4.2. *Let Γ be a mapping from T into $K(X)$. Then a necessary and sufficient condition that Γ be p.u.s.c. at each point of T is that the graph $\gamma(\Gamma)$ be closed.*

We are more interested in the following result.

Theorem 1.4.3. *Let T be a locally compact separable metric space, X a separable metric space, and $\Gamma: T \to \mathcal{P}(X)$ a set-valued function with complete values [i.e., $\Gamma(t)$ is complete for each $t \in T$]. Then the following statements are equivalent.*

i. *Γ is measurable.*
ii. *Γ is weakly measurable.*
iii. *$\phi_\Gamma(t, x) \equiv \rho(x, \Gamma(t))$ is a measurable function of t on T for each $x \in X$.*
iv. *For each $\varepsilon > 0$ there exists a closed set $T_\varepsilon \subset T$ such that $\mu(T \setminus T_\varepsilon) < \varepsilon$ and ϕ_Γ restricted to $T_\varepsilon \times X$ is continuous.*
v. *For each $\varepsilon > 0$ there exists a closed (souslin) set $T_\varepsilon \subset T$ such that $\mu(T \setminus T_\varepsilon) < \varepsilon$ and the graph $\gamma(\Gamma_\varepsilon)$ of Γ_ε ($\equiv$restriction of Γ to T_ε) is a closed (souslin) subspace of $T_\varepsilon \times X$.*
vi. *There exists a souslin subspace T_0 of T such that $\mu(T \setminus T_0) = 0$ and $\gamma(\Gamma_0)$ is souslin where Γ_0 is the restriction of Γ to T_0.*

If X is a souslin space [Bo.1, p. 197] the values of Γ need only be closed, rather than complete, to obtain the equivalence of each of (i)–(vi) with each of the following:

vii. *For each $\varepsilon > 0$ there exists a closed subspace T_ε of T such that $\mu(T \setminus T_\varepsilon) < \varepsilon$ and Γ_ε ($\equiv$restriction of Γ to T_ε) is p.u.s.c.*
viii. *For each $\varepsilon > 0$ there exists a set $T_\varepsilon \subset T$ such that $\mu(T \setminus T_\varepsilon) < \varepsilon$ and $\gamma(\Gamma_\varepsilon)$ is a closed subspace of $T_\varepsilon \times X$.*

1.4.4. Selection Theorems

Let Γ be a measurable multifunction from T into $K(X)$. A mapping $\xi: T \to X$ is called *a selection of* Γ if $\xi(t) \in \Gamma(t)$, $t \in T$. A mapping $\xi: T \to X$ is called a *measurable selection of* Γ if

i. ξ is a measurable function from T into X and
ii. $\xi(t) \in \Gamma(t)$ a.e. on T.

Theorem 1.4.4. *Let T be a locally compact metric space, $X=(X, \rho)$ a separable metric space, and $G: T \to \mathcal{P}(X)$ a measurable multifunction with complete values. Then there exists a measurable function $\xi: T \to X$ such that $\xi(t) \in G(t)$ for almost all $t \in T$.*

This result was obtained by Himmelberg–Jacobs–Van Vleck [HJV.1] under conditions weaker than those stated here. For our purpose these conditions are sufficient. An earlier selection theorem due to Kuratowski and Ryll-Nardzewski [KR.1] uses stronger conditions. This result is given below.

Theorem 1.4.5. *Let X be a complete separable metric space, T Lebesgue measurable set, and $G: T \to K(X)$ a measurable multifunction. Then there exists a Lebesgue measurable selection.*

The following result is a consequence of Theorem 1.4.4.

Theorem 1.4.6. *Let T be a locally compact metric space, X a separable metric space, and $G: T \to \mathcal{P}(X)$ a measurable multifunction with complete values. Let Y be another metric space and $f: T \times X \to Y$ a function measurable in t on T for each $x \in X$ and locally uniformly continuous in x on X for each $t \in T$. Let $y: T \to Y$ be a measurable function such that $y(t) \in f(t, G(t))$ for all $t \in T$. Then there exists a measurable function $x: T \to X$ such that $x(t) \in G(t)$ and $y(t) = f(t, x(t))$ for all $t \in T$.*

We close this section by stating the following result.

Proposition 1.4.1. Let l and s be positive integers, Q be an open subset of R^l with compact closure $\bar{Q}$, M a nonempty bounded subset of R^s, and $KC(M)$ the class of nonempty, closed convex subsets of M. Let $U: \bar{Q} \to KC(M)$ be a measurable multifunction on $\bar{Q}$ with values $U(q) \in KC(M)$, and let $\mathcal{U}$ denote the class of measurable functions $\{u\}$ defined on $\bar{Q}$ with values $u(q) \in U(q)$ a.e. on $\bar{Q}$. Then $\mathcal{U}$ is a w*-compact subset of $L_\infty(\bar{Q}, R^s)$.

REMARK 1.4.1. For an excellent survey of measurable set valued maps and selection theorems see [Io.1, W.1].

CHAPTER TWO

PARTIAL DIFFERENTIAL EQUATIONS, INTEGRAL EQUATIONS, AND EVOLUTION EQUATIONS

In this chapter, we discuss the question of existence and uniqueness, and study the properties of solutions of several classes of dynamical systems that are of direct concern in later chapters. We consider two major classes of systems. The first major class consists of first boundary value problems of second-order partial differential equations. The second major class consists of abstract evolution equations in Banach spaces.

2.1. First Boundary Value Problems for Parabolic Partial Differential Equations in General Form

In this section, a class of systems governed by linear second-order parabolic partial differential equations in general form with first boundary condition is studied. This class of problems is sometimes referred to as first boundary value problems (in general form).

Consider the following system:

$$\begin{aligned} L\phi(x,t)&=f(x,t), \qquad (x,t)\in Q,\\ \phi(x,0)&=\phi_0(x), \qquad x\in\Omega,\\ \phi(x,t)&=0, \qquad (x,t)\in\partial\Omega\times[0,T], \end{aligned} \tag{2.1.1}$$

where T is a fixed positive number, Ω an open bounded subset of n-dimensional euclidean space R^n, $\partial\Omega$ the boundary of Ω, $I\equiv(0,T)$, $Q\equiv\Omega\times I$, and the operator L given by

$$\begin{aligned} L\psi(x,t)\equiv\psi_t(x,t)&-\sum_{i,j=1}^{n} a_{ij}(x,t)\psi_{x_ix_j}(x,t)\\ &-\sum_{i=1}^{n} b_i(x,t)\psi_{x_i}(x,t)-c(x,t)\psi(x,t), \end{aligned} \tag{2.1.2}$$

where

$$\psi_t(x,t)=\frac{\partial\psi(x,t)}{\partial t},\quad \psi_{x_i}(x,t)=\frac{\partial\psi(x,t)}{\partial x_i},\quad \text{and}\quad \psi_{x_ix_j}(x,t)=\frac{\partial^2\psi(x,t)}{\partial x_i\partial x_j}.$$

We introduce the following definition.

Definition 2.1.1. A function $\phi:\overline{Q}\to R$ is said to be a *classical solution* of system (2.1.1) if it is continuous in $\overline{Q}$, has continuous derivatives $\phi_t, \phi_{x_i}, \phi_{x_ix_j}$ $(i, j=1,\ldots,n)$ in Q, and it satisfies all the equations of the system everywhere in their domains of definition.

The following theorem is the well-known maximum principle for system (2.1.1).

Theorem 2.1.1. *Consider system* (2.1.1). *Suppose that*

$$\sum_{i,j=1}^{n} a_{ij}(x,t)\xi_i\xi_j>0 \tag{2.1.3}$$

on Q for any nonzero vector $\xi\in R^n$. Let ϕ be the classical solution of the system, and let M be some constant such that $c(x,t)<M$ for all $(x,t)\in\overline{Q}$. Then $\phi(x,t)\geq 0$ in $\overline{Q}$ if $\phi_0(x)\geq 0$ in $\overline{\Omega}$ and $f(x,t)\geq 0$ on Q.

PROOF. The proof is divided into two parts, (i) $M\leq 0$ and (ii) $M>0$. First, we consider the case when $M\leq 0$. Then $c(x,t)<0$ in $\overline{Q}$. If ϕ takes a negative value at some point of $\overline{Q}$, then there exists a point $(x^0,t^0)\in\overline{Q}$ such that $\phi(x^0,t^0)$ is the minimum value of ϕ on $\overline{Q}$. From the hypotheses of the theorem, (x^0,t^0) must lie either in Q or in $\Omega\times\{T\}$. Consequently,

$$\left(\frac{\partial\phi}{\partial x_i}\right)(x^0,t^0)=0,\qquad i=1,\ldots,n,$$

$$\left(\frac{\partial\phi}{\partial t}\right)(x^0,t^0)\leq 0,$$

and

$$c(x^0,t^0)\phi(x^0,t^0)>0.$$

Next we note that at a minimum point the pure second-order derivatives $(\partial^2\phi/\partial y_\kappa^2)(x^0,t^0)$ are nonnegative in any direction $y_\kappa=\sum_{l=1}^n k_{\kappa l}(x_l-x_l^0)$, where the matrix $K=(k_{\kappa l})$ is a nondegenerate matrix. Choose the matrix K to be such that $KAK'=\operatorname{diag}(\lambda_{11},\lambda_{22},\ldots,\lambda_{nn})$, where $A=(a_{ij})$, K' denotes the transpose of K, and

$$\lambda_{\kappa\kappa}(x,t)=\sum_{i,j=1}^{n} k_{\kappa j}a_{ij}(x,t)k_{\kappa i},\qquad \kappa=1,\ldots,n.$$

Thus, it follows from (2.1.3) that $\lambda_{\kappa\kappa}(x,t)>0$ in $\overline{Q}$. Next, it is easily deduced that

$$\sum_{i,j=1}^{n} a_{ij}(x,t)\frac{\partial^2\phi(x,t)}{\partial x_i \partial x_j} = \sum_{\kappa=1}^{n} \lambda_{\kappa\kappa}(x,t)\frac{\partial^2\phi(x,t)}{\partial y_\kappa^2}.$$

Since $(\partial^2\phi/\partial y_\kappa^2)(x^0, t^0)$ are nonnegative in any direction, $y_\kappa = \sum_{l=1}^{n} k_{\kappa l}(x_l - x_l^0)$, and $\lambda_{\kappa\kappa}(x,t)>0$ in $\overline{Q}$, it follows readily that

$$\sum_{i,j=1}^{n} a_{ij}(x^0, t^0)\left(\frac{\partial^2\phi}{\partial x_i \partial x_j}\right)(x^0, t^0) \geq 0.$$

Thus, $f(x^0, t^0)<0$. This is a contradiction and hence the proof is complete for the case $M \leq 0$.

It remains to prove the second case, $M>0$. Define the function ψ by

$$\psi(x,t) \equiv \phi(x,t)e^{\lambda t},$$

where λ is for the time being an arbitrary number. Clearly, the function satisfies the differential equation

$$\psi_t(x,t) - \sum_{i,j=1}^{n} a_{ij}(x,t)\psi_{x_i x_j}(x,t) - \sum_{i=1}^{n} b_i(x,t)\psi_{x_i}(x,t)$$
$$-(c(x,t)+\lambda)\psi(x,t) = f(x,t)e^{\lambda t}$$

for all $(x,t)\in Q$. Choose $-\lambda > M$. Then $c(x,t)+\lambda < M+\lambda < 0$. Consequently, by what has been proved, $\psi(x,t)\geq 0$ in $\overline{Q}$. Thus, $\phi(x,t) = \psi(x,t)e^{-\lambda t}$ is also nonnegative in $\overline{Q}$. The proof of the theorem is now complete. □

Before we present a result concerning the existence and uniqueness of classical solutions of system (2.1.1.), we need the concept of the compatibility condition (of order $m \geq 0$) for system (2.1.1). For this, we shall introduce the following notation.

$$Z^{(k)}(x) = \left.\frac{\partial^k Z(x,t)}{\partial t^k}\right|_{t=0}. \tag{2.1.4}$$

$$A\left(x,t,\frac{\partial}{\partial x}\right)\phi(x,t) = \sum_{i,j=1}^{n} a_{ij}(x,t)\frac{\partial^2\phi(x,t)}{\partial x_i \partial x_j} + \sum_{i=1}^{n} b_i(x,t)\frac{\partial\phi(x,t)}{\partial x_i} + c(x,t)\phi(x,t). \tag{2.1.5}$$

The functions $\phi^{(k)}(x)$, $k=0,1$, are determined as follows:

$$\phi^{(0)}(x) = \phi_0(x), \tag{2.1.6}$$

$$\phi^{(1)}(x) = A\left(x, 0, \frac{\partial}{\partial x}\right)\phi_0(x) + f(x,0)). \tag{2.1.7}$$

The remaining functions are found from the recursion relation

$$\phi^{(k+1)}(x) = \left[\frac{\partial^k}{\partial t^k} \left(A\left(x, t, \frac{\partial}{\partial x}\right) \phi(x,t) \right) + \frac{\partial^k f(x,t)}{\partial t^k} \right]_{t=0}$$
$$= \sum_{l=1}^{n} \binom{k}{l} A^{(l)}\left(x, 0, \frac{\partial}{\partial x}\right) \phi^{(k-l)}(x) + f^{(k)}(x), \qquad (2.1.8)$$

where the operator $A^{(l)}(x, t, \partial/\partial x)$ is defined by

$$A^{(l)}\left(x, t, \frac{\partial}{\partial x}\right)\psi(x,t) = \sum_{i,j=1}^{n} \frac{\partial^l a_{ij}(x,t)}{\partial t^l} \frac{\partial^2 \psi(x,t)}{\partial x_i \partial x_j}$$
$$+ \sum_{i=1}^{n} \frac{\partial^l b_i(x,t)}{\partial t^l} \frac{\partial \psi(x,t)}{\partial x_i} + \frac{\partial^l c(x,t)}{\partial t^l} \psi(x,t). \qquad (2.1.9)$$

At this stage, the concept of the compatibility condition (of order $m \geq 0$) for system (2.1.1) can be defined.

Definition 2.1.2. The compatibility condition of order $m \geq 0$ is said to be satisfied for system (2.1.1) if

$$\phi^{(k)}(x)|_{x \in \partial\Omega} = 0, \qquad k = 0, 1, \ldots, m.$$

Intuitively, the above compatibility condition ensures that the classical solution has a similar order of smoothness as that of the given data (f, ϕ_0).

The next theorem contains a fundamental result concerning the existence and uniqueness of classical solutions of system (2.1.1). This theorem is a classical result; the proof can be found in Ladyžhenskaja, Solonikov, and Ural'ceva [LSU.1, Theorem 5.2, Chapter IV, p. 320]. For this, we need the following assumption.

Assumption (2.1.A1). Ω is open and connected with compact closure $\bar{\Omega} = \Omega \cup \partial\Omega$, where the boundary $\partial\Omega$ of Ω is a compact manifold of class C^2.

Theorem 2.1.2. *Consider system* (2.1.1). *Suppose that Assumption* (2.1.A1) *is satisfied and that the coefficients and free term of the system belong to the Hölder space* $\mathcal{H}^{\iota, \iota/2}(\bar{Q})$ *(see Section* 1.2.1), *where* ι *is a nonintegral positive number. Furthermore, let there exist a positive constant* α_1 *such that*

$$\sum_{i,j=1}^{n} a_{ij}(x,t)\xi_i\xi_j \geq \alpha_1 \sum_{i=1}^{n} (\xi_i)^2$$

for all $\xi \in R^n$ *uniformly on* $\bar{Q}$ *and let* $\phi_0 \in \mathcal{H}^{\iota+2}(\bar{\Omega})$. *In addition, let the compatibility condition of order* $[\iota/2]+1$ $([\gamma] \equiv$ *largest integer* $\leq \gamma)$ *be satisfied for system* (2.1.1). *Then, the system has a unique classical solution* $\phi \in \mathcal{H}^{\iota+2, (\iota/2)+1}(Q)$.

System (2.1.1) can also be solved in the space $W_p^{2,1}(\bar{Q})$, $3/2<p<\infty$ (see Section 1.2.3). In this situation, the conditions on the coefficients and data of the system can be substantially relaxed. These assumptions are given below.

Assumption (2.1.A2). a_{ij}, $i, j=1,\ldots,n$, are continuous functions on $\bar{Q}$.

Assumption (2.1.A3). There exist positive constants α_1 and α_2 such that

$$\alpha_1 \sum_{i=1}^{n} (\xi_i)^2 \leq \sum_{i,j=1}^{n} a_{ij}(x,t)\xi_i\xi_j \leq \alpha_2 \sum_{i=1}^{n} (\xi_i)^2$$

for all $\xi \in R^n$ uniformly on $\bar{Q}$.

Assumption (2.1.A4). b_i, $i=1,\ldots,n$, c, and f are bounded measurable functions on $\bar{Q}$.

Assumption (2.1.A5). $\phi_0 \in C_0^2(\Omega)$.

Presentation of the results on the existence and uniqueness of solution from the class $W_p^{2,1}(Q)$, $3/2<p<\infty$, of system (2.1.1) requires the following definition.

Definition 2.1.3. A function $\phi: \bar{Q} \to R$ is said to be a *solution from the class* $W_p^{2,1}(Q)$, $3/2<p<\infty$, *of system* (2.1.1) if it satisfies the following conditions:

i. $\phi \in W_p^{2,1}(Q)$, $3/2<p<\infty$;
ii. ϕ satisfies the differential equation of the system a.e. in Q;
iii. ϕ satisfies the initial and boundary conditions of the system everywhere in its domain of definition.

Note that all the partial derivatives in the above definition are understood in the generalized sense (see Section 1.2.2).

Theorem 2.1.3. *Consider system* (2.1.1). *Suppose that Assumptions* (2.1.A1)–(2.1.A5) *are satisfied. Then the system has a unique solution* $\phi \in W_p^{2,1}(Q)$, $3/2<p<\infty$. *In addition, for given second-order coefficients and* Q, ϕ *satisfies the following a priori estimate*:

$$\|\phi\|_{p,Q}^{(2,1)} \leq K_1\{\|f\|_{p,Q} + \|\phi_0\|_{p,\Omega}^{(2)}\} \tag{2.1.10}$$

for all $p \geq 3/2$, *where* $\|\cdot\|_{p,\Omega}^{(2)}$, $\|\cdot\|_{p,Q}$, *and* $\|\cdot\|_{p,Q}^{(2,1)}$ *are the norms in the Banach spaces* $W_p^2(\Omega)$, $L_p(Q)$, *and* $W_p^{2,1}(Q)$ (*see Sections* 1.2.2 *and* 1.2.3), *respectively, and the constant* K_1 *depends only on bounds for* b_i, $i=1,\ldots,n$, *and* c *on* $\bar{Q}$.

The result of the above theorem is contained in Ladyžhenskaja, Solonikov, and Ural'ceva [LSU.1, Theorem 9.1, Chapter IV, pp. 341–342].

If p is large enough, then it can be proved [LSU.1, Lemma 3.3, Chapter II, p. 80] that the a priori estimate (2.1.10) implies a Hölder estimate as given in the following result.

Theorem 2.1.4. *Consider system* (2.1.1). *Suppose that the hypotheses of Theorem* 2.1.3 *are satisfied and that* $p>n+2$. *Then*

$$|\phi|_{Q}^{(1+\mu,(1+\mu)/2)} \le K_2 \|\phi\|_{p,Q}^{(2,1)}, \tag{2.1.11}$$

where $\mu = 1-[(n+2)/p]$, $|\cdot|_Q^{(\lambda,\lambda/2)}$ *denotes the norm in the Hölder space* $\mathcal{H}^{\lambda,\lambda/2}(\overline{Q})$ (*see Section* 1.2.1), *and the constant* K_2 *depends only on* Q *and* p.

REMARK 2.1.1. Since Q is bounded it is clear that, under Assumptions (2.1.A1)–(2.1.A5), the estimate (2.1.10) reduces to a stronger version,

$$\|\phi\|_{p,Q}^{(2,1)} \le K_3\{\|f\|_{\infty,Q} + \|\phi_0\|_{\infty,\Omega}^{(2)}\} \tag{2.1.12}$$

for all $p>3/2$, where $K_3 \equiv K_1 \max\{(|Q|)^{1/p}, (|\Omega|)^{1/p}\}$ and $|G|$ denotes the Lebesgue measure of G. On the other hand, we recall that the estimate (2.1.11) holds for all $p>n+2$. Thus, it follows from (2.1.12) that

$$|\phi|_{Q}^{(1+\mu,(1+\mu)/2)} \le K_4\{\|f\|_{\infty,Q} + \|\phi_0^{(2)}\|_{\infty,\Omega}\} \tag{2.1.13}$$

for all $\mu \in (0,1)$, where $K_4 \equiv K_2 K_3$. □

For the proof of the next two theorems, we need to take the integral averages [for definition, see formula (1.2.1)] of the coefficients and the data. Thus, it is convenient to extend the definition of system (2.1.1) to all of R^{n+1}. Let

$$L\psi(x,t) \equiv \psi_t(x,t) - \sum_{i=1}^{n} \psi_{x_i x_i}(x,t), \qquad (x,t) \in R^{n+1} \backslash Q,$$

and

$$f(x,t) \equiv 0, \qquad (x,t) \in R^{n+1} \backslash Q.$$

These conventions are used throughout the rest of this section without further mention.

REMARK 2.1.2. For each positive integer σ, let a_{ij}^{σ}, $i,j=1,\dots,n$; b_i^{σ}, $i=1,\dots,n$; c^{σ}; and f^{σ} denote, respectively, the integral averages of the functions a_{ij}, $i,j=1,\dots,n$; b_i, $i=1,\dots,n$; c; and f on R^{n+1}. Further, let $\{\Omega^{\sigma}\}$ be a sequence of open domains such that $\overline{\Omega^{\sigma}} \subset \Omega^{\sigma+1} \subset \overline{\Omega^{\sigma+1}} \subset \Omega$ for all integers $\sigma \ge 1$ and $\lim_{\sigma\to\infty} \Omega^{\sigma} = \Omega$. Let $Q^{\sigma} \equiv \Omega^{\sigma} \times I^{\sigma}$, where $\{I^{\sigma}\}$ is a sequence of intervals in $(0,T)$ for which $\lim_{\sigma\to\infty} I^{\sigma} = (0,T)$. For each integer $\sigma \ge 1$, let $g^{\sigma} \in C_0^{\infty}(Q)$, $g^{\sigma}(x,t)=1$ on $Q^{\sigma-1}$ and $0 \le g^{\sigma}(x,t) \le 1$ elsewhere. □

We consider the following sequence of first boundary value problems:

$$\begin{aligned} &L^\sigma\phi(x,t)=g^\sigma(x,t)f^\sigma(x,t), \qquad (x,t)\in Q,\\ &\phi(x,t)=0, \qquad (x,t)\in\partial\Omega\times[0,T], \\ &\phi(x,0)=\phi_0(x), \qquad x\in\Omega, \end{aligned} \tag{2.1.14}$$

where, for each positive integer σ, the operator L^σ is defined by

$$L^\sigma\psi(x,t)\equiv\psi_t(x,t)-\sum_{i,j=1}^{n} a_{ij}^\sigma(x,t)\psi_{x_ix_j}(x,t)$$
$$-\sum_{i=1}^{n} b_i^\sigma(x,t)\psi_{x_i}(x,t)-c^\sigma(x,t)\psi(x,t). \tag{2.1.15}$$

Suppose that Assumptions (2.1.A1)–(2.1.A5) are satisfied. Then we observe from Theorem 1.2.1 that, for each positive integer σ, the coefficients and the free term of the system are all C^∞-functions. Further, it is clear from the definition of the function g^σ and Assumption (2.1.A5) that the compatibility condition of order 1 is satisfied. At this stage, we note that all the hypotheses of Theorem 2.1.2 are satisfied. Thus, it follows that, for each integer $\sigma\geq 1$, system (2.1.14) admits a unique classical solution $\phi^\sigma\in\mathcal{H}^{\iota+2,(\iota/2)+1}(\overline{Q})$, where $\iota\in(0,1)$. However, by Theorem 2.1.3, ϕ^σ is also the unique solution of system (2.1.14) in the sense of Definition 2.1.3 for each integer $\sigma\geq 1$.

Theorem 2.1.5. *Consider systems* (2.1.1) *and* (2.1.14). *Suppose that Assumptions* (2.1.A1)–(2.1.A5) *are satisfied. Moreover let* $\phi\in W_p^{2,1}(Q)$, $3/2<p<\infty$, *be the solution of system* (2.1.1), *and let* $\{\phi^\sigma\}$ *be the sequence of classical solutions of system* (2.1.14). *Then there exists a subsequence of the sequence* $\{\phi^\sigma\}$, *which is again denoted by* $\{\phi^\sigma\}$, *such that, as* $\sigma\to\infty$,

$$\phi^\sigma\overset{\text{u}}{\to}\phi, \qquad \phi^\sigma_{x_i}\overset{\text{u}}{\to}\phi_{x_i}, \qquad i=1,\ldots,n,$$

(*uniformly*) *on* $\overline{Q}$, *and*

$$\phi^\sigma_t\overset{\text{w}}{\to}\phi_t, \qquad \phi^\sigma_{x_ix_j}\overset{\text{w}}{\to}\phi_{x_ix_j}, \qquad i,j=1,\ldots,n,$$

(*weakly*) *in* $L_p(Q)$, $3/2<p<\infty$.

PROOF. By Assumptions (2.1.A2) and (2.1.A4), and the definition of g^σ in Remark 2.1.2, we deduce readily that

$$\begin{aligned} &|a_{ij}^\sigma|_Q^{(0)}\equiv\max\{|a_{ij}^\sigma(x,t)|:(x,t)\in\overline{Q}\}\leq|a_{ij}|_Q^{(0)}, \qquad i,j=1,\ldots,n,\\ &|b_i^\sigma|_Q^{(0)}\leq|b_i|_Q^{(0)}, \qquad i=1,\ldots,n,\\ &|c^\sigma|_Q^{(0)}\leq|c|_Q^{(0)},\\ &\|g^\sigma f^\sigma\|_{\infty,Q}\leq\|f\|_{\infty,Q}. \end{aligned} \tag{2.1.16}$$

Thus, from Remark 2.1.1, it follows that

$$\|\phi^\sigma\|_{p,Q}^{(2,1)} \le K_3\{\|f\|_{\infty,Q} + \|\phi_0\|_{\infty,\Omega}^{(2)}\} \quad \text{for } p>3/2, \tag{2.1.17}$$

and

$$|\phi^\sigma|_{\bar{Q}}^{(1+\mu,(1+\mu)/2)} \le K_4\{\|f\|_{\infty,Q} + \|\phi_0\|_{\infty,\Omega}^{(2)}\} \quad \text{for } \mu, 0<\mu<1, \tag{2.1.18}$$

where σ is any positive integer and the constants K_3 and K_4 are as given in Remark 2.1.1.

Clearly, the estimate (2.1.18) implies that, if $\xi^\sigma = \phi^\sigma$ or $\phi^\sigma_{x_i}$, $i=1,\ldots,n$, then

$$|\xi^\sigma(x,t)| + \frac{|\xi^\sigma(x',t') - \xi^\sigma(x,t)|}{\left[|x'-x|^2 + |t'-t|\right]^{\mu/2}} \le \hat{K}_4\{\|f\|_{\infty,Q} + \|\phi_0\|_{\infty,\Omega}^{(2)}\} \equiv \tilde{K}_4 \tag{2.1.19}$$

for all (x,t), $(x',t') \in \bar{Q}$, where $\mu \in (0,1)$, while $\hat{K}_4$ and hence $\tilde{K}_4$ are independent of σ. Thus, from the Ascoli–Arzelà theorem, there exists a system of subsequences $\{\phi^{\sigma(k)}, \phi^{\sigma(k)}_{x_i}, i=1,\ldots,n\} \subset \{\phi^\sigma, \phi^\sigma_{x_i}, i=1,\ldots,n\}$ and a system of functions $\{\phi, \phi^i, i=1,\ldots,n\}$ such that, as $k \to \infty$,

$$\phi^{\sigma(k)} \overset{u}{\to} \phi \quad \text{and} \quad \phi^{\sigma(k)}_{x_i} \overset{u}{\to} \phi^i, \qquad i=1,\ldots,n,$$

(uniformly) on $\bar{Q}$. Clearly, the limit functions ϕ and ϕ^i, $i=1,\ldots,n$, satisfy the estimate (2.1.19).

Since $\{\phi^{\sigma(k)}\}$ is a subsequence of the sequence $\{\phi^\sigma\}$, it follows from the estimate (2.1.17) that

$$\|\phi^{\sigma(k)}\|_{p,Q}^{(2,1)} \le K_3\{\|f\|_{\infty,Q} + \|\phi_0\|_{\infty,\Omega}^{(2)}\} \equiv \tilde{K}_3 \tag{2.1.20}$$

for all $p \in (3/2, \infty)$. Thus, there exists a common system of subsequences

$$\left\{\phi^{\sigma(k(l))}, \phi^{\sigma(k(l))}_t, \phi^{\sigma(k(l))}_{x_i}, \phi^{\sigma(k(l))}_{x_i x_j}, i, j=1,\ldots,n\right\}$$

and a system of functions $\{\phi, \psi, \phi^i, \psi^{ij}, i, j=1,\ldots,n\}$ such that, as $l \to \infty$,

$$\phi^{\sigma(k(l))} \overset{w}{\to} \phi, \qquad \phi^{\sigma(k(l))}_t \overset{w}{\to} \psi, \qquad \phi^{\sigma(k(l))}_{x_i} \overset{w}{\to} \phi^i, \qquad i=1,\ldots,n,$$

and

$$\phi^{\sigma(k(l))}_{x_i x_j} \overset{w}{\to} \psi^{ij}, \qquad i, j=1,\ldots,n,$$

(weakly) in $L_p(Q)$, $3/2<p<\infty$, where the limit functions satisfy the estimate

$$\|\phi\|_{p,Q} + \|\psi\|_{p,Q} + \sum_{i=1}^n \|\phi^i\|_{p,Q} + \sum_{i,j=1}^n \|\psi^{ij}\|_{p,Q} \le \tilde{K}_3 \tag{2.1.21}$$

for all $p>3/2$.

Next, for any $Z \in C_0^1(Q)$, it follows from integration by parts that

$$\iint_Q \{\phi_{x_i}^{\sigma(k(l))}(x,t)Z(x,t)\}\,dx\,dt = -\iint_Q \{\phi^{\sigma(k(l))}(x,t)Z_{x_i}(x,t)\}\,dx\,dt,$$

$$i=1,\dots,n,$$

$$\iint_Q \{\phi_t^{\sigma(k(l))}(x,t)Z(x,t)\}\,dx\,dt = -\iint_Q \{\phi^{\sigma(k(l))}(x,t)Z_t(x,t)\}\,dx\,dt,$$

$$\iint_Q \{\phi_{x_i x_j}^{\sigma(k(l))}(x,t)Z(x,t)\}\,dx\,dt = -\iint_Q \{\phi_{x_i}^{\sigma(k(l))}(x,t)Z_{x_j}(x,t)\}\,dx\,dt,$$

$$i, j=1,\dots,n.$$

Thus, by letting $l \to \infty$, we have

$$\iint_Q \{\phi^i(x,t)Z(x,t)\}\,dx\,dt = -\iint_Q \{\phi(x,t)Z_{x_i}(x,t)\}\,dx\,dt,$$

$$i=1,\dots,n,$$

$$\iint_Q \{\psi(x,t)Z(x,t)\}\,dx\,dt = -\iint_Q \{\phi(x,t)Z_t(x,t)\}\,dx\,dt,$$

$$\iint_Q \{\psi^{ij}(x,t)Z(x,t)\}\,dx\,dt = -\iint_Q \{\phi^i(x,t)Z_{x_j}(x,t)\}\,dx\,dt,$$

$$i, j=1,\dots,n.$$

This implies that ϕ^i, ψ, and ψ^{ij} are respectively, the generalized derivatives of, ϕ with respect to x_i, ϕ with respect to t, and ϕ^i with respect to x_j. Therefore, they can be written as $\phi_{x_i}, \phi_t, \phi_{x_i x_j}$, respectively. Thus, it follows from (2.1.21) that $\phi \in W_p^{2,1}(Q)$, $3/2 < p < \infty$.

At this point, we have, as $l \to \infty$,

$$\phi^{\sigma(k(l))} \xrightarrow{u} \phi, \qquad \phi_{x_i}^{\sigma(k(l))} \xrightarrow{u} \phi_{x_i}, \qquad \text{on } \bar{Q},\ i=1,\dots,n$$

$$\phi_t^{\sigma(k(l))} \xrightarrow{w} \phi_t, \qquad \phi_{x_i x_j}^{\sigma(k(l))} \xrightarrow{w} \phi_{x_i x_j}, \qquad \text{in } L_p(Q),\quad 3/2 < p < \infty,\ i, j=1,\dots,n. \tag{2.1.22}$$

Now we shall show that the function ϕ satisfies the differential equation of system (2.1.1) a.e. on Q. For this, first note that, for each positive integer l, $\phi^{\sigma(k(l))}$ satisfies the differential equation of system (2.1.14) everywhere on

Q. Thus, for almost all $(x,t)\in Q$,

$$\begin{aligned}
\phi_t(x,t) &- \sum_{i,j=1}^{n} a_{ij}(x,t)\phi_{x_ix_j}(x,t) - \sum_{i=1}^{n} b_i(x,t)\phi_{x_i}(x,t) - c(x,t)\phi(x,t) \\
&- f(x,t) = \left(\phi_t(x,t) - \phi_t^{\sigma(k(l))}(x,t)\right) \\
&\qquad - \sum_{i,j=1}^{n} \left(a_{ij}(x,t)\phi_{x_ix_j}(x,t)\right. \\
&\qquad \left. - a_{ij}^{\sigma(k(l))}(x,t)\phi_{x_ix_j}^{\sigma(k(l))}(x,t)\right) - \sum_{i=1}^{n} \left(b_i(x,t)\phi_{x_i}(x,t)\right. \\
&\qquad \left. - b_i^{\sigma(k(l))}(x,t)\phi_{x_i}^{\sigma(k(l))}(x,t)\right) - \left(c(x,t)\phi(x,t)\right. \\
&\qquad \left. - c^{\sigma(k(l))}(x,t)\phi^{\sigma(k(l))}(x,t)\right) - \left(f(x,t)\right. \\
&\qquad \left. - g^{\sigma(k(l))}(x,t)f^{\sigma(k(l))}(x,t)\right).
\end{aligned} \tag{2.1.23}$$

Recall that ϕ satisfies the estimate (2.1.21) and, by Assumptions (2.1.A2) and (2.1.A4), the coefficients and free term of system (2.1.1) are bounded measurable on $\overline{Q}$. Thus, the expression on the left-hand side of (2.1.23) is an element of $L_p(Q)$, $3/2<p<\infty$. In view of (2.1.16) and (2.1.20), we observe that

$$\begin{aligned}
&\phi_t^{\sigma(k(l))}(\cdot,\cdot) - \sum_{i,j=1}^{n} a_{ij}^{\sigma(k(l))}(\cdot,\cdot)\phi_{x_ix_j}^{\sigma(k(l))}(\cdot,\cdot) \\
&- \sum_{i=1}^{n} b_i^{\sigma(k(l))}(\cdot,\cdot)\phi_{x_i}^{\sigma(k(l))}(\cdot,\cdot) - c^{\sigma(k(l))}(\cdot,\cdot)\phi^{\sigma(k(l))}(\cdot,\cdot) \\
&\qquad - g^{\sigma(k(l))}(\cdot,\cdot)f^{\sigma(k(l))}(\cdot,\cdot)
\end{aligned}$$

is also an element of $L_p(Q)$, $3/2<p<\infty$, for each positive integer l. Consequently,

$$\begin{aligned}
&\iint_Q \left(\phi_t(x,t) - \sum_{i,j=1}^{n} a_{ij}(x,t)\phi_{x_ix_j}(x,t) - \sum_{i=1}^{n} b_i(x,t)\phi_{x_i}(x,t)\right. \\
&\qquad \left. - c(x,t)\phi(x,t) - f(x,t)\right) Z(x,t)\,dx\,dt \\
&= \iint_Q \left\{ \left(\phi_t(x,t) - \phi_t^{\sigma(k(l))}(x,t)\right) Z(x,t)\right. \\
&\qquad - \sum_{i,j=1}^{n} \left(a_{ij}(x,t)\phi_{x_ix_j}(x,t) - a_{ij}^{\sigma(k(l))}(x,t)\phi_{x_ix_j}^{\sigma(k(l))}(x,t)\right) Z(x,t) \\
&\qquad - \sum_{i=1}^{n} \left(b_i(x,t)\phi_{x_i}(x,t) - b_i^{\sigma(k(l))}(x,t)\phi_{x_i}^{\sigma(k(l))}(x,t)\right) Z(x,t)
\end{aligned}$$

$$-\big(c(x,t)\phi(x,t)-c^{\sigma(k(l))}(x,t)\phi^{\sigma(k(l))}(x,t)\big)Z(x,t)$$

$$\left.-\big(f(x,t)-g^{\sigma(k(l))}(x,t)f^{\sigma(k(l))}(x,t)\big)Z(x,t)\right\}dx\,dt \tag{2.1.24}$$

for any $Z\in L^{p'}(Q)$, where $(1/p)+(1/p')=1$.

Recall (Theorem 1.2.1) that if $\xi\in L_\lambda(Q)$, $1\le\lambda<\infty$, and if ξ^σ is the integral average of ξ, then $\xi^\sigma\to\xi$ strongly in $L_\lambda(Q)$, $1\le\lambda<\infty$, as $\sigma\to\infty$. In particular, $\xi^{\sigma(k(l))}\to\xi$ strongly in $L_\lambda(Q)$, $1\le\lambda<\infty$, as $l\to\infty$.

By virtue of the fact that Q is bounded, it follows from Assumptions (2.1.A2) and (2.1.A4) that $a_{ij}\in L_\lambda(Q)$, $b_i\in L_\lambda(Q)$, $i,j=1,\dots,n$, $c\in L_\lambda(Q)$ and $f\in L_\lambda(Q)$ for any $\lambda\in[1,\infty]$. Thus, as $l\to\infty$,

$$\begin{aligned}
&a_{ij}^{\sigma(k(l))}\xrightarrow{s}a_{ij}, \qquad i,j=1,\dots,n,\\
&b_i^{\sigma(k(l))}\xrightarrow{s}b_i, \qquad i=1,\dots,n,\\
&c^{\sigma(k(l))}\xrightarrow{s}c,\\
&f^{\sigma(k(l))}\xrightarrow{s}f \quad \text{(strongly) in } L_\lambda(Q), \quad 1\le\lambda<\infty.
\end{aligned} \tag{2.1.25}$$

Next, in view of the definition of $g^{\sigma(k(l))}$ that appeared in Remark 2.1.3, we observe that $g^{\sigma(k(l))}\to1$ everywhere in Q and $|g^{\sigma(k(l))}(x,t)|\le1$ on Q for all positive integers l. Thus, we obtain from the (fourth) inequality (2.1.16) and Lebesgue's dominated convergence theorem that

$$g^{\sigma(k(l))}f^{\sigma(k(l))}\to f \tag{2.1.26}$$

strongly in $L_\lambda(Q)$, $1\le\lambda<\infty$, as $l\to\infty$.

Using (2.1.22), (2.1.25), and (2.1.26), we observe that the relation (2.1.24), in the limit with respect l, reduces to

$$\iint_Q\left\{\left(\phi_t(x,t)-\sum_{i,j=1}^n a_{ij}(x,t)\phi_{x_ix_j}(x,t)-\sum_{i=1}^n b_i(x,t)\phi_{x_i}(x,t)\right.\right.$$

$$\left.\left.-c(x,t)\phi(x,t)-f(x,t)\right)Z(x,t)\right\}dx\,dt=0, \tag{2.1.27}$$

which holds for all $Z\in L^{p'}(Q)$, where p' is such that $(1/p)+(1/p')=1$ and $p\in(3/2,\infty)$. Thus,

$$\phi_t(x,t)-\sum_{i,j=1}^n a_{ij}(x,t)\phi_{x_ix_j}(x,t)-\sum_{i=1}^n b_i(x,t)\phi_{x_i}(x,t)$$

$$-c(x,t)\phi(x,t)=f(x,t) \tag{2.1.28}$$

for almost all $(x,t)\in Q$. This implies that ϕ satisfies the differential equation of system (2.1.1) a.e. in Q.

It remains to show that ϕ satisfies the boundary and initial conditions of system (2.1.1). For this, recall that (i) $\phi^{\sigma(k(l))} \to \phi$ uniformly on $\overline{Q}$ as $l \to \infty$, (ii) $\phi^{\sigma(k(l))}(x,t)=0$ on $\partial\Omega \times [0,T]$ for all integers l, and (iii) $\phi^{\sigma(k(l))}(x,0)=\phi_0(x)$ on Ω. Thus, we have $\phi(x,t)=0$ on $\partial\Omega \times [0,T]$ and $\phi(x,0)=\phi_0(x)$ on Ω. Therefore, we conclude that $\phi \in W_p^{2,1}(Q)$, $3/2<p<\infty$, and is a solution of system (2.1.1). However, by virtue of Theorem 2.1.3, system (2.1.1) has only a unique solution. This completes the proof. □

Theorem 2.1.6. *Consider system* (2.1.1). *Suppose that assumptions* (2.1.A1)–(2.1.A5) *are satisfied. Then, the solution* $\phi \in W_p^{2,1}(Q)$, $3/2<p<\infty$, *of system* (2.1.1) *is nonnegative in* $\overline{Q}$ *if* $\phi_0(x) \geq 0$ *for all* $x \in \Omega$ *and* $f(x,t) \geq 0$ *for all* $(x,t) \in Q$.

PROOF. From Theorem 2.1.5, there exists a subsequence $\{\phi^{\sigma(l)}\} \subset \{\phi^{\sigma}\}$ such that $\phi^{\sigma(l)} \to \phi$ uniformly on $\overline{Q}$ as $l \to \infty$. By virtue of the definitions of ϕ_0^{σ} and f^{σ}, we observe that the signs of $\phi^{\sigma(l)}$ and $f^{\sigma(l)}$ depend, respectively, on those of ϕ_0 and f. Thus, the conclusion of the theorem follows readily from Theorem 2.1.1. This completes the proof. □

Consider the sequence of first boundary value problems

$$
\begin{aligned}
L^m\phi(x,t) &= f^m(x,t), && (x,t) \in Q, \\
\phi(x,0) &= \phi_0(x,t), && x \in \Omega, \\
\phi(x,t) &= 0, && (x,t) \in \partial\Omega \times [0,T],
\end{aligned}
\tag{2.1.29}
$$

where, for each positive integer m, the operator L^m is defined by

$$
\begin{aligned}
L^m\psi(x,t) \equiv \psi_t(x,t) - &\sum_{i,j=1}^{n} a_{ij}(x,t)\psi_{x_i x_j}(x,t) \\
&- \sum_{i=1}^{n} b_i^m(x,t)\psi_{x_i}(x,t) - c^m(x,t)\psi(x,t).
\end{aligned}
\tag{2.1.30}
$$

For the system (2.1.29), it is assumed that the first- and zeroth-order coefficients and the free term satisfy the Assumption (2.1.A4) uniformly with respect to the integers $m \geq 1$. For convenience, this assumption will be referred to as Assumption (2.1.A6).

Under Assumptions (2.1.A1)–(2.1.A3), (2.1.A5), and (2.1.A6), it follows from Theorem 2.1.3 that system (2.1.29) admits a unique solution $\phi^m \in W_p^{2,1}(Q)$, $3/2<p<\infty$, for each integer $m \geq 1$.

By adapting the proof of Theorem 2.1.5, it is not difficult to obtain the following result.

Theorem 2.1.7. *Consider system* (2.1.29). *Suppose that Assumptions* (2.1.A1)–(2.1.A3), (2.1.A5) *and* (2.1.A6) *are satisfied. Then there exists a system of subsequences* $\{b_i^{m(k)}, c^{m(k)}, f^{m(k)}, i=1,\ldots,n\}$ *of the sequence* $\{b_i^m, c^m, f^m,$

$i=1,\ldots,n\}$ and a system of functions $\{b_i, c, f,\ i=1,\ldots,n\}$ satisfying the corresponding conditions of Assumptions (2.1.A6) *such that, as $k\to\infty$,*

$$b_i^{m(k)} \overset{w^*}{\to} b_i, \qquad i=1,\ldots,n,$$

$$c^{m(k)} \overset{w^*}{\to} c,$$

$$f^{m(k)} \overset{w^*}{\to} f \quad \textit{in (the weak* topology of) } L_\infty(Q).$$

Moreover, there exists a further subsequence

$$\left\{\phi^{m(k(l))}, \phi_{x_i}^{m(k(l))}, \phi_t^{m(k(l))}, \phi_{x_i x_j}^{m(k(l))},\ i, j=1,\ldots,n\right\}$$

such that, as $l\to\infty$,

$$\phi^{m(k(l))} \overset{u}{\to} \phi, \qquad \phi_{x_i}^{m(k(l))} \overset{u}{\to} \phi_{x_i}, \qquad \text{on } \bar{Q},\ i=1,\ldots,n,$$

and

$$\phi_t^{m(k(l))} \overset{w}{\to} \phi_t, \qquad \phi_{x_i x_j}^{m(k(l))} \overset{w}{\to} \phi_{x_i x_j}, \textit{ in } L_p(Q), \quad 3/2<p<\infty,\ i, j=1,\ldots,n$$

where $\phi\in W_p^{2,1}(Q)$, $3/2<p<\infty$, is the solution of the limit system (2.1.1).

PROOF. Note that any closed, bounded, and convex subset of $L_\infty(Q)$ is compact (and closed) in the weak* topology of $L_\infty(Q)$. Thus, by Assumption (2.1.A6), the first part of the theorem follows easily.

It remains to prove the second part. From Remark 2.1.1 and Assumptions (2.1.A5) and (2.1.A6), there exists a constant K_5, independent of k, such that

$$\|\phi^{m(k)}\|_{p,Q}^{(2,1)} \le K_5 \quad \text{for } p>3/2, \tag{2.1.31}$$

and

$$|\phi^{m(k)}|_{\bar{Q}}^{(1+\mu,(1+\mu)/2)} \le K_5 \quad \text{for } \mu, \quad 0<\mu<1. \tag{2.1.32}$$

Using the estimates (2.1.31) and (2.1.32) instead of the estimates (2.1.17) and (2.1.18), the rest of the proof is similar to that given for Theorem 2.1.5. This completes the proof. □

2.2. First Boundary Value Problems for Parabolic Partial Differential Equations in Divergence Form

In the previous section, we considered a class of systems governed by linear second-order parabolic partial differential equations in general form with a first boundary condition. The second-order coefficients of the differential operator were assumed to be continuous, because without this assumption it is not possible to show the existence and uniqueness of solutions for the system.

In this section, a class of systems governed by linear second-order parabolic partial differential equations in *divergence form* with first boundary conditions is studied. Since the differential operator is in divergence form, it is no longer essential for the second-order coefficients to be continuous. However, the system admits only "weak solutions" rather than "classical solutions" or "solutions" as considered in the previous section. The basic reference of this section is Ladyžhenskaja, Solonikov, and Ural'ceva [LSU.1].

Let Ω be a connected bounded open subset of R^n, and let $T>0$, $I\equiv(0,T)$, $Q\equiv\Omega\times I$, and $\overline{Q}$ denote the closure of Q.

To proceed further, we recall the notation of Sections 1.2.1–1.2.3.

Throughout this section, it is assumed that the domain Ω has piecewise smooth boundary (Definition 1.2.1).

Consider the following first boundary value problem:

$$
\begin{aligned}
&L\phi(x,t)=\sum_{j=1}^{n}\left(F_j(x,t)\right)_{x_j}+f(x,t), \qquad (x,t)\in Q,\\
&\phi(x,0)=\phi_0(x), \qquad x\in\Omega,\\
&\phi(x,t)=0, \qquad (x,t)\in\partial\Omega\times[0,T],
\end{aligned}
\tag{2.2.1}
$$

where the operator L is given by

$$
\begin{aligned}
L\psi(x,t)\equiv\psi_t(x,t)-\Bigg\{&\sum_{j=1}^{n}\left(\sum_{i=1}^{n}a_{ij}(x,t)\psi_{x_i}(x,t)+a_j(x,t)\psi(x,t)\right)_{x_j}\\
&+\sum_{j=1}^{n}b_j(x,t)\psi_{x_j}(x,t)+c(x,t)\psi(x,t)\Bigg\}.
\end{aligned}
\tag{2.2.2}
$$

For brevity, we introduce the following notation.

$$
\begin{aligned}
\mathfrak{L}(\psi,\zeta)(t)\equiv\int_{\Omega}\Bigg\{&\sum_{j=1}^{n}\left(\sum_{i=1}^{n}a_{ij}(x,t)\psi_{x_i}(x,t)\right.\\
&\left.+a_j(x,t)\psi(x,t)\right)\zeta_{x_j}(x,t)\\
&-\sum_{j=1}^{n}b_j(x,t)\psi_{x_j}(x,t)\zeta(x,t)\\
&-c(x,t)\psi(x,t)\zeta(x,t)\Bigg\}\,dx
\end{aligned}
\tag{2.2.3}
$$

$$
\tilde{F}(x,t)\equiv\sum_{j=1}^{n}\left(F_j(x,t)\right)_{x_j}+f(x,t),
\tag{2.2.4}
$$

$$
(\tilde{F},\zeta)(t)\equiv-\int_{\Omega}\left\{\sum_{j=1}^{n}F_j(x,t)\zeta_{x_j}(x,t)-f(x,t)\zeta(x,t)\right\}dx.
\tag{2.2.5}
$$

With reference to the differential equation of system (2.2.1), if a_{ij}, $i, j=1,\ldots,n$, a_i, $i=1,\ldots,n$, and F_j, $j=1,\ldots,n$, are nondifferentiable functions, then the differential equation must be put in a form that does not contain derivatives of these functions. This is done as follows.

The differential equation of system (2.2.1) is multiplied by an arbitrary function $\eta \in \mathring{W}_2^{1,1}(Q)$ (see Section 1.2.3); then both sides are integrated over $\Omega \times (0, \tau)$ with $\tau \in (0, T)$, and for terms containing the a_{ij}, a_i, and F_j, a single integration by parts is carried out. This gives rise to the following identity:

$$\int_\Omega \{\phi(x,\tau)\eta(x,\tau)\}\,dx - \int_0^\tau \int_\Omega \{\phi(x,t)\eta_t(x,t)\}\,dx\,dt + \int_0^\tau \{\mathfrak{L}(\phi,\eta)(t) - (\tilde{F},\eta)(t)\}\,dt = \int_\Omega \{\phi_0(x)\eta(x,0)\}\,dx. \quad (2.2.6)$$

Clearly, (2.2.6) is equivalent to the differential equation of system (2.2.1) if the a_{ij}, a_i, and F_j are sufficiently smooth. However, if these functions are not differentiable, then (2.2.6) has meaning (for appropriate functions ϕ and η) while the differential equation of system (2.2.1) does not. On this basis, even if a_{ij}, a_i, and F_j are nondifferentiable functions, we can still determine a function ϕ from a certain class of functions that solves Equation (2.2.6). Such a function is called a weak solution of the differential equation of system (2.2.1).

Definition 2.2.1. A function $\phi: \bar{Q} \to R$ is said to be a *weak solution of the differential equation of system* (2.2.1) if it satisfies the following conditions:

i. $\phi \in V_2(Q)$, (see Section 1.2.3), and
ii. for almost all $\tau \in [0, T]$,

$$\int_\Omega \{\phi(x,\tau)\eta(x,\tau)\}\,dx - \int_0^\tau \int_\Omega \{\phi(x,t)\eta_t(x,t)\}\,dx\,dt + \int_0^\tau \{\mathfrak{L}(\phi,\eta)(t) - (\tilde{F},\eta)(t)\}\,dt = 0$$

for all $\eta \in \mathring{W}_2^{1,1}(Q)$ that equal 0 for $t=0$.

Definition 2.2.2. *A weak solution from* $V_2^{1,0}(Q)$ (see Section 1.2.3) of the differential equation of system (2.2.1) *is as given in Definition* 2.2.1 *except that condition* (*i*) *is replaced by* $\phi \in V_2^{1,0}(Q)$ *and condition* (*ii*) *holds* for all $\tau \in [0, T]$.

For weak solutions of system (2.2.1) instead of merely the differential equation, the initial and boundary conditions of the system are also required to be satisfied. This is done by modifying appropriately both conditions (i) and (ii) of Definition 2.2.1 (or Definition 2.2.2). The precise statement of a weak solution from $V_2(Q)$ of system (2.2.1) is given in the following definition.

Definition 2.2.3. A function $\phi\colon \bar{Q}\to R$ is said to be a *weak solution from* $V_2(Q)$ *of system* (2.2.1) if the following conditions are satisfied:

i. $\phi\in \mathring{V}_2(Q)$ *(see Section* 1.2.3) *and*
ii. *for almost all* $\tau\in[0,T]$,

$$\int_\Omega \{\phi(x,\tau)\eta(x,\tau)\}dx - \int_0^\tau \int_\Omega \{\phi(x,t)\eta_t(x,t)\}dx\,dt$$

$$+\int_0^\tau \{\mathcal{L}(\phi,\eta)(t)-(\tilde{F},\eta)(t)\}dt = \int_\Omega \{\phi_0(x)\eta(x,0)\}dx$$

for any $\eta\in \mathring{W}_2^{1,1}(Q)$.

Definition 2.2.4. A *weak solution* ϕ *from* $V_2^{1,0}(Q)$ *of system* (2.2.1) is as given in Definition 2.2.3 except that condition (i) is replaced by $\phi\in \mathring{V}_2^{1,0}(Q)$ (see Section 1.2.3) and condition (ii) holds for all $\tau\in[0,T]$.

Throughout most of this section, we need the following assumptions.

Assumption (2.2.A1). a_{ij}, $i, j=1,\ldots,n$, are bounded and measurable functions defined on $\bar{Q}$.

Assumption (2.2.A2). There exist positive constants α_l and α_u such that

$$\alpha_l \sum_{i=1}^n (\xi_i)^2 \le \sum_{i,j=1}^n a_{ij}(x,t)\xi_i\xi_j \le \alpha_u \sum_{i=1}^n (\xi_i)^2$$

for all $\xi\in R^n$ uniformly on $\bar{Q}$.

Assumption (2.2.A3). The coefficients a_j, b_j, $j=1,\ldots,n$, and c are measurable functions defined on $\bar{Q}$. Furthermore, there exists a positive constant $M_1\equiv M_1(q,r)$ such that

$$\left\| \sum_{j=1}^n (a_j)^2 \right\|_{q,r,Q} \le M_1,$$

$$\left\| \sum_{j=1}^n (b_j)^2 \right\|_{q,r,Q} \le M_1,$$

and

$$\|c\|_{q,r,Q} \le M_1$$

for arbitrary pair of constants q and r satisfying

$$\frac{1}{r}+\frac{n}{2q}=1,$$

$$q\in\left(\frac{n}{2},\infty\right],\qquad r\in[1,\infty)\quad \text{for } n\geq 2, \tag{2.2.7}$$
$$q\in[1,\infty],\qquad r\in[1,2]\quad \text{for } n=1,$$

where $\|\cdot\|_{q,r,Q}$ is the norm in the Banach space $L_{q,r}(Q)$ (see Section 1.2.3).

Assumption (2.2.A4). There exists a positive constant $M_2\equiv M_2(q_1,r_1)$ such that

$$\left(\iint_Q\left\{\sum_{j=1}^{n}(F_j(x,t))^2\right\}dx\,dt\right)^{1/2}\leq M_2$$

and

$$\|f\|_{q_1,r_1,Q}\leq M_2,$$

where q_1 and r_1 are arbitrary pair of constants satisfying

$$\frac{1}{r_1}+\frac{n}{2q_1}=1+\frac{n}{4},$$

$$\begin{aligned} &q_1\in\left[\frac{2n}{n+2},2\right], && r_1\in[1,2] && \text{for } n\geq 3,\\ &q_1\in(1,2], && r_1\in[1,2) && \text{for } n=2,\\ &q_1\in[1,2], && r_1\in[1,4/3] && \text{for } n=1. \end{aligned} \tag{2.2.8}$$

Assumption (2.2.A5). $\phi_0\in L_2(\Omega)$.

Lemma 2.2.1. *Let the coefficients and free terms of the differential equation of system* (2.2.1) *satisfy the corresponding Assumptions* (2.2.A1), (2.2.A3), *and* (2.2.A4). *If $\phi\in \overset{\circ}{V}_2(Q)$, then the following three statements are valid.*

i. $\{\sum_{i=1}^{n}a_{ij}(\cdot,\cdot)\phi_{x_i}(\cdot,\cdot)+a_j(\cdot,\cdot)\phi(\cdot,\cdot)+F_j(\cdot,\cdot)\}$, $j=1,\dots,n$, *are elements of* $L_2(Q)$.
ii. $c(\cdot,\cdot)\phi(\cdot,\cdot)$ *and* $\sum_{j=1}^{n}b_j(\cdot,\cdot)\phi_{x_j}(\cdot,\cdot)$ *are elements of* $L_{\lambda_1,\lambda_2}(Q)$, *where* $\lambda_1=2q/(q+1)$ *and* $\lambda_2=2r/(r+1)$.
iii. f *is an element of* $L_{q_1,r_1}(Q)$.

PROOF. The third statement follows immediately from Assumption (2.2.A4). Thus, it remains to verify the first two statements.

From Minkowski's inequality and Assumption (2.2.A1), we obtain

$$\left\|\sum_{i=1}^{n}a_{ij}\phi_{x_i}+a_j\phi+F_j\right\|_{2,Q}\leq \nu_1\sum_{i=1}^{n}\|\phi_{x_i}\|_{2,Q}+\|a_j\phi\|_{2,Q}+\|F_j\|_{2,Q}, \tag{2.2.9}$$

for each $j\in\{1,\dots,n\}$, where $\nu_1\equiv\max\{\|a_{ij}\|_{\infty,Q}: i,\, j=1,\dots,n\}$.

Now, by virtue of Hölder's inequality and Assumption (2.2.A3), it follows that

$$\iint_Q \left\{ \left(a_j(x,t)\phi(x,t) \right)^2 \right\} dx\, dt \le M_1 \left(\|\phi\|_{2q',2r',Q} \right)^2 \tag{2.2.10}$$

for all $j \in \{1, \ldots, n\}$, where the constant M_1 is as defined in Assumption (2.2.A3), $(1/q)+(1/q')=1$, $(1/r)+(1/r')=1$, and the numbers q and r satisfy condition (2.7) of Assumption (2.2.A3). Obviously, $q'=q/(q-1)$ and $r'=r/(r-1)$. Thus, in view of Theorem 1.2.6, we obtain from the inequality (2.2.10) that

$$\iint_Q \left\{ \left(a_j(x,t)\phi(x,t) \right)^2 \right\} dx\, dt \le M_1 \left(\beta |\phi|_Q \right)^2 \tag{2.2.11}$$

for all $j \in \{1, \ldots, n\}$, where the constant β is as defined in Theorem 1.2.6 and $|\cdot|_Q$ denotes the norm in the Banach space $V_2(Q)$.

By hypothesis, $\phi \in \mathring{V}_2(Q)$. Thus, it follows readily from the inequalities (2.2.9), (2.2.11), and Assumptions (2.2.A1), (2.2.A3), and (2.2.A4) that, for each $j \in \{1, \ldots, n\}$,

$$\sum_{i=1}^{n} a_{ij}(\cdot,\cdot)\phi_{x_i}(\cdot,\cdot) + a_j(\cdot,\cdot)\phi(\cdot,\cdot) + F_j(\cdot,\cdot)$$

belongs to $L_2(Q)$. This proves the first statement.

It remains to prove the second statement. From Hölder's inequality, Theorem 1.2.6, Assumption (2.2.A3), and the hypothesis $\phi \in \mathring{V}_2(Q)$, we deduce that

$$\|c\phi\|_{2q/(q+1),2r/(r+1),Q} \le \beta M_1 |\phi|_Q. \tag{2.2.12}$$

On the other hand, by using Cauchy's inequality, Hölder's inequality, Assumption (2.2.A3), and the hypothesis $\phi \in \mathring{V}_2(Q)$, it follows that for $\lambda_1 = 2q/(q+1)$ and $\lambda_2 = 2r/(r+1)$,

$$\left\| \sum_{j=1}^{n} b_j \phi_{x_j} \right\|_{\lambda_1,\lambda_2,Q} \le (M_1)^{1/2} \|\phi_x\|_{2,Q} \le (M_1)^{1/2} |\phi|_Q < \infty, \tag{2.2.13}$$

where $\|\phi_x\|_{2,Q} \equiv \left(\iint_Q \{ \sum_{j=1}^{n} (\phi_{x_j}(x,t))^2 \} dx\, dt \right)^{1/2}$.

Combining (2.2.12) and (2.2.13), we obtain the second statement. This completes the proof. □

REMARK 2.2.1. Denote $2q/(q+1)$ and $2r/(r+1)$ by λ_1 and λ_2, respectively, with q and r satisfying condition (2.2.7). Then we can easily verify that the quantities λ_1 and λ_2 satisfy condition (2.2.8).

Note that Definition 2.2.3 makes sense only if all the integrals in condition (ii) of this definition are finite for some $\phi \in \mathring{V}_2(Q)$ and all $\eta \in \mathring{W}_2^{1,1}(Q)$. Thus, we need the following lemma.

Lemma 2.2.2. *Let the coefficients and data of system* (2.2.1) *satisfy the corresponding Assumptions* (2.2.A1) *and* (2.2.A3)–(2.2.A5). *Then all the integrals appearing in condition* (ii) *of Definition* 2.2.3 *are finite for any pair of functions* $\phi \in \mathring{V}_2(Q)$ *and* $\eta \in \mathring{W}_2^{1,1}(Q)$.

PROOF. Obviously, the first two integrals in condition (ii) of Definition 2.2.3 are finite. We shall itemize the rest of our proof. The hypotheses, $\phi \in \mathring{V}_2(Q)$ and $\eta \in \mathring{W}_2^{1,1}(Q)$, are used without further mention.

i. From Cauchy's inequality, Hölder's inequality, and Assumption (2.2.A1), we obtain

$$\iint_Q \left\{ \left| \sum_{i,j=1}^{n} a_{ij}(x,t)\phi_{x_i}(x,t)\eta_{x_j}(x,t) \right| \right\} dx\,dt \le \nu_1 n |\phi|_Q \|\eta_x\|_{2,Q} < \infty, \tag{2.2.14}$$

where $\nu_1 \equiv \max\{\|a_{ij}\|_{\infty,Q} : i, j = 1, \ldots, n\}$.

ii. From Cauchy's inequality, Hölder's inequality, Theorem 1.2.6, and Assumption (2.2.A3), we deduce that

$$\iint_Q \left\{ \left| \sum_{j=1}^{n} a_j(x,t)\phi(x,t)\eta_{x_j}(x,t) \right| \right\} dx\,dt \le (M_1)^{1/2}\beta |\phi|_Q \|\eta_x\|_{2,Q} < \infty, \tag{2.2.15}$$

where the constant β is as defined in Theorem 1.2.6.

iii. From Hölder's inequality, inequality (2.2.13), Theorem 1.2.6, and Assumption (2.2.A3), we get

$$\iint_Q \left\{ \left| \sum_{j=1}^{n} b_j(x,t)\phi_{x_j}(x,t)\eta(x,t) \right| \right\} dx\,dt \le \beta (M_1)^{1/2} |\phi|_Q |\eta|_Q < \infty. \tag{2.2.16}$$

iv. From Hölder's inequality, Theorem 1.2.6, Assumption (2.2.A3), and inequality (2.2.12), we obtain

$$\iint_Q \{|c(x,t)\phi(x,t)\eta(x,t)|\} dx\,dt \le (\beta)^2 M_1 |\phi|_Q |\eta|_Q < \infty. \tag{2.2.17}$$

v. From Cauchy's inequality, Hölder's inequality, and Assumption (2.2.A4), it follows that

$$\iint_Q \left\{ \left| \sum_{j=1}^{n} F_j(x,t)\eta_{x_j}(x,t) \right| \right\} dx\,dt \le M_2 \|\eta_x\|_{2,Q} < \infty. \tag{2.2.18}$$

vi. By Hölder's inequality, Theorem 1.2.6, and Assumption (2.2.A4), we deduce that

$$\iint_Q \{|f(x,t)\eta(x,t)|\} dx\,dt \le \beta M_2 |\eta|_Q < \infty. \tag{2.2.19}$$

vii. Since $W_2^{1,1}(Q)$ is contained in $V_2^{1,0}(Q)$, it follows from Assumption (2.2.A5) that

$$\int_\Omega \{\phi_0(x)\eta(x,0)\}dx < \infty.$$

Combining all the results established above, the proof of the lemma is complete. □

REMARK 2.2.2. Since $V_2^{1,0}(Q) \subset V_2(Q)$, the conclusions of Lemmas 2.2.1 and 2.2.2 are valid with $\phi \in \mathring{V}_2^{1,0}(Q)$ instead of $\phi \in \mathring{V}_2(Q)$.

Lemma 2.2.3. *Let the coefficients and data of system* (2.2.1) *satisfy the corresponding Assumptions* (2.2.A1) *and* (2.2.A3)–(2.2.A5). *If* $\phi \in \mathring{V}_2(Q)$, *then condition* (ii) *of Definition* 2.2.3 *is equivalent to condition* (*ii*)′:

$$-\iint_Q \{\phi(x,t)\eta_t(x,t)\}dx\,dt$$

$$+\int_0^T \{\mathfrak{L}(\phi,\eta)(t) - (\tilde{F},\eta)(t)\}dt = \int_\Omega \{\phi_0(x)\eta(x,0)\}dx$$

for all $\eta \in \mathring{W}_2^{1,1}(Q)$ *that are equal to zero for* $t = T$.

PROOF. First, we show that condition (ii) implies condition (ii)′. For this, let $\varepsilon > 0$ be a small number. Take $\eta \in \mathring{W}_2^{1,1}(Q)$ such that $\eta(x,t) = 0$ for all $(x,t) \in \Omega \times [T-\varepsilon, T)$. Then, by taking $\tau > T - \varepsilon$ in condition (ii) of Definition 2.2.3, condition (ii)′ holds for such η, and hence for all η required in condition (ii)′. Conversely, we shall show that condition (ii)′ implies condition (ii) of Definition 2.2.3. For this, let $\varepsilon > 0$ be a small number and let μ_ε be the function defined by

$$\mu_\varepsilon(t,\tau) = \begin{cases} 1, & t \in [0, \tau-\varepsilon] \\ (\tau - t)/\varepsilon, & t \in [\tau-\varepsilon, \tau] \\ 0, & t \in [\tau, T]. \end{cases}$$

Let $\hat{\eta}$ be an arbitrary element of $\mathring{W}_2^{1,1}(Q)$. Then, it is clear that $\eta_\varepsilon(\cdot,\cdot) \equiv \mu_\varepsilon(\cdot,\tau)\eta(\cdot,\cdot) \in \mathring{W}_2^{1,1}(Q)$ and is equal to zero for $t \geq \tau$. Thus, by substituting this η_ε into condition (ii)′, we have

$$-\iint_Q \{\phi(x,t)(\mu_\varepsilon(t,\tau)\hat{\eta}_t(x,t) + (\mu_\varepsilon)_t(t,\tau)\hat{\eta}(x,t))\}dx\,dt$$

$$+\int_0^T \{\mathfrak{L}(\phi,\eta_\varepsilon)(t) - (\tilde{F},\eta_\varepsilon)(t)\}dt = \int_\Omega \{\phi_0(x)\mu_\varepsilon(0,\tau)\hat{\eta}(x,0)\}dx. \tag{2.2.20}$$

From the definition of μ_ε and the above equality, it is clear that

a. the right-hand side of the above equality is precisely that of condition (ii) and
b. the first integral on the left-hand side of the equality is

$$-\iint_Q \{\phi(x,t)(\mu_\varepsilon(t,\tau)\hat{\eta}_t(x,t)+(\mu_\varepsilon)_t(t,\tau)\hat{\eta}(x,t))\}\,dx\,dt$$

$$=-\int_0^T\int_\Omega \{\phi(x,t)\mu_\varepsilon(t,\tau)\hat{\eta}_t(x,t)\}\,dx\,dt$$

$$+\frac{1}{\varepsilon}\int_{\tau-\varepsilon}^{\tau}\int_\Omega \{\phi(x,t)\hat{\eta}(x,t)\}\,dx\,dt. \tag{2.2.21}$$

Note that the function

$$\psi(\cdot)\equiv\int_\Omega \{\phi(x,\cdot)\hat{\eta}(x,\cdot)\}\,dx$$

is integrable on $[0,T]$. Thus,

$$\lim_{\varepsilon\downarrow 0}\frac{1}{\varepsilon}\int_{\tau-\varepsilon}^{\tau}\int_\Omega \{\phi(x,t)\hat{\eta}(x,t)\}\,dx\,dt=\int_\Omega \{\phi(x,\tau)\hat{\eta}(x,\tau)\}\,dx \tag{2.2.22}$$

for almost all $\tau\in[0,T]$.

For the first integral on the right-hand side of (2.2.21), it follows from Lebesgue's dominated convergence theorem that

$$\lim_{\varepsilon\downarrow 0}\int_0^T\int_\Omega \{\phi(x,t)\mu_\varepsilon(t,\tau)\hat{\eta}_t(x,t)\}\,dx\,dt$$

$$=\int_0^\tau\int_\Omega \{\phi(x,t)\hat{\eta}_t(x,t)\}\,dx\,dt. \tag{2.2.23}$$

Letting $\varepsilon\to 0$ in (2.2.21), we obtain from (2.2.22) and (2.2.23) that

$$\lim_{\varepsilon\downarrow 0}\left[-\iint_Q \{\phi(x,t)(\mu_\varepsilon(x,t)\hat{\eta}_t(x,t)+(\mu_\varepsilon)_t(x,t)\hat{\eta}(x,t))\}\,dx\,dt\right]$$

$$=-\int_0^\tau\int_\Omega \{\phi(x,t)\hat{\eta}_t(x,t)\}\,dx\,dt+\int_\Omega \{\phi(x,\tau)\hat{\eta}(x,\tau)\}\,d\tau \tag{2.2.24}$$

for almost all $\tau\in[0,T]$. This gives us the first two integrals on the left-hand side of condition (ii).

It remains to show that the second integral on the left-hand side of (2.2.20) tends to that of condition (ii) as $\varepsilon\downarrow 0$. In fact, this conclusion follows easily from the definition of the function η_ε and the hypotheses of the lemma. Thus, the proof is complete. □

REMARK 2.2.3. Using an argument similar to that given for Lemma 2.2.3, we can show that condition (ii) of Definition 2.2.4 is equivalent to condition (ii)′:

$$-\iint_Q \{\phi(x,t)\eta_t(x,t)\}\,dx\,dt + \int_0^T \{\mathcal{L}(\phi,\eta)(t) - (\tilde{F},\eta)(t)\}\,dt$$

$$= \int_\Omega \{\phi_0(x)\eta(x,0)\}\,dx$$

for all $\eta \in \mathring{W}_2^{1,1}(Q)$ that are equal to zero for $t=T$. The only difference is that the function ψ [Equation (2.2.22)] is now a continuous function on $[0,T]$. Thus, $(1/\varepsilon)\int_{\tau-\varepsilon}^{\tau}\{\psi(t)\}\,dt$ converges to $\psi(\tau)$ *for all* $\tau \in [0,T]$ instead of *for almost all* $\tau \in [0,T]$.

In order to show that any weak solution from $V_2(Q)$ satisfies an a priori estimate, we need the following two lemmas.

Lemma 2.2.4. *Let the coefficients and free terms of the differential equation of system* (2.2.1) *satisfy Assumptions* (2.2.A1)–(2.2.A4). *Suppose that, for* $\phi \in V_2(Q)$, *the inequality,*

$$\frac{1}{2}\int_\Omega \{(\phi(x,t))^2\}\,dx\Big|_{t=t_1}^{t=t_2} + \int_{t_1}^{t_2} \{\mathcal{L}(\phi,\phi)(t) - (\tilde{F},\phi)(t)\}\,dt \le 0, \tag{2.2.25}$$

holds for almost all t_1, t_2 *from* $[0,T]$, *including* $t_1 = 0$. *Then*

$$|\phi|_Q \le K\{\|\phi(\cdot,0)\|_{2,\Omega} + \|F\|_{2,Q} + \|f\|_{q_1,r_1,Q}\}, \tag{2.2.26}$$

where

$$\|F\|_{2,Q} \equiv \left(\iint_Q \left\{\sum_{j=1}^n (F_j(x,t))^2\right\}dx\,dt\right)^{1/2} \tag{2.2.27}$$

and the constant K *depends only on* n, α_1, α_2, M_1, *and the quantities* q *and* r *satisfying condition* (2.2.7).

PROOF. Throughout the proof, all the inequalities are understood to be valid for almost all $t_1, t_2 \in [0,T]$.

From inequality (2.2.25) and Assumption (2.2.A2), we obtain

$$\frac{1}{2}\int_\Omega \{(\phi(x,t))^2\}\,dx\Big|_{t=t_1}^{t=t_2} + \alpha_l \int_{t_1}^{t_2}\int_\Omega \left\{\sum_{j=1}^n (\phi_{x_j}(x,t))^2\right\}dx\,dt$$

$$< \int_{t_1}^{t_2}\int_\Omega \{\mathcal{F}_1(x,t)\}\,dx\,dt, \tag{2.2.28}$$

where

$$\mathcal{F}_1(x,t) \equiv \sum_{j=1}^{n} \{|a_j(x,t)\phi(x,t)\phi_{x_j}(x,t)| + |b_j(x,t)\phi_{x_j}(x,t)\phi(x,t)| + |F_j(x,t)\phi_{x_j}(x,t)|\} + |c(x,t)(\phi(x,t))^2| + |f(x,t)\phi(x,t)|. \quad (2.2.29)$$

Using the Cauchy's inequality,

$$ab \le \frac{\varepsilon}{2}(a)^2 + \frac{1}{2\varepsilon}(b)^2, \qquad \varepsilon > 0, \quad a, b, \in R,$$

with $\varepsilon = \alpha_l/2$, in the first and second terms on the right-hand side of (2.2.28), we obtain

$$\frac{1}{2}\int_{\Omega}\{(\phi(x,t))^2\}dx\Big|_{t=t_1}^{t=t_2} + \alpha_l\int_{t_1}^{t_2}\int_{\Omega}\left\{\sum_{j=1}^{n}(\phi_{x_j}(x,t))^2\right\}dx\,dt$$

$$< \int_{t_1}^{t_2}\int_{\Omega}\left\{\frac{\alpha_l}{2}\sum_{j=1}^{n}\left[\phi_{x_j}(x,t)\right]^2 + \mathcal{C}(x,t)(\phi(x,t))^2 + \sum_{j=1}^{n}|F_j(x,t)\phi_{x_j}(x,t)| + |f(x,t)\phi(x,t)|\right\}dx\,dt, \quad (2.2.30)$$

where

$$\mathcal{C}(x,t) \equiv \frac{1}{\alpha_l}\sum_{j=1}^{n}\left((a_j(x,t))^2 + (b_j(x,t))^2\right) + |c(x,t)|. \quad (2.2.31)$$

Define

$$Q(t_1,t_2) \equiv \Omega \times (t_1,t_2), \quad (2.2.32)$$

$$\mathcal{B}(t_1,t_2) \equiv 2(\beta)^2 \|\mathcal{C}\|_{q,r,Q(t_1,t_2)}, \quad (2.2.33)$$

$$\mathcal{P}(t_1,t_2) \equiv 2\left(\|F\|_{2,Q(t_1,t_2)} + \beta\|f\|_{q_1,r_1,Q(t_1,t_2)}\right), \quad (2.2.34)$$

where q, r satisfy condition (2.2.7), q_1, r_1 satisfy condition (2.2.8), the constant β is as defined in Theorem 1.2.6, and $\|F\|_{2,Q(t_1,t_2)}$ is as defined by (2.2.27) with Q replaced by $Q(t_1,t_2)$.

From (2.2.30) and the Cauchy–Schwarz inequality, it follows that

$$\frac{1}{2}\int_{\Omega}\{[\phi(x,t)]^2\}dx\Big|_{t=t_1}^{t=t_2} + \alpha_l\iint_{Q(t_1,t_2)}\left\{\sum_{j=1}^{n}(\phi_{x_j}(x,t))^2\right\}dx\,dt$$

$$\le \iint_{Q(t_1,t_2)}\{\mathcal{F}_2(x,t)\}dx\,dt, \quad (2.2.35)$$

where

$$\mathcal{F}_2(x,t) \equiv \frac{\alpha_l}{2} \sum_{j=1}^{n} \left(\phi_{x_j}(x,t)\right)^2 + \mathcal{Q}(x,t)(\phi(x,t))^2$$
$$+ \left(\sum_{j=1}^{n} \left(F_j(x,t)\right)^2 \right)^{1/2} \left(\sum_{j=1}^{n} \left(\phi_{x_j}(x,t)\right)^2 \right)^{1/2}$$
$$+ (|f(x,t)|)(|\phi(x,t)|).$$

By using Assumptions (2.2.A3)–(2.2.A4), Hölder's inequality, Theorem 1.2.6, and the expressions (2.2.33) and (2.2.34), the inequality (2.2.35) reduces to

$$\int_\Omega \left\{(\phi(x,t_2))^2\right\} dx + \alpha_l \iint_{Q(t_1,t_2)} \left\{ \sum_{j=1}^{n} \left(\phi_{x_j}(x,t)\right)^2 \right\} dx\,dt$$
$$\leq \int_\Omega \left\{(\phi(x,t_1))^2\right\} dx + \mathcal{B}(t_1,t_2)\left(|\phi|_{Q(t_1,t_2)}\right)^2 + \mathcal{P}(t_1,t_2)|\phi|_{Q(t_1,t_1)}. \tag{2.2.36}$$

Since $\alpha_l > 0$, this implies that

$$\int_\Omega \left\{(\phi(x,t_2))^2\right\} dx \leq \int_\Omega \left\{(\phi(x,t_1))^2\right\} dx + \mathcal{B}(t_1,t_2)\left(|\phi|_{Q(t_1,t_2)}\right)^2$$
$$+ \mathcal{P}(t_1,t_2)|\phi|_{Q(t_1,t_2)} \tag{2.2.37}$$

and

$$\alpha_l \iint_{Q(t_1,t_2)} \left\{ \sum_{j=1}^{n} \left(\phi_{x_j}(x,t)\right)^2 \right\} dx\,dt$$
$$\leq \int_\Omega \left\{(\phi(x,t_1))^2\right\} dx + \mathcal{B}(t_1,t_2)\left(|\phi|_{Q(t_1,t_2)}\right)^2$$
$$+ \mathcal{P}(t_1,t_2)|\phi|_{Q(t_1,t_2)}. \tag{2.2.38}$$

From the definitions of $Q(t_1,t_2)$, $\mathcal{B}(t_1,t_2)$, and $\mathcal{P}(t_1,t_2)$, it follows from the inequality (2.2.37) that, for almost all $t \in [t_1,t_2]$,

$$\int_\Omega \left\{(\phi(x,t))^2\right\} dx < \int_\Omega \left\{(\phi(x,t_1))^2\right\} dx + \mathcal{B}(t_1,t_2)\left(|\phi|_{Q(t_1,t_2)}\right)^2$$
$$+ \mathcal{P}(t_1,t_2)|\phi|_{Q(t_1,t_2)}. \tag{2.2.39}$$

Let $\alpha_1 \equiv \frac{1}{2}\min\{1,\alpha_l\}$. Then it follows from (2.2.39) and (2.2.38) that

$$\alpha_1 |\phi|^2_{Q(t_1,t_2)} \leq \int_\Omega \left\{(\phi(x,t_1))^2\right\} dx + \mathcal{B}(t_1,t_2)\left(|\phi|_{Q(t_1,t_2)}\right)^2$$
$$+ \mathcal{P}(t_1,t_2)|\phi|_{Q(t_1,t_2)}. \tag{2.2.40}$$

From (2.2.40), we shall show that $|\phi|_{Q(t_1,t_2)}$ is bounded from above in terms of known data. To start with, let $\varepsilon \in (0, T)$ be arbitrary and let T_1 be a point in $(T-\varepsilon, T]$ such that it can be taken as an upper limit of integration in (2.2.25). Now, we divide the interval $[0, T]$ into a finite number of subintervals $[t_0 = 0, t_1], [t_1, t_2], \ldots, [t_{s-1}, t_s = T_1]$, where the $\{t_k\}_{k=1}^{s}$ are selected in such a way that they can also be taken as limits. For this, we note, however, that there are two possibilities:

i. there exists a $t' \in (0, T_1)$ such that $\mathcal{B}(0, t') > \alpha_1/2$;
ii. $\mathcal{B}(0, T_1) \leq \alpha_1/2$.

For case (i), the subdivisions of the interval $[0, T_1]$ are such that

$$\frac{\alpha_1}{4} \leq \mathcal{B}(t_{k-1}, t_k) \leq \frac{\alpha_1}{2}. \tag{2.2.41}$$

The first part of the above inequality is needed only to ensure that the number of subintervals s is finite. For case (ii), it is not necessary to partition the interval $[0, T]$. Thus, the first part of inequality (2.2.41) is redundant. Therefore, the proof is required only for case (i). This is done as follows.

In view of condition (2.2.7), we observe that $r < \infty$. Thus, it follows from (2.2.33) that

$$\sum_{k=1}^{s} (\mathcal{B}(t_{k-1}, t_k))^r \leq (2(\beta)^2)^r (\|\mathcal{A}\|_{q,r,Q})^r. \tag{2.2.42}$$

From the first part of inequality (2.2.41), the above inequality reduces to

$$s\left(\frac{\alpha_1}{4}\right)^r \leq (2(\beta)^2)^r (\|\mathcal{A}\|_{q,r,Q})^r. \tag{2.2.43}$$

Thus, the number s is finite and is given by

$$s \leq \left(\frac{8(\beta)^2}{\alpha_1} \|\mathcal{A}\|_{q,r,Q}\right)^r. \tag{2.2.44}$$

For each of these subintervals, we obtain from inequality (2.2.40) and the second part of inequality (2.2.41) that

$$\alpha_1 \left(|\phi|_{Q(t_{k-1},t_k)}\right)^2 \leq (\|\phi(\cdot, t_{k-1})\|_{2,\Omega})^2 + \frac{\alpha_1}{2}\left(|\phi|_{Q(t_{k-1},t_k)}\right)^2$$
$$+ \mathcal{P}(t_{k-1}, t_k)|\phi|_{Q(t_{k-1},t_k)}. \tag{2.2.45}$$

Consequently,

$$\frac{\alpha_1}{2}\left(|\phi|_{Q(t_{k-1},t_k)}\right)^2 \leq (\|\phi(\cdot, t_{k-1})\|_{2,\Omega})^2 + \mathcal{P}(t_{k-1}, t_k)|\phi|_{Q(t_{k-1},t_k)}, \tag{2.2.46}$$

and, therefore, it follows from Cauchy's inequality with $\varepsilon=\alpha_1/2$ that

$$\frac{\alpha_1}{2}\left(|\phi|_{Q(t_{k-1},t_k)}\right)^2\leq\left(\|\phi(\cdot,t_{k-1})\|_{2,\Omega}\right)^2+\frac{\alpha_1}{4}\left(|\phi|_{Q(t_{k-1},t_k)}\right)^2 + \frac{1}{\alpha_1}\left(\mathcal{P}(t_{k-1},t_k)\right)^2. \tag{2.2.47}$$

Thus,

$$\left(|\phi|_{Q(t_{k-1},t_k)}\right)^2\leq(\alpha_2)^2\left(\|\phi(\cdot,t_{k-1})\|_{2,\Omega}+\hat{\mathcal{P}}\right)^2, \tag{2.2.48}$$

where

$$(\alpha_2)^2\equiv 4\max\left\{\frac{1}{\alpha_1},\frac{1}{(\alpha_1)^2}\right\} \quad\text{and}\quad \hat{\mathcal{P}}\equiv 2\max\{\beta,1\}\left(\|F\|_{2,Q}+\|f\|_{q_1,r_1,Q}\right)$$

Consequently,

$$\operatorname*{ess\,sup}_{t_{k-1}\leq t\leq t_k}\|\phi(\cdot,t)\|_{2,\Omega}\leq\alpha_2\left\{\|\phi(\cdot,t_{k-1})\|_{2,\Omega}+\hat{\mathcal{P}}\right\} \tag{2.2.49}$$

and

$$\iint_{Q(t_{k-1},t_k)}\left\{\sum_{i=1}^{n}\left(\phi_{x_i}(x,t)\right)^2\right\}dx\,dt\leq(\alpha_2)^2\left(\|\phi(\cdot,t_{k-1})\|_{2,\Omega}+\hat{\mathcal{P}}\right)^2. \tag{2.2.50}$$

Using (2.2.49) and the definition of the division points t_k, we can show, by induction, that

$$\|\phi(\cdot,t_k)\|_{2,\Omega}\leq(\alpha_2)^k\|\phi(\cdot,0)\|_{2,\Omega}+\sum_{l=1}^{k}(\alpha_2)^l\hat{\mathcal{P}} \tag{2.2.51}$$

for $k=1,\ldots,s$.

Substituting (2.2.51) into (2.2.49), we obtain

$$\operatorname*{ess\,sup}_{t_{k-1}\leq t\leq t_k}\|\phi(\cdot,t)\|_{2,\Omega}\leq(\alpha_2)^k\|\phi(\cdot,0)\|_{2,\Omega}+\sum_{l=1}^{k}(\alpha_2)^l\hat{\mathcal{P}} \tag{2.2.52}$$

for $k=1,\ldots,s$.

This, in turn, implies that

$$\operatorname*{ess\,sup}_{0\leq t\leq T_1}\|\phi(\cdot,t)\|_{2,\Omega}\leq\alpha_3\left(\|\phi(\cdot,0)\|_{2,\Omega}+\hat{\mathcal{P}}\right), \tag{2.2.53}$$

where $\alpha_3\equiv\{\sum_{k=1}^{s}(s-k+1)(\alpha_2)^k\}$.

Substituting (2.2.51) into (2.2.50), we have

$$\iint_{Q(t_{k-1},t_k)}\left\{\sum_{i=1}^{n}\left(\phi_{x_i}(x,t)\right)^2\right\}dx\,dt\leq\left((\alpha_2)^k\|\phi(\cdot,0)\|_{2,\Omega}+\sum_{l=1}^{k}(\alpha_2)^l\hat{\mathcal{P}}\right)^2 \tag{2.2.54}$$

for $k=1,\ldots,s$.

Summing (2.2.54) over k from 1 to s and then taking the square root of both sides, it follows that

$$\|\phi_x\|_{2,\Omega\times(0,T_1)} \leq \alpha_3\big(\|\phi(\cdot,0)\|_{2,\Omega} + \hat{\mathcal{P}}\big), \tag{2.2.55}$$

where α_3 is as defined for (2.2.53).

Combining (2.2.53) and (2.2.55), we obtain

$$|\phi|_{2,\Omega\times(0,T_1)} \leq 2\alpha_3\big(\|\phi(\cdot,0)\|_{2,\Omega} + \hat{\mathcal{P}}\big).$$

Since $T_1 \in (T-\varepsilon, T)$ and $\varepsilon > 0$ is arbitrary, the conclusion of the lemma follows from the above inequality. □

REMARK 2.2.4. Let η be an element of $W_2^{1,1}(\Omega\times(-h,T))$ that is equal to zero for $t \geq T-h$ and for $t \leq 0$. Define

$$\eta_{\bar{h}}(x,t) \equiv \frac{1}{h}\int_{t-h}^{t} \{\eta(x,\tau)\}\,d\tau. \tag{2.2.56}$$

Then it is easily verified that

$$(\eta_{\bar{h}})_t(x,t) = (\eta_t)_{\bar{h}}(x,t), \tag{2.2.57}$$

where

$$(\eta_t)_{\bar{h}}(x,t) \equiv \frac{1}{h}\int_{t-h}^{t} \{\eta_\tau(x,\tau)\}\,d\tau. \tag{2.2.58}$$

REMARK 2.2.5. Let ζ and $\hat{\eta}$ be square integrable functions on $[-h,T]$ such that on each of the intervals $[-h,0]$ and $[T-h,T]$ at least one of them is identically zero. Define

$$\zeta_h(t) \equiv \frac{1}{h}\int_{t}^{t+h} \{\zeta(\tau)\}\,d\tau \tag{2.2.59}$$

and

$$\hat{\eta}_{\bar{h}}(t) \equiv \frac{1}{h}\int_{t-h}^{t} \{\hat{\eta}(\tau)\}\,d\tau. \tag{2.2.60}$$

Then, by interchanging the order of integration with respect to t and τ, we obtain

$$\int_0^T \{\zeta(t)\hat{\eta}_{\bar{h}}(t)\}\,dt = \int_0^{T-h} \{\zeta_h(t)\hat{\eta}(t)\}\,dt. \quad \square \tag{2.2.61}$$

Lemma 2.2.5. *Consider system* (2.2.1). *Suppose that Assumptions* (2.2.A1)–(2.2.A5) *are satisfied and that* ϕ *is a weak solution from* $V_2^{1,0}(Q)$ *of system* (2.2.1). *Then the equality*

$$\frac{1}{2}\int_\Omega \{(\phi(x,t))^2\}\,dx\Big|_{t=0}^{t=t_1} + \int_0^{t_1} \{\mathfrak{L}(\phi,\phi)(t) - (\tilde{F},\phi)(t)\}\,dt = 0, \tag{2.2.62}$$

holds for all $t_1 \in [0,T]$.

PROOF. By hypothesis, ϕ is a weak solution from $V_2^{1,0}(Q)$ of system (2.2.1). Let $\hat{\eta}$ be an element of $\mathring{W}_2^{1,1}(\Omega\times(-h,T))$ that is equal to zero for $t\geq T-h$ and for $t\leq 0$. Then, we can construct the corresponding function $\hat{\eta}_{\bar{h}}$ according to the definition (2.2.56). For each real number $h>0$, $\hat{\eta}_{\bar{h}}\in \mathring{W}_2^{1,1}(Q)$ and is equal to zero for $t=0$ and $t=T$. Thus, by virtue of Remark 2.2.3, we have

$$-\iint_Q\{\phi(x,t)(\hat{\eta}_{\bar{h}})_t(x,t)\}\,dx\,dt+\int_0^T\{\mathfrak{L}(\phi,\hat{\eta}_{\bar{h}})(t)-(\tilde{F},\hat{\eta}_{\bar{h}})(t)\}\,dt=0. \tag{2.2.63}$$

By Remarks 2.2.4 and 2.2.5 and the properties of the function $\hat{\eta}$, the first integral on the left-hand side of the above equality gives

$$\begin{aligned}-\iint_Q\{\phi(x,t)(\hat{\eta}_{\bar{h}})_t(x,t)\}\,dx\,dt&=-\iint_Q\{\phi(x,t)(\hat{\eta}_t)_{\bar{h}}(x,t)\}\,dx\,dt\\&=-\int_0^{T-h}\int_\Omega\{\phi_h(x,t)\hat{\eta}_t(x,t)\}\,dx\,dt\\&=\int_0^{T-h}\int_\Omega\{(\phi_h)_t(x,t)\hat{\eta}(x,t)\}\,dx\,dt.\end{aligned} \tag{2.2.64}$$

In each term of the second integral on the left-hand side of (2.2.63), the averaging in t, $(\cdot)_{\bar{h}}$, can be transferred from $\hat{\eta}$ to each of its coefficients, because differentiation with respect to x commutes with averaging operation in t. Therefore, by virtue of (2.2.64), Equation (2.2.63) becomes

$$\begin{aligned}\int_0^{T-h}\int_\Omega\Bigg\{&(\phi_h)_t(x,t)\hat{\eta}(x,t)+\sum_{j=1}^n\left(\sum_{i=1}^n a_{ij}\phi_{x_i}+a_j\phi\right)_h(x,t)\hat{\eta}_{x_j}(x,t)\\&-\sum_{j=1}^n\left(b_j\phi_{x_j}\right)_h(x,t)\hat{\eta}(x,t)-(c\phi)_h(x,t)\hat{\eta}(x,t)\\&+\sum_{j=1}^n(F_j)_h(x,t)\hat{\eta}_{x_j}(x,t)-(f)_h(x,t)\hat{\eta}(x,t)\Bigg\}\,dx\,dt=0.\end{aligned} \tag{2.2.65}$$

The above equality can be rearranged as

$$\begin{aligned}\int_0^{T-h}\int_\Omega\Bigg\{&(\phi_h)_t(x,t)\hat{\eta}(x,t)+\sum_{j=1}^n\left(\sum_{i=1}^n a_{ij}\phi_{x_i}+a_j\phi+F_j\right)_h(x,t)\hat{\eta}_{x_j}(x,t)\\&-\sum_{j=1}^n\left(b_j\phi_{x_j}+c\phi+f\right)_h(x,t)\hat{\eta}(x,t)\Bigg\}\,dx\,dt=0,\end{aligned} \tag{2.2.66}$$

where $\hat{\eta}$ is an arbitrary element of $\mathring{W}_2^{1,1}(\Omega\times(-h,T))$ that is equal to zero for $t\geq T-h$ and $t\leq 0$.

In what follows, we shall show that the above equality remains valid for any function $\hat{\eta}$ defined by

$$\hat{\eta}(x,t)\equiv\begin{cases}0, & (x,t)\in\Omega\times(t_1,T)\\ \eta(x,t), & (x,t)\in\Omega\times[0,t_1],\end{cases} \tag{2.2.67}$$

where $t_1\le T-h$ and η is an element of $\mathring{V}_2^{1,0}(\Omega\times(-h,T))$. For this, we note, first of all, that the set $\mathring{W}_2^{1,1}(\Omega\times(-h,T))$ is dense in $\mathring{V}_2^{1,0}(\Omega\times(-h,T))$. Thus, for any η from $\mathring{V}_2^{1,0}(\Omega\times(-h,T))$, there is a sequence of functions $\{\eta_m\}$ from $\mathring{W}_2^{1,1}(\Omega\times(-h,T))$, so that $\eta_m\xrightarrow{s}\eta$ (strongly) in $V_2^{1,0}(\Omega\times(-h,T))$ as $m\to\infty$. Let $\chi_\kappa(\cdot)$, $\kappa>0$, be a continuous piecewise linear function defined by

$$\chi_\kappa(t)\equiv\begin{cases}0, & t\le 0\\ \kappa t, & 0\le t\le 1/\kappa\\ 1, & 1/\kappa\le t\le t_1-(1/\kappa)\\ \kappa(t_1-t), & t_1-(1/\kappa)\le t\le t_1\\ 0, & t\ge t_1.\end{cases} \tag{2.2.68}$$

If $t_1\le T-h$, then, for each pair of positive integers m and κ, $\eta_m(\cdot,\cdot)\chi_\kappa(\cdot)$ belongs to $\mathring{W}_2^{1,1}(\Omega\times(-h,T))$ and is equal to zero for $t\le 0$ and $t\ge t_1$. Thus, for $\hat{\eta}\equiv\eta_m\chi_\kappa$, it follows from (2.2.66) that

$$\begin{aligned}\int_0^{t_1}\int_\Omega\Bigg\{&(\phi_h)_t(x,t)\eta_m(x,t)\chi_\kappa(t)+\sum_{j=1}^n\left(\sum_{i=1}^n a_{ij}\phi_{x_i}+a_j\phi+F_j\right)_h(x,t)\\ &\times(\eta_m)_{x_j}(x,t)\chi_\kappa(t)-\left(\sum_{j=1}^n b_j\phi_{x_j}+c\phi+f\right)_h(x,t)\\ &\times\hat{\eta}_m(x,t)\chi_\kappa(t)\Bigg\}dx\,dt=0\end{aligned} \tag{2.2.69}$$

for any pair of positive integers m and κ, where $t_1\le T-h$.

Since $\phi\in\mathring{V}_2^{1,0}(Q)$, it is easy to check that $\phi_h\in\mathring{W}_2^{1,1}(\Omega\times(0,T-h))$. Also, from Lemma 2.2.1 we observe that

$$\begin{cases}\text{(i)} & \left(\sum_{i=1}^n a_{ij}\phi_{x_i}+a_j\phi+F_j\right)_h(\cdot,\cdot),\qquad j=1,\dots,n,\\ & \text{are elements of } L_2(Q),\\ \text{(ii)} & (c\phi)_h(\cdot,\cdot)\ \text{ and }\ \left(\sum_{j=1}^n b_j\phi_{x_j}\right)_h(\cdot,\cdot)\\ & \text{are elements of } L_{\lambda_1,\lambda_2}(Q),\\ & \text{where } \lambda_1=2q/(q+1) \text{ and } \lambda_2=2r/(r+1),\\ \text{(iii)} & f_h\in L_{q_1,r_1}(Q).\end{cases} \tag{2.2.70}$$

Recall that

$$\eta_{m,\kappa}(\cdot,\cdot)\equiv\eta_m(\cdot,\cdot)\chi_\kappa(\cdot)\in \mathring{W}_2^{1,1}(\Omega\times(-h,T))\subset \mathring{V}_2^{1,0}(\Omega\times(-h,T))$$

and is equal to zero for $t\le 0$ and $t\ge t_1$, where $t_1\le T-h$. This in turn, implies that $\eta_{m,\kappa}\in L_2(Q)$ and $(\eta_{m,\kappa})_{x_j}=(\eta_m)_{x_j}\chi_\kappa\in L_2(Q)$, $j=1,\dots,n$. Moreover, by Theorem 1.2.6 we have

$$\|\eta_{m,\kappa}\|_{\lambda_3,\lambda_4,Q}\le\beta|\eta_{m,\kappa}|_Q, \tag{2.2.71}$$

where $\lambda_3=2q/(q-1)$, $\lambda_4=2r/(r-1)$, the constant β is as defined in Theorem 1.2.6, and the quantities q and r satisfy condition (2.2.7). In addition, it follows again from Theorem 1.2.6 that

$$\|\eta_{m,\kappa}\|_{q_1',r_1',Q}\le\beta|\eta_{m,\kappa}|_Q, \tag{2.2.72}$$

where q_1' and r_1' are such that $1/q_1+1/q_1'=1$ and $1/r_1+1/r_1'=1$ with the quantities q_1 and r_1 satisfying condition (2.2.8). Obviously, $q_1'=q_1/(q_1-1)$ and $r_1'=r_1/(r_1-1)$.

Since $\eta_{m,\kappa}\overset{s}{\to}\eta$ (strongly) in $V_2^{1,0}(\Omega\times(0,t_1))$, it follows from (2.2.71) and (2.2.72), respectively, that $\eta_{m,\kappa}\overset{s}{\to}\eta$ (strongly) in $L_{\lambda_3,\lambda_4}(\Omega\times(0,t_1))$ and in $L_{q_1',r_1'}(\Omega\times(0,t_1))$. From these facts, including (2.2.70), and the fact that $(\phi_h)_t\in L_2(Q)$, we deduce easily that the equality (2.2.69), in the limit with respect to m and κ (irrespective of order), reduces to

$$\int_0^{t_1}\int_\Omega\left\{(\phi_h)_t(x,t)\eta(x,t)+\sum_{j=1}^n\left(\sum_{i=1}^n a_{ij}\phi_{x_i}+a_j\phi+F_j\right)_h\eta_{x_j}(x,t)\right.$$
$$\left.-\left(\sum_{j=1}^n b_j\phi_{x_j}+c\phi+f\right)_h(x,t)\eta(x,t)\right\}dx\,dt=0 \tag{2.2.73}$$

for any $\eta\in\mathring{V}_2^{1,0}(\Omega\times(0,T))$, where $t_1\le T-h$.
Taking $\eta=\phi_h$, it follows from the first term of the above equality that

$$\int_0^{t_1}\int_\Omega\{(\phi_h)_t(x,t)\phi_h(x,t)\}dx\,dt=\frac{1}{2}\int_\Omega\{(\phi_h(x,t))^2\}dx\Big|_{t=0}^{t=t_1}. \tag{2.2.74}$$

Replacing η by ϕ_h in (2.2.73) and using (2.2.74), we obtain

$$\frac{1}{2}\int_\Omega\{(\phi_h(x,t))^2\}dx\Big|_{t=0}^{t=t_1}$$
$$+\int_0^{t_1}\int_\Omega\left\{\sum_{j=1}^n\left(\sum_{i=1}^n a_{ij}\phi_{x_j}+a_j\phi+F_j\right)_h(x,t)(\phi_{x_j})_h(x,t)\right.$$
$$\left.-\left(\sum_{j=1}^n b_j\phi_{x_j}+c\phi+f\right)_h(x,t)\phi_h(x,t)\right\}dx\,dt=0. \tag{2.2.75}$$

Now let us consider the following integrals:

$$\int_0^{t_1}\int_\Omega\left\{\sum_{j=1}^n\left(\sum_{i=1}^n a_{ij}\phi_{x_i}+a_j\phi+F_j\right)_h(x,t)(\eta_{x_j})_h(x,t)\right\}dx\,dt \tag{2.2.76}$$

and

$$\int_0^{t_1}\int_\Omega\left\{\left(\sum_{j=1}^n b_j\phi_{x_j}+c\phi+f\right)_h(x,t)\eta_h(x,t)\right\}dx\,dt. \tag{2.2.77}$$

For any $q, r\in[1,\infty)$, let $g\in L_{q,r}(Q)$, and let g_h be its integral average with respect to t as defined by (2.2.59). Then, it is clear that $g_h \overset{s}{\to} g$ (strongly) in $L_{q,r}(\Omega\times(0,T-\delta))$, where δ is some positive number. On the other hand, if $g\in V_2^{1,0}(Q)$, then we observe from Theorem 1.2.7 that $g_h \overset{s}{\to} g$ (strongly) in $V_2^{1,0}(\Omega\times(0,T-\delta))$. Thus, by virtue of Theorem 1.2.7, it is clear that $g_h \overset{s}{\to} g$ (strongly) in $L_{q,r}(\Omega\times(0,T-\delta))$ with q, r satisfying condition (1.2.9). In particular, $g_h \overset{s}{\to} g$ (strongly) in the norms of the Banach spaces

$$L_{\lambda_3,\lambda_4}(\Omega\times(0,T-\delta)),\qquad \lambda_3=\frac{2q}{q-1},\quad \lambda_4=\frac{2r}{r-1},$$

and

$$L_{q_1',r_1'}(\Omega\times(0,T-\delta)),\qquad q_1'=\frac{q_1}{q_1-1},\quad r_1'=\frac{r_1}{r_1-1},$$

where q and r (q_1 and r_1) satisfy condition (2.2.7) [condition (2.2.8)]. Using these relations, the fact that $\phi\in \overset{\circ}{V}{}_2^{1,0}(Q)$, and Lemma 2.2.1, we conclude that, for any $\eta\in \overset{\circ}{V}{}_2^{1,0}(Q)$, the integrals (2.2.76) and (2.2.77), in the limit with respect to h, converge, respectively, to

$$\int_0^{t_1}\int_\Omega\left\{\sum_{j=1}^n\left(\sum_{i=1}^n a_{ij}(x,t)\phi_{x_i}(x,t)+a_j(x,t)\phi(x,t)\right.\right.$$
$$\left.\left.+F_j(x,t)\right)\eta_{x_j}(x,t)\right\}dx\,dt \tag{2.2.78}$$

and

$$\int_0^{t_1}\int_\Omega\left\{\left(\sum_{j=1}^n b_j(x,t)\phi_{x_j}(x,t)+c(x,t)\phi(x,t)\right.\right.$$
$$\left.\left.+f(x,t)\right)\eta(x,t)\right\}dx\,dt, \tag{2.2.79}$$

where $t_1\in[0,T-\delta]$ and δ is some positive number. In particular, these are true for $\eta=\phi$. Thus,

$$\lim_{h\downarrow 0}\int_0^{t_1}\int_\Omega\left\{\sum_{j=1}^n\left(\sum_{i=1}^n a_{ij}\phi_{x_i}+a_j\phi+F_j\right)_h(x,t)(\phi_{x_j})_h(x,t)\right.$$
$$\left.-\left(\sum_{j=1}^n b_j\phi_{x_j}+c\phi+f\right)_h(x,t)\phi_h(x,t)\right\}dx\,dt$$
$$=\int_0^{t_1}\{\mathcal{L}(\phi,\phi)(t)-(\tilde{F},\phi)(t)\}dt. \tag{2.2.80}$$

On the other hand, since ϕ depends continuously on t in the $L_2(\Omega)$-norm, it follows that

$$\lim_{h\downarrow 0}\int_\Omega\{(\phi_h(x,t))^2\}dx=\int_\Omega\{(\phi(x,t))^2\}dx \tag{2.2.81}$$

for all $t\in[0,T-\delta]$.

Therefore, from (2.2.80) and (2.2.81), the limit of (2.2.75) is

$$\frac{1}{2}\int_\Omega\{(\phi(x,t))^2\}dx\Big|_{t=0}^{t=t_1}+\int_0^{t_1}\{\mathcal{L}(\phi,\phi)(t)-(\tilde{F},\phi)(t)\}dt=0 \tag{2.2.82}$$

for any $t_1\leq T-\delta$, where δ is some positive number. To complete the proof, it suffices to show that ϕ and all the coefficients and free terms of system (2.2.1) and hence the differential equation of system (2.2.1) can be extended beforehand onto $\Omega\times(0,T+\delta)$ with all the properties of these functions preserved. For this, let $\hat{\phi}$, $\hat{L}$, $\hat{f}$, and $\hat{F}_j$, $j=1,\ldots,n$, be defined, respectively, by

$$\hat{\phi}(x,t)\equiv\begin{cases}\phi(x,t), & (x,t)\in\bar{\Omega}\times[0,T]\\ 0, & (x,t)\in\bar{\Omega}\times(T,T+\delta),\end{cases}$$

$$\tilde{L}\psi(x,t)\equiv\begin{cases}L\psi(x,t), & (x,t)\in Q\\ \Delta\psi(x,t), & (x,t)\in\Omega\times[T,T+\delta),\end{cases}$$

where Δ is the Laplacian operator,

$$\hat{f}(x,t)=\begin{cases}f(x,t), & (x,t)\in Q\\ 0, & (x,t)\in\Omega\times[T,T+\delta),\end{cases}$$

and, for each $j\in\{1,\ldots,n\}$,

$$\hat{F}_j(x,t)=\begin{cases}F_j(x,t), & (x,t)\in Q\\ 0, & (x,t)\in\Omega\times[T,T+\delta).\end{cases}$$

Now it is easy to check that these extended functions satisfy their original properties. This completes the proof. □

The next theorem shows that any weak solution from $V_2^{1,0}(Q)$ satisfies an a priori estimate.

Theorem 2.2.1. *Consider system* (2.2.1). *Suppose that Assumptions* (2.2.A1)–(2.2.A5) *are satisfied. If* ϕ *is a weak solution from* $V_2^{1,0}(Q)$, *then it satisfies the estimate*

$$\|\phi\|_Q \leq K\{\|\phi_0\|_{2,\Omega} + \|F\|_{2,Q} + \|f\|_{q_1,r_1,Q}\}, \tag{2.2.83}$$

where $\|\cdot\|_Q$ *is the norm in the Banach space* $V_2^{1,0}(Q)$ [*given by* (1.2.8)], $\|F\|_{2,Q}$ *is as defined by* (2.2.27) *and the constant* K *depends only on* n, α_l, α_u, *and* M_1, *and the quantities* q, q_1 *and* r, r_1 *satisfy conditions* (2.2.7) *and* (2.2.8).

PROOF. The proof follows from Lemmas 2.2.4 and 2.2.5.

REMARK 2.2.6. Consider system (2.2.1) with $f = \sum_{j=1}^{\nu} f_j$. Suppose that a_{ij}, $i, j = 1, \ldots, n$, a_j, b_j, $j = 1, \ldots, n$, c, F_j, $j = 1, \ldots, n$, and ϕ_0 satisfy the corresponding Assumptions (2.2.A1)–(2.2.A5). Furthermore, assume that $\|f_j\|_{q_j,r_j,Q} \leq M_2$, $j = 1, \ldots, \nu$, where the quantities q_j and r_j satisfy condition (2.2.8). Then, by an approach similar to that used in Lemma 2.2.4, Lemma 2.2.5, and Theorem 2.2.1 (with some obvious modifications), we can show that their conclusions remain valid with the term $\|f\|_{q_1,r_1,Q}$ on the right-hand side of the estimate (2.2.26) replaced by $\sum_{j=1}^{\nu} \|f_j\|_{q_j,r_j,Q}$.

The next theorem shows that if ϕ is a weak solution from $V_2(Q)$, then ϕ belongs to $\overset{\circ}{V}{}_2^{1,0}(Q)$.

Theorem 2.2.2. *Consider system* (2.2.1). *Suppose that all the hypotheses of Theorem* 2.2.1 *are satisfied. If* ϕ *is a weak solution from* $V_2(Q)$, *then* ϕ *belongs to* $\overset{\circ}{V}{}_2^{1,0}(Q)$.

PROOF. Since ϕ is a weak solution from $V_2(Q)$ of system (2.2.1), it follows from Lemma 2.2.3 that

$$-\iint_Q \{\phi(x,t)\eta_t(x,t)\} dx\, dt + \int_0^T \{\mathfrak{L}(\phi,\eta)(t) - (\tilde{F},\eta)(t)\} dt$$

$$= \int_\Omega \{\phi_0(x)\eta(x,0)\} dx \tag{2.2.84}$$

for any $\eta \in \overset{\circ}{W}{}_2^{1,1}(Q)$ that is equal to zero for $t = T$.

Define

$$G_j(x,t) \equiv \sum_{i=1}^{n} a_{ij}(x,t)\phi_{x_i}(x,t) + a_j(x,t)\phi(x,t) + F_j(x,t), \tag{2.2.85}$$

and

$$g(x,t) \equiv \sum_{j=1}^{n} b_j(x,t)\phi_{x_j}(x,t) + c(x,t)\phi(x,t). \tag{2.2.86}$$

Substituting (2.2.85) and (2.2.86) into (2.2.84), we obtain

$$\iint_Q \{\phi(x,t)\eta_t(x,t)\}\,dx\,dt + \int_\Omega \{\phi_0(x)\eta(x,0)\}\,dx$$
$$= \iint_Q \left\{ \sum_{j=1}^{n} G_j(x,t)\eta_{x_j}(x,t) - (g(x,t)+f(x,t))\eta(x,t) \right\} dx\,dt \tag{2.2.87}$$

for any $\eta \in \mathring{W}_2^{1,1}(Q)$ that is equal to zero for $t=T$.

In view of Lemma 2.2.2, we note that the $G_j, j=1,\ldots,n$, belong to $L_2(Q)$ and that g belongs to $L_{\lambda_1,\lambda_2}(Q)$, where $\lambda_1 = 2q/(q+1)$ and $\lambda_2 = 2r/(r+1)$.

To show that $\phi \in \mathring{V}_2^{1,0}(Q)$, we extend the definitions of the functions ϕ, $G_j, j=1,\ldots,n$, g, and f onto $\Omega \times R$ as follows:

$$\phi^*(x,t) = \begin{cases} \phi(x,t), & (x,t) \in \Omega \times [0,T] \\ \phi(x,-t), & (x,t) \in \Omega \times [-T,0) \\ 0, & |t| > T, \end{cases} \tag{2.2.88}$$

$$G_j^*(x,t) = \begin{cases} G_j(x,t), & (x,t) \in \Omega \times [0,T] \\ G_j(x,-t), & (x,t) \in \Omega \times [-T,0) \\ 0, & |t| > T, \end{cases} \tag{2.2.89}$$

$$g^*(x,t) = \begin{cases} g(x,t), & (x,t) \in \Omega \times [0,T] \\ g(x,-t), & (x,t) \in \Omega \times [-T,0) \\ 0, & |t| > T, \end{cases} \tag{2.2.90}$$

$$f^*(x,t) = \begin{cases} f(x,t), & (x,t) \in \Omega \times [0,T] \\ f(x,-t), & (x,t) \in \Omega \times [-T,0) \\ 0, & |t| > T. \end{cases} \tag{2.2.91}$$

With these functions, the corresponding version of equality (2.2.87) takes the following form:

$$\int_{-\infty}^{\infty} \int_\Omega \{\phi^*(x,t)\eta_t(x,t)\}\,dx\,dt$$
$$= \int_{-\infty}^{\infty} \int_\Omega \left\{ \sum_{j=1}^{n} G_j^*(x,t)\eta_{x_j}(x,t) - (g^*(x,t)+f^*(x,t))\eta(x,t) \right\} dx\,dt, \tag{2.2.92}$$

where η is an arbitrary element of $\mathring{W}_2^{1,1}(\Omega \times (-\infty,\infty))$ that is equal to zero for $|t|>T$. Now, let ω be an element of $C^1((-\infty,\infty))$ such that $\omega(t)=1$ for all $t\in[-T+\delta, T-\delta]$, where δ is some positive number, and $\omega(t)=0$ for $|t|\geq T$. Let $\hat{\eta}$ be an element of $\mathring{W}_2^{1,1}(\Omega\times(-\infty,\infty))$ and let $\eta(x,t)\equiv \omega(t)\hat{\eta}(x,t)$. Substituting this function into equality (2.2.92), we have

$$\int_{-\infty}^{\infty}\int_{\Omega}\{\psi(x,t)\hat{\eta}_t(x,t)\}\,dx\,dt$$
$$=\int_{-\infty}^{\infty}\int_{\Omega}\Bigg\{\sum_{j=1}^{n} G_j^*(x,t)\omega(t)\hat{\eta}_{x_j}(x,t)$$
$$-(g^*(x,t)\omega(t)+f^*(x,t)\omega(t)+\phi^*(x,t)\omega_t(t))\hat{\eta}(x,t)\Bigg\}\,dx\,dt, \tag{2.2.93}$$

where $\psi(x,t)\equiv\phi^*(x,t)\omega(t)$.

Let $\hat{\eta}$ be an arbitrary element of $\mathring{V}_2^{1,0}(\Omega\times(-\infty,\infty))$ and $\hat{\eta}_{\bar{h}}$ as defined by (2.2.56). Then, it is clear that $\hat{\eta}_{\bar{h}}\in\mathring{W}_2^{1,1}(\Omega\times(-\infty,\infty))$. Thus, by taking $\hat{\eta}$ as $\hat{\eta}_{\bar{h}}$ in Equation (2.2.93) and noting that the averaging operation in t, $(\cdot)_{\bar{h}}$, commutes with differential operator with respect to x, we obtain

$$\int_{-\infty}^{\infty}\int_{\Omega}\{\psi(x,t)(\hat{\eta}_{\bar{h}})_t(x,t)\}\,dx\,dt$$
$$=\int_{-\infty}^{\infty}\int_{\Omega}\Bigg\{\sum_{j=1}^{n} \hat{G}_j^*(x,t)(\hat{\eta}_{\bar{h}})_{x_j}(x,t)$$
$$-\left[\hat{g}^*(x,t)+\hat{f}^*(x,t)+\phi^*(x,t)\omega_t(t)\right]\hat{\eta}_{\bar{h}}(x,t)\Bigg\}\,dx\,dt$$
$$=\int_{-\infty}^{\infty}\int_{\Omega}\Bigg\{\sum_{j=1}^{n} \hat{G}_j^*(x,t)(\hat{\eta}_{x_j})_{\bar{h}}(x,t)$$
$$-\left(\hat{g}^*(x,t)+\hat{f}^*(x,t)-\phi^*(x,t)\omega_t(t)\right)\hat{\eta}_{\bar{h}}(x,t)\Bigg\}\,dx\,dt, \tag{2.2.94}$$

where $\hat{G}_j^*(x,t)\equiv G_j^*(x,t)\omega(t)$, $j=1,\ldots,n$, $\hat{g}^*(x,t)\equiv g^*(x,t)\omega(t)$, and $\hat{f}^*(x,t)\equiv f^*(x,t)\omega(t)$.

In view of Remarks 2.2.4 and 2.2.5, it is clear that the averaging in t, $(\cdot)_{\bar{h}}$, can be transferred from the second factor of each term of the equality (2.2.94) to its corresponding first factor. Thus,

$$\int_{-\infty}^{\infty}\int_{\Omega}\{\psi_h(x,t)\hat{\eta}_t(x,t)\}\,dx\,dt$$
$$=\int_{-\infty}^{\infty}\int_{\Omega}\Bigg\{\sum_{j=1}^{n} (\hat{G}_j^*)_h(x,t)\hat{\eta}_{x_j}(x,t)$$
$$-\left[\hat{g}_h^*(x,t)+\hat{f}_h^*(x,t)+(\phi^*\omega_t)_h(x,t)\right]\hat{\eta}(x,t)\Bigg\}\,dx\,dt. \tag{2.2.95}$$

Let Ξ be an element of $\mathring{W}_2^1(\Omega)$, γ an element of $C_0^1((-\infty,\infty))$, and $\hat{\eta}(x,t)\equiv\Xi(x)\gamma(t)$. Then, by substituting this $\hat{\eta}$ into the above equation, we have

$$\int_{-\infty}^{\infty}\left\{\gamma_t(t)\left[\int_\Omega\{\psi_h(x,t)\Xi(x)\}dx\right]\right\}dt$$

$$=\int_{-\infty}^{\infty}\left\{\gamma(t)\left[\sum_{j=1}^{n}\int_\Omega\left\{\left(\hat{G}_j^*\right)_h(x,t)\Xi_{x_j}(x)\right\}dx\right.\right.$$

$$\left.\left.-\int_\Omega\left[\left(\hat{g}_h^*(x,t)+\hat{f}_h^*(x,t)+(\phi^*\omega_t)_h(x,t)\right]\Xi(x)dx\right]\right\}dt, \quad (2.2.96)$$

where $(\cdot)_h$ is as defined in (2.2.59).

In view of the definition of the generalized derivative d/dt, it is clear that, for almost all t,

$$-\frac{d}{dt}\left(\int_\Omega\{\psi_h(x,t)\Xi(x)\}dx\right)=\sum_{j=1}^{n}\int_\Omega\left\{\left(\hat{G}_j^*\right)_h(x,t)\Xi_{x_j}(x)\right\}dx$$

$$-\int_\Omega\left\{\left(\hat{g}_h^*(x,t)+\hat{f}_h^*(x,t)\right.\right.$$

$$\left.\left.+(\phi^*\omega_t)_h(x,t)\right)\Xi(x)\right\}dx \quad (2.2.97)$$

for any $\Xi\in\mathring{W}_2^1(\Omega)$.

Note that the function ψ_h is strongly continuous in t in the norm of $L_2(\Omega)$. Thus $\int_\Omega\psi_h(x,\cdot)\Xi(x)\,dx$ is continuous in t. On the other hand, it is known that the generalized derivative $(\psi_h)_t$ of ψ_h belongs to $L_2(\Omega\times(-\infty,\infty))$. Thus, for almost all t,

$$\frac{d}{dt}\left(\int_\Omega\{\psi_h(x,t)\Xi(x)\}dx\right)=\int_\Omega\{(\psi_h)_t(x,t)\Xi(x)\}dx. \quad (2.2.98)$$

From (2.2.97) and (2.2.98), it follows that, for almost all t,

$$-\int_\Omega\{(\psi_h)_t(x,t)\Xi(x)\}dx$$

$$=\sum_{j=1}^{n}\int_\Omega\left\{\left(\hat{G}_j^*\right)_h(x,t)\Xi_{x_j}(x)\right\}dx$$

$$-\int_\Omega\left\{\left(\hat{g}_h^*(x,t)+\hat{f}_h^*(x,t)+(\phi^*\omega_t)_h(x,t)\right)\Xi(x)\right\}dx. \quad (2.2.99)$$

Therefore, for any pair of positive real numbers h_1 and h_2, it follows from the above equality that, for almost all t,

$$\int_\Omega \{((\psi_{h_1})_t(x,t)-(\psi_{h_2})_t(x,t))\Xi(x)\}dx$$

$$=-\sum_{i=1}^{n}\int_\Omega\{((\hat{G}_j^*)_{h_1}(x,t)-(\hat{G}_j^*)_{h_2}(x,t))\Xi_{x_j}(x)\}dx$$

$$+\int_\Omega\{(\hat{g}_{h_1}^*(x,t)-\hat{g}_{h_2}^*(x,t))\Xi(x)\}dx$$

$$+\int_\Omega\{(\hat{f}_{h_1}^*(x,t)-\hat{f}_{h_2}^*(x,t))\Xi(x)\}dx$$

$$+\int_\Omega\{((\phi^*\omega_t)_{h_1}(x,t)-(\phi^*\omega_t)_{h_2}(x,t))\Xi(x)\}dx. \tag{2.2.100}$$

Since $\psi_{h_1}-\psi_{h_2}\in \mathring{W}_2^{1,1}(\Omega\times(-\infty,\infty))$, it is clear that, for each $t\in(-\infty,\infty)$, $\psi_{h_1}(\cdot,t)-\psi_{h_2}(\cdot,t)\in\mathring{W}_2^1(\Omega)$. Therefore, by taking $\Xi=\psi_{h_1}-\psi_{h_2}$ in the above equality, we have, for almost all t,

$$\int_\Omega\{((\psi_{h_1})_t(x,t)-(\psi_{h_2})_t(x,t))(\psi_{h_1}(x,t)-\psi_{h_2}(x,t))\}dx$$

$$=-\sum_{j=1}^{n}\int_\Omega\{((\hat{G}_j^*)_{h_1}(x,t)-(\hat{G}_j^*)_{h_2}(x,t))(\psi_{h_1}-\psi_{h_2})_{x_j}(x,t)\}dx$$

$$+\int_\Omega\{(\hat{g}_{h_1}^*(x,t)-\hat{g}_{h_2}^*(x,t))(\psi_{h_1}(x,t)-\psi_{h_2}(x,t))\}dx$$

$$+\int_\Omega\{(\hat{f}_{h_1}^*(x,t)-\hat{f}_{h_2}^*(x,t))(\psi_{h_1}(x,t)-\psi_{h_2}(x,t))\}dx$$

$$+\int_\Omega\{((\hat{\phi}^*\omega_t)_{h_1}(x,t)-(\hat{\phi}^*\omega_t)_{h_2}(x,t))(\psi_{h_1}(x,t)-\psi_{h_2}(x,t))\}dx. \tag{2.2.101}$$

Since the averaging operation in t, $(\cdot)_h$, commutes with the differential operator with respect to x, the above equality can be written as

$$\int_\Omega \{((\psi_{h_1})_t(x,t)-(\psi_{h_2})_t(x,t))(\psi_{h_1}(x,t)-\psi_{h_2}(x,t))\}dx$$

$$=-\sum_{j=1}^n \int_\Omega \{((\hat{G}_j^*)_{h_1}(x,t)-(\hat{G}_j^*)_{h_2}(x,t))$$

$$\times((\psi_{x_j})_{h_1}(x,t)-(\psi_{x_j})_{h_2}(x,t))\}dx$$

$$+\int_\Omega \{(\hat{g}_{h_1}^*(x,t)-\hat{g}_{h_2}^*(x,t))(\psi_{h_1}(x,t)-\psi_{h_2}(x,t))\}dx$$

$$+\int_\Omega \{(\hat{f}_{h_1}^*(x,t)-\hat{f}_{h_2}(x,t))(\psi_{h_1}(x,t)-\psi_{h_2}(x,t))\}dx$$

$$+\int_\Omega \{((\phi^*\omega_t)_{h_1}(x,t)-(\phi^*\omega_t)_{h_2}(x,t))(\psi_{h_1}(x,t)-\psi_{h_2}(x,t))\}dx. \tag{2.2.102}$$

On the other hand, we observe that

$$\int_\Omega \{((\psi_{h_1})_t(x,t)-(\psi_{h_2})_t(x,t))(\psi_{h_1}(x,t)-\psi_{h_2}(x,t))\}dx$$

$$=\frac{1}{2}\frac{d}{dt}(\|\psi_{h_1}(\cdot,t)-\psi_{h_2}(\cdot,t)\|_{2,\Omega})^2 \tag{2.2.103}$$

for almost all t.

Thus, from (2.2.102) and (2.2.103), it follows that

$$\frac{1}{2}\frac{d}{dt}(\|\psi_{h_1}(\cdot,t)-\psi_{h_2}(\cdot,t)\|_{2,\Omega})^2$$

$$=-\sum_{j=1}^n \int_\Omega \{((\hat{G}_j^*)_{h_1}(x,t)-(\hat{G}_j^*)_{h_2}(x,t))$$

$$\times((\psi_{x_j})_{h_1}(x,t)-(\psi_{x_j})_{h_2}(x,t))\}dx$$

$$+\int_\Omega \{(\hat{g}_{h_1}^*(x,t)-\hat{g}_{h_2}^*(x,t))(\psi_{h_1}(x,t)-\psi_{h_2}(x,t))\}dx$$

$$+\int_\Omega \{(\hat{f}_{h_1}^*(x,t)-\hat{f}_{h_2}^*(x,t))(\psi_{h_1}(x,t)-\psi_{h_2}(x,t))\}dx$$

$$+\int_\Omega \{((\phi^*\omega_t)_{h_1}(x,t)-(\phi^*\omega_t)_{h_2}(x,t))(\psi_{h_1}(x,t)-\psi_{h_2}(x,t))\}dx \tag{2.2.104}$$

for almost all t.

Integrating both sides of the above equality with respect to t from t_1 to t_2, we obtain

$$-\frac{1}{2}\left(\|\psi_{h_1}(\cdot,t)-\psi_{h_2}(\cdot,t)\|_{2,\Omega}\right)\Big|_{t=t_1}^{t=t_2}$$
$$=\int_{t_1}^{t_2}\Bigg[\sum_{j=1}^{n}\int_{\Omega}\Big[\left((\hat{G}_j^*)_{h_1}(x,t)-(\hat{G}_j^*)_{h_2}(x,t)\right)$$
$$\times\left((\psi_{x_j})_{h_1}(x,t)-(\psi_{x_j})_{h_2}(x,t)\right)\Big]dx$$
$$-\int_{\Omega}\left\{\left(\hat{g}_{h_1}^*(x,t)-\hat{g}_{h_2}^*(x,t)\right)\left(\psi_{h_1}(x,t)-\psi_{h_2}(x,t)\right)\right\}dx$$
$$-\int_{\Omega}\left\{\left(\hat{f}_{h_1}^*(x,t)-\hat{f}_{h_2}^*(x,t)\right)\left(\psi_{h_1}(x,t)-\psi_{h_2}(x,t)\right)\right\}dx$$
$$-\int_{\Omega}\left\{\left((\phi^*\omega_t)_{h_1}(x,t)-(\phi^*\omega_t)_{h_2}(x,t)\right)\left(\psi_{h_1}(x,t)-\psi_{h_2}(x,t)\right)\right\}dx\Bigg]dt \tag{2.2.105}$$

for all t_1 and t_2.

Note that the function ϕ^* defined by (2.2.88) is an even function with respect to the variable t. Thus, by virtue of the definition of the function ω and Theorem 1.2.6, it follows that

$$\|\phi^*(\cdot,\cdot)\omega_t(\cdot)\|_{\lambda_3,\lambda_4,\Omega\times(-\infty,\infty)}\le M\beta|\phi|_Q, \tag{2.2.106}$$

where

$$\lambda_3=\frac{2q}{q-1},\qquad \lambda_4=\frac{2r}{r-1},\qquad M\equiv\max_{-T\le t\le T}|\omega_t(t)|;$$

$|\cdot|_Q$ is the norm in the Banach space $V_2(Q)$ and the constant β is as defined in Theorem 1.2.6. Therefore, it is clear from Theorem 1.2.7 that

$$(\phi^*\omega_t)_h(\cdot,\cdot)\overset{s}{\to}\phi^*(\cdot,\cdot)\omega_t(\cdot) \tag{2.2.107}$$

(strongly) in $L_{\lambda_3,\lambda_4}(Q)$.

Observe that ϕ^*, $\hat{G}_j^*$, $j=1,\ldots,n$, $\hat{g}^*$, and $\hat{f}^*$ are even functions with respect to the variable t and equal to zero for $|t|>T$. Thus, by using the fact that $\phi\in\overset{\circ}{V}_2(Q)$, it follows from Lemma 2.2.1, Theorem 1.2.7 and the relation (2.2.107) that

$$\Big\{\left(\|(\hat{G}_j^*)_{h_1}-(\hat{G}_j^*)_{h_2}\|_{2,\Omega\times(-\infty,\infty)}\times\|(\psi_{h_1})_x-(\psi_{h_2})_x\|_{2,\Omega\times(-\infty,\infty)}\right)$$
$$+\left(\|\hat{g}_{h_1}^*-\hat{g}_{h_2}^*\|_{\lambda_1,\lambda_2,\Omega\times(-\infty,\infty)}\times\|\psi_{h_1}-\psi_{h_2}\|_{\lambda_3,\lambda_4,\Omega\times(-\infty,\infty)}\right)$$
$$+\left(\|\hat{f}_{h_1}^*-\hat{f}_{h_2}^*\|_{q_1,r_1,\Omega\times(-\infty,\infty)}\times\|\psi_{h_1}-\psi_{h_2}\|_{q_1',r_1',\Omega\times(-\infty,\infty)}\right)$$
$$+\Big(\|(\phi^*\omega_t)_{h_1}-(\phi^*\omega_t)_{h_2}\|_{\lambda_3,\lambda_4,\Omega\times(-\infty,\infty)}$$
$$\times\|\psi_{h_1}-\psi_{h_2}\|_{\lambda_1,\lambda_2,\Omega\times(-\infty,\infty)}\Big)\Big\}\to 0 \tag{2.2.108}$$

as $h_1\downarrow 0$ and $h_2\downarrow 0$, where

$$\lambda_1=\frac{2q}{q+1},\qquad \lambda_2=\frac{2r}{r+1},\qquad \lambda_3=\frac{2q}{q-1},\qquad \lambda_4=\frac{2r}{r-1},$$

$$q_1'=\frac{q_1}{q_1-1},\qquad r_1'=\frac{r_1}{r_1-1},$$

and

$$\|Z_x\|_{2,\Omega\times(-\infty,\infty)}\equiv\left(\int_{-\infty}^{\infty}\int_{\Omega}\left\{\sum_{i=1}^{n}\left(Z_{x_i}(x,t)\right)^2\right\}dx\,dt\right)^{1/2}.$$

Setting $t_1=-\infty$ and t_2 arbitrary in (2.2.105), we get

$$-\frac{1}{2}\left(\|\psi_{h_1}(\cdot,t_2)-\psi_{h_2}(\cdot,t_2)\|_{2,\Omega}\right)$$
$$=\int_{-\infty}^{t_2}\left[\sum_{j=1}^{n}\int_{\Omega}\left\{\left((\hat{G}_j^*)_{h_1}(x,t)-(\hat{G}_j^*)_{h_2}(x,t)\right)\right.\right.$$
$$\left.\times\left((\psi_{x_j})_{h_1}(x,t)-(\psi_{x_j})_{h_2}(x,t)\right)\right\}dx$$
$$-\int_{\Omega}\left\{\left(\hat{g}_{h_1}^*(x,t)-\hat{g}_{h_2}^*(x,t)\right)\left(\psi_{h_1}(x,t)-\psi_{h_2}(x,t)\right)\right\}dx$$
$$-\int_{\Omega}\left\{\left(\hat{f}_{h_1}^*(x,t)-\hat{f}_{h_2}^*(x,t)\right)\left(\psi_{h_1}(x,t)-\psi_{h_2}(x,t)\right)\right\}dx$$
$$\left.-\int_{\Omega}\left\{\left((\phi^*\omega_t)_{h_1}(x,t)-(\phi^*\omega_t)_{h_2}(x,t)\right)\left(\psi_{h_1}(x,t)-\psi_{h_2}(x,t)\right)\right\}dx\right]dt.$$

(2.2.109)

Applying Hölder's inequality to all those integrals on the right-hand side of the above equality and then using relation (2.2.108), we conclude that

$$\|\psi_{h_1}(\cdot,t)-\psi_{h_2}(\cdot,t)\|_{2,\Omega}\to 0$$

as $h_1\downarrow 0$ and $h_2\downarrow 0$ uniformly in $t\in(-\infty,\infty)$. This implies that $\{\psi_h(\cdot,t)\}$ is a Cauchy sequence in $L_2(\Omega)$ uniformly in $t\in(-\infty,\infty)$. Thus, $\{\psi_h(\cdot,t)\}$ converges strongly in $L_2(\Omega)$ uniformly in $t\in(-\infty,\infty)$. Therefore, the limit function $\psi(x,t)\equiv\phi^*(x,t)\omega(t)$ is equivalent in $\Omega\times(-\infty,\infty)$ to a function that is strongly continuous in t in the norm topology of $L_2(\Omega)$. With this relation, we conclude from the definition of ω and ϕ^* that $\phi\in\mathring{V}_2^{1,0}(\Omega\times(0,T-\delta))$. However, since ϕ and all the coefficients and free terms of system (2.2.1) and hence the differential equation of the system can be extended (beforehand) onto $\Omega\times(0,T+\delta)$ with all the properties of these functions preserved, the proof of the theorem is complete. □

In the next theorem, we shall show that system (2.2.1) has a weak solution from $V_2(Q)$.

Theorem 2.2.3. *Consider system* (2.2.1). *Suppose that Assumptions* (2.2.A1)–(2.2.A5) *are satisfied. Then the system has a weak solution* ϕ *from* $V_2(Q)$.

PROOF. Let $\{\psi_k\}$ be a fundamental system of functions in $\mathring{W}_2^1(\Omega)$ so that

$$\langle\psi_k,\psi_l\rangle_\Omega \equiv \int_\Omega \{\psi_k(x)\psi_l(x)\}dx \equiv \delta_{kl}, \tag{2.2.110}$$

where

$$\delta_{kl} = \begin{cases} 1 & \text{if } k=l \\ 0 & \text{if } k\neq l. \end{cases} \tag{2.2.111}$$

We look for an approximate solution in the form

$$\phi^N(x,t) = \sum_{k=1}^{N} \gamma_k^N(t)\psi_k(x). \tag{2.2.112}$$

The functions γ_k^N are determined by the conditions

$$\frac{d}{dt}\langle\phi^N(\cdot,t),\psi_k(\cdot)\rangle_\Omega + \mathfrak{L}(\phi^N,\psi_k)(t) - (\tilde{F},\psi_k)(t) = 0 \tag{2.2.113}$$

and

$$\gamma_k^N(0) = \langle\phi_0,\psi_k\rangle_\Omega, \qquad k=1,\ldots,N. \tag{2.2.114}$$

Since $\{\psi_k\}$ is orthonormal in $L_2(\Omega)$, the conditions (2.2.113) give rise to

$$\frac{d\gamma_k^N(t)}{dt} = \sum_{l=1}^{N} A_{kl}(t)\gamma_l^N(t) + B_k(t), \qquad k=1,\ldots,N, \tag{2.2.115}$$

where

$$A_{kl}(t) \equiv -\int_\Omega \Bigg\{ \sum_{i,j=1}^{n} a_{ij}(x,t)(\psi_l)_{x_i}(x)(\psi_k)_{x_j}(x) + \sum_{j=1}^{n} a_j(x,t)\psi_l(x)(\psi_k)_{x_j}(x) - \sum_{j=1}^{n} b_j(x,t)(\psi_l)_{x_j}(x)\psi_k(x) - c(x,t)\psi_l(x)\psi_k(x) \Bigg\} dx \tag{2.2.116}$$

and

$$B_k(t) \equiv -\int_\Omega \Bigg\{ \sum_{j=1}^{n} F_j(x,t)(\psi_k)_{x_j}(x) - f(x,t)\psi_k(x) \Bigg\} dx \tag{2.2.117}$$

From Lemma 2.2.2, it follows that the functions A_{kl}, $k, l=1,\ldots, N$, and B_k, $k=1,\ldots, N$, are integrable on $[0, T]$. Thus, for each positive integer N, the ordinary differential equation (2.2.115) has a unique solution corresponding to the initial condition given by (2.2.114). This, in turn, implies that, for each positive integer N, the function ϕ^N is uniquely determined by Equation (2.2.112).

Multiplying each equation of (2.2.115) by γ_k^N, summing the obtained equalities over all k from 1 to N, and then integrating the result with respect to t from 0 to t_1, we obtain

$$\frac{1}{2}\left(\|\phi^N(\cdot,t)\|_{2,\Omega}\right)^2\Big|_{t=0}^{t=t_1}+\int_0^{t_1}\{\mathfrak{L}(\phi^N,\phi^N)(t)-(\tilde{F},\phi^N)(t)\}dt=0. \tag{2.2.118}$$

Thus, it follows from Lemma 2.2.4 that

$$|\phi^N|_Q \leq K\left\{\|\phi^N(\cdot,0)\|_{2,\Omega}+\|F\|_{2,Q}+\|f\|_{q_1,r_1,Q}\right\}, \tag{2.2.119}$$

where the constant K is independent of N.

From (2.2.112),

$$(\phi^N(x,0))^2=\sum_{k,l=1}^{N}\gamma_k^N(0)\gamma_l^N(0)\psi_k(x)\psi_l(x); \tag{2.2.120}$$

thus, it follows from (2.2.110) that

$$\left(\|\phi^N(\cdot,0)\|_{2,\Omega}\right)^2=\sum_{k=1}^{N}(\gamma_k^N(0))^2. \tag{2.2.121}$$

On the other hand, by virtue of (2.2.114) and Bessel's inequality, we have

$$\sum_{k=1}^{N}(\gamma_k^N(0))^2\leq(\|\phi_0\|_{2,\Omega})^2. \tag{2.2.122}$$

Therefore, it is clear from (2.2.121) and (2.2.122) that

$$\|\phi^N(\cdot,0)\|_{2,\Omega}\leq\|\phi_0\|_{2,\Omega}. \tag{2.2.123}$$

Using the above inequality, we observe that the estimate (2.2.119) reduces to

$$|\phi^N|_Q\leq K_1, \tag{2.2.124}$$

where the constant K_1 is independent of N.

In view of the above estimate and Hölder's inequality, we have

$$\|\phi^N\|_{2,Q}\leq(T)^{1/2}\|\phi^N\|_{2,\infty,Q}\leq(T)^{1/2}K_1\equiv K_2. \tag{2.2.125}$$

It follows from the inequalities (2.2.124) and (2.2.125) that there exists a subsequence of the sequence $\{\phi^N\}$, again denoted by $\{\phi^N\}$, such that

$$\phi^N\xrightarrow{w}\phi$$

$$\phi^N_{x_i}\xrightarrow{w}\phi_{x_i},\qquad i=1,\ldots,n, \tag{2.2.126}$$

(weakly) in $L_2(Q)$ as $N\to\infty$. According to the Banach–Saks theorem (Corollary 1.1.1), we can extract a subsequence $\{\phi^{N(k)}\}$ so that the averages $\hat{\phi}^l\equiv(1/l)\Sigma_{k=1}^{l}\phi^{N(k)}$ converge to ϕ strongly in $W_2^{1,0}(Q)$ as $l\to\infty$. However, ϕ^N and hence $\hat{\phi}^l$ belong to the Banach space $\mathring{W}_2^{1,0}(Q)$. Thus, it follows that ϕ also belongs to $\mathring{W}_2^{1,0}(Q)$. On the other hand, by using (2.2.124) and the first part of the relation (2.2.126), we obtain from Theorem 1.2.12 that $\phi\in L_{2,\infty}(Q)$. Therefore, we conclude that $\phi\in\mathring{V}_2(Q)$.

At this stage, we observe that if ϕ also satisfies condition (ii) of Definition 2.2.3, then it is a weak solution from $V_2(Q)$. Thus, we consider, first of all, a class of functions $\tilde{W}\equiv\{\zeta\in C^\infty([0,T]):\zeta$ is equal to zero for $t=T\}$. Multiplying each equation of (2.2.113) by a function ζ_κ from $\tilde{W}$, summing the obtained equalities over all κ from 1 to $N'\le N$, and then integrating the result with respect to t from 0 to T, we obtain

$$\int_0^T\{\langle\phi_t^N(\cdot,t),\Psi^{N'}(\cdot,t)\rangle_\Omega+\mathfrak{L}(\phi^N,\Psi^{N'})(t)-(\tilde{F},\Psi^{N'})(t)\}dt=0 \tag{2.2.127}$$

where $\Psi^{N'}(x,t)\equiv\Sigma_{\kappa=1}^{N'}\zeta_\kappa(t)\psi_\kappa(x)$. Thus, performing integration by parts, we obtain

$$\int_0^T\{-\langle\phi^N(\cdot,t),\Psi_t^{N'}(\cdot,t)\rangle_\Omega+\mathfrak{L}(\phi^N,\Psi^{N'})(t)$$
$$-(\tilde{F},\Psi^{N'})(t)\}dt=\langle\phi^N(\cdot,0),\Psi^{N'}(\cdot,0)\rangle_\Omega. \tag{2.2.128}$$

In what follows, we shall show that (2.2.128) converges to condition (ii)′ of Lemma 2.2.3 in the limit with respect to N and N'. For this, we need the following.

i. $\Psi_t^{N'}\in L_2(Q)$.
ii. From Lemma 2.2.1, it follows that $a_{ij}\Psi_{x_j}^{N'}\in L_2(Q)$, $i,j=1,\dots,n$, $a_j\Psi_{x_j}^{N'}\in L_2(Q)$, $j=1,\dots,n$, $b_i\Psi^{N'}\in L_{\lambda_1,\lambda_2}(Q)$, $i=1,\dots,n$, and $c\Psi^{N'}\in L_{\lambda_1,\lambda_2}(Q)$, where $\lambda_1=2q/(q+1)$, $\lambda_2=2r/(r+1)$, and the quantities q and r satisfy condition (2.2.7). Clearly, the conjugate space of $L_{\lambda_1,\lambda_2}(Q)$ is $L_{\lambda_1',\lambda_2'}(Q)$, where $\lambda_1'=2q/(q-1)$ and $\lambda_2'=2r(r-1)$.
iii. Since the quantities q and r satisfy condition (2.2.7), we obtain from Remark 2.2.1 and Theorem 1.2.6 that $V_2(Q)$ is continuously embedded in $L_{\lambda_1',\lambda_2'}(Q)$. Thus, by virtue of the estimate (2.2.124), $\{\phi^N\}$ can be considered as a bounded set in $L_{\lambda_1',\lambda_2'(Q)}$. Therefore, it is clear that there exists a subsequence of the sequence $\{\phi^N\}$, again denoted by $\{\phi^N\}$, such that

$$\iint_Q\{\phi^N(x,t)\eta(x,t)\}dx\,dt\to\iint_Q\{\phi(x,t)\eta(x,t)\}dx\,dt \tag{2.2.129}$$

for any $\eta\in L_{\lambda_1,\lambda_2}(Q)$ as $N\to\infty$.

iv. From (2.2.112) and (2.2.114), we observe that $\phi^N(\cdot,0) \overset{s}{\to} \phi_0$ (strongly) in $L_2(\Omega)$ as $N\to\infty$. Thus, in particular, we have

$$\langle\phi^N(\cdot,0),\Psi^{N'}(\cdot,0)\rangle_\Omega\to\langle\phi_0,\Psi^{N'}(\cdot,0)\rangle_\Omega \qquad (2.2.130)$$

as $N\to\infty$.

v. Consider a set of functions $\{\Psi^{N'}\}$ such that $\Psi^{N'}(x,t)\equiv \sum_{\kappa=1}^{N'}\zeta_\kappa(t)\psi_\kappa(x)$ and $\zeta_\kappa\in\tilde{W}$. Then it follows from Theorem 1.2.8 that $\{\Psi^{N'}\}$ is dense in the space of all those $\{\eta\}$ required in condition (ii)′ of Lemma 2.2.3 (i.e., $\mathring{W}_2^{1,1}$). Thus, for any $\Psi\in\mathring{W}_2^{1,1}$, there exists a sequence $\{\Psi^{N'}\}$ in the above class that converges strongly to Ψ.

vi. From (iii), we recall that $V_2(Q)$ is continuously embedded in $L_{\lambda_1',\lambda_2'}(Q)$. Now, let $\{\Psi^{N'}\}$ be as defined in (v). Then, it follows that

$$\Psi^{N'}\overset{s}{\to}\Psi \qquad (2.2.131)$$

(strongly) also in $L_{\lambda_1',\lambda_2'}(Q)$ as $N'\to\infty$.

vii. From Lemma 2.2.1, it follows that $a_{ij}\phi_{x_i}\in L_2(Q)$, $i,j=1,\ldots,n$, $a_j\phi\in L_2(Q)$, $b_j\phi_{x_j}\in L_{\lambda_1,\lambda_2}(Q)$, $j=1,\ldots,n$, $c\phi\in L_{\lambda_1,\lambda_2}(Q)$, $F_j\in L_2(Q)$, $j=1,\ldots,n$, $f\in L_{q_1,r_1}(Q)$, where $\lambda_1=2q/(q+1)$, $\lambda_2=2r/(r+1)$, and the numbers q and r (q_1 and r_1) satisfy condition (2.2.7) [condition (2.2.8)]. Obviously, the conjugate spaces of $L_{\lambda_1,\lambda_2}(Q)$ and $L_{q_1,r_1}(Q)$ are $L_{\lambda_1',\lambda_2'}(Q)$ and $L_{q_1',r_1'}(Q)$, respectively, where

$$\lambda_1'=\frac{2q}{q-1},\qquad \lambda_2'=\frac{2r}{r-1},\qquad q_1'=\frac{q_1}{q_1-1},\qquad r_1'=\frac{r_1}{r_1-1}.$$

viii. Since q_1 and r_1 satisfy condition (2.2.8), we obtain from Theorem 1.2.6 that $V_2(Q)$ is continuously embedded in $L_{q_1',r_1'}(Q)$. Thus,

$$\Psi^{N'}\overset{s}{\to}\Psi \qquad (2.2.132)$$

(strongly) also in $L_{q_1',r_1'}(Q)$ as $N'\to\infty$.

Letting $N\to\infty$ in (2.2.128), we obtain from (2.2.129), (2.2.130), and the results given in (i) and (ii) that

$$\int_0^T\{-\langle\phi(\cdot,t),\Psi_t^{N'}(\cdot,t)\rangle_\Omega+\mathfrak{L}(\phi,\Psi^{N'})(t)-(\tilde{F},\Psi^{N'})(t)\}dt$$

$$=\langle\phi_0,\Psi^{N'}(\cdot,0)\rangle_\Omega. \qquad (2.2.133)$$

By using (2.2.131), (2.2.132), and the results given in (v) and (vii), we can easily show that (2.2.133), in the limit with respect to N', reduces to condition (ii)′ of Lemma 2.2.3. Thus, by virtue of that lemma, condition (ii) of Definition 2.2.3 is satisfied. Hence the proof is complete. □

In the next theorem, we shall show that system (2.2.1) admits only a unique solution from $V_2^{1,0}(Q)$.

Theorem 2.2.4. *Consider system* (2.2.1). *Suppose that Assumptions* (2.2.A1)–(2.2.A5) *are satisfied. Then the system has only a unique solution* ϕ *from* $V_2^{1,0}(Q)$.

PROOF. From Theorems 2.2.3 and 2.2.2, it follows that system (2.2.1) has a weak solution from $V_2^{1,0}(Q)$.

It remains to prove the uniqueness. For this, let $\tilde{\phi}$ be another weak solution from $V_2^{1,0}(Q)$. Then $\psi \equiv \phi - \tilde{\phi}$ is a weak solution from $V_2^{1,0}(Q)$ of the system

$$\begin{aligned} L\psi(x,t) &= 0, \qquad (x,t)\in Q \\ \psi(x,0) &= 0, \qquad x\in\Omega \\ \psi(x,t) &= 0, \qquad (x,t)\in\partial\Omega\times[0,T]. \end{aligned} \tag{2.2.134}$$

Thus, it follows from Theorem 2.2.1 that $\|\psi\|_Q = 0$, where $\|\cdot\|_Q$ is the norm in the Banach space $V_2^{1,0}(Q)$. Therefore, $\psi = 0$ and hence $\phi = \tilde{\phi}$. This completes the proof. □

REMARK 2.2.7. Consider system (2.2.1) with $f = \sum_{j=1}^{\nu} f_j$. Suppose that a_{ij}, $i, = 1, \dots, n$, a_j, b_j, $j = 1, \dots, n$, c, F_j, $j = 1, \dots, n$, and ϕ_0 satisfy Assumptions (2.2.A1)–(2.2.A5). Furthermore, assume that $\|f_j\|_{q_j, r_j, Q} \le M_2$, $j = 1, \dots, \nu$, with q_j and r_j satisfying condition (2.2.8). Then, by an argument similar to that given in Theorems 2.2.2, 2.2.3, and 2.2.4 (with some obvious modifications), we can show that their conclusions remain valid.

In the next two theorems, we shall investigate the properties of the continuity of the weak solutions with respect to the coefficients and data of the system.

Theorem 2.2.5. *Consider the sequence of first boundary value problems*

$$\begin{aligned} L^m\phi(x,t) &= \sum_{j=1}^{n} \left(F_j^m(x,t)\right)_{x_j} + f^m(x,t), \qquad (x,t)\in Q \\ \phi(x,0) &= \phi_0^m(x), \qquad x\in\Omega \\ \phi(x,t) &= 0, \qquad (x,t)\in\partial\Omega\times[0,T], \end{aligned} \tag{2.2.135}$$

where, for each positive integer m, the operator L^m *is defined by*

$$\begin{aligned} L^m\zeta(x,t) \equiv \zeta_t(x,t) - \sum_{j=1}^{n} &\left(\sum_{i=1}^{n} a_{ij}^m(x,t)\zeta_{x_i}(x,t) \right. \\ &\left. + a_j^m(x,t)\zeta(x,t) \right)_{x_j} + \sum_{j=1}^{n} b_j^m(x,t)\zeta_{x_j}(x,t) \\ &- c^m(x,t)\zeta(x,t). \end{aligned} \tag{2.2.136}$$

Suppose that the following hypotheses are satisfied:

i. *The coefficients and data of the system satisfy Assumptions* (2.2.A1)–(2.2.A5) *uniformly with respect to integers* $m \geq 1$.
ii. *The functions* a_{ij}^m, $i=1,\dots,n$, *converge to* a_{ij}, $i, j=1,\dots,n$, *a.e. on* Q, *where* a_{ij}, $i, j=1,\dots,n$, *are assumed to satisfy conditions* (2.2.A1) *and* (2.2.A2).
iii. *The functions* a_j^m, b_j^m, $j=1,\dots,n$, c^m, F_j^m, $j=1,\dots,n$, f^m, *and* ϕ_0^m *converge, respectively, to* a_j, b_j, $j=1,\dots,n$, c, F_j, $j=1,\dots,n$, f, *and* ϕ_0 *in the norms of the spaces to which they belong, where the limit functions* a_j, b_j, $j=1,\dots,n$, c, F_j, $j=1,\dots,n$, f, *and* ϕ_0 *are assumed to satisfy* (2.2.A3)–(2.2.A5).

Then $\phi^m \overset{s}{\to} \phi$ *in* $V_2^{1,0}(Q)$, *where, for each integer* $m \geq 1$, ϕ^m *is the weak solution from* $V_2^{1,0}(Q)$ *of system* (2.2.135) *and* ϕ *is the weak solution from* $V_2^{1,0}(Q)$ *of the limit system* (2.2.1).

PROOF. Let $\mathfrak{L}^m(\cdot,\cdot)(\cdot)$ be as defined by (2.2.3) with the coefficients a_{ij}, $i, j=1,\dots,n$, a_j, b_j, $j=1,\dots,n$, and c replaced, respectively by a_{ij}^m, $i, j=1,\dots,n$, a_j^m, b_j^m, $j=1,\dots,n$, and c^m. Furthermore, let $\tilde{F}^m$ be as defined by (2.2.4) with F_j, $j=1,\dots,n$, and f replaced, respectively, by F_j^m, $j=1,\dots,n$, and f^m. Then, we observe from Definition 2.2.4 that, for all $\tau \in [0, T]$,

$$\int_\Omega \{\phi^m(x,\tau)\eta(x,\tau)\}dx - \int_0^\tau \int_\Omega \{\phi^m(x,t)\eta_t(x,t)\}dx\,dt$$
$$+ \int_0^\tau \{\mathfrak{L}^m(\phi^m,\eta)(t) - (\tilde{F}^m,\eta)(t)\}dt = \int_\Omega \{\phi_0^m(x)\eta(x,0)\}dx \quad (2.2.137)$$

for all $\eta \in \mathring{W}_2^{1,1}(Q)$.

Consider the limit system (2.2.1). Then, it is also clear from Definition 2.2.4 that, for all $\tau \in [0, T]$,

$$\int_\Omega \{\phi(x,\tau)\eta(x,\tau)\}dx - \int_0^\tau \int_\Omega \{\phi(x,t)\eta_t(x,t)\}dx\,dt$$
$$+ \int_0^\tau \{\mathfrak{L}(\phi,\eta)(t) - (\tilde{F},\eta)(t)\}dt = \int_\Omega \{\phi_0(x)\eta(x,0)\}dx. \quad (2.2.138)$$

Subtracting (2.2.138) from (2.2.137) and then denoting $\phi^m - \phi$ by ψ^m, we obtain

$$\int_\Omega \{\psi^m(x,\tau)\eta(x,\tau)\}dx - \int_0^\tau \int_\Omega \{\psi^m(x,t)\eta(x,t)\}dx\,dt$$
$$+ \int_0^\tau \{\mathfrak{L}^m(\psi^m,\eta)(t) - (\tilde{G}^m,\eta)(t)\}dx\,dt$$
$$= \int_\Omega \{(\phi_0^m(x) - \phi_0(x))\eta(x,0)\}dx, \quad (2.2.139)$$

where

$$(\tilde{G}^m,\eta)(t)\equiv-\int_\Omega\left\{\sum_{j=1}^n g_j^m(x,t)\eta_{x_j}(x,t)\right.$$
$$\left.-\left(g^m(x,t)+\tilde{f}^m(x,t)\right)\eta(x,t)\right\}dx, \tag{2.2.140}$$

$$g_j^m(x,t)\equiv\sum_{i=1}^n\left(a_{ij}^m(x,t)-a_{ij}(x,t)\right)\phi_{x_i}(x,t)+\left(a_j^m(x,t)-a_j(x,t)\right)\phi(x,t)$$
$$+\left(F_j^m(x,t)-F_j(x,t)\right),\qquad i,j=1,\ldots,n, \tag{2.2.141}$$

$$g^m(x,t)\equiv\sum_{j=1}^n\left(b_j^m(x,t)-b_j(x,t)\right)\phi_{x_j}(x,t)+\left(c^m(x,t)-c(x,t)\right)\phi(x,t), \tag{2.2.142}$$

$$\tilde{f}^m(x,t)\equiv f^m(x,t)-f(x,t). \tag{2.2.143}$$

By following arguments similar to those given for Lemma 2.2.1, we can easily show that $g_j^m\in L_2(Q)$ and $g^m\in L_{\lambda_1,\lambda_2}(Q)$, where $\lambda_1=2q/(q+1)$, $\lambda_2=2r/(r+1)$, and the quantities q and r satisfy condition (2.2.7). In view of Remark 2.2.1, we observe that λ_1 and λ_2 satisfy condition (2.2.8). Thus, it follows from Theorem 2.2.4 that, for each positive integer m, ψ^m is the weak solution from $V_2^{1,0}(Q)$ of system (2.2.135) with F_j^m, $j=1,\ldots,n$, f^m, and ϕ_0 replaced, respectively, by g_j^m, $j=1,\ldots,n$, $g^m+\tilde{f}^m$, and $\phi_0^m-\phi_0$. By using Theorem 2.2.1, we obtain

$$\|\psi^m\|_Q\le K\left\{\|\phi_0^m-\phi_0\|_{2,\Omega}+\left(\iint_Q\left\{\sum_{j=1}^n\left(g_j^m(x,t)\right)^2\right\}dx\,dt\right)^{1/2}\right.$$
$$\left.+\|g^m\|_{\lambda_1,\lambda_2,Q}+\|\tilde{f}^m\|_{q_1,r_1,Q}\right\}, \tag{2.2.144}$$

where $\lambda_1=2q/(q+1)$ and $\lambda_2=2r/(r+1)$. Thus, it follows that $\|\psi^m\|_Q\to0$ as $m\to\infty$. This completes the proof. □

The following theorem is similar to the above, except that the coefficients here are assumed to converge to their limits a.e. (on Q) rather than strongly. This result will be found more useful for our purpose.

Theorem 2.2.6. *Consider the following sequence of first boundary value problems*:

$$L^m\phi(x,t)=\sum_{j=1}^n\left(F_j^m(x,t)\right)_{x_j}+f^m(x,t),\qquad (x,t)\in Q$$
$$\phi(x,0)=\phi_0(x),\qquad x\in\Omega \tag{2.2.145}$$
$$\phi(x,t)=0,\qquad (x,t)\in\partial\Omega\times[0,T],$$

where, for each positive integer m, the operator L^m is as defined by (2.2.136). *Suppose that the following hypotheses are satisfied*:

i. *The coefficients of the system satisfy Assumptions* (2.2.A1)–(2.2.A5) *uniformly with respect to the integers $m \geq 1$.*
ii. $\| F_j^m \|_{2,Q}$, $j=1,\ldots,n$, *and* $\| f^m \|_{2,r,Q}$ *are bounded uniformly with respect to the integers $m \geq 1$, where $r \in (1,\infty)$.*
iii. $\phi_0 \in L_2(\Omega)$.
iv. a_{ij}^m, $i,j=1,\ldots,n$, a_j^m, b_j^m, $j=1,\ldots,n$, c^m, F_j^m, $j=1,\ldots,n$, *and* f^m *converge, respectively, to* a_{ij}, $i,j=1,\ldots,n$, a_j, b_j, $j=1,\ldots,n$ c, F_j, $j=1,\ldots,n$, *and f a.e. on Q.*

Then there exists a subsequence $\{\phi^{m(\kappa)}\}$ of the sequence $\{\phi^m\}$ such that $\phi^{m(\kappa)} \overset{w}{\rightarrow} \phi$ and $\phi_{x_i}^{m(\kappa)} \overset{w}{\rightarrow} \phi_{x_i}$, $i=1,\ldots,n$, (weakly) in $L_2(Q)$ as $\kappa \to \infty$, where, for each integer $m \geq 1$, ϕ^m is the weak solution from $V_2^{1,0}(Q)$ of system (2.2.145) and ϕ is the weak solution from $V_2^{1,0}(Q)$ of the limit system (2.2.1).

PROOF. Since $r>1$, it follows from Hölder's inequality that

$$\| f^m \|_{2,1,Q} \leq (T)^{r/(r-1)} \| f^m \|_{2,r,Q}. \tag{2.2.146}$$

Thus, by virtue of the above inequality and conditions (i)–(iii) of the hypotheses, we deduce from inequality (2.2.83) that

$$\| \phi^m \|_Q \leq K K_1 \equiv K_2 \tag{2.2.147}$$

for all positive integers m, where $\| \cdot \|_Q$ is the norm in the Banach space $V_2^{1,0}(Q)$, the constant K is as defined for the estimate (2.2.83), and the constant K_1 is given by

$$K_1 \equiv \| \phi_0 \|_{2,\Omega} + \sup_{m \geq 1} \| F^m \|_{2,Q} + (T)^{r/(r-1)} \sup_{m \geq 1} \| f^m \|_{2,r} \tag{2.2.148}$$

with

$$\| F^m \|_{2,Q} \equiv \left(\iint_Q \left\{ \sum_{j=1}^n (F_j^m(x,t))^2 \right\} dx\, dt \right)^{1/2}. \tag{2.2.149}$$

Since, for each integer $m \geq 1$, $\phi^m \in V_2^{1,0}(Q)$, it follows from Hölder's inequality that

$$\| \phi^m \|_{2,Q} \leq (T)^{1/2} \max_{t \in [0,T]} \| \phi^m(\cdot,t) \|_{2,\Omega} \leq (T)^{1/2} K_2 \equiv K_3. \tag{2.2.150}$$

From the estimates (2.2.147) and (2.2.150), we observe that $\{\phi^m, \phi_{x_1}^m, \ldots, \phi_{x_n}^m\}$ can be considered as a bounded sequence in $[L_2(Q)]^{n+1}$, where $[L_2(Q)]^{n+1}$ denotes the $(n+1)$-copies of $L_2(Q)$. Thus, there exists a subsequence of the sequence $\{\phi^m, \phi^m_{x_i}, \ldots, \phi^m_{x_m}\}$, which is denoted by the original sequence, such that

$$\phi^m \xrightarrow{w} \phi$$

$$\phi^m_{x_i} \xrightarrow{w} \phi_i, \qquad i=1,\ldots,n, \tag{2.2.151}$$

(weakly) in $L_2(Q)$ as $m\to\infty$. According to the Banach–Saks theorem (Corollary 1.1.1), there exists a further subsequence $\{\phi^{m(\kappa)}\}$ such that $\hat{\phi}^l \equiv (1/l)\Sigma^l_{\kappa=1}\phi^{m(\kappa)} \xrightarrow{s} \phi$ (strongly) in $W_2^{1,0}(Q)$ as $l\to\infty$. Since ϕ^m and hence $\hat{\phi}^l$ belong to the Banach space $\mathring{W}_2^{1,0}(Q)$, it follows that $\phi\in \mathring{W}_2^{1,0}(Q)$. On the other hand, by virtue of the inequalities (2.2.147) and (2.2.150) and the first part of relation (2.2.151), we conclude from Theorem 1.2.12 that $\phi\in L_{2,\infty}(Q)$. Therefore, we have $\phi\in \mathring{V}_2(Q)$.

Now we shall show ϕ is a weak solution from $V_2^{1,0}(Q)$. For this, let η be an arbitrary element of $\mathring{W}_2^{1,1}(Q)$ that is equal to 0 for $t=T$. Since, for each integer $m\geq 1$, ϕ^m is the weak solution from $V_2^{1,0}(Q)$ of system (2.2.145), it follows from condition (ii)′ of Remark 2.2.3 that

$$-\iint_Q \{\phi^m(x,t)\eta_t(x,t)\}\,dx\,dt + \int_0^T \{\mathfrak{L}^m(\phi^m,\eta)(t) - (\tilde{F}^m,\eta)(t)\}\,dt$$

$$= \int_\Omega \{\phi_0(x)\eta(x,0)\}\,dx, \tag{2.2.152}$$

where $\mathfrak{L}^m(\cdot,\cdot)(\cdot)$ and $\tilde{F}^m$ are as defined for Equation (2.2.137). Thus by virtue of conditions (i), (ii), and (iv) of the hypotheses, the estimates (2.2.147) and (2.2.150) and relation (2.2.151), we obtain from Theorems 1.2.9 and 1.2.10 that the expression (2.2.152), in the limit with respect to m (through an appropriate subsequence, if necessary), reduces to

$$-\iint_Q \{\phi(x,t)\eta_t(x,t)\}\,dx\,dt + \int_0^T \{\mathfrak{L}(\phi,\eta)(t) - (\tilde{F},\eta)(t)\}\,dt$$

$$= \int_\Omega \{\phi_0(x)\eta(x,0)\}\,dx. \tag{2.2.153}$$

Since η is an arbitrary element of $\mathring{W}_2^{1,1}(Q)$ that is equal to zero for $t=T$, it follows from Remark 2.2.3 that condition (ii) of Definition 2.2.3 is satisfied. Therefore, ϕ is a weak solution from $V_2(Q)$ of the limit system (2.2.1). However, by virtue of Theorem 2.2.2, ϕ belongs to $\mathring{V}_2^{1,0}(Q)$ and hence is a weak solution from $V_2^{1,0}(Q)$. The uniqueness of this weak solution follows from Theorem 2.2.4. Thus, the proof is complete. □

REMARK 2.2.8. The conclusion of Theorem 2.2.6 remains valid if the assumptions for F_j^m, $j=1,\ldots,n$, and f^m given in condition (iv) of the hypotheses are replaced by condition (iv)′: F_j^m, $j=1,\ldots,n$, converge, respectively, to F_j, $j=1,\ldots,n$, weakly in $L_2(Q)$, and f^m converges to f weakly in $L_{2,r}(Q)$. The proof is similar to that given for Theorem 2.2.6.

In the rest of this section, we shall present a list of results concerning the properties of weak solutions of the differential equation of system (2.2.1). Their proofs are rather lengthy. Due to space limitation, we omit these proofs. Interested readers are referred to [LSU.1].

To present these results, stronger assumptions on the free terms and all but the second-order coefficients of the system are required.

Assumption (2.2.A6). There exists a positive constant $M_2 \equiv M_2(q,r)$ such that

$$\left\| \sum_{j=1}^{n} (a_j)^2 \right\|_{q,r,Q} \le M_2,$$

$$\left\| \sum_{j=1}^{n} (b_j)^2 \right\|_{q,r,Q} \le M_2,$$

$$\| c \|_{q,r,Q} \le M_2,$$

$$\left\| \sum_{j=1}^{n} (F_j)^2 \right\|_{q,r,Q} \le M_2,$$

$$\| f \|_{q,r,Q} \le M_2,$$

where q and r are arbitrary pair of positive numbers satisfying the condition

$$\frac{1}{r} + \frac{n}{2q} = 1 - \delta$$

with

$$q \in \left[\frac{n}{2(1-\delta)}, \infty \right], \qquad r \in \left[\frac{1}{1-\delta}, \infty \right], \qquad 0<\delta<1, \quad \text{for } n \ge 2,$$

$$q \in [1,\infty], \qquad r \in \left[\frac{1}{1-\delta}, \frac{2}{1-2\delta} \right], \qquad 0<\delta<\frac{1}{2}, \quad \text{for } n=1. \tag{2.2.154}$$

Theorem 2.2.7. *Let Assumptions* (2.2.A1), (2.2.A2), *and* (2.2.A6) *be satisfied.*

If ϕ *is a weak solution from* $V_2^{1,0}(Q)$ *of the differential equation of system*

(2.2.1), *then the following statements are valid*:

i. *If* $\phi(x,t)\le\hat{K}$ *for all* $(x,t)\in\Gamma\equiv\{\partial\Omega\times[0,T]\}\cup\{(x,t): x\in\Omega, t=0\}$, *then there exists a constant* $M_3\equiv M_3(n,\alpha_l,M_2,q,r)$ *such that*

$$\operatorname*{ess\,sup}_{(x,t)\in Q}\phi(x,t)\le M_3K_1,$$

where $K_1\equiv\max\{1,\hat{K}\}$.

ii. *If* $\phi(x,t)\ge\hat{K}$ *for all* $(x,t)\in\Gamma$, *then*

$$M_3K_1\le\operatorname*{ess\,inf}_{(x,t)\in Q}\phi(x,t)$$

Note that, the above theorem is valid only if $\phi(x,t)\le\hat{K}$ or $\phi(x,t)\ge\hat{K}$ for all $(x,t)\in\Gamma$. However, if we are only interested in the local boundedness of a weak solution, then the above mentioned condition is not essential, as shown in

Theorem 2.2.8. *Suppose Assumptions* (2.2.A1), (2.2.A2), *and* (2.2.A6) *are satisfied. If* ϕ *is a weak solution from* $V_2^{1,0}(Q)$ *of the differential equation of system* (2.2.1), *then* $M\equiv\text{ess sup}\{|\phi(x,t)|:(x,t)\in\hat{Q}\}<\infty$ *for any domain* $\hat{Q}\subset Q$ *separated from* Γ *by a positive distance* d, *where* Γ *is as defined in Theorem* 2.2.7. *Moreover*, M *is bounded from above by a constant that depends only on* n, $\|\phi\|_{2,Q}$, *the distance* d, *and the parameters* α_l, α_u, m_2, q, *and* r [*from Assumptions* (2.2.A1), (2.2.A2) *and* (2.2.A6)].

The next theorem presents two results: (i) ϕ belongs to the Hölder space $\mathcal{H}^{\alpha,\alpha/2}(Q)$ with some $\alpha\in(0,1)$ (for definition see Section 1.2.1); and (ii) $|\phi|_{\hat{Q}}^{(\alpha,\alpha/2)}$ is bounded from above by a constant depending only on $\|\phi\|_{2,Q}$ and the known parameters found in Assumptions (2.2.A1), (2.2.A2), and (2.2.A6), where $\hat{Q}$ is a closed subset of Q and $|\cdot|_{\hat{Q}}^{(\alpha,\alpha/2)}$ is as defined in (1.2.3).

Theorem 2.2.9. *Suppose that Assumptions* (2.2.A1), (2.2.A2), *and* (2.2.A6) *are satisfied and that* ϕ *is a weak solution from* $V_2^{1,0}(Q)$ *of the differential equation of system* (2.2.1). *If* $M\equiv\text{ess sup}\{|\phi(x,t)|:(x,t)\in Q\}<\infty$, *then* $\phi\in\mathcal{H}^{\alpha,\alpha/2}(Q)$. *Furthermore, if* $\hat{Q}$ *is a closed subset of* Q *separated from* Γ *by a positive distance* d, *then* $|\phi|_{\hat{Q}}^{(\alpha,\alpha/2)}$ *is bounded from above by a constant that depends only on* n, M, d, *and the parameters* a_l, α_u, M_2, q, *and* r. *The exponent* $\alpha\in(0,1)$ *is determined by* n, α_l, α_u, q, *and* r.

We shall close this section by making the following two remarks.

REMARK 2.2.9. Note that all the coefficients and free terms of system (2.2.1) and hence the differential equation can be extended onto $\Omega\times(0,T+\varepsilon)$ with all the properties of these functions preserved, where ε is some positive number. Thus, in particular, Theorems 2.2.7–2.2.9 remain valid with Q replaced by $\Omega\times(0,T]$.

REMARK 2.2.10. For the same reasons given in Remark 2.2.9, it is clear that a weak solution from $V_2^{1,0}(Q)$ of system (2.2.1) is also a weak solution from $V_2^{1,0}(\hat{Q})$ of the differential equation of the system with Q replaced by $\hat{Q}$, where $\hat{Q}$ is any compact subset of $\Omega\times(0,T]$.

2.3. A Class of Volterra Integral Equations

In this section we wish to consider the question of existence and uniqueness of solutions for a special class of Volterra integral equations arising in hyperbolic problems (Section 4.3).

Let $Q\equiv\{(x,t): x\in I_1\equiv[a_1,a_2], t\in I_2\equiv[b_1,b_2]\}$, where a_1, a_2, b_1, and b_2 are finite real numbers. Consider the following class of integral equation of Volterra type:

$$\phi(x,t)=\xi(x,t)+\alpha\int_{a_1}^{x} f(x;\eta,t;\phi(\eta,t))\,d\eta$$

$$+\beta\int_{b_1}^{t} g(t;x,\tau;\phi(x,\tau))\,d\tau, \qquad (x,t)\in Q, \tag{2.3.1}$$

where $f\colon I_1\times Q\times R^n\to R^n$, $g\colon I_2\times Q\times R^n\to R^n$, and $\xi\colon Q\to R^n$ are given functions and α,β are given scalars. We seek a function $\phi\colon Q\to R^n$ that satisfies Equation (2.3.1). For a precise definition of the solution we need the following notation. For a vector $v\in R^m$ we denote its norm by $|v|\equiv\Sigma_{1\le i\le m}|v_i|$, where m is any finite positive integer. Let G be a measurable subset of R^s and define $L_p(G,R^n)$, $1\le p\le\infty$, to be the n-copies of $L_p(G)\equiv L_p(G,R)$. Equipped with the norm

$$\|z\|_{p,G}\equiv\sum_{1\le i\le n}\|z_i\|_{L_p(G)}, \tag{2.3.2}$$

$L_p(G,R^n)$ is a Banach space.

Definition 2.3.1. A function $\phi\colon Q\to R^n$ is said to be a *solution of the integral equation* (2.3.1) if $\phi\in L_p(Q,R^n)$ for some p, $1\le p\le\infty$, and satisfies the equality (2.3.1) for almost all $(x,t)\in Q$.

For existence and uniqueness of solutions of (2.3.1) we introduce the following assumptions:

Assumption (2.3.A1). $\xi\in L_p(Q,R^n)$, $1\le p\le\infty$.

Assumption (2.3.A2). The function $f(x;\eta,t;\theta)\colon I_1\times Q\times R^n\to R^n$ is continuous in x on I_1 for fixed $(\eta,t,\theta)\in Q\times R^n$ and measurable in (η,t) on Q

for each $(x, \theta) \in I_1 \times R^n$; the function $g(t; x, \tau; \theta)$: $I_2 \times Q \times R^n \to R^n$ is continuous in t on I_2 for fixed $(x, \tau, \theta) \in Q \times R^n$ and measurable in (x, τ) on Q for each $(t, \theta) \in I_2 \times R^n$.

Assumption (2.3.A2). There exist constants $K_1, K_2 \geq 0$ and $h_1, h_2 \in L_p(Q, R_0)$, $R_0 = [0, \infty)$, $1 \leq p \leq \infty$, such that

$$|f(x; \eta, t; \theta_1) - f(x; \eta, t; \theta_2)| \leq K_1 |\theta_1 - \theta_2|, \qquad (x, \eta, t) \in I_1 \times Q,$$

$$|g(t; x, \tau; \theta_1) - g(t; x, \tau; \theta_2)| \leq K_2 |\theta_1 - \theta_2|, \qquad (t, x, \tau) \in I_2 \times Q,$$

$$|f(x; \eta, t; 0)| \leq h_1(\eta, t), \qquad x \in I_1,$$

$$|g(t; x, \tau; 0)| \leq h_2(x, \tau), \qquad t \in I_2.$$

Now we can consider the existence problem. Let (X, ρ) be a complete metric space and T a mapping of X into itself. T is said to be a *contraction map* if and only if there exists a positive number $a < 1$ such that for all $x, y \in X$, $\rho(Tx, Ty) \leq a\rho(x, y)$. An element x of X is said to be a *fixed point of* T if x is invariant under T, that is, $x = Tx$.

Lemma 2.3.1. *Let $(X, \rho) \equiv X$ be a complete metric space and T a continuous mapping of X into itself. If there exists a positive integer $m \geq 1$ such that T^m ($\equiv T \cdot T \cdots T$, m times) is a contraction in X, then T has a unique fixed point.*

In case X is a Banach space, the above result is known as the *Banach fixed point theorem*.

Let n_1, n_2, n_3 be positive integers and Γ a measurable subset of R^{n_1}. Let g be a Carathéodory function from $\Gamma \times R^{n_2}$ into R^{n_3}: that is, for each $y \in R^{n_2}$, $\nu \to g(\nu, y)$ is measurable on Γ with values in R^{n_3}; and for almost all $\nu \in \Gamma$, $y \to g(\nu, y)$ is continuous on R^{n_2} with values in R^{n_3}. Then it is clear that for any measurable function w: $\Gamma \to R^{n_2}$, the function $\nu \to g(\nu, w(\nu))$ is a measurable function on Γ. Let G denote the operator defined by $(Gw)(\nu) = g(\nu, w(\nu))$, $\nu \in \Gamma$. The operator G is called the *Nemytsky operator*. Basic properties of this operator are well known [Kr.1]. For the operator G we introduce the following assumption:

Assumption (2.3.A4). There exist numbers $p_1, p_2 \in [1, \infty)$, a constant $K_3 \geq 0$, and an $a \in L_{p_2}(\Gamma, R_0)$ such that

$$|g(\nu, y)| \leq a(\nu) + K_3 |y|^{p_1/p_2} \quad \text{a.e. on } \Gamma \tag{2.3.3}$$

for all $y \in R^{n_2}$.

It is clear from this inequality that if $w \in L_{p_1}(\Gamma, R^{n_2})$, then $Gw \in L_{p_2}(\Gamma, R^{n_3})$. Thus under Assumption (2.3.A4), the Nemytsky operator maps $L_{p_1}(\Gamma, R^{n_2})$ into $L_{p_2}(\Gamma, R^{n_3})$. In fact the following result is well known in the theory of nonlinear integral equations [Kr.1].

Lemma 2.3.2. *If the function g satisfies Assumption* (2.3.A4), *then the corresponding Nemytsky operator G is continuous and bounded from* $L_{p_1}(\Gamma, R^{n_2})$ *into* $L_{p_2}(\Gamma, R^{n_3})$, *and conversely.*

In other words, the inequality (2.3.3) is both a necessary and sufficient condition for G to be a continuous and bounded operator from $L_{p_1}(\Gamma, R^{n_2})$ into $L_{p_2}(\Gamma, R^{n_3})$. A detailed proof of this result is given in [Kr.1, p. 22].

With the help of the above results, we shall prove the existence and uniqueness of solutions of the integral equation (2.3.1). We introduce the operator T defined as

$$(T\phi)(x,t)=\xi(x,t)+\alpha\int_{a_1}^{x} f(x;\eta,t;\phi(\eta,t))\,d\eta$$

$$+\beta\int_{b_1}^{t} g(t;x,\tau;\phi(x,\tau))\,d\tau \tag{2.3.4}$$

and show that the operator T has a fixed point in $L_p(Q, R^n)$; that is, there exists a $\phi\in L_p(Q, R^n)$ such that $T\phi=\phi$. More precisely, we shall prove the following result.

Theorem 2.3.1. *Consider the integral equation* (2.3.1) *and suppose Assumptions* (2.3.A1)–(2.3.A3) *hold for an arbitrary but fixed* $p\in[1,\infty]$ *and* α, β *any finite real numbers. Then the integral equation* (2.3.1) *has a unique solution* $\phi\in L_p(Q, R^n)$, *and there exists a number* r_0 (≥ 0), *dependent on* α, β, K_1, K_2, *and the norms of* h_1, h_2, *and* ξ, *such that* $\|\phi\|_{p,Q}<r_0$.

PROOF. By virtue of Assumption (2.3.A3), it is clear that, for any $\theta\in R^n$,

$$|f(x;\eta,t;\theta)|\leq h_1(\eta,t)+K_1|\theta| \quad \text{for all } x\in I_1$$

and

$$|g(t;x,\tau;\theta)|\leq h_2(x,\tau)+K_2|\theta| \quad \text{for all } t\in I_2.$$

Therefore, for fixed but arbitrary $x\in I_1[t\in I_2]$, $f(x;\cdots;\cdot)$ [$g(t;\cdots;\cdot)$] is a Nemytsky operator and maps $L_p(Q, R^n)$ into $L_p(Q, R^n)$ whenever $h_1(h_2)$ $\in L_p(Q, R_0)$; consequently, by Lemma 2.3.2, it is continuous and bounded on $L_p(Q, R^n)$. Note that, by virtue of Fubini's theorem, for any $\phi\in L_p(Q, R^n)$

$$(x,t)\to\int_{a_1}^{x} f(x;\eta,t;\phi(\eta,t))\,d\eta$$

and

$$(x,t)\to\int_{b_1}^{t} g(t;x,\tau;\phi(x,\tau))\,d\tau$$

are measurable functions on Q to R^n. Therefore, it is clear that, for

$\xi \in L_p(Q, R^n)$ and $\phi \in L_p(Q, R^n)$, $(x, t) \to (T\phi)(x, t)$ is a measurable function on Q to R^n and that $T\phi \in L_p(Q, R^n)$. We show that T is a continuous operator in $L_p(Q, R^n)$ and that there exists an integer $m_0 \geq 1$ and a positive number r_0 such that for all $m \geq m_0$, T^m is a contraction in $L_p(Q, R^n)$ and the ball $B_{r_0} \equiv \{\phi \in L_p(Q, R^n) : \|\phi\|_{p,Q} \leq r_0\}$ is invariant under T, that is, $TB_{r_0} \subseteq B_{r_0}$.

Define

$$a \equiv |\alpha| K_1(a_2 - a_1), \qquad b = |\beta| K_2(b_2 - b_1).$$

For $\phi_1, \phi_2 \in L_p(Q, R^n)$, with $p \in [1, \infty]$, it is easily verified that

$$\|T\phi_1 - T\phi_2\|_{p,Q} \leq (a+b)\|\phi_1 - \phi_2\|_{p,Q}, \tag{2.3.5}$$

and, for arbitrary integer $m \geq 0$,

$$\|T^m\phi_1 - T^m\phi_2\|_{p,Q} \leq \lambda_m \|\phi_1 - \phi_2\|_{p,Q}, \tag{2.3.6}$$

where $\{\lambda_m, m \geq 0\}$ are nonnegative numbers satisfying the bounds

$$\lambda_m \leq \left((a+b)^m \Big/ \left[\frac{m}{2}\right]!\right) \quad \text{for } p = 1, \infty,$$

$$\lambda_m \leq \left((2a+2b)^{pm} / p^m \left[\frac{m}{2}\right]!\right) \quad \text{for } p \in (1, \infty). \tag{2.3.7}$$

(Note: we have used $[m/2]$ to denote the greatest integer less than or equal to $m/2$). It is clear from (2.3.5) that T is a continuous and bounded operator in $L_p(Q, R^n)$, $1 \leq p \leq \infty$. It follows from (2.3.7) that as $m \to \infty$, $\lambda_m \to 0$; consequently, there exists an integer $m_0 > 0$ such that $0 \leq \lambda_m < 1$ for all $m \geq m_0$, and hence, due to (2.3.6), T^m is a contraction for all $m \geq m_0$. Further, we note that, due to (2.3.7), the infinite series $\sum_{m=0}^{\infty} \lambda_m$ converges and has a limit, which we denote by λ:

$$\lambda = \sum_{m=0}^{\infty} \lambda_m. \tag{2.3.8}$$

Define

$$\sigma \equiv |\alpha|(a_2 - a_1)\|h_1\|_{p,Q} + |\beta|(b_2 - b_1)\|h_2\|_{p,Q}.$$

Then, for $\phi \equiv 0$,

$$\|T\phi\|_{p,Q} = \|T0\|_{p,Q} \leq (\|\xi\|_{p,Q} + \sigma). \tag{2.3.9}$$

For arbitrary $\phi \in L_p(Q, R^n)$ and integer $m \geq 1$,

$$\begin{aligned}\|T^m\phi\|_{p,Q} &\leq \|\phi\|_{p,Q} + \sum_{i=0}^{m-1} \|T^{i+1}\phi - T^i\phi\|_{p,Q} \\ &\leq \|\phi\|_{p,Q} + \left(\sum_{i=0}^{m-1} \lambda_i\right)\|T\phi - \phi\|_{p,Q},\end{aligned}$$

and consequently, due to (2.3.8), we have $\|T^m\phi\|_{p,Q} \le \|\phi\|_{p,Q} + \lambda\|T\phi - \phi\|_{p,Q}$: in particular,

$$\|T^m 0\|_{p,Q} \le \lambda(\|\xi\|_{p,Q} + \sigma) \quad \text{for all integers } m \ge 1. \tag{2.3.10}$$

Thus, for $m = m_0$ and $\phi \in L_p(Q, R^n)$, $p \in [1, \infty]$,

$$\begin{aligned}\|T^{m_0}\phi\|_{p,Q} &\le \|T^{m_0}\phi - T^{m_0}0\|_{p,Q} + \|T^{m_0}0\|_{p,Q} \\ &\le \lambda_{m_0}\|\phi\|_{p,Q} + \lambda(\|\xi\|_{p,Q} + \sigma).\end{aligned} \tag{2.3.11}$$

Define $r_0 \equiv (\lambda/(1-\lambda_{m_0}))(\|\xi\|_{p,Q} + \sigma)$; then

i. $T^{m_0}(B_{r_0}) \subseteq B_{r_0}$ and, for $\phi_1, \phi_2 \in B_{r_0}$,
ii. $\|T^{m_0}\phi_1 - T^{m_0}\phi_2\|_{p,Q} \le \lambda_{m_0}\|\phi_1 - \phi_2\|_{p,Q}$, where $0 \le \lambda_{m_0} < 1$.

Therefore, by the fixed point theorem (Lemma 2.3.1), there exists a unique $\phi \in B_{r_0}$ such that $T^{m_0}\phi = \phi$ and hence $T\phi = \phi$. Thus ϕ is the unique solution of the integral equation (2.3.1). This completes the proof of the theorem. □

REMARK 2.3.1. For much more general linear and nonlinear integral equations on Banach spaces the reader is referred to Section 2.4 and [Ah.5].

For basic materials on linear and nonlinear integral equations see [Kr.1, Tr.1].

2.4. Semigroup Approach to a Class of Linear Evolution Equations

In this section we wish to study the question of existence and uniqueness of solutions for a class of linear evolution equations in a Hilbert space. In Section 1.3 we have investigated in some details the properties of semigroups and their generators. We have seen that the Hille–Yosida theorem gives us the necessary and sufficient conditions for the existence of a semigroup. According to this fundamental result, for an operator A acting within a Banach space X to be the generator of a strongly continuous semigroup $\{T(t), t \ge 0\}$ of bounded linear operators in X, it is necessary and sufficient that

i. A be a closed linear operator in X with domain $D(A)$ dense in X, and
ii. there exists a pair of numbers $\{M, w_0\}$, $M > 0$ such that $\|(\lambda I - A)^{-n}\| \le M/(\lambda - w_0)^n$ for all $\lambda > w_0$, $n = 1, 2, 3, \ldots$.

For convenience we shall denote by $\zeta(M, w_0)$ the class of operators $\{A\}$ satisfying properties (i) and (ii).

It is known (Section 1.3.7) that if $A \in \zeta(M, w_0)$, then the corresponding semigroup $T_A(t)$ is strongly continuous and

$$\|T_A(t)\| \le Me^{w_0 t}, \qquad t \ge 0, \quad \text{and} \quad T_A(0) = I. \tag{2.4.1}$$

A semigroup $T_A(t)$, $t \ge 0$ corresponding to the operator A is said to be *of type* w_0 if $w_0 = \inf\{w: A \in \zeta(M,w)\}$. Thus for $A \in \zeta(1,0)$, the corresponding semigroup is a family of contraction operators in X; for $A \in \zeta(M,0)$, the corresponding semigroup is a family of bounded (linear) operators in X; and for $w_0 < 0$, the corresponding semigroup is a family of dissipative operators in X. There are various classes of such semigroups. For details the reader is referred to the book by Hille and Phillips [HP.1].

Lemma 2.4.1. *Let $A \in \zeta(M, w_0)$ and $x_0 \in D(A) \subset X$. Then the abstract Cauchy (initial value) problem*

$$\mathrm{S} \qquad \begin{cases} \dfrac{d}{dt}x(t) = Ax(t), & t>0, \\ x(0) = x_0 \end{cases}$$

has a unique solution $x(t)$, $t>0$, satisfying

i. $x(t) \in D(A)$, $t \geq 0$, *and is continuous,*
ii. $x(t)$ *is strongly differentiable for* $t>0$, *and*
iii. $\operatorname{s\,lim}_{t \downarrow 0} x(t) = x_0$.

PROOF. We have seen in Section 1.3 that under the given assumptions the operator A admits a unique strongly continuous semigroup $T(t)$, $t \geq 0$. Further, for any $h \in D(A)$, $T(t)h$ is strongly differentiable for all $t>0$ and

$$\frac{d}{dt}T(t)h = T(t)Ah = AT(t)h, \qquad t>0. \tag{2.4.2}$$

Thus for $h = x_0$ we have

$$\frac{d}{dt}(T(t)x_0) = A(T(t)x_0),$$

and consequently, the function $x(t) \equiv T(t)x_0$ satisfies the differential equation $\dot{x}(t) = Ax(t)$, $t>0$. Further, $T(t)x_0 \in D(A)$ and $x(t)$ is strongly continuous since the semigroup is, and consequently, $\operatorname{s\,lim} x(t) = x_0$. For the proof of uniqueness we note that if there are two solutions, then their difference must satisfy the equation $d\xi/dt = A\xi$, $\xi(0) = 0$. Let $0 < s < t$ and define $z(s) = T(t-s)\xi(s)$. Clearly z is absolutely continuous on $[0,t]$, $z(0) = 0$, $z(t) = \xi(t)$, and $dz/ds = -T(t-s)A\xi(s) + T(t-s)A\xi(s) = 0$ for $0 < s < t$. Thus, z is continuous and constant, and consequently, $\xi(t) = 0$. Since t is arbitrary, this completes the proof. □

With the help of the above result we can prove the existence of solutions of the inhomogeneous differential equation

$$\begin{aligned} \frac{dx}{dt} &= Ax + u, \\ x(0) &= x_0, \end{aligned} \tag{2.4.3}$$

where $x_0 \in D(A)$ and u is a given function with values in X and is assumed to be strongly continuous for $t \geq 0$. If x is a solution of (2.4.3), then

$$x(t) = T(t)x_0 + \int_0^t T(t-s)u(s)ds \tag{2.4.4}$$

with $x(0) = x_0$.

Consider the function

$$T(t-s)x(s) \quad \text{for } 0 < s < t. \tag{2.4.5}$$

Differentiating this function, we have

$$\begin{aligned} \frac{d}{ds}(T(t-s)x(s)) &= -T(t-s)Ax(s) + T(t-s)\dot{x}(s) \\ &= -T(t-s)Ax(s) + T(t-s)(Ax(s) + u(s)) \\ &= T(t-s)u(s). \end{aligned} \tag{2.4.6}$$

Integrating this on $(0, t)$, we obtain

$$x(t) = T(t)x_0 + \int_0^t T(t-s)u(s)ds, \qquad x(0) = x_0. \tag{2.4.7}$$

In particular, this implies that the solution of (2.4.3) is uniquely determined by x_0. We are more interested in the converse.

Theorem 2.4.1. *Let $A \in \zeta(M, w_0)$ and let $u(t)$ be a given (strongly) continuously differentiable function for $t \geq 0$. For any $x_0 \in D(A)$, the function $x(t)$ given by* (2.4.7) *is continuously differentiable for $t \geq 0$ and is a solution of* (2.4.3) *with the initial condition $x(0) = x_0$.*

PROOF. Define

$$z(t) = \int_0^t T(t-s)u(s)ds = \int_0^t T(s)u(t-s)ds.$$

Since u is strongly continuously differentiable, it is clear that z is also (strongly) continuously differentiable and

$$\frac{dz}{dt} = \int_0^t T(s)\dot{u}(t-s)ds + T(t)u(0) \tag{2.4.8}$$

with

$$\left|\frac{dz}{dt}\right|_X \leq Me^{w_0 t}|u(0)|_X + M\int_0^t e^{w_0 s}|\dot{u}(t-s)|_X ds < \infty. \tag{2.4.9}$$

On the other hand, we can write

$$\frac{z(t+h) - z(t)}{h} = \frac{1}{h}\left\{\int_0^t (T(h) - I)T(t-s)u(s)ds + \int_t^{t+h} T(t+h-s)u(s)ds\right\}. \tag{2.4.10}$$

Thus,

$$\frac{z(t+h)-z(t)}{h}=\left(\frac{T(h)-I}{h}\right)z(t)+\frac{1}{h}\int_t^{t+h}T(t+h-s)u(s)ds. \tag{2.4.11}$$

By virtue of (2.4.8) and (2.4.9), the limit of the expression on the left-hand side of (2.4.11) exists and belongs to X for each $t\geq 0$. Consequently, the (strong) limits of the expressions on the right-hand side of (2.4.11) also exist. Since A is the infinitesimal generator of the strongly continuous semigroup $T(t)$, it follows from the first term in the right of (2.4.11) that

$$\operatorname{s\,lim}\left(\frac{T(h)-I}{h}\right)z(t)=Az(t). \tag{2.4.12}$$

Consequently, by definition of the set $D(A)$, $z(t)\in D(A)$ for all $t\geq 0$. Since $T(t)$ is strongly continuous and u is strongly continuously differentiable, it is easy to verify that

$$\operatorname{s\,lim}\frac{1}{h}\int_t^{t+h}T(t+h-s)u(s)ds=u(t). \tag{2.4.13}$$

Therefore, it follows from (2.4.11)–(2.4.13) that $dz/dt=Az+u$. That is, $z=z(t)$, $t>0$, satisfies the differential equation for $z(0)=0$. Since u is strongly continuously differentiable, it is clear from (2.4.8) that z is also strongly continuously differentiable. Consequently, x, defined as

$$x(t)=T(t)x_0+z(t), \qquad t\geq 0, \tag{2.4.14}$$

is strongly continuously differentiable since obviously

$$\dot{x}(t)=T(t)(Ax_0)+\dot{z}(t) \tag{2.4.15}$$

is strongly continuous for $t>0$. It follows from this relation and the fact that: $T(t)Av=AT(t)v$ for $v\in D(A)$ (Theorem 1.3.2) that

$$\dot{x}(t)=A(T(t)x_0+z(t))+u(t)=Ax(t)+u(t) \quad \text{for } t>0.$$

Thus the function defined by (2.4.7) is a solution of the evolution equation (2.4.3) with initial condition $x(0)=x_0$. □

REMARK 2.4.1. The solution is unique. This follows from the fact that if there are two solutions then their difference must satisfy the homogeneous equation $dx/dt=Ax$, $x(0)=0$. In that case, $x\equiv 0$, as shown in Lemma 2.4.1.

In Theorem 2.4.1, it was assumed that the initial state $x_0\in D(A)$ and that u is a (strongly) continuously differentiable function with values in X. Both conditions are rather too strong for application. For application to control theory, we wish to admit x_0 from X and $u\in L_p(I, X)$, $1\leq p\leq\infty$, with I any bounded interval. For this purpose, we introduce the following definition.

Definition 2.4.1. A function $x: I \rightarrow X$ is called a *mild solution of the problem* (2.4.3) if it admits the integral representation

$$x(t)=T(t)x_0+\int_0^t T(t-\theta)u(\theta)d\theta. \tag{2.4.16}$$

We note that (2.4.16) need not give a solution of (2.4.3) for every $x_0 \in X$ and $u \in L_p(I, X)$. The existence of dx/dt and $Ax(t)$ can be proved only under certain stronger assumptions on x_0 and u. On the other hand, if we assume that X is a reflexive Banach space—in particular, a Hilbert space—we can prove the existence of a solution of problem (2.4.3) in certain weak sense.

Definition 2.4.2. A function x with values $x(t) \in X$, $t \in I$, is said to be a *weak solution of the evolution equation* (2.4.3) if (i) $(x(t), y)$ is absolutely continuous for $y \in D(A^*)$ and (ii) $(d/dt)(x(t), y)=(x(t), A^*y)+(u(t), y)$ *a.e.* on I, for all $y \in D(A^*)$.

Theorem 2.4.2. *Let X be a reflexive Banach space, $A \in \zeta(M, w_0)$, $x_0 \in X$, and $u \in L_p(I, X)$, $1 \leq p \leq \infty$. Then the evolution equation* (2.4.3) *has a unique weak solution.*

PROOF. Since $A \in \zeta(M, w_0)$, it admits a unique strongly continuous semigroup $T(t)$, $t \geq 0$, of operators in X with the properties $T(0)=I$ and $\|T(t)\| \leq Me^{w_0 t}$, $t \geq 0$. Define

$$x(t)=T(t)x_0+\int_0^t T(t-\theta)u(\theta)d\theta, \qquad t \geq 0. \tag{2.4.17}$$

Since

$$|x(t)|_X \leq Me^{w_0 t}\left\{|x_0|_X+\left(\int_0^t e^{-qw_0 s}ds\right)^{1/q}\left(\int_0^t |u(s)|_X^p ds\right)^{1/p}\right\}$$

and for $t \in I$ (a finite interval) and $(1/p)+(1/q)=1$, $1<p, q<\infty$,

$$|x(t)|_X \leq \begin{cases} Me^{w_0 t}\left\{|x_0|_X+\int_0^t e^{-w_0\theta}|u(\theta)|_X d\theta\right\} & \text{for } p=1 \\ Me^{w_0 t}\left\{|x_0|_X+\left(\int_0^t e^{-w_0\theta}d\theta\right)\operatorname*{ess\,sup}_{0\leq\theta\leq t}|u(\theta)|_X\right\} & \text{for } p=\infty, \end{cases}$$

it is clear that x, given by (2.4.17), is well defined and that $|x(t)|_X < \infty$ for t finite. Further, since $T(t)$ is strongly continuous and

$$\int_I |f(\theta+h)-f(\theta)|_X^p d\theta \rightarrow 0 \quad \text{as } h \rightarrow 0 \quad \text{(Theorem 1.1.18)}$$

for any $f \in L_p(I, X)$, it follows from the inequality

$$
\begin{aligned}
|x(t+h)-x(t)|_X \le & |(T(t+h)-T(t))x_0|_X \\
& + M\left(\int_0^t e^{qw_0\theta}\,d\theta\right)^{1/q}\left(\int_0^t |u(\theta+h)-u(\theta)|_X^p\,d\theta\right)^{1/p} \\
& + M\left(\int_t^{t+h} e^{qw_0\theta}\,d\theta\right)^{1/q}\left(\int_0^h |u(\theta)|^p\,d\theta\right)^{1/p}
\end{aligned}
$$

that $x \in C^0(I, X)$. In fact, x, as given by (2.4.17), is absolutely continuous in the sense that, for any $y \in D(A^*) \subset X^*$, the scalar function $t \to (x(t), y)$ is absolutely continuous and differentiable almost everywhere on I. Indeed, for $y \in D(A^*) \subset X^*$,

$$
\begin{aligned}
\frac{d}{dt}(x(t), y) &= \frac{d}{dt}\left\{(T(t)x_0, y) + \left(\int_0^t T(t-\theta)u(\theta)\,d\theta, y\right)\right\} \\
&= \frac{d}{dt}\left\{(x_0, T^*(t)y) + \int_0^t (u(\theta), T^*(t-\theta)y)\,d\theta\right\}, \qquad (2.4.18)
\end{aligned}
$$

where $(z, w) \equiv (w, z) = w(z)$ denotes the duality pairing between $w \in X^*$ and $z \in X$. Since X is a reflexive Banach space, the dual A^* of the infinitesimal generator A of the c_0-semigroup $T(t)$, $t \ge 0$, is itself the infinitesimal generator of a c_0-semigroup that equals the dual $T^*(t)$ of the semigroup $T(t)$ [Theorem 1.3.11(i), (iii)]. Further,

$$
\frac{d}{dt}(T^*(t)y) = A^*T^*(t)y = T^*(t)A^*y \qquad (2.4.19)
$$

for all $y \in D(A^*)$ and $D(A^*)$ is dense in X^*.

Using (2.4.19) in Equation (2.4.18) and recalling that $T^*(t)$ is strongly continuous and $T^*(0) = I^*$, we obtain

$$
\begin{aligned}
\frac{d}{dt}(x(t), y) &= (x_0, T^*(t)A^*y) + (u(t), y) \\
&\quad + \int_0^t (u(\theta), T^*(t-\theta)A^*y)\,d\theta \quad \text{a.e.} \\
&= \left(T(t)x_0 + \int_0^t T(t-\theta)u(\theta)\,d\theta, A^*y\right) + (u(t), y) \quad \text{a.e.} \\
&= (x(t), A^*y) + (u(t), y) \quad \text{a.e.} \qquad (2.4.20)
\end{aligned}
$$

This shows that $(x(t), y)$ is absolutely continuous and differentiable a.e. on I and that, by Definition 2.4.2, x, as given by (2.4.17), is a weak solution of problem (2.4.3). Clearly, since x is an element of $C^0(I, X)$, $x(0)$ is defined, and due to strong continuity of $T(t)$, $x(0) = x_0$. Using (2.4.17) and the fact that $\|T(t)\| \le Me^{w_0 t}$, it is easy to verify that the mapping $\{x_0, u\} \to x$ from $X \times L_p(I, X)$ into $C^0(I, X)$ is continuous. The uniqueness of the solution is a consequence of this. □

In dealing with linear evolution equations, occasionally we encounter certain specific Volterra integral equations in Banach spaces. Specifically, the integral equation given by

$$x(t)=g(t)+\int_0^t T(t-\theta)\Gamma(\theta)x(\theta)d\theta, \qquad t\in I, \tag{2.4.21}$$

occurs frequently while considering the perturbation of systems governed by differential equations of the form (2.4.3). We are interested primarily in the perturbation induced by linear state feedback replacing u by Γx. Let $\mathfrak{L}(X)$ denote the space of bounded linear operators in an arbitrary Banach space X and $C_s(I, \mathfrak{L}(X))$ denote the class of strongly continuous functions on I with values in $\mathfrak{L}(X)$ equipped with the strong operator topology. Let $L_\infty(I, \mathfrak{L}(X))$ denote the class of strongly measurable essentially bounded $\mathfrak{L}(X)$-valued functions on I.

Theorem 2.4.3. *Let A be the generator of a strongly continuous semigroup $\{T(t),\ t\geq 0\}$ in X [that is, $A\in\zeta(M, w_0)$], $\Gamma(t)$, $t\geq 0$, a given family of bounded linear operators in X, and $g(t)$, $t\geq 0$, a given function with values in X. Then the integral equation* (2.4.21) *has*

i. *a unique solution $x\in C(I, X)$ whenever $g\in C(I, X)$ and $\Gamma\in C_s(I, \mathfrak{L}(X))$ and*

ii. *a unique solution $x\in L_p(I, X)$ whenever $g\in L_p(I, X)$ and $\Gamma\in L_\infty(I, \mathfrak{L}(X))$, $1\leq p<\infty$.*

PROOF. The proof is entirely similar to that for the corresponding finite-dimensional case. □

With the help of the above result we can prove the existence and uniqueness of a solution for the perturbed problem:

$$\begin{aligned} &\frac{d}{dt}x=Ax+\Gamma(t)x,\\ &x(s)=h\in X, \qquad t\in(s,T), \quad s\geq 0. \end{aligned} \tag{2.4.22}$$

Theorem 2.4.4. *Under the assumptions of Theorem* 2.4.3, *the perturbed system* (2.4.22) *has a unique strongly continuous solution x given by $x(t)=\phi(t,s)h$, where $\phi(t,s)$ is the transition (evolution) operator satisfying the properties*

$$\begin{cases} \text{(a)} \quad \phi(t,t)=I, \qquad t\in[0,\tau], \\ \text{(b)} \quad \phi(t,s)\phi(s,\theta)=\phi(t,\theta), \qquad 0\leq\theta\leq s\leq t\leq\tau, \\ \text{(c)} \quad (\partial/\partial s)\phi(t,s)h=-\phi(t,s)(A+\Gamma(s))h, \quad h\in D(A). \end{cases} \tag{2.4.23}$$

PROOF. By definition, the "mild" solution of problem (2.4.22) is given by the solution of the integral equation in X

$$x(t)=T(t-s)h+\int_s^t T(t-\theta)\Gamma(\theta)x(\theta)d\theta, \qquad t\in(s,\tau). \tag{2.4.24}$$

By Theorem 2.4.3, this equation has a unique solution $x \in C(s,\tau;X)$. We show that there exists a linear (transition) operator $\phi(t,s)$, $0 \le s \le t \le \tau$, with values in $\mathcal{L}(X)$ such that $x(t) = \phi(t,s)h$ and ϕ satisfies the properties (a)–(c). If such an operator-valued function exists, it must satisfy the abstract integral equation

$$\phi(t,s)h = T(t-s)h + \int_s^t T(t-\theta)\Gamma(\theta)\phi(\theta,s)h\,d\theta,$$
$$0 \le s \le t \le \tau, \tag{2.4.25}$$

for $h \in X$. We show that this equation has a unique solution ϕ that is strongly continuous in the first argument and satisfies (a)–(c). Define, for $h \in X$ and $0 \le s \le t \le \tau$,

$$\begin{cases} \text{(i)} & \phi_0(t,s)h = T(t-s)h, \\ \text{(ii)} & K^{(1)}(t,s)h = T(t-s)\Gamma(s)h, \\ \text{(iii)} & K^{(l+1)}(t,s)h = \int_s^t K^{(l)}(t,\theta)K^{(1)}(\theta,s)h\,d\theta \quad \text{for } l = 1,2,\ldots, \\ \text{(iv)} & \phi_n(t,s)h = \phi_0(t,s)h + \int_s^t \sum_{l=1}^{n} K^{(l)}(t,\theta)\phi_0(\theta,s)h\,d\theta \quad \text{for } n = 1,2,\ldots. \end{cases} \tag{2.4.26}$$

We show that the sequence (of operator-valued functions) $\{\phi_n\}$ has a strong limit uniformly on $\Delta \equiv \{(s,t): 0 < s \le t \le \tau\}$ and that the limit satisfies the integral equation (2.4.25). Since $\Gamma \in C_s(I, \mathcal{L}(X))$ and $A \in \zeta(M, w_0)$, there exists a constant M_0 such that

$$\|T(t-s)\Gamma(s)\| \le M_0 e^{w_0(t-s)} \quad \text{for } 0 \le s \le t \le \tau. \tag{2.4.27}$$

Using this inequality and (iii) of (2.4.26), we obtain

$$|K^{(l+1)}(t,s)h|_X \le M_0 e^{w_0(t-s)} \cdot \left((M_0(t-s))^l / l!\right)|h|_X, \qquad l = 0,1,2,\ldots. \tag{2.4.28}$$

Using the estimate (2.4.28), it is easy to verify that

$$\sup_{(s,t)\in\Delta} \|\phi_{n+1}(t,s) - \phi_n(t,s)\|_{\mathcal{L}(X)} \le M_0 e^{|w_0|\tau} \cdot \frac{(M_0\tau)^n}{n!}. \tag{2.4.29}$$

Thus, $\{\phi_n(t,s): (s,t) \in \Delta\}$ is a Cauchy sequence of bounded linear operators in X uniformly on Δ. Since $\mathcal{L}(X)$ is a Banach space, $\phi_n(t,s)$ has a limit $\phi(t,s)$ for $(s,t) \in \Delta$; and for arbitrary $h \in X$

$$\phi_n(t,s)h \to \phi(t,s)h \quad \text{strongly in } X \text{ for } (s,t) \in \Delta. \tag{2.4.30}$$

Similarly, defining

$$R_n(t,s) = \sum_{l=1}^{n} K^{(l)}(t,s) \quad \text{for } (s,t) \in \Delta \tag{2.4.31}$$

and using the estimate (2.4.28), one can verify that $\{R_n(t,s):(s,t)\in\Delta\}$ is also a Cauchy sequence of bounded linear operators in X and that

$$\|R_n(t,s)\|\le M_0\exp(w_0+M_0)(t-s),\qquad (s,t)\in\Delta,\qquad (2.4.32)$$

independently of n. Therefore, it follows from the Lebesgue dominated convergence theorem that

$$\int_s^t\sum_{l=1}^{n}K^{(l)}(t,\theta)\phi_0(\theta,s)h\,d\theta=\int_s^t R_n(t,\theta)\phi_0(\theta,s)h\,d\theta$$
$$\to\int_s^t R(t,\theta)\phi_0(\theta,s)h\,d\theta\quad\text{(strongly)},\qquad (2.4.33)$$

where $R_n(t,\theta)z\to R(t,\theta)z$ for $z\in X$, $(\theta,t)\in\Delta$. Letting $n\to\infty$ in (iv) of (2.4.26) and using the limits (2.4.30) and (2.4.33), we obtain

$$\phi(t,s)h=\phi_0(t,s)h+\int_s^t R(t,\theta)\phi_0(\theta,s)h\,d\theta.\qquad (2.4.34)$$

By substituting (2.4.34) in the expression on the right-hand side of (2.4.25) and recalling that $\phi_0(t,s)=T(t-s)$, we obtain

$$\begin{aligned}\text{R.H.S.}(2.4.25)&=\phi_0(t,s)h+\int_s^t K^{(1)}(t,\theta)\phi_0(\theta,s)h\,d\theta\\&\quad+\int_s^t\sum_{l=2}^{\infty}K^{(l)}(t,\theta)\phi_0(\theta,s)h\,d\theta\\&=\phi_0(t,s)h+\int_s^t R(t,\theta)\phi_0(\theta,s)h\,d\theta=\phi(t,s)h\\&=\text{L.H.S.}(2.4.25).\end{aligned}$$

Thus ϕ, as given by (2.4.34), satisfies the integral equation (2.4.25). In fact, ϕ is the unique solution. Indeed, if ψ is another solution, then, for any $h\in X$, one can show by successive iteration that

$$|(\phi(t,s)-\psi(t,s))h|_X\le\frac{(M_0t)^n}{n!}\left(M_0e^{|w_0|t}\right)\int_s^t|(\phi(t,\theta)-\psi(t,\theta))h|_X\,d\theta$$

for all integers n. This implies that $\phi=\psi$ for $(s,t)\in\Delta$. Subtracting (2.4.25) from (2.4.24) and using the above argument, we obtain $x(t)=\phi(t,s)h$. From (2.4.34), it follows that there is a constant k, so that

$$\sup_{(s,t)\in\Delta}\|\phi(t,s)\|_{\mathcal{L}(X)}\le k.\qquad (2.4.35)$$

For the strong continuity of x, we show that $t\to\phi(t,s)$ is strongly continuous on $[s,\tau]$. For arbitrary $h\in X$ it follows from (2.4.25) that

$$\begin{aligned}&(\phi(t+\Delta t,s)-\phi(t,s))h\\&\quad=(T(\Delta t)-I)\phi(t,s)h+\int_t^{t+\Delta t}T(t+\Delta t-\theta)\Gamma(\theta)\phi(\theta,s)h\,d\theta.\end{aligned}\qquad (2.4.36)$$

Since Γ and ϕ are bounded and T is a c_0-semigroup, it follows from (2.4.36) that $t \to \phi(t,s)$ is strongly continuous for $t \in [s,\tau]$. Thus x is strongly continuous. Since the integrand in (2.4.25) is bounded and T is strongly continuous, (a) follows on letting $s \uparrow t$. For (b), we note that, due to uniqueness of the solution of (2.4.24) or (2.4.25), we have, for $s \le \theta \le t$, $x(\theta) = \phi(\theta,s)h$, $x(t) = \phi(t,\theta)x(\theta)$, and also $x(t) = \phi(t,s)h$. Thus, $\phi(t,s)h = \phi(t,\theta)\phi(\theta,s)h$ for all $h \in X$, proving (b). For the proof of (c) we use (2.4.25) to compute $(1/\Delta s)\{(\phi(t,s+\Delta s) - \phi(t,s))h\}$ for $t \ge s + \Delta s$. Thus,

$$\begin{aligned}
&\frac{1}{\Delta s}\{\phi(t,s+\Delta s)h - \phi(t,s)h\} \\
&\qquad = -T(t-s-\Delta s)\left(\frac{T(\Delta s)-I}{\Delta s}\right)h \\
&\qquad\quad + \int_s^t T(t-\theta)\Gamma(\theta)\left(\frac{\phi(\theta,s+\Delta s)-\phi(\theta,s)}{\Delta s}\right)h\,d\theta \\
&\qquad\quad - \frac{1}{\Delta s}\int_s^{s+\Delta s} T(t-\theta)\Gamma(\theta)\phi(\theta,s+\Delta s)h\,d\theta. \qquad (2.4.37)
\end{aligned}$$

Letting $\Delta s \to 0$ in (2.4.37), we obtain the integral equation

$$\begin{aligned}
\left[\frac{\partial}{\partial s}\phi(t,s)h\right] &= -T(t-s)(A+\Gamma(s))h \\
&\quad + \int_s^t T(t-\theta)\Gamma(\theta)\left[\frac{\partial}{\partial s}\phi(\theta,s)h\right]d\theta. \qquad (2.4.38)
\end{aligned}$$

Recalling that the integral equation

$$y(t) = T(t-s)z + \int_s^t T(t-\theta)\Gamma(\theta)y(\theta)\,d\theta, \qquad z \in X,$$

has the unique solution $y(t) = \phi(t,s)z$, we obtain, by comparison with (2.4.38),

$$\frac{\partial}{\partial s}\phi(t,s)h = -\phi(t,s)(A+\Gamma(s))h. \qquad (2.4.39)$$

That this is the solution of (2.4.38) is also verified by direct substitution into (2.4.38). This completes the proof of the theorem. □

REMARK 2.4.2. Without additional assumptions on h and Γ we cannot, in general, prove that $(\partial/\partial t)\phi(t,s)h = (A+\Gamma(t))\phi(t,s)h$ unless X is a finite-dimensional space. However, since x is a mild solution of (2.4.22), we have

$$\frac{d}{dt}(x(t),z) = (x(t),(A+\Gamma(t))^*z) \quad \text{for } z \in D(A^*), \quad t \in [0,\tau). \qquad (2.4.40)$$

But $x(t) = \phi(t,s)h$ for $h \in X$ and $t \ge s \ge 0$, and consequently,

$$\frac{d}{dt}(\phi(t,s)h,z) = (\phi(t,s)h,(A+\Gamma(t))^*z) \qquad (2.4.41)$$

for $t \geq s$ and $z \in D(A^*)$. Thus, $t \to (\phi(t,s)h, z)$ is an absolutely continuous scalar-valued function on $[s, \tau]$ for $h \in X$ and $z \in D(A^*)$. In other words, $\phi(\cdot, s): [s, \tau] \to \mathcal{L}(X)$ is absolutely continuous in the weak operator topology in the restricted sense, that is, the weak operator topology induced on $\mathcal{L}(X)$ by taking finite subsets of $D(A^*) \subset X^*$ instead of X^*. □

REMARK 2.4.3. By integrating (2.4.41), we have

$$(\phi(t,s)h - h, z) = \int_s^t (\phi(\theta, s)h, (A + \Gamma(\theta))^* z)\, d\theta, \qquad h \in X, z \in D(A^*), \tag{2.4.42}$$

and integrating (c) of (2.4.23), we have

$$(\phi(t,s)h - h, z) = \int_s^t (\phi(t, \theta)(A + \Gamma(\theta))h, z)\, d\theta, \qquad h \in D(A), z \in X^*. \tag{2.4.43}$$

In fact, these relations can be generalized to characterize mild evolution operators and their corresponding generators.

2.5. A Class of Nonlinear Evolution Equations

In this section, we wish to study the question of existence of solutions of a very general class of nonlinear equations on a Banach space. These results are used in the sequel to prove the existence and uniqueness of solutions of nonlinear evolution equations on a reflexive Banach space. In Chapter 5, we study optimal control problems involving these nonlinear evolution equations.

Let Y be a reflexive Banach space, with Y^* its dual and (x, y) the duality pairing between an element x of Y^* and an element y of Y. In general, we assume that Y is a complex Banach space, so that Y^* is the space of bounded conjugate linear functionals on Y while (x, y) is linear in x and conjugate linear in y. We shall also write (y, x), to denote $(x, y)^*$, where * denotes complex conjugate in this case.

By a *mapping T from Y to Y** we mean a function T (not necessarily linear) whose domain $D(T)$ is a dense linear subset of Y and whose range $R(T) \subset Y^*$. Such a mapping is said to be *demicontinuous* if it is continuous from the strong topology of Y to the weak topology of Y^*. That is, if $y_n \xrightarrow{s} y$ (strongly) in Y, then $Ty_n \xrightarrow{w} Ty$ (weakly) in Y^*. If L is a densely defined linear mapping from Y to Y^*, then $D(L)$ is a dense linear subset of Y. L is said to be *closed* if its graph $G(L) = \{[y, x]: y \in D(L), x = Ly\}$ is a closed subset of $Y \times Y^*$. If L is not closed, it is said to be *closable* if $L \subseteq L_0$ for some closed linear operator L_0 from Y to Y^*, that is, $D(L) \subset D(L_0)$, $Ly = L_0 y$ for $y \in D(L)$. If L is closable let L_s, the closure of L, be the smallest closed linear operator containing L. Then

$$D(L_s)=\left\{\begin{matrix} y: y\in Y; \text{ there exists } w\in Y^* \text{ and a sequence} \\ y_\kappa \in D(L) \text{ such that } y_\kappa \overset{s}{\to} y \text{ and } Ly_\kappa \overset{s}{\to} w \text{ in } Y^* \end{matrix}\right\} \quad \text{and} \quad L_s y=w. \tag{2.5.1}$$

For every densely defined linear operator L from Y to Y^*, its adjoint L^* is the linear operator with domain in Y given by

$$D(L^*)\equiv\left\{\begin{matrix} y: y\in Y; \text{ there exists a constant } c_y \text{ such that} \\ |(y, Lx)|\leq c_y \|x\|_Y \text{ for all } x\in D(L) \end{matrix}\right\}. \tag{2.5.2}$$

For each $y\in D(L^*)$, there exists an unique element $x\in Y^*$ such that $(y, Lv)=(x, v)$ for all $v\in D(L)$. We set

$$L^*y=x. \tag{2.5.3}$$

The operator L is *closable* if and only if $D(L^*)$ is dense in Y, that is, if and only if L^* itself is a densely defined linear operator from Y to Y^*. L^* is always a closed linear operator. If L is closable, so that L^* is densely defined, we may form $L^{**}=(L^*)^*$. Then $L_s=L^{**}$.

Definition 2.5.1. Let L be a densely defined linear operator from Y to Y^* such that $D(L)\subset D(L^*)$. Let L' be the restriction of L^* to $D(L)$, that is, $L'\equiv L^*|_{D(L)}$. Then

i. L_s is said to be the *strong extension of* L from Y to Y* and
ii. $L_w\equiv(L')^*$ is said to be the *weak extension of* L from Y to Y*.

Since $L'\subseteq L^*$ we have $L_s\subseteq L_w$.

Definition 2.5.2. The strong and weak extensions of L are said to be *equal* if $L_s=L_w$.

Theorem 2.5.1. *Let Y be a reflexive complex Banach space and T a densely defined mapping from Y to Y^* such that $T=L+F$, where L and F satisfy the following conditions*:

i. *L is a densely defined closable linear operator from Y to Y^* with $D(L)\subset D(L^*)$ and $L_s=L_w$.*
ii. *F: $Y\to Y^*$ is demicontinuous and maps bounded sets of Y into bounded sets of Y^*.*
iii. *For all $x, y\in D(T)$*

$$\text{Re}(Tx-Ty, x-y)\geq 0,$$

that is, T is monotone on $D(T)$.

iv. *There exists a continuous real valued function* η *on* $(0,\infty)$ *with* $\eta(\xi)\to\infty$ *as* $\xi\to\infty$ *such that*

$$\operatorname{Re}(Tx,x)\geq\eta(\|x\|)\|x\| \quad \text{for all } x\in D(T).$$

Then, if $T_s=L_s+F$, *the range of* T_s *is all of* Y^*.

PROOF. Since Y is a reflexive Banach space and L is a densely defined closable linear operator from Y to Y^*, L^* is closed and densely defined and $L_s=L^{**}$ [K.2, Theorem 5.29, p. 168]. By hypothesis, $L_s=L_w$, that is, $L_s=(L')^*$, and consequently, $L_s^*=(L')^{**}$. Since $L^*=L_s^*$, this implies that L_s^* or L^* is the closure of L'. Thus, L_s^* is the closure of its restriction to $D(L_s)\cap D(L_s^*)$. By hypotheses (iii) and (iv)

$$\operatorname{Re}(Tx-Ty,x-y)\geq 0 \quad \text{for } x,y\in D(T)=D(L),$$

$$\operatorname{Re}(Tx,x)\geq\eta(\|x\|)\|x\| \quad \text{for } x\in D(T)=D(L).$$

Suppose $x,y\in D(T_s)=D(L_s)$. Then, by the definition of L_s, there exist two sequences $\{x_n\}$ and $\{y_n\}$ from $D(L)=D(T)$ such that

$$x_n\xrightarrow{s}x, \qquad Lx_n\xrightarrow{s}L_sx,$$

$$y_n\xrightarrow{s}y, \qquad Ly_n\xrightarrow{s}L_sy.$$

Since F is assumed to be demicontinuous, we have

$$Fx_n\xrightarrow{w}Fx, \qquad Fy_n\xrightarrow{w}Fy.$$

Thus,

$$Tx_n\xrightarrow{w}T_sx, \qquad Ty_n\xrightarrow{w}T_sy.$$

Since $Tx_n-Ty_n\xrightarrow{w}T_sx-T_sy$ and $x_n-y_n\xrightarrow{s}x-y$, it follows that

$$\operatorname{Re}(Tx_n-Ty_n,x_n-y_n)\to\operatorname{Re}(T_sx-T_sy,x-y) \quad \text{as } n\to\infty.$$

On the other hand, for all n,

$$\operatorname{Re}(Tx_n-Ty_n,x_n-y_n)\geq 0.$$

Therefore,

$$\operatorname{Re}(T_sx-T_sy,x-y)\geq 0 \quad \text{for all } x,y\in D(T_s).$$

Similarly,

$$\operatorname{Re}(Tx_n,x_n)\to\operatorname{Re}(T_sx,x) \quad \text{as } n\to\infty,$$

and, due to continuity of the function η,

$$\eta(\|x_n\|)\|x_n\|\to\eta(\|x\|)\|x\| \quad \text{as } n\to\infty.$$

Hence

$$\operatorname{Re}(T_sx,x)\geq\eta(\|x\|)\|x\| \quad \text{for all } x\in D(T_s).$$

By hypothesis, F is demicontinuous and hence it is hemicontinuous in the sense that $F(x+\alpha_n y) \overset{w}{\to} F(x)$ whenever $\alpha_n \to 0$, where $x, y \in Y$ and $\alpha_n \in R$. Summarizing the above results, we have,

a. $T_s \equiv L_s + F$ is a monotone operator mapping $D(T_s)$ ($\subset Y$) into Y^*,
b. L_s^* is the closure of its restriction to $D(L_s) \cap D(L_s^*)$, and
c. there exists a real-valued continuous function η defined on $[0, \infty]$ such that

$$\operatorname{Re}(T_s x, x) \geq \eta(\|x\|)\|x\| \quad \text{for all } x \in D(T_s).$$

We show that, under the above conditions, $R(T_s) = Y^*$. This is equivalent to showing that $0 \in R(T_s)$. Indeed, for an arbitrary $w \in Y^*$, we can define $T_1\xi = T_s\xi - w$ and observe that $D(T_1) = D(T_s)$ and T_1 satisfies all the above properties, in particular, $\operatorname{Re}\langle T_1 x - T_1 y, x-y\rangle \geq 0$ and $\operatorname{Re}\langle T_1 x, x\rangle \geq \eta_1(\|x\|)\|x\|$, where $\eta_1(r) = \eta(r) - \|w\|$. Thus, if $0 \in R(T_1)$, then there exists an $x \in D(T_1)$ such that $0 = T_1 x \equiv T_s x - w$, and consequently, $w \in R(T_s)$. Therefore, it suffices to show that, under the given conditions, $0 \in R(T_s)$. Define $\Lambda \equiv \{G : G \subset Y$ and G is a finite dimensional subspace of $D(L)\}$. Ordered by inclusion, Λ is a directed set. Let j denote the injection map of G into Y and its adjoint j^* the injection of Y^* onto G^*. Define $T_G \equiv j^* T_s j$ for $G \in \Lambda$. Clearly, T_G maps G into G^* and both G and G^* are finite-dimensional subspaces of Y and Y^*, respectively. Further, for $x, y \in G$,

$$\begin{aligned} \operatorname{Re}\langle T_G x - T_G y, x-y\rangle &= \operatorname{Re}\langle T_s x - T_s y, x-y\rangle \geq 0, \\ \operatorname{Re}\langle T_G x, x\rangle &= \operatorname{Re}\langle T_s x, x\rangle \geq \eta(\|x\|)\|x\|, \end{aligned} \tag{2.5.4}$$

and, since T_s is hemicontinuous, T_G is continuous and $D(T_G) = G$. Thus, it follows from Browder [Br.1, Lemma 1.2, p. 488] that $R(T_G) = G^*$, or equivalently, there exists an $x_G \in G$ such that $T_G x_G = 0$. Clearly $0 = \langle T_G x_G, x_G\rangle$ and $0 = \operatorname{Re}\langle T_G x_G, x_G\rangle \geq \eta(\|x_G\|)\|x_G\|$. Since η is continuous on R_0 and $\eta(r) \to +\infty$ as $r \to +\infty$, either $\|x_G\| = 0$ or $\eta(\|x_G\|) \leq 0$. In either case, there exists a finite number $r_0 > 0$ such that $\|x_G\| \leq r_0$ for all $G \in \Lambda$. Since $x_G \in G$, for $y \in G \cap D(L_s^*)$, we have

$$\begin{aligned} 0 &= \langle T_G x_G, y\rangle = \langle j^* T_s j x_G, y\rangle = \langle T_s j x_G, jy\rangle = \langle T_s x_G, y\rangle \\ &= \langle (L_s + F) x_G, y\rangle. \end{aligned}$$

Thus,

$$\langle L_s x_G, y\rangle = -\langle F x_G, y\rangle \quad \text{for all } y \in G. \tag{2.5.5}$$

Since F maps bounded subsets of Y into bounded subsets of Y^*, there exists a number $\beta > 0$ such that $\|Fx\| \leq \beta$ for all $x \in B_{r_0} = \{x \in Y : \|x\| \leq r_0\}$. Consequently, it follows from (2.5.5) that

$$|\langle L_s x_G, y\rangle| = |\langle x_G, L_s^* y\rangle| \leq \beta\|y\| \quad \text{for all } y \in G \cap D(L_s^*). \tag{2.5.6}$$

Finally, since $T_G x_G = 0$, for $y \in G$,

$$0 \le \operatorname{Re}\langle T_G x_G - T_G y, x_G - y\rangle_{G^*-G}$$
$$= \operatorname{Re}\langle -T_G y, x_G - y\rangle_{G^*-G} = -\operatorname{Re}\langle T_s y, x_G - y\rangle_{Y^*-Y},$$

and consequently,

$$\operatorname{Re}\langle T_s y, x_G - y\rangle \le 0 \quad \text{for all } y \in G \text{ and } G \in \Lambda. \tag{2.5.7}$$

The mapping $G \to x_G$, denoted by h $[x_G = h(G)]$, is a mapping from the directed set Λ into the ball B_{r_0} of Y. Since Y is reflexive, B_{r_0} is weakly compact and hence there exists an $x_0 \in B_{r_0}$ such that for each $G_0 \in \Lambda$, each $\varepsilon > 0$, and each finite set $\{y_1^*, y_2^*, \ldots, y_m^*\} \subset Y^*$, there exists a $G \in \Lambda$ with $G \supset G_0$ such that $|y_i^*(x_0 - x_G)| < \varepsilon$ for $1 \le i \le m$, that is, $x_G \in N(x_0; y_2^*, \ldots, y_m^*; \varepsilon)$ (see Section 1.1.6). To complete the proof, we show that $x_0 \in D(T_s) = D(L_s)$ and $T_s x_0 = 0$. Suppose $y \in D(L_s) \cap D(L_s^*)$ with $\|y\| = \varepsilon$, let G_0 denote the 1-dimensional subspace of $D(L_s)$ generated by the single element $\{y\}$, and consider the element $\{L_s^* y\} \in Y^*$. Then clearly there exists a $G \in \Lambda$ with $G \supset G_0$ such that

$$|\langle x_0 - x_G, L_s^* y\rangle| \le \|y\|. \tag{2.5.8}$$

Since $y \in G$, it follows from (2.5.6) that

$$|\langle x_G, L_s^* y\rangle| \le \beta \|y\|. \tag{2.5.9}$$

Combining (2.5.8) and (2.5.9), we have

$$|\langle x_0, L_s^* y\rangle| \le (1+\beta)\|y\| \quad \text{for all } y \in D(L_s) \cap D(L_s^*). \tag{2.5.10}$$

Let L' denote the restriction of L_s^* to $D(L_s) \cap D(L_s^*)$. Since L_s^* is the closure of L', the graph $\gamma(L_s^*)$ of L_s^* is equal to the closure of the graph $\gamma(L')$ of L'. Let y be any element of $D(L_s^*)$. Then the function $\langle x_0, L_s^* y\rangle - (1+\beta)\|y\|$ is continuous on the graph $\gamma(L_s)$ of L_s and $\langle x_0, L_s^* y\rangle - (1+\beta)\|y\| \le 0$ on $\gamma(L')$, a dense subset of the graph of L_s^*, and hence ≤ 0 on $\gamma(L_s)$. This implies that $\langle x_0, L_s^* y\rangle \le (1+\beta)\|y\|$ for all $y \in D(L_s^*)$, and consequently, $x_0 \in D(L_s) = D(T_s)$. It remains to show that $T_s x_0 = 0$. By (2.5.7) we know that, for each $G \in \Lambda$ and all $y \in G$, $\operatorname{Re}\langle T_s y, x_G - y\rangle \le 0$. Let y be a given element of $D(L_s)$, and let $G_0 \equiv \{y\}$. Then we can choose $G \in \Lambda$ with $G \supset G_0$ such that for given $\varepsilon > 0$

$$|\langle T_s y, x_G - x_0\rangle| = |\langle T_s y, x_G - y\rangle - \langle T_s y, x_0 - y\rangle| < \varepsilon. \tag{2.5.11}$$

From (2.5.7) and (2.5.11), we obtain, for all $y \in D(L_s) = D(T_s)$,

$$\operatorname{Re}\langle T_s y, y - x_0\rangle \ge -\varepsilon. \tag{2.5.12}$$

Since ε (>0) is arbitrary, this implies that

$$\operatorname{Re}\langle T_s y, y - x_0\rangle \ge 0 \quad \text{for all } y \in D(T_s). \tag{2.5.13}$$

We claim that $T_s x_0 = 0$. Suppose to the contrary; then, since $D(L_s)$ is dense in Y, there exists a $\xi \in D(L_s) = D(T_s)$ such that $\operatorname{Re}\{\langle T_s x_0, \xi\rangle\} > 0$. Define $y \equiv x_0 - \theta\xi$, $\theta > 0$. Clearly $y \in D(T_s)$, and hence, it follows from

(2.5.13) that

$$\mathrm{Re}\langle T_s(x_0-\theta\xi), \theta\xi\rangle \le 0 \quad \text{for all } \theta>0. \tag{2.5.14}$$

Dividing (2.5.14) by θ and letting $\theta\to 0^+$ and recalling that T_s is hemicontinuous, we obtain $\mathrm{Re}\langle T_s(x_0), \xi\rangle \le 0$, which is a contradiction. Thus $T_s x_0 = 0$. This completes the proof of the theorem. □

Under an additional assumption, one can prove that the mapping T_s has a continuous inverse from Y^* into Y. Since we do not make use of this result, we state the theorem without proof.

Theorem 2.5.2. *Suppose that the hypotheses of Theorem* 2.5.1 *hold and that in addition there exist two continuous real-valued functions* η_1 *and* η_2 *on* $(0,\infty)$ *with* $\eta_1(\xi)>0$ *for* $\xi\in(0,\infty)$, η_2 *strictly increasing, and* $\eta_2(0)=0$, *such that*

(*v*) $\mathrm{Re}(Tx-Ty, x-y)\ge \eta_1(\|x\|\wedge\|y\|)\cdot\eta_2(\|x-y\|), \qquad x,y\in D(T),$

or, for $x\ne y$,

$$\mathrm{Re}(T_s x - T_s y, x-y)>0.$$

Then T_s *is one-to-one and, under the assumption* (*v*), *has a continuous inverse from* Y^* *into* Y.

For control problems, we are interested in the class of general evolution equations of the form

$$\frac{dy}{dt}=A(t)y+f(t,y), \qquad t\in I=[t_0,t_1],$$

$$y(t_0)=y_0, \qquad -\infty<t_0<t_1<+\infty, \tag{2.5.15}$$

where $\{A(t), t\in I\}$ is, in general, a family of unbounded linear operators and f a suitable nonlinear function.

Let H be a Hilbert space and E a dense linear subset of H carrying the structure of a reflexive Banach space. Let E^* be the dual of E. Identifying H with its dual H^*, we have $E\subset H\subset E^*$ with the injection map $E\hookrightarrow H$ assumed continuous. We introduce the following assumptions for A and f.

Assumption (2.5.A1). $\{A(t), t\in I\}$ is a family of densely defined linear operators in H with domain $D(A(t))\subset E$ and range $R(A(t))\subset E^*$, $t\in I$.

Assumption (2.5.A2). $\mathrm{Re}(A(t)e,e)\le 0$ for all $e\in D(A(t))\subset E$, $t\in I$.

Assumption (2.5.A3). f: $I\times E\to E^*$, $e\to f(t,e)$, is continuous from E to E^* for almost all $t\in I$, and $t\to(f(t,e),w)$ is measurable on I for $e,w\in E$.

Assumption (2.5.A4). $\mathrm{Re}(f(t,e_1)-f(t,e_2), e_1-e_2)\le 0$ for all $e_1,e_2\in E$, $t\in I$.

For $p>1$, $q=p/(p-1)$,

Assumption (2.5.A5). There exist a constant $\alpha \geq 0$ and $h \in L_q(I, R_+)$ such that $|f(t,e)|_{E^*} \leq h(t)+\alpha|e|_E^{p-1}$ for all $e \in E$ and $t \in I$.

Assumption (2.5.A6). There exist a constant $\beta>0$ and $h_1 \in L_1(I, R)$ such that $\operatorname{Re}(f(t,e),e) \leq h_1(t)-\beta|e|_E^p$ for all $e \in E$ and $t \in I$.

Definition 2.5.3. For $p>1$, $q=p/(p-1)$, let $Y \equiv L_p(I, E)$, let Y^* ($=L_q(I, E^*)$) be the conjugate space of Y, and let L be given by

$$(Lx)(t)=\left(\frac{d}{dt}x(t)-A(t)x(t)\right), \qquad x \in D(L), \quad t \in I, \quad (2.5.16)$$

where

$$D(L) \equiv \left\{ \begin{array}{l} x \in Y: x \in C^1(I,E),\, x(t) \in D(A(t)) \cap D(A^*(t)), \\ A(t)x(t),\, A^*(t)x(t) \text{ are continuous from } I \text{ to } E^* \end{array} \right\}. \quad (2.5.17)$$

Then the strong and weak extensions of L from Y to Y^* are said to *coincide* if (i) $D(L)$ is dense in Y and (ii) setting $(L'y)(t)=-(d/dt)y-A^*(t)y$, we know that for any pair $x \in Y$ and $z \in Y^*$ for which

$$\int_I (x(t),(L'y)(t))\,dt=\int_I (z(t),y(t))\,dt$$

$$\text{for all } y \in D(L), \quad (2.5.18)$$

there exists a sequence $\{x_n\} \in D(L)$ such that $x_n \overset{w}{\to} x$ in Y and $Lx_n \overset{w}{\to} z$ in Y^*.

Definition 2.5.4. Let L_s be the (strong) closure of L from Y to Y^*, and $D(L)$, as given in the definition above, is dense in Y, so that L is closable.

Definition 2.5.5. Let F be the mapping of Y into Y^* given by

$$(Fx,y)=\int_I (-f(\theta,x(\theta)),y(\theta))\,d\theta \quad \text{for } x,y \in Y. \quad (2.5.19)$$

Note that the integral in (2.5.19) is well defined due to Assumption (2.5.A5).

Definition 2.5.6. An element $x \in Y$ is said to be a strong solution of the initial value problem

$$\frac{dx}{dt}=A(t)x+f(t,x), \qquad t \in I,$$

$$x(t_0)=0 \quad (2.5.20)$$

if $x \in D(L_s)$ and

$$L_s x + Fx = 0. \tag{2.5.21}$$

We note that, for $y_0 \in \bigcap_{t\in I} D(A(t))$, the initial value problem (2.5.20) is entirely equivalent to the original problem (2.5.15). Indeed, writing $x = y - y_0$, problem (2.5.15) reduces to

$$\frac{dx}{dt} = A(t)x + f_1(t, x), \qquad t \in I,$$
$$x(t_0) = 0,$$

where $f_1(t, x) = A(t)y_0 + f(t, y_0 + x)$.

To show that this problem is equivalent to the original problem, it is only essential to verify that f_1 satisfies all the properties (2.5.A3)–(2.5.A6). Verification of (2.5.A3)–(2.5.A5) is straightforward; we show only (2.5.A6).

For $x \in Y$, we have

$$\begin{aligned} &\operatorname{Re}(f_1(t, x), x) \\ &= \operatorname{Re}\{(A(t)y_0 + f(t, y_0 + x), x)\} \\ &= \operatorname{Re}\{(A(t)y_0, x) + (f(t, y_0 + x), y_0 + x) - (f(t, y_0 + x), y_0)\} \\ &\le |A(t)y_0|_{E^*}|x|_E + (h_1(t) - \beta|x + y_0|_E^p) + |f(t, x + y_0)|_{E^*}|y_0|_E. \end{aligned} \tag{2.5.22}$$

Using property (2.5.A5) in the last term of the above inequality, we obtain

$$\begin{aligned} &\operatorname{Re}(f_1(t, x), x) \le \{h_2(t) - \beta_1|x|_E^p + \beta_2|x|_E^{p-1} + h_3(t)|x|_E\}, \\ &\qquad h_2(t) \equiv \{h_1(t) + |y_0|h(t) - 2^p(\beta - \alpha/2)|y_0|^p\}, \\ &\qquad h_3(t) \equiv |A(t)y_0|_{E^*}, \qquad \beta_1 = 2^p\beta, \quad \beta_2 = 2^{p-1}\alpha|y_0|_E. \end{aligned} \tag{2.5.23}$$

Using Cauchy's inequality,

$$a \cdot b \le \frac{\delta^p}{p}a^p + \frac{\delta^{-q}}{q}b^q, \qquad a, b > 0, \quad p > 1, \quad q = p/(p-1), \tag{2.5.24}$$

which is valid for any $\delta > 0$, we have

$$|x|_E^{p-1} \le \frac{\delta^p}{p} + \frac{\delta^{-q}}{q}|x|_E^p, \qquad \delta > 0 \tag{2.5.25}$$

and

$$h_3(t)|x|_E \le \frac{\varepsilon^p}{p}|x|^p + \frac{\varepsilon^{-q}}{q}|h_3(t)|^q, \qquad \varepsilon > 0. \tag{2.5.26}$$

Combining (2.5.23), (2.5.25), and (2.5.26), we can rewrite (2.5.23) as

$$\operatorname{Re}(f_1(t, x), x) \le h^*(t) - (\beta_1 - \gamma(\varepsilon, \delta))|x|_E^p = h^*(t) - \beta^*|x|_E^p, \tag{2.5.27}$$

where (i) h^* depends on h_2, h_3, β_2, ε, and δ, and $h^* \in L_1(I, R)$ for $\varepsilon, \delta > 0$, and (ii) $\gamma(\varepsilon, \delta)$ depends on β_2, ε, and δ. Since $\varepsilon, \delta > 0$ but otherwise are arbitrary, we can choose these numbers in such a way that $0 < \gamma(\varepsilon, \delta) < \beta_1$

and hence $\beta^*>0$. This shows that f_1 satisfies property (2.5.A6). Thus we conclude that, for $y_0 \in \cap D(A(t))$, there is no loss of generality if we consider the problem (2.5.20) or (2.5.21) instead of (2.5.15). We shall prove the existence and uniqueness of solutions of the initial value problem (2.5.20) or the functional equation (2.5.21). In preparation for this, we present the following lemma.

Lemma 2.5.1. *Suppose that f satisfies Assumptions* (2.5.A3)–(2.5.A6) *and let F denote the mapping given by* (2.5.19) *in Definition* 2.5.5. *Then F is a well-defined continuous mapping of Y into Y^* and carries bounded sets of Y into bounded sets of Y^*. Moreover, F satisfies the inequalities*

$$\text{(a)} \quad \operatorname{Re}(Fx, x) \geq \beta \|x\|_Y^p - c \quad \text{for all } x \in Y \equiv L_p(I, E). \tag{2.5.28}$$

and

$$\text{(b)} \quad \operatorname{Re}(Fx - Fy, x - y) \geq 0 \quad \text{for all } x, y \in Y. \tag{2.5.29}$$

PROOF. For $x, y \in Y$ define $N(x, y) \equiv \int_I (f(\theta, x(\theta)), y(\theta))\, d\theta$. Using Hölder's inequality and properties (2.5.A3) and (2.5.A5), it is easy to verify that

$$|N(x, y)| \leq \left(\|h\|_{L_q} + \alpha \|x\|_Y^{p-1}\right) \|y\|_Y \quad \text{for all } x, y \in Y. \tag{2.5.30}$$

Since, for a fixed but arbitrary $x \in Y$, $y \to N(x, y)$ is a conjugate linear functional on Y, there exists a $z \in Y^*$ dependent on x alone such that $-N(x, y) = (z, y)$. Setting $z = Fx$, we conclude that F maps Y into Y^* and, by virtue of (2.5.30),

$$\|z\|_{Y^*} = \|Fx\|_{Y^*} \leq \|h\|_{L_q} + \alpha \|x\|_Y^{p-1} \quad \text{for all } x \in Y. \tag{2.5.31}$$

Therefore, F carries bounded sets of Y into bounded sets of Y^*. By using property (2.5.A6) and the definition for f given in (2.5.19), we obtain (a), where the constant c is given by

$$c = \int_I h_1(\theta)\, d\theta.$$

Similarly, (b)=(2.5.29) follows from the definition of F and property (2.5.A4). It remains now to verify the continuity of F. For that purpose, it suffices to show that for any sequence $\{x_n\} \in Y$ converging strongly to x_0 and any $y \in Y$ and $\varepsilon > 0$, there exist a finite number $\gamma > 0$, and an integer $n_0 = n_0(\varepsilon)$ such that

$$|N(x_n, y) - N(x_0, y)| \leq (\gamma \|y\|)\varepsilon \quad \text{for all } n \geq n_0. \tag{2.5.32}$$

First, we show this for I with finite Lebesgue measure, that is, $l(I) < \infty$, and then indicate the necessary modification for $l(I) = \infty$. We can write

$$\begin{aligned} |(Fx_n - Fx_0, y)| &= |N(x_n, y) - N(x_0, y)| \\ &\leq \int_I |f(t, x_n(t)) - f(t, x_0(t))|_{E^*} |y|_E\, dt, \end{aligned} \tag{2.5.33}$$

where $x_n \xrightarrow{s} x_0$ in Y. For fixed $\delta > 0$, define

$$I_{n,\delta} \equiv \{t \in I : |x_n(t) - x_0(t)|_E \geq \delta\}, \qquad n = 1, 2, \ldots.$$

Since $x_n \xrightarrow{s} x_0$ and

$$\delta^p l(I_{n,\delta}) \leq \int_I |x_n(t) - x_0(t)|_E^p \, dt, \tag{2.5.34}$$

it is clear that, for $\delta > 0$, $\lim_{n\to\infty} l(I_{n,\delta}) = 0$. This shows that $x_n \to x_0$ strongly (norm topology of E) in measure. For each integer m and a given $\varepsilon > 0$, define

$$J_{m,\varepsilon} = \left\{ \begin{array}{l} t \in I : |f(t, x_0(t)) - f(t, e)|_{E^*} \leq \varepsilon \text{ for all } e \\ \text{for which } |x_0(t) - e|_E \leq 1/m \end{array} \right\}. \tag{2.5.35}$$

For a fixed $\varepsilon > 0$, it may be observed that $\{J_{m,\varepsilon}\}_m$ is a nondecreasing sequence of measurable subsets of I. Define

$$J_\varepsilon = \bigcup_m J_{m,\varepsilon}. \tag{2.5.36}$$

Since, by Assumption (2.5.A3), the map $e \to f(t, e)$ from E to E^* is continuous for almost all $t \in I$, we have

$$l(I \backslash J_\varepsilon) = 0 \quad \text{for all } \varepsilon > 0. \tag{2.5.37}$$

Therefore, for any given number $\nu > 0$, there exists an integer $m_0 = m_0(\nu)$ such that

$$l(J_{m_0,\varepsilon}) > l(I) - \nu/2. \tag{2.5.38}$$

For the given $\varepsilon > 0$, define

$$K_{n,\varepsilon} \equiv \{t \in I : |f(t, x_0(t)) - f(t, x_n(t))|_{E^*} \geq \varepsilon\}. \tag{2.5.39}$$

Choosing $\delta = 1/m_0$, we note that

$$K_{n,\varepsilon} \subset \{J_{m_0,\varepsilon} \cap I \backslash I_{n,1/m_0}\}',$$

or equivalently,

$$K_{n,\varepsilon} \subset (I \backslash J_{m_0,\varepsilon}) \cup I_{n,1/m_0},$$

and consequently, it follows from (2.5.34) and (2.5.38) that there exists an integer $n_0 = n_0(\nu)$ such that

$$l(K_{n,\varepsilon}) \leq l(I \backslash J_{m_0,\varepsilon}) + l(I_{n,1/m_0}) \leq \nu \quad \text{for all } n \geq n_0.$$

Clearly

$$\begin{aligned} &|N(x_n, y) - N(x_0, y)| \\ &\quad \leq \int_I |f(t, x_n(t)) - f(t, x_0(t))|_{E^*} |y(t)|_E \, dt \\ &\quad \leq \left(\int_{K_{n,\varepsilon}} + \int_{I \backslash K_{n,\varepsilon}} \right) \left[|f(t, x_n(t)) - f(t, x_0(t))|_{E^*} |y(t)|_E \right] dt, \end{aligned} \tag{2.5.40}$$

where

$$\int_{I\setminus K_{n,\varepsilon}} |f(t,x_n)-f(t,x_0)|_{E^*}|y(t)|_E\,dt$$
$$\le \varepsilon\int_{I\setminus K_{n,\varepsilon}} |y(t)|_E\,dt \le \varepsilon l^{1/q}(I\setminus K_{n,\varepsilon})\|y\|_Y$$
$$\le \varepsilon l^{1/q}(I)\|y\|_Y, \tag{2.5.41}$$

and, by virtue of (2.5.A5),

$$\int_{K_{n,\varepsilon}} |f(t,x_n)-f(t,x_0)|_{E^*}|y(t)|_E\,dt$$
$$\le \int_{K_{n,\varepsilon}} (2h(t)+\alpha|x_n(t)|_E^{p-1}+\alpha|x_0(t)|_E^{p-1})|y(t)|_E\,dt$$
$$\le \left\{2\left(\int_{K_{n,\varepsilon}} |h(t)|^q\,dt\right)^{1/q}+\alpha 2^{p-1}\left(\int_{K_{n,\varepsilon}} |x_n(t)-x_0(t)|_E^p\,dt\right)^{1/q}\right.$$
$$\left.+\alpha 2^{p-1}\left(\int_{K_{n,\varepsilon}} |x_0(t)|_E^p\,dt\right)^{1/q}+\alpha\left(\int_{K_{n,\varepsilon}} |x_0(t)|_E^p\,dt\right)^{1/q}\right\}\|y\|_Y. \tag{2.5.42}$$

In arriving at the final expression in (2.5.42), we have used Hölder's inequality and the fact that $(p-1)q=p$. Since $h\in L_q(I, R_+)$ and $x_0\in Y\equiv L_p(I, E)$, for the given $\varepsilon>0$, there exists an integer $n_1=n_1(\varepsilon)$ such that, for all $n\ge n_1$,

$$\int_{K_{n,\varepsilon}} |h(t)|^q\,dt\le\varepsilon^q,$$
$$\int_{K_{n,\varepsilon}} |x_0(t)|_E^p\,dt\le\varepsilon^q. \tag{2.5.43}$$

Similarly, due to strong convergence of x_n to x_0 in Y, there exists an integer $n_2=n_2(\varepsilon)$ such that

$$\int_{K_{n,\varepsilon}\subset I} |x_n(t)-x_0(t)|_E^p\,dt\le\varepsilon^q \quad \text{for all } n\ge n_2. \tag{2.5.44}$$

Thus, for $n\ge n_1\vee n_2$, we have

$$\int_{K_{n,\varepsilon}} |f(t,x_n(t))-f(t,x_0(t))|_{E^*}|y(t)|_E\,dt\le\varepsilon[2+\alpha(1+2^p)]\|y\|. \tag{2.5.42$'$}$$

Therefore, it follows from (2.5.40), (2.5.41), and (2.5.42)′ that for every $\varepsilon>0$

there exists an integer $n_0 = n_0(\varepsilon)$ such that for all $n \geq n_0$

$$|N(x_n, y) - N(x_0, y)| \leq \varepsilon \cdot \{l^{1/q}(I) + 2 + \alpha(1+2^p)\} \|y\|$$

$$\leq \varepsilon \cdot (\gamma \|y\|). \tag{2.5.45}$$

This proves the continuity of F in the sense that, whenever $x_n \overset{s}{\to} x_0$ in Y, $Fx_n \overset{w}{\to} Fx_0$ in Y^*. The last statement follows from the fact that weak and weak* convergence are equivalent in reflexive Banach spaces. For the proof of continuity in case $l(I) = \infty$, we define, for each number $a > 0$, $I_a \equiv I \cap \{t \in I : |t| \leq a\}$ and $I'_a = I \cap \{t \in I : |t| \geq a\}$. Then for any $y \in Y$,

$$|N(x_n, y) - N(x_0, y)| \leq \int_{I_a} |(f(t, x_n(t)) - f(t, x_0(t)), y(t))| \, dt$$

$$+ \int_{I'_a} |(f(t, x_n(t)) - f(t, x_0(t)), y(t))| \, dt. \tag{2.5.46}$$

Considering the last term in the above inequality and taking into account property (2.5.A5), we obtain

$$\int_{I'_a} |(f(t, x_n(t)) - f(t, x_0(t)), y(t))| \, dt$$

$$\leq \left\{ 2 \left(\int_{I'_a} |h(t)|^q \right)^{1/q} + \alpha \left(\int_{I'_a} |x_n(t)|_E^p \, dt \right)^{1/q} \right.$$

$$\left. + \alpha \left(\int_{I'_a} |x_0(t)|_E^p \, dt \right)^{1/q} \right\} \|y\|_Y. \tag{2.5.47}$$

Since $h \in L_q(I, R_+)$, $x_0 \in L_p(I, E)$, $x_n \in L_p(I, E)$, and $x_n \overset{s}{\to} x_0$ in Y, for every $\varepsilon > 0$ there exists a finite number $a_0 = a(\varepsilon) > 0$ such that

$$\int_{I'_{a_0}} |h(t)|^q \, dt \leq \varepsilon^q,$$

$$\int_{I'_{a_0}} |x_n(t)|_E^p \, dt \leq \varepsilon^q, \qquad \int_{I'_{a_0}} |x_0(t)|_E^p \, dt \leq \varepsilon^q \tag{2.5.48}$$

for all n. Replacing I by I_{a_0} in (2.5.45), it follows from (2.5.46)–(2.5.48) that, for any $\varepsilon > 0$ there exists an integer $n_0 = n(\varepsilon)$ such that

$$|N(x_n, y) - N(x_0, y)| \leq (\gamma_0 \|y\|)\varepsilon \quad \text{for all } n \geq n_0,$$

where $\gamma_0 \equiv \{l^{1/q}(I_{a_0}) + 4 + \alpha(3 + 2^p)\}$. This completes the proof of the lemma. □

We are now prepared to consider the question of existence and uniqueness of solutions of the evolution equation

$$\frac{dy}{dt}=A(t)y+f(t,y), \qquad t\in I,$$
$$y(t_0)=y_0.$$

We have observed that for $y_0\in\cap D(A(t))$, this equation is equivalent to (2.5.20), and consequently, under this condition we can consider the later equation.

Theorem 2.5.3. *Consider the evolution equation*

$$\begin{aligned}\frac{dx}{dt}&=A(t)x+f(t,x)\\ x(t_0)&=0,\end{aligned} \tag{2.5.49}$$

where f maps $I\times E$ into E^* and satisfies conditions (2.5.A3)–(2.5.A6) *and $\{A(t),\ t\in I\}$ is a family of densely defined linear operators in H with domain $D(A(t))\subset E$ and range $R(A(t))\subset E^*$. Suppose that A satisfies conditions* (2.5.A1) *and* (2.5.A2) *and that for $Y=L_p(I,E)$, the strong and weak extensions of $L\equiv((d/dt)-A(t))$ from Y to Y^* coincide in the sense of Definition* 2.5.3. *Then system* (2.5.49) *has a unique strong solution $x\in Y$.*

PROOF. Under the given assumption, the operator L, defined by $L=((d/dt)-A(t))$, is a densely defined linear operator from Y to Y^* and $L_s=L_w$. By Definition 2.5.6, a strong solution of Equation (2.5.49) is an element of $D(L_s)\subset Y$ such that $L_s x+Fx=0$, where F denotes the operator given by

$$(Fx,y)=-\int_I(f(\theta,x(\theta)),y(\theta))d\theta, \qquad x,y\in Y.$$

By Lemma 2.5.1, the mapping F from Y into Y^* is demicontinuous and carries bounded sets of Y into bounded sets of Y^*. Defining $T=L+F$, we have for $x,y\in D(T)\subset Y$

$$\mathrm{Re}(Tx-Ty,x-y)=\mathrm{Re}(L(x-y),x-y)+\mathrm{Re}(Fx-Fy,x-y). \tag{2.5.50}$$

By Lemma 2.5.1(b),

$$\mathrm{Re}(Fx-Fy,x-y)\geq 0 \quad \text{for all } x,y\in Y, \tag{2.5.51}$$

and due to Assumption (2.5.A2),

$$\mathrm{Re}(Lx,x)=\mathrm{Re}\int_I\left\{\left(\frac{dx}{dt},x(t)\right)-(A(t)x(t),x(t))\right\}dt\geq 0 \tag{2.5.52}$$

for $x\in D(L)=D(T)$. Therefore, it follows from (2.5.50)–(2.5.52) that

$$\mathrm{Re}(Tx-Ty,x-y)\geq 0 \quad \text{for all } x,y\in D(T). \tag{2.5.53}$$

Again, for $x \in D(T)$, by Lemma 2.5.1(a),

$$\operatorname{Re}(Tx, x) = \operatorname{Re}(Lx, x) + \operatorname{Re}(Fx, x) \geq \operatorname{Re}(Fx, x) \geq \beta \| x \|_Y^p - c. \tag{2.5.54}$$

Hence T satisfies all the conditions [(i)–(iv)] of Theorem 2.5.1, and consequently, T_s ($\equiv L_s + F$) maps $D(T_s)$ onto Y^*. Thus, for every $y^* \in Y^*$, there exists an $x^* \in D(T_s)$ such that $T_s x^* = L_s x^* + Fx^* = y^*$. In particular, there exists $x \in D(L_s) = D(T_s)$ such that $L_s x + Fx = 0$. This proves the existence of a strong solution. For uniqueness, we note that if x and $y \in Y$ are any two strong solutions of the problem (2.5.49), then

$$L_s(x-y) + Fx - Fy = 0. \tag{2.5.55}$$

Let $C_t(\cdot)$ denote the characteristic function of the set $[t_0, t]$; then, clearly, $(x-y)C_t \in Y$ and

$$(L_s(x-y) + Fx - Fy, (x-y)C_t)_{Y^* - Y} = 0 \quad \text{for all } t \in I. \tag{2.5.56}$$

Hence

$$\int_{t_0}^{t} \left(\frac{d}{d\theta}(x-y), x-y \right)_{E^* - E} d\theta - \int_{t_0}^{t} (A(\theta)(x-y), x-y)_{E^* - E} d\theta$$
$$- \int_0^t (f(\theta, x(\theta)) - f(\theta, y(\theta)), x(\theta) - y(\theta))_{E^* - E} d\theta = 0, \qquad t \in I. \tag{2.5.57}$$

By virtue of the hypotheses (2.5.A2) and (2.5.A4), we have

$$- \int_{t_0}^{t} (A(\theta)(x-y)(\theta), (x-y)(\theta))_{E^* - E} d\theta \geq 0, \qquad t \in I,$$

and

$$- \int_{t_0}^{t} (f(\theta, x(\theta)) - f(\theta, y(\theta)), x(\theta) - y(\theta))_{E^* - E} d\theta \geq 0, \qquad t \in I.$$

Using these conditions in (2.5.57), we obtain

$$\int_{t_0}^{t} \left(\frac{d}{d\theta}(x-y), x-y \right) d\theta \leq 0 \quad \text{for all } t \in I. \tag{2.5.58}$$

Since $x \in Y$ and $Lx \in Y^*$, by Theorem 1.2.15 we have $x \in C(\bar{I}, H)$, and consequently, we can write (2.5.58) as

$$\frac{1}{2} \int_{t_0}^{t} \frac{d}{d\theta} (|x(\theta) - y(\theta)|_H^2) d\theta \leq 0, \qquad t \in I.$$

This implies that $|x(t) - y(t)|_H = 0$ for all $t \in I$, or equivalently, $x = y$. Thus, we have uniqueness.

REMARK 2.5.1. Suppose $A(t)$, $t \geq 0$, admits an evolution operator $K(t, \tau)$, $(t, \tau) \in \Delta \equiv \{(t, \tau) : 0 \leq \tau \leq t < \infty\}$, of bounded linear operators from E^* into

E. Then the evolution equation (2.5.15) can be written as an abstract integral equation

$$y(t)=K(t,t_0)y_0+\int_{t_0}^{t}K(t,\tau)f(\tau,y(\tau))d\tau \tag{2.5.59}$$

in E. A solution of this equation is called a *mild* solution of the evolution equation (2.5.15). One of the authors has studied [Ah.5] the question of existence of solutions of a more general class of nonlinear integral equations than (2.5.59) on reflexive Banach spaces with application to evolution equations of the form (2.5.15).

REMARK 2.5.2. The basic references for the results of Section 2.4 are Kato [K2], Hille and Phillips [HP.1], and Yosida [Yo.1]. The results of Section 2.5 are taken from the recent literature and are due mainly to Bowder [Br.1,2].

CHAPTER THREE

OPTIMAL CONTROL OF PARABOLIC PARTIAL DIFFERENTIAL SYSTEMS WITH CONTROLS IN THE COEFFICIENTS

In recent years, significant emphasis has been given to the study of optimal control of systems governed by parabolic partial differential equations (PPDE) with first boundary conditions or with Cauchy conditions (for details, see the articles mentioned in Section 3.4). In these studies, the differential equations are either in *general form* or in *divergence form*. It is known [F1.2, TA.1] that a general class of optimal control problems of systems governed by Ito stochastic differential equations with Markov (fixed) terminal time can be converted into a class of optimal control problems of systems governed by linear second-order PPDE with first boundary condition (Cauchy condition).

Questions concerning necessary conditions for optimality and existence of optimal controls for these problems have been investigated in [AT.1, AT.4, AT.5, Fl.1, Fl.2, RT.2, Te.1, Z.1, Z.2,]. Moreover, a few results [Boy.1, Fl.1, R.1, RT.3, TRB.1] on the computational methods for finding optimal controls are also available in the literature.

This chapter is divided into four sections. In Section 3.1, we consider a class of systems governed by linear second-order PPDE in general form with first boundary condition. In Section 3.2, we consider a similar class of systems but with the differential equation given in divergence form. In each of these two sections necessary conditions for optimality are derived and several results on the existence of optimal controls are proved. A few results on the computational methods are discussed in Section 3.3. Due to space limitation, questions concerning optimal control of systems governed by linear second-order PPDE with Cauchy condition are not considered. Instead, we refer the reader to the literature [Te.1, AT.4, AT.5]. Section 3.4 contains brief notes in which certain related results are discussed and several open problems are posed.

3.1. First Boundary Value Problems in General Form

3.1.1. Introduction

In this section, we consider a class of systems governed by second-order linear parabolic partial differential equations in general form with a first boundary condition. Here, the controls appear in the first- and zeroth-order coefficients and the free term of the system.

In Section 3.1.2, we describe the system and impose certain basic assumptions. Results concerning the existence and uniqueness of solutions of the system and certain important properties of the solutions are given in Section 3.1.3. In Sections 3.1.4 and 3.1.5, an optimal control problem is solved giving the necessary conditions for optimality. The question of existence of optimal controls is discussed in Section 3.1.6.

3.1.2. System Description

Let $\Omega \subset R^n$ be open and connected with compact closure $\bar{\Omega} = \Omega \cup \partial\Omega$, where the boundary $\partial\Omega$ of Ω is assumed to satisfy the condition (2.1.A1). Let T be a fixed positive constant and $Q \equiv \Omega \times (0, T)$, and let $\mathcal{U}$ be the class of admissible controls to be defined later. In addition, we recall the notation of Section 1.2.1.

We consider the system

$$\begin{aligned} L(u)\phi(x,t) &= f(x,t,u(x,t)), && (x,t) \in Q, \\ \phi(x,0) &= \phi_0(x), && x \in \Omega, \\ \phi(x,t) &= 0, && (x,t) \in \partial\Omega \times [0,T], \end{aligned} \tag{3.1.1}$$

where $u \in \mathcal{U}$ and, for each $u \in \mathcal{U}$, $L(u)$ is a second-order partial differential operator given by

$$\begin{aligned} L(u)\psi(x,t) \equiv \psi_t(x,t) &- \sum_{i,j=1}^{n} a_{ij}(x,t)\psi_{x_i x_j}(x,t) \\ &- \sum_{i=1}^{n} b_i(x,t,u(x,t))\psi_{x_i}(x,t) - c(x,t,u(x,t))\psi(x,t). \end{aligned} \tag{3.1.2}$$

Let V be a given nonempty compact subset of R^m and $U: \bar{Q} \to R^m$ a measurable multifunction (see Section 1.4.1) with values from the class of nonempty closed convex subsets of V. A measurable function $u: \bar{Q} \to V$ is called an *admissible control* if $u(x,t) \in U(x,t)$ a.e. in $\bar{Q}$. Let $\mathcal{U}$ denote the *class of admissible controls*.

The coefficients and data of system (3.1.1) are assumed (throughout this section) to satisfy the following conditions.

Assumption (3.1.A1). a_{ij}, $i, j = 1, \ldots, n$, are continuous on $\bar{Q}$.

Assumption (3.1.A2). There exist numbers α_l, $\alpha_u > 0$ such that

$$\alpha_l \sum_{i=1}^{n} (\xi_i)^2 \leq \sum_{i,j=1}^{n} a_{ij}(x,t)\xi_i\xi_j \leq \alpha_u \sum_{i=1}^{n} (\xi_i)^2$$

for all $\xi \in R^n$ uniformly on $\bar{Q}$.

Assumption (3.1.A3). b_i, $i=1,\dots,n$, c, and f are bounded measurable on $\bar{Q}\times V$ and continuous on V for almost all $(x,t)\in\bar{Q}$.

Assumption (3.1.A4). $\phi_0 \in C_0^2(\Omega)$.

We shall see in the next subsection that Assumptions (2.1.A1) and (3.1.A1)–(3.1.A4) are sufficient conditions under which the system has a unique solution corresponding to each $u\in\mathcal{U}$.

3.1.3. *Existence and Uniqueness of Solutions*

With reference to system (3.1.1), we introduce the following definition.

Definition 3.1.1. For each $u\in\mathcal{U}$, a function $\phi(u)$: $\bar{Q}\to R$ is said to be a *solution of system* (3.1.1) if $\phi(u)\in W_p^{2,1}(Q)$, $3/2<p<\infty$, and it satisfies the differential equation a.e. on Q and the boundary conditions everywhere in their corresponding domains of definition.

REMARK 3.1.1. In view of Remark 2.1.1, we observe that if Assumptions (2.1.A1) and (3.1.A1)–(3.1.A4) are satisfied, then, for each $u\in\mathcal{U}$, the system (3.1.1) has a unique solution $\phi(u)$. Furthermore, $\phi(u)$ satisfies the following estimates:

$$\|\phi(u)\|_{p,Q}^{(2,1)} \leq K_1\{\|f(\cdot,\cdot,u(\cdot,\cdot))\|_{\infty,Q} + \|\phi_0\|_{\infty,\Omega}^{(2)}\} \tag{3.1.3}$$

for all $p>3/2$, and

$$|\phi(u)|_{Q}^{(1+\mu,(1+\mu)/2)} \leq K_2\{\|f(\cdot,\cdot,u(\cdot,\cdot))\|_{\infty,Q} + \|\phi_0\|_{\infty,\Omega}^{(2)}\} \tag{3.1.4}$$

for all $\mu\in(0,1)$ (for notation, see Sections 1.2.2–1.2.3). Here, for given second-order coefficients, the constants K_1 and K_2 depend only on Q, Ω, p, and the bounds for b_i, $i=1,\dots,n$, and c.

3.1.4. *Formulation of Some Control Problems*

Given the system (3.1.1), we can formulate an optimal control problem by including a functional that gives a measure of performance of the system corresponding to each control. Such a functional is called a *cost functional* or *objective functional*. In this subsection, several cost functionals will be defined. The most general one is

$$J(u)=\int_{\Omega}\{\gamma_0(x,\phi(u)(x,T),(\phi(u))_x(x,T))\}\,dx$$
$$+\iint_{Q}\{\gamma_1(x,t,\phi(u)(x,t),(\phi(u))_x(x,t),u(x,t))\}\,dx\,dt. \tag{3.1.5}$$

where $\phi(u)\in W_p^{2,1}(Q)$, $3/2<p<\infty$, is the response of system (3.1.1) corresponding to the control $u\in\mathcal{U}$ and $(\phi(u))_x\equiv((\phi(u))_{x_1},\dots,(\phi(u))_{x_n})$. In what follows, we shall impose certain conditions on the real-valued functions γ_0 and γ_1.

Assumption (3.1.A5)

i. γ_0 is a Carathéodory function defined on $\bar{\Omega}\times R^{n+1}$.
ii. There exists a nonnegative constant δ_1 and a nonnegative function $g_1\in L_1(\Omega)$ such that

$$|\gamma_0(x,\theta_0,\theta)|\le g_1(x)+\delta_1\sum_{j=0}^{n}|\theta_j| \quad \text{a.e. on } \Omega$$

for all $\theta_0\in R$ and $\theta\equiv(\theta_1,\dots,\theta_n)\in R^n$.

Assumption (3.1.A6)

i. $\gamma_1$is a Carathéodory function defined on $\bar{Q}\times\{R^{n+1}\times V\}$.
ii. There exists a nonnegative function $g_2\in L_1(Q)$ such that

$$|\gamma_1(x,t,\theta_0^1,\theta^1,v)-\gamma_1(x,t,\theta_0^2,\theta^2,v)|\le g_2(x,t)\left(\sum_{j=0}^{n}|\theta_j^1-\theta_j^2|\right)$$
$$\text{a.e. on } \bar{Q}$$

for all $\theta_0^1,\theta_0^2\in R$, $\theta^1,\theta^2\in R^n$, and $v\in V$.
iii. There exist functions $G_i\in L_1(Q)$, $i=1,\dots,m$, such that

$$-\sum_{i=1}^{m}G_i(x,t)(v_i^1-v_i^2)\le\gamma_1(x,t,\theta_0,\theta,v^1)-\gamma_1(x,t,\theta_0,\theta,v^2)$$
$$\le\sum_{i=1}^{m}G_i(x,t)(v_i^1-v_i^2) \quad \text{a.e. on } Q,$$

for all $\theta_0\in R$, $\theta\in R^n$, and $v^1\equiv(v_1^1,\dots,v_m^1)$, $v^2\equiv(v_1^2,\dots,v_m^2)\in R^m$.
iv. $\gamma_1(\cdot,\cdot,0,0,0)\in L_1(Q)$.

With reference to the cost functional (3.1.5), we may state a general optimal control problem subject to system (3.1.1) as follows:

Problem (3.1.P1). Subject to system (3.1.1), find a control $u\in\mathcal{U}$ such that the cost functional (3.1.5) is minimized.

The next problem is a special case of Problem (3.1.P1), in which the cost functional depends only on the terminal values of the state.

Problem (3.1.P2). Subject to system (3.1.1), find a control $u \in \mathfrak{U}$ such that the cost functional

$$J(u) = \int_{\Omega} \{\gamma(x, \phi(u)(x, T))\}\, dx \tag{3.1.6}$$

is minimized.

Considering problem (3.1.P2), we impose certain conditions on the real-valued function γ, namely,

Assumption (3.1.A7)

i. There exists a Carathéodory function $g: \bar{\Omega} \times R \to R$ such that

$$\gamma(x, \theta^2) \leq \gamma(x, \theta^1) + g(x, \theta^1)(\theta^2 - \theta^1)$$

for almost all $x \in \Omega$ and for every $\theta^1, \theta^2 \in R$.

ii. There exist a nonnegative bounded measurable function g_3 defined on $\bar{\Omega}$ and a nonnegative constant δ_3 such that

$$|g(x, \theta)| \leq g_3(x) + \delta_3 |\theta|$$

for almost all $x \in \bar{\Omega}$ and for all $\theta \in R$.

Assumption (3.1.A8). $\gamma(\cdot, \theta) \in L_1(\Omega)$ for *some* $\theta \in R$.

3.1.5. Necessary Conditions For Optimality

In this subsection, we shall derive necessary conditions for optimality for the terminal control problem (3.1.P2). For this, we assume that an optimal control exists and is denoted by u^*. Before we present the main result of this subsection (Theorem 3.1.3), we need some preparation.

To start with, we introduce a system that is called the *adjoint system of system* (3.1.1). The function a_i, $i = 1, \ldots, n$, defined by

$$a_i(x, t, v) \equiv \sum_{j=1}^{n} \frac{\partial a_{ij}(x, t)}{\partial x_j} - b_i(x, t, v), \tag{3.1.7}$$

will appear in the differential operator of the adjoint system. For each $u \in \mathfrak{U}$, this operator is the formal adjoint of the operator $L(u)$ [which is defined by (3.1.2)], and is denoted by $L^*(u)$. For simplicity of presentation, we assume that the second-order coefficients a_{ij}, $i, j = 1, \ldots, n$, are such that a_i, $i = 1, \ldots, n$, as given by (3.1.7), are bounded and measurable on Q. More precisely, we assume the following:

Assumption (3.1.A9). There exists a positive constant K_4 such that, for all $i, j \in \{1, \ldots, n\}$,

$$\frac{|a_{ij}(x,t) - a_{ij}(x',t')|}{|x-x'|+|t-t'|} \leq K_4 \tag{3.1.8}$$

for all (x,t), $(x',t') \in \overline{Q}$.

For each $u \in \mathcal{U}$, the operator $L^*(u)$ is given by

$$L^*(u)\psi(x,t) \equiv -\psi_t(x,t) - \sum_{i=1}^{n}\left(\sum_{j=1}^{n} a_{ij}(x,t)\psi_{x_j}(x,t) + a_i(x,t,u(x,t))\psi(x,t)\right)_{x_i} - c(x,t,u(x,t))\psi(x,t), \tag{3.1.9}$$

and the adjoint system is defined as

$$\begin{aligned} L^*(u)z(x,t) &= 0, \qquad (x,t) \in Q, \\ z(x,t) &= 0, \qquad (x,t) \in \partial\Omega \times [0,T], \\ z(x,T) &= g(x, \phi(u)(x,T)), \quad x \in \Omega, \end{aligned} \tag{3.1.10}$$

where the function g is as defined in Assumption (3.1.A7) and $\phi(u)$ is the solution of system (3.1.1) corresponding to the control $u \in \mathcal{U}$.

In order to derive a necessary condition for optimality for the terminal control problem (3.1.P2), we must show that the adjoint system (3.1.10) has a unique (weak) solution in the sense of the following definition.

Definition 3.1.2. For each $u \in \mathcal{U}$, a function $z(u)$ is said to be a *weak solution of system* (3.1.10) if $z(u) \in \overset{\circ}{V}{}_2^{1,0}(Q)$ and, for all $\tau \in [0, T]$,

$$\begin{aligned} &\int_\Omega \{z(u)(x,\tau)\eta(x,\tau)\}\,dx + \int_\tau^T \int_\Omega \Bigg\{ z(u)(x,t)\eta_t(x,t) \\ &\quad + \sum_{i=1}^{n}\left(\sum_{j=1}^{n} a_{ij}(x,t)(z(u))_{x_j}(x,t) + a_i(x,t,u(x,t))z(u)(x,t)\right) \\ &\quad \times \eta_{x_i}(x,t) - c(x,t,u(x,t))z(u)(x,t)\eta(x,t)\Bigg\}\,dx\,dt \\ &= \int_\Omega \{g(x,\phi(u)(x,T))\eta(x,T)\}\,dx \end{aligned}$$

for all $\eta \in \overset{\circ}{W}_2^{1,1}(Q)$, where the Banach spaces $\overset{\circ}{V}_2^{1,0}(Q)$ and $\overset{\circ}{W}_2^{1,1}(Q)$ are defined in Section 1.2.3.

In view of Assumption (3.1.A3), there exists a constant K_5 such that

$$|b_i(x,t,v)| \le K_5, \qquad i=1,\dots,n, \tag{3.1.11}$$

and

$$|c(x,t,v)| \le K_5 \tag{3.1.12}$$

for all $(x,t,v) \in \bar{Q} \times V$. Since Q is a bounded subset of R^{n+1} and V a compact subset of R^m, it is clear that

$$\|c(\cdot,\cdot,u(\cdot,\cdot))\|_{q,r,Q} \le K_6 \tag{3.1.13}$$

uniformly with respect to $u \in \mathcal{U}$, for any $q, r \ge 1$ and hence, for any q, r satisfying condition (2.2.7) or the condition given in Assumption (2.2.A6), where $K_6 \equiv K_5((|\Omega|)^{1/q}(T)^{1/r})$ and $|\Omega|$ denotes the Lebesgue measure of Ω. On the other hand, by using the definition of a_i given in (3.1.7), Assumption (3.1.A9), and the inequality (3.1.11), we can verify that

$$\left\| \sum_{i=1}^{n} (a_i(\cdot,\cdot,u(\cdot,\cdot)))^2 \right\|_{q,r,Q} \le K_7 \tag{3.1.14}$$

uniformly with respect to $u \in \mathcal{U}$, for any $q, r \ge 1$ and hence, for any q, r satisfying condition (2.2.7) or the condition given in Assumption (2.2.A6), where $K_7 \equiv n(nK_4 + K_5)^2(|\Omega|)^{1/q}(T)^{1/r}$.

REMARK 3.1.2. In view of Remark 3.1.1, we observe that, for each $u \in \mathcal{U}$, system (3.1.1) has a unique solution $\phi(u)$ that satisfies the estimate (3.1.4). Thus, $\phi(u)(\cdot,T)$ is defined on $\bar{\Omega}$ and belongs to $\mathcal{H}^{1+\mu}(\bar{\Omega})$, $0<\mu<1$ (for definition, see Section 1.2.1). Therefore, it follows from the second inequality of Assumption (3.1.A7) that $g(\cdot,\phi(u)(\cdot,T))$ is bounded on $\bar{\Omega}$ uniformly with respect to $u \in \mathcal{U}$. Since Ω is bounded, it is clear that $g(\cdot,\phi(u)(\cdot,T)) \in L_2(\Omega)$ for any $u \in \mathcal{U}$.

The next theorem presents a result concerning the existence and uniqueness of weak solutions of the adjoint system.

Theorem 3.1.1. *Consider the adjoint system* (3.1.10). *Suppose that Assumptions* (2.1.A1), (3.1.A2)–(3.1.A4), (3.1.A7), *and* (3.1.A9) *are satisfied. Then, for each* $u \in \mathcal{U}$, *the adjoint system has a unique weak solution* $z(u)$. *Further, there exists a constant* $K_8 \equiv K_8(n, \alpha_l, \alpha_u, K_5, \Omega, T, q, r)$ *such that*

$$\|z(u)\|_Q \le K_8\{\|g(\cdot,\phi(u)(\cdot,T))\|_{2,\Omega}\}, \tag{3.1.15}$$

where $\|\cdot\|_Q$ *is given in* (1.2.8), $\|\cdot\|_{2,\Omega}$ *is the norm in the Banach space* $L_2(\Omega)$, *and the constant* K_5 *is as defined for the inequalities* (3.1.11) *and* (3.1.12).

PROOF. Letting $t=T-t'$ and then setting $z(x,T-t)\equiv\hat{z}(x,t)$, system (3.1.10) can be written as

$$\begin{aligned}\hat{L}^*(u)\hat{z}(x,t)&=0, && (x,t)\in Q,\\ \hat{z}(x,t)&=0, && (x,t)\in\partial\Omega\times[0,T],\\ \hat{z}(x,0)&=g(x,\phi(u)(x,T)), && x\in\Omega,\end{aligned}\tag{3.1.16}$$

where, for each $u\in\mathcal{U}$,

$$\begin{aligned}\hat{L}^*(u)\psi(x,t)\equiv\psi_t(x,t)-\sum_{i=1}^{n}\Bigg(\sum_{j=1}^{n}a_{ij}(x,T-t)\psi_{x_j}(x,t)\\ +a_i(x,T-t,u(x,T-t))\psi(x,t)\Bigg)_{x_i}\\ -c(x,T-t,u(x,T-t))\psi(x,t).\end{aligned}\tag{3.1.17}$$

Obviously, for each $u\in\mathcal{U}$, if system (3.1.16) possesses a unique weak solution $\hat{z}(u)$, then $\hat{z}(u)(x,T-t)\equiv z(u)(x,t)$ is the unique weak solution of the system (3.1.10). Furthermore, if $\hat{z}(u)$ satisfies the estimate (3.1.15), then $z(u)$ satisfies also the same estimate. Thus, it suffices to prove that system (3.1.16) has a unique weak solution $\hat{z}(u)$ that satisfies the estimate (3.1.16). First, we note that the free terms of system (3.1.16) are equal to zero identically on Q. Further, we recall from Remark 3.1.2 that $g(\cdot,\phi(u)(\cdot,T))\in L_2(\Omega)$. Thus, by using these facts and Assumptions (2.1.A1), (3.1.A2)–(3.1.A3), and (3.1.A9), we can verify that all the hypotheses of Theorems 2.2.4 and 2.2.1 are satisfied. Consequently, the proof is complete. □

REMARK 3.1.3. Recall that $g(\cdot,\phi(u)(\cdot,T))$ is bounded on $\bar{\Omega}$ uniformly with respect to $u\in\mathcal{U}$ and that estimates (3.1.13) and (3.1.14) are satisfied uniformly with respect to $u\in\mathcal{U}$, for any pair of quantities q and r satisfying condition (2.2.7) or the condition stated in Assumption (2.2.A6). Thus, under Assumptions (2.1.A1), (3.1.A2)–(3.1.A4), (3.1.A7), and (3.1.A9), we obtain from Theorem 2.2.7 that $\hat{z}(u)$ is bounded on $\bar{Q}$ uniformly with respect to $u\in\mathcal{U}$. This, in turn, implies that $z(u)$ is bounded on $\bar{Q}$ uniformly with respect to $u\in\mathcal{U}$.

In our later analysis, we need to take integral averages [for definition, see (1.2.4)] of the coefficients and data of the adjoint system. Thus, for each $u\in\mathcal{U}$, we shall adopt the following convention:

$$L^*(u)\psi(x,t)\equiv-\psi_t(x,t)-\sum_{i=1}^{n}\psi_{x_ix_i}(x,t)\tag{3.1.18}$$

for all $(x,t)\in R^{n+1}\setminus Q$, and

$$g(x,\phi(u)(x,T))\equiv 0\tag{3.1.19}$$

for all $x\in R^n\setminus\Omega$.

For each $u \in \mathfrak{U}$ and for each integer $\sigma \geq 1$, let $\beta^\sigma(\cdot,\cdot,u(\cdot,\cdot))$ denote the integral average of $\beta(\cdot,\cdot,u(\cdot,\cdot))$, where β stands for any of the functions a_{ij}, $i=1,\ldots,n$, a_i, $i=1,\ldots,n$, and c (on R^{n+1}). Further, let $g^\sigma(\cdot,\phi(u)(\cdot,T))$ denote the integral average of $g(\cdot,\phi(u)(\cdot,T))$ (on R^n).

Let $\{\Omega^\sigma\}$ be a sequence of open domains with sufficiently smooth boundaries such that $\overline{\Omega^\sigma} \subset \Omega^{\sigma+1} \subset \overline{\Omega^{\sigma+1}} \subset \Omega$ for all integers $\sigma \geq 1$ and $\lim_{\sigma\to\infty} \Omega^\sigma = \Omega$. For positive integer σ, let d_σ be an element in $C^\infty(\Omega)$ with compact support in Ω^σ so that $d_\sigma(x)=1$ on $\Omega^{\sigma-1}$ and $0 \leq d_\sigma(x) \leq 1$ elsewhere.

We now consider the following sequence of first boundary value problems

$$
\begin{aligned}
L^{*\sigma}(u)z(x,t)&=0, \qquad (x,t)\in Q,\\
z(x,t)&=0, \qquad (x,t)\in \partial\Omega\times[0,T],\\
z(x,T)&=g^\sigma(x,\phi(u)(x,T))d_\sigma(x), \qquad x\in\Omega,
\end{aligned}
\tag{3.1.20}
$$

where $\phi(u)$ is the solution of system (3.1.1) corresponding to $u \in \mathfrak{U}$ and, for each $u \in \mathfrak{U}$ and for each positive integer σ, the operator $L^{*\sigma}(u)$ is defined by

$$
\begin{aligned}
L^{*\sigma}(u)\psi(x,t) \equiv -\psi_t(x,t) - \sum_{i=1}^{n}\Bigg(\sum_{j=1}^{n} a_{ij}^\sigma(x,t)\psi_{x_j}(x,t)&\\
+a_i^\sigma(x,t,u(x,t))\psi(x,t)\Bigg)_{x_i}&\\
-c^\sigma(x,t,u(x,t))\psi(x,t).&
\end{aligned}
\tag{3.1.21}
$$

Definition 3.1.3. For each $u \in \mathfrak{U}$ and for each integer $\sigma \geq 1$, a function $z^\sigma(u)\colon \overline{Q} \to R$ is said to be a *classical solution of system* (3.1.20) if it is continuous on $\overline{Q}$, has continuous derivatives $(z^\sigma(u))_t$, $(z^\sigma(u))_{x_i}$, $(z^\sigma(u))_{x_ix_j}$, $i,j=1,\ldots,n$, on Q, and satisfies all the equations everywhere in their domains of definition.

REMARK 3.1.4. For each $u \in \mathfrak{U}$ and for each integer $\sigma \geq 1$, it is clear that a classical solution of the system (3.1.20) is also a weak solution.

Note that the integral averages have derivatives of arbitrary order [see Theorem 1.2.1(i)]. Thus, for each $u \in \mathfrak{U}$ and for each integer $\sigma \geq 1$, the system (3.1.20) with t replaced by $T-t$ may be written in the form of system (3.1.1) in which the free term is identically zero. By virtue of the definition of d_σ, it is easily observed that the compatibility condition of order 1 is satisfied for the reduced system (see Definition 2.1.2). Thus, for the reduced system, all the required conditions of Theorem 2.1.2 are fulfilled, and therefore, it has a unique classical solution $\hat{z}^\sigma(u) \in \mathcal{H}^{2+\iota,(2+\iota)/2}(\overline{Q})$, $0<\iota<1$, where $\mathcal{H}^{2+\iota,(2+\iota)/2}(\overline{Q})$ is defined in Section

1.2.1. This, in turn, implies that $z^\sigma(u)(x,t) \equiv \hat{z}^\sigma(x, T-t)$ belongs to $\mathcal{H}^{2+\iota,(2+\iota)/2}(\bar{Q})$, $0<\iota<1$, and is the unique classical solution of system (3.1.20). From Remark 3.1.4 and Theorem 3.1.1, it follows that $z^\sigma(u)$ is also the (unique) weak solution. Since the integral averages do not increase norm [see Theorem 1.2.1(iii)], it is clear from the estimate (3.1.15) that

$$\|z^\sigma(u)\|_Q \leq K_8\{\|g(\cdot, \phi(u)(\cdot, T))\|_{2,\Omega}\}. \tag{3.1.22}$$

Theorem 3.1.2. *Consider systems* (3.1.10) *and* (3.1.20). *Suppose that Assumptions* (2.1.A1), (3.1.A2)–(3.1.A4), (3.1.A7), *and* (3.1.A9) *are satisfied. Then* $z^\sigma(u) \xrightarrow{s} z(u)$ *(strongly) in* $V_2^{1,0}(Q)$, *where, for each integer* $\sigma \geq 1$, $z^\sigma(u) \in \mathcal{H}^{2+\iota,(2+\iota)/2}(\bar{Q})$, $0<\iota<1$, *is the classical solution of system* (3.1.20), *and* $z(u)$ *is the weak solution of system* (3.1.10), *both corresponding to* $u \in \mathcal{U}$.

PROOF. Recall that, for each $u \in \mathcal{U}$ and for each integer $\sigma \geq 1$, the classical solution of the system (3.1.20) is also the weak solution. Thus, by replacing t with $T-t$ in systems (3.1.20) and (3.1.10), the proof of the theorem follows easily from Theorem 2.2.5. □

In the sequel, we need several lemmas, the first of which is the following.

Lemma 3.1.1. *Consider system* (3.1.1) *and the adjoint system* (3.1.10). *Suppose that Assumptions* (2.1.A1), (3.1.A2)–(3.1.A4), (3.1.A7), *and* (3.1.A9) *are satisfied. Then, for each pair of admissible controls* u^1 *and* u^2,

$$\langle L(u^1)\Phi, z(u^1)\rangle = \int_\Omega \{\Phi(x,T) g(x, \phi(u^1)(x,T))\}\, dx, \tag{3.1.23}$$

where $\Phi \equiv \phi(u^1) - \phi(u^2)$ *and*

$$\langle \xi, \eta\rangle \equiv \iint_Q \{\xi(x,t)\eta(x,t)\}\, dx\, dt. \tag{3.1.24}$$

PROOF. From the estimate (3.1.3), we observe that $L(u^1)\Phi$ belongs to $L_2(Q)$. Using this fact, including the fact that T is finite, it follows from Theorem 3.1.2 that

$$\lim_{\sigma\to\infty} \langle L(u^1)\Phi, z^\sigma(u^1)\rangle = \langle L(u^1)\Phi, z(u^1)\rangle. \tag{3.1.25}$$

By virtue of the definition of the operator $L(u^1)$, we obtain

$$\begin{aligned}
&\langle L(u^1)\Phi, z^\sigma(u^1)\rangle \\
&\quad = \iint_Q \{\Phi_t(x,t) z^\sigma(u^1)(x,t)\}\, dx\, dt \\
&\qquad - \sum_{i,j=1}^n \iint_Q \{a_{ij}(x,t)\Phi_{x_i x_j}(x,t) z^\sigma(u^1)(x,t)\}\, dx\, dt
\end{aligned}$$

$$-\sum_{i=1}^{n}\iint_Q\left\{b_i(x,t,u^1(x,t))\Phi_{x_i}(x,t)z^\sigma(u^1)(x,t)\right\}dx\,dt$$

$$-\iint_Q\left\{c(x,t,u^1(x,t))\Phi(x,t)z^\sigma(u^1)(x,t)\right\}dx\,dt. \qquad (3.1.26)$$

Performing integration by parts and noting that $\Phi(x,0)\equiv\phi(u^1)(x,0)-\phi(u^2)(x,0)=0$ on Ω and $\Phi(x,t)=0$, $z^\sigma(u^1)(x,t)=0$ on $\partial\Omega\times[0,T]$, it follows that

$$\langle L(u^1)\Phi, z^\sigma(u^1)\rangle=\int_\Omega\left\{\Phi(x,T)z^\sigma(u^1)(x,T)\right\}dx+\langle\Phi, L^*(u^1)z^\sigma\rangle. \qquad (3.1.27)$$

Write

$$\langle\Phi, L^*(u^1)z^\sigma(u^1)\rangle=\langle\Phi,\left(L^*(u^1)-L^{*\sigma}(u^1)\right)z^\sigma(u^1)\rangle$$
$$+\langle\Phi, L^{*\sigma}(u^1)z^\sigma(u^1)\rangle. \qquad (3.1.28)$$

Then, it is clear from the differential equation of the system (3.1.20) that the second term on the right-hand side of the above equality is zero. Thus, by performing integration by parts and using the facts that $\Phi(x,t)=0$, $z^\sigma(u^1)(x,t)=0$ on $\partial\Omega\times[0,T]$, it follows from (3.1.28) that

$$\langle\Phi, L^*(u^1)z^\sigma(u^1)\rangle=\sum_{i=1}^{n}\iint_Q\left\{\Phi_{x_i}(x,t)\left[\sum_{j=1}^{n}(a_{ij}(x,t)-a_{ij}^\sigma)\right.\right.$$
$$\times(x,t)\left(z^\sigma(u^1)\right)_{x_j}(x,t)+\left(a_i(x,t,u^1(x,t))\right.$$
$$\left.\left.\left.-a_i^\sigma(x,t,u^1(x,t))\right)z^\sigma(u^1)(x,t)\right]\right\}dx\,dt$$
$$-\iint_Q\left\{\Phi(x,t)\left(c(x,t,u^1(x,t))\right.\right.$$
$$\left.\left.-c^\sigma(x,t,u^1(x,t))\right)z^\sigma(u^1)(x,t)\right\}dx\,dt. \qquad (3.1.29)$$

From Theorem 1.2.1(iii), $\xi^\sigma\overset{s}{\to}\xi$ (strongly) in $L_2(Q)$ as $\sigma\to\infty$, where ξ^σ denotes the integral average of ξ and ξ denotes any of the functions $a_{ij}(\cdot,\cdot)$, $i,j=1,\ldots,n$, $a_i(\cdot,\cdot,u^1(\cdot,\cdot))$, $i=1,\ldots,n$, and $c(\cdot,\cdot,u^1(\cdot,\cdot))$. Thus by virtue of the definition of Φ, the estimates (3.1.4) and (3.1.22), and Hölder's inequality it follows that the relation (3.1.29), in the limit with respect to σ, reduces to

$$\lim_{\sigma\to\infty}\langle\Phi, L^*(u^1)z^\sigma(u^1)\rangle=0. \qquad (3.1.30)$$

Further, by using Theorem 3.1.2, we have

$$\lim_{\sigma\to\infty}\int_\Omega\{\Phi(x,T)z^\sigma(u^1)(x,T)\}\,dx=\int_\Omega\{\Phi(x,T)z(u^1)(x,T)\}\,dx$$
$$=\int_\Omega\{\Phi(x,T)g(x,\phi(u^1)(x,T))\}\,dx. \tag{3.1.31}$$

Thus, by virtue of the above two equalities, the relation (3.1.27), in the limit with respect to σ, reduces to

$$\lim_{\sigma\to\infty}\langle L(u^1)\Phi, z^\sigma(u^1)\rangle=\int_\Omega\{\Phi(x,T)g(x,\phi(u^1)(x,T))\}\,dx. \tag{3.1.32}$$

Combining (3.1.25) and the above equality, we obtain the proof of the lemma. □

Consider system (3.1.1). Let u^0 be an element in $\mathfrak{U}$, $(x',t')\in Q$ a regular point for all the coefficients and the free term of the system corresponding to the control u^0, $\{G_l\}$ a sequence of cubes in R^n with center x' so that $\lim_{l\to\infty}|G_l|=0$, and $\{I_l\}$ a sequence of intervals in (0,T) with midpoint t' so that $\lim_{l\to\infty}|I_l|=0$. Here $|E|$ denotes the Lebesgue measure of the set E. Define $\Omega_l\equiv G_l\cap\Omega$ and, for each $v\in V$,

$$u^l(x,t)\equiv(1-\chi_l(x,t))u^0(x,t)+\chi_l(x,t)v, \tag{3.1.33}$$

where χ_l is the characteristic function of $\Omega_l\times I_l\equiv Q_l$.

Let $\phi(u^0)$, $\phi(u^l)$ be the solutions of system (3.1.1) corresponding to the controls $u^0,u^l\in\mathfrak{U}$, respectively, where u^l is given by (3.1.33). Define

$$\Phi_l(x,t)\equiv\phi(u^0)(x,t)-\phi(u^l)(x,t). \tag{3.1.34}$$

Then

$$\begin{aligned}L(u^0)\Phi_l(x,t)&=L(u^0)\phi(u^0)(x,t)-L(u^l)\phi(u^l)(x,t)\\&\quad-(L(u^0)-L(u^l))\phi(u^l)(x,t)\\&=\sum_{i=1}^n\{(b_i(x,t,u^0(x,t))-b_i(x,t,u^l(x,t)))\\&\quad\times(\phi(u^l))_{x_i}(x,t)\}+\{(c(x,t,u^0(x,t))\\&\quad-c(x,t,u^l(x,t)))\phi(u^l)(x,t)\}\\&\quad+f(x,t,u^0(x,t))-f(x,t,u^l(x,t))\\&\equiv Y_l(x,t).\end{aligned} \tag{3.1.35}$$

Further,

$$\begin{aligned} \Phi_l(x,t) &= 0 \quad \text{for} \quad (x,t)\in\partial\Omega\times[0,T], \\ \Phi_l(x,0) &= 0 \quad \text{for} \quad x\in\Omega. \end{aligned} \tag{3.1.36}$$

We consider the following sequence of the first boundary value problems:

$$\begin{aligned} L(u^0)\Phi_l(x,t) &= Y_l(x,t), \qquad (x,t)\in Q, \\ \Phi_l(x,t) &= 0, \qquad (x,t)\in\partial\Omega\times[0,T], \\ \Phi_l(x,0) &= 0, \qquad x\in\Omega. \end{aligned} \tag{3.1.37}$$

Under Assumptions (2.1.A1) and (3.1.A1)–(3.1.A4), it follows from the definition of Φ_l given in (3.1.34) and Equations (3.1.35) and (3.1.36) that, for each integer $l\geq 1$, Φ_l is a solution of the system (3.1.37). On the other hand, it is clear that the coefficients of system (3.1.37) satisfy the corresponding Assumptions (3.1.A1)–(3.1.A3). Furthermore, $\phi(u^0)$ and $\phi(u^l)$ satisfy the estimate (3.1.4). Thus, we can find a constant K_9, independent of the integer $l\geq 1$, such that

$$|Y_l(x,t)|\leq K_9 \tag{3.1.38}$$

for almost all $(x,t)\in\bar{Q}$. Therefore, for each integer $l\geq 1$, it follows from Remark 3.1.1 that the system (3.1.37) admits a unique solution. This implies that, for each integer $l\geq 1$, Φ_l is the *only* solution of the system (3.1.37). Furthermore, it follows from Theorem 2.1.3 that there exists a constant K_{10}, independent of integers $l\geq 1$, such that

$$\|\Phi_l\|_{p,Q}^{(2,1)}\leq K_{10}\{\|Y_l\|_{p,Q}\}. \tag{3.1.39}$$

for any $p>3/2$.

Lemma 3.1.2. *Consider system* (3.1.37). *Suppose that Assumptions* (2.1.A1) *and* (3.1.A1)–(3.1.A4) *are satisfied. Then,*

$$\iint_Q\Big\{(\Phi_l(x,t))^2+((\Phi_l)_t(x,t))^2+\sum_{i=1}^n((\Phi_l)_{x_i}(x,t))^2 + \sum_{i,j=1}^n((\Phi_l)_{x_ix_j}(x,t))^2\Big\}\,dx\,dt\leq K_{11}|Q_l|, \tag{3.1.40}$$

where the constant K_{11} *is independent of* l.

PROOF. Using the definitions of Y_l and u^l, the inequalities (3.1.11) and (3.1.12), the estimate (3.1.4), and Assumption (3.1.A3), it follows from (3.1.39) with $p=2$ that

$$\|\Phi_l\|_{2,Q}^{(2,1)}\leq N_0(|Q_l|)^{1/2}, \tag{3.1.41}$$

where

$$N_0 \equiv 2(n+1)K_5K_2\left(N_1 + \|\phi_0\|^{(2)}_{\infty,\Omega}\right) + 2N_1$$

and

$$N_1 \equiv \max\left\{|f(x,t,v)| : (x,t,v) \in \bar{Q} \times V\right\}.$$

Define $K_{11} \equiv (n^2+n+2)(N_0)^2$. Then we can easily verify from the above inequality that

$$\iint_Q \left\{ (\Phi_l(x,t))^2 + ((\Phi_l)_t(x,t))^2 + \sum_{i=1}^{n} ((\Phi_l)_{x_i}(x,t))^2 + \sum_{i,j=1}^{n} ((\Phi_l)_{x_i x_j}(x,t))^2 \right\} dx\,dt \le K_{11}|Q_l|. \tag{3.1.42}$$

This completes the proof. □

Lemma 3.1.3. *Consider systems* (3.1.37) *and* (3.1.10). *Suppose that Assumptions* (2.1.A1), (3.1.A2)–(3.1.A4), (3.1.A7), *and* (3.1.A9) *are satisfied. Then, as* $l \to \infty$,

$$\alpha_l^i \equiv \frac{1}{|Q_l|} \iint_Q \left\{ \left(b_i(x,t,u^l(x,t)) - b_i(x,t,u^0(x,t))\right) \times (\Phi_l)_{x_i}(x,t) z(u^0)(x,t) \right\} dx\,dt \to 0, \quad i = 1,\ldots,n, \tag{3.1.43}$$

and

$$\beta_l \equiv \frac{1}{|Q_l|} \iint_Q \left\{ \left(c(x,t,u^l(x,t)) - c(x,t,u^0(x,t))\right) \times \Phi_l(x,t) z(u^0)(x,t) \right\} dx\,dt \to 0 \tag{3.1.44}$$

PROOF. The proofs of convergence of α_l^i and β_l are similar. Thus, a detailed proof of only the first part is given.

From Remark 3.1.3, the definition of the control u^l, and Assumption (3.1.A3), we obtain

$$\operatorname*{ess\,sup}_Q \left|\left(b_i(x,t,u^l(x,t)) - b_i(x,t,u^0(x,t))\right) z(u^0)(x,t)\right| = \operatorname*{ess\,sup}_{Q_l} \left|\left(b_i(x,t,v) - b_i(x,t,u^0(x,t))\right) z(u^0)(x,t)\right| \le N_0 \tag{3.1.45}$$

for all $i = 1,\ldots,n$, where the constant N_0 is independent of integers $l \ge 1$.

Therefore, by using the definition of α_l^i, it follows that

$$|\alpha_l^i| \leq \frac{1}{|Q_l|} \iint_{Q_l} \{|N_0(\Phi_l)_{x_i}(x,t)|\}\,dx\,dt \tag{3.1.46}$$

for all integers $l \geq 1$.

Let p and p' be any pair of real numbers such that (i) $p>2$ if $n=1$; (ii) $2<p<2n/(n-2)$ if $n\geq 2$; and (iii) $(1/p)+(1/p')=1$. Then, there exists a real number $\delta>0$ such that $1/p'=(1/2)+\delta$. Thus, it follows from Hölder's inequality that

$$\iint_{Q_l} \{|N_0(\Phi_l)_{x_i}(x,t)|\}\,dx\,dt$$

$$\leq N_0\left[\int_{I_l}\left\{\left[\int_{\Omega_l}\{|(\Phi_l)_{x_i}(x,t)|^p\}\,dx\right]^{1/p}(|\Omega_l|)^{1/p'}\right\}dt\right]. \tag{3.1.47}$$

Using Theorem 1.2.5, we can find a constant N_1, independent of integers $l\geq 1$, such that

$$\left(\int_{\Omega_l}\{|(\Phi_l)_{x_i}(x,t)|^p\}\,dx\right)^{1/p} \leq N_1\left[\left[\int_{\Omega}\{|(\Phi_l)_{x_i}(x,t)|^2\}\,dx\right.\right.$$
$$\left.\left.+\sum_{j=1}^{n}\int_{\Omega}\{|(\phi_l)_{x_ix_j}(x,t)|^2\}\,dx\right]^{1/2}\right]. \tag{3.1.48}$$

Substituting the above inequality into the inequality (3.1.47), we obtain

$$\iint_{Q_l}\{|N_0(\Phi_l)_{x_i}(x,t)|\}\,dx\,dt \leq N_2(|\Omega_l|)^{1/p'}\left[\int_{I_l}\left\{\left[\int_{\Omega}\{|(\Phi_l)_{x_i}(x,t)|^2\}\,dx\right.\right.\right.$$
$$\left.\left.\left.+\sum_{j=1}^{n}\int_{\Omega}\{|(\Phi_l)_{x_ix_j}(x,t)|^2\}\,dx\right]^{1/2}\right\}dt\right], \tag{3.1.49}$$

where $N_2 \equiv N_0 N_1$. Thus, it follows from Hölder's inequality that

$$\iint_{Q_l}\{|N_0(\Phi_l)_{x_i}(x,t)|\}\,dx\,dt$$

$$\leq N_2(|\Omega_l|)^{1/p'}(|I_l|)^{1/2}\times\left[\left[\iint_{Q}\{|(\Phi_l)_{x_i}(x,t)|^2\}\,dx\,dt\right.\right.$$
$$\left.\left.+\sum_{j=1}^{n}\iint_{Q}\{|(\Phi_l)_{x_ix_j}(x,t)|^2\}\,dx\,dt\right]^{1/2}\right], \tag{3.1.50}$$

and consequently, from (3.1.40), we have

$$\iint_{Q_l} \{|N_0(\Phi_l)_{x_i}(x,t)|\}\, dx\, dt \le N_3(|\Omega_l|)^{1/p'}(|I_l|)^{1/2}(|Q_l|)^{1/2}, \tag{3.1.51}$$

where $N_3 \equiv N_2(K_{11})^{1/2}$. Therefore, by using the fact that $1/p' = (1/2)+\delta$, we obtain

$$\iint_{Q_l} \{|N_0(\Phi_l)_{x_i}(x,t)|\}\, dx\, dt \le N_3(|Q_l|)(|\Omega_l|)^{\delta}. \tag{3.1.52}$$

From inequality (3.1.46) and the above inequality, it follows that

$$|\alpha_l^i| \le N_3(|\Omega_l|)^{\delta}. \tag{3.1.53}$$

This implies that $\alpha_l^i \to 0$ as $l \to \infty$ for $i = 1, \ldots, n$.

By similar arguments, we can show that $\beta_l \to 0$ as $l \to \infty$. Thus, the proof is complete. □

REMARK 3.1.5. From the properties of γ given in Assumptions (3.1.A7) and (3.1.A8), it is easily shown that $\gamma(\cdot, \phi(\cdot)) \in L_1(\Omega)$ for every $\phi \in L_2(\Omega)$. This, in turn, imples that $\gamma(\cdot, \phi(u)(\cdot, T)) \in L_1(\Omega)$ for every $u \in \mathfrak{U}$, where $\phi(u)$ is the solution of the system (3.1.1) corresponding to $u \in \mathfrak{U}$.

We are now in a position to derive a necessary condition for optimality for the terminal control problem (3.1.P2).

Theorem 3.1.3. *Consider Problem* (3.1.P2). *Suppose that all the hypotheses of Lemma* 3.1.3 *are satisfied and that* $U(x,t) = V$ *for all* $(x,t) \in \overline{Q}$. *Then, if* $u^* \in \mathfrak{U}$ *is an optimal control,*

$$\Bigg[\sum_{i=1}^{n} (b_i(x,t,v) - b_i(x,t,u^*(x,t)))(\phi(u^*))_{x_i}(x,t) + (c(x,t,v) - c(x,t,u^*(x,t)))\phi(u^*)(x,t) + (f(x,t,v) - f(x,t,u^*(x,t)))\Bigg] z(u^*)(x,t) \ge 0 \tag{3.1.54}$$

for almost all $(x,t) \in Q$ *and for all* $v \in V$.

PROOF. Let $v \in V$ be arbitrary and let $(x', t') \in Q$ be a regular point for all the functions involved on the left hand side of (3.1.54). Let u^l be as defined by Equation (3.1.33), with u^0 in this equation replaced by u^*, and let $\Phi_l \equiv \phi(u^*) - \phi(u^l)$. Then it is easily verified that, for almost all $(x,t) \in Q$,

$$L(u^*)\Phi_l(x,t) = L(u^*)\phi(u^*)(x,t) - L(u^l)\phi(u^l)(x,t)$$

$$
\begin{aligned}
&+\big(L(u^*)-L(u^l)\big)\Phi_l(x,t)+\big(L(u^l)-L(u^*)\big)\phi(u^*)(x,t)\\
&=\big(f(x,t,u^*(x,t))-f(x,t,u^l(x,t))\big)+\big(L(u^*)-L(u^l)\big)\Phi_l(x,t)\\
&\quad+\sum_{i=1}^{n}\big(b_i(x,t,u^*(x,t))-b_i(x,t,u^l(x,t))\big)\big(\phi(u^*)\big)_{x_i}(x,t)\\
&\quad+\big(c(x,t,u^*(x,t))-c(x,t,u^l(x,t))\big)\phi(u^*)(x,t). \qquad (3.1.55)
\end{aligned}
$$

Further, from Lemma 3.1.1, we have

$$\langle L(u^*)\Phi_l, z(u^*)\rangle=\int_\Omega\{\Phi_l(x,T)g(x,\phi(u^*)(x,T))\}\,dx. \qquad (3.1.56)$$

Since $u^*\in\mathcal{U}$ is an optimal control, it is clear that

$$\int_\Omega\{\gamma(x,\phi(u^*)(x,T))-\gamma(x,\phi(u^l)(x,T))\}\,dx\le 0 \qquad (3.1.57)$$

for all integers $l\ge 1$. Thus, it follows from the first inequality of Assumption (3.1.A7) that

$$
\begin{aligned}
&\int_\Omega\{(\phi(u^*)(x,T)-\phi(u^l)(x,T))g(x,\phi(u^*)(x,T))\}\,dx\\
&\qquad\le\int_\Omega\{\gamma(x,\phi(u^*)(x,T))-\gamma(x,\phi(u^l)(x,T))\}\,dx\le 0 \qquad (3.1.58)
\end{aligned}
$$

for all integers $l\ge 1$.

Combining the equality (3.1.56) and the above inequality, we obtain

$$\langle L(u^*)\Phi_l, z(u^*)\rangle\le 0 \qquad (3.1.59)$$

for all integers $l\ge 1$.

Substituting the expression (3.1.55) into the above inequality and dividing the obtained inequality by $|Q_l|$, we have

$$
\begin{aligned}
&\frac{1}{|Q_l|}\langle L(u^*)\Phi_l, z(u^*)\rangle\\
&\quad=\frac{1}{|Q_l|}\iint_Q\{f(x,t,u^*(x,t))-f(x,t,u^l(x,t))\}\\
&\qquad\times z(u^*)(x,t)\,dx\,dt+\sum_{i=1}^{n}\alpha_l^i+\beta_l\\
&\qquad+\frac{1}{|Q_l|}\iint_Q\Bigg\{\Bigg[\sum_{i=1}^{n}\big(b_i(x,t,u^*(x,t))-b_i(x,t,u^l(x,t))\big)\big(\phi(u^*)\big)_{x_i}(x,t)\\
&\qquad+\big(c(x,t,u^*(x,t))-c(x,t,u^l(x,t))\big)\phi(u^*)(x,t)\Bigg]z(u^*)(x,t)\Bigg\}\,dx\,dt\\
&\quad\le 0 \qquad (3.1.60)
\end{aligned}
$$

for all integers $l \geq 1$, where α_l^i, $(i=1,\ldots,n)$, β_l are as defined in (3.1.43) and (3.1.44) with u^0 replaced by u^*. Thus, it follows from Lemma 3.1.3 and Theorem 1.1.19, specialized to the finite-dimensional case, that the inequality (3.1.60), in the limit with respect to l, reduces to

$$\left[\sum_{i=1}^{n} (b_i(x',t',u^*(x',t')) - b_i(x',t',v))(\phi(u^*))_{x_i}(x',t') \right.$$
$$+ (c(x',t',u^*(x',t')) - c(x',t',v))\phi(u^*)(x',t')$$
$$\left. + (f(x',t',u^*(x',t')) - f(x',t',v))\right] z(u^*)(x',t') \leq 0, \qquad v \in V. \tag{3.1.61}$$

Since almost all $(x,t) \in Q$ are regular points, V is separable and the functions b_i, $(i=1,\ldots,n)$, c and f are continuous on V, the conclusion of the theorem follows. □

3.1.6. *Existence of Optimal Controls*

In the previous subsection, we presented a necessary condition for optimality. In the derivation of the necessary condition, it was assumed that an optimal control exists. In this subsection, we wish to present sufficient conditions ensuring the existence of optimal controls.

Here, we present several existence theorems. These results are essentially due to Fleming [Fl.2], Ahmed and Teo [AT.1], and Teo [TE.3] with certain modification. More general results will be found in Chapter 5.

As a consequence of Theorem 2.1.7, we have the following result.

Theorem 3.1.4. *Consider system* (3.1.1). *Suppose that Assumptions* (2.1.A1) *and* (3.1.A1)–(3.1.A4) *are satisfied. Let* $\{u^k\} \subset \mathcal{U}$, *and let* $\{\phi(u^k)\}$ *be the corresponding sequence of solutions. Then there exist a subsequence* $\{u^{k(l)}\}$ *of sequence* $\{u^k\}$ *and functions* b_i^*, $i=1,\ldots,n$, c^*, *and* f^* *satisfying the corresponding Assumption* (3.1.A3) *such that*

$$b_i(\cdot,\cdot,u^{k(l)}(\cdot,\cdot)) \xrightarrow{w^*} b_i^*(\cdot,\cdot), \qquad i=1,\ldots,n,$$
$$c(\cdot,\cdot,u^{k(l)}(\cdot,\cdot)) \xrightarrow{w^*} c^*(\cdot,\cdot),$$
$$f(\cdot,\cdot,u^{k(l)}(\cdot,\cdot)) \xrightarrow{w^*} f^*(\cdot,\cdot),$$

in (the weak topology of) $L_\infty(Q)$ as $l \to \infty$. Furthermore,*

$$\left.\begin{aligned} &\phi(u^{k(l)}) \overset{\mathrm{u}}{\to} \phi^* \\ &(\phi(u^{k(l)}))_{x_i} \overset{\mathrm{u}}{\to} \phi^*_{x_i}, \qquad i=1,\dots,n \end{aligned}\right\} \text{(uniformly) on } \bar{Q},$$

$$\left.\begin{aligned} &(\phi(u^{k(l)}))_t \overset{\mathrm{w}}{\to} \phi^*_t \\ &(\phi(u^{k(l)}))_{x_i x_j} \overset{\mathrm{w}}{\to} \phi^*_{x_i x_j}, \qquad i,j=1,\dots,n \end{aligned}\right\} \text{(weakly) in } L_p(Q),$$

as $l \to \infty$, where $p \in (3/2, \infty)$ and ϕ^* is the unique solution of the limit system

$$\begin{aligned} \phi_t(x,t) - \sum_{i,j=1}^{n} a_{ij}(x,t)\phi_{x_i x_j}(x,t) - \sum_{i=1}^{n} b_i^*(x,t)\phi_{x_i}(x,t) & \\ - c^*(x,t)\phi(x,t) = f^*(x,t), \qquad & (x,t) \in Q, \\ \phi(x,t) = 0, \qquad & (x,t) \in \partial\Omega \times [0,T], \\ \phi(x,0) = \phi_0(x), \qquad & x \in \Omega. \end{aligned} \tag{3.1.62}$$

For the proof of existence of optimal controls, it is often convenient to formulate the problem as a sort of contingent problem. Define a multifunction F on Q with values

$$F(x,t) \equiv \{\xi \in R^{n+2} : \xi = g(x,t,v), v \in U(x,t)\}, \qquad (x,t) \in Q, \tag{3.1.63}$$

where

$$g(x,t,v) = (b_1(x,t,v), \dots, b_n(x,t,v), c(x,t,v), f(x,t,v))$$

and U is a measurable multifunction with values $U(x,t) \subset R^m$.

As an immediate consequence of Theorem 1.4.5, we have the following result.

Lemma 3.1.4. *Suppose U is a measurable multifunction on Q with values from the class of nonempty closed convex subsets of the compact set V. Let b_i, $i=1,\dots,n$, c, and f satisfy assumption* (3.1.A3), *and let the multifunction F, as given by* (3.1.63), *be measurable. If y is a measurable function on Q with values $y(x,t) \in F(x,t)$ a.e. in Q, then there exists a measurable function u^0 on Q such that $u^0(x,t) \in U(x,t)$ and $y(x,t) = g(x,t,u^0(x,t))$ a.e. on Q.*

Lemma 3.1.5. *Let the hypotheses of Theorem* 3.1.4 *hold. Suppose that the multifunction F satisfies the assumptions of Lemma* 3.1.4 *and has convex* *values $F(x,t)$ for each $(x,t)\in\overline{Q}$. Then there exists a control $u^0\in\mathfrak{U}$ so that*

$$b_i^*(x,t)=b_i(x,t,u^0(x,t)), \qquad i=1,\dots,n,$$
$$c^*(x,t)=c(x,t,u^0(x,t)),$$
$$f^*(x,t)=f(x,t,u^0(x,t)),$$

a.e. *on* $\overline{Q}$.

PROOF. Let $\{u^{k(l)}\}$ be a subsequence of the sequence $\{u^k\}$ as given in Theorem 3.1.4. Define the function ξ^l with values

$$\xi^l(x,t)\equiv\big(b_1(x,t,u^l(x,t)),\dots,b_n(x,t,u^l(x,t)),c(x,t,u^l(x,t)),$$
$$f(x,t,u^l(x,t))\big), \qquad (x,t)\in Q,$$

and the set

$$\mathfrak{N}\equiv\{\omega:\omega \text{ measurable on } \overline{Q}, \omega(x,t)\in F(x,t), \text{ a.e. in } \overline{Q}\}.$$

By Theorem 3.1.4, the sequence $\xi^l \overset{w^*}{\to} \xi^*$ in $L_\infty(\overline{Q},R^{n+2})$. Since, by Proposition 1.4.1, $\mathfrak{N}$ is a w*-compact subset of $L_\infty(\overline{Q},R^{n+2})$, $\xi^*\in\mathfrak{N}$. Thus, it follows from Lemma 3.1.4 that there exists a control $u^0\in\mathfrak{U}$ such that $\xi^*(x,t)=g(x,t,u^0(x,t))$ a.e. on Q. This completes the proof. □

Combining Theorem 3.1.4 and Lemma 3.1.5, we have the following result.

Theorem 3.1.5. *Consider system* (3.1.1). *Suppose that the hypotheses of Lemma 3.1.5 are satisfied and that $\{u^{k(l)}\}$ is the sequence as defined in Theorem 3.1.4. Then there exists a control $u^0\in\mathfrak{U}$ such that*

$$\left.\begin{aligned}&\phi(u^{k(l)})\overset{u}{\to}\phi(u^0)\\&(\phi(u^{k(l)}))_{x_i}\overset{u}{\to}(\phi(u^0))_{x_i}, \qquad i=1,\dots,n\end{aligned}\right\} \text{ on } \overline{Q},$$

$$\left.\begin{aligned}&(\phi(u^{k(l)}))_t\overset{w}{\to}(\phi(u^0))_t\\&(\phi(u^{k(l)}))_{x_ix_j}\overset{w}{\to}(\phi(u^0))_{x_ix_j}, \qquad i,j=1,\dots,n,\end{aligned}\right\} \text{ in } L_p(Q), 3/2<p<\infty,$$

as $l\to\infty$, where $\phi(u^0)$ is the solution of system (3.1.1) *corresponding to the control $u^0\in\mathfrak{U}$.*

We are now in a position to present few results on the existence of optimal controls for Problem (3.1.P1), which includes Problem (3.1.P2) as a special case.

Theorem 3.1.6. *Suppose that Assumptions* (2.1.A1) *and* (3.1.A1)–(3.1.A6) *are satisfied. Let the multifunction U satisfy the corresponding hypotheses of Lemma* 3.1.4. *If the first- and zeroth-order coefficients and the free term of the system* (3.1.1) *are linear in the control variable, then Problem* (3.1.P1) *has a solution.*

PROOF. By virtue of Assumptions (3.1.A5)–(3.1.A6), the estimate (3.1.4), and the facts that Q and V are bounded, it can be verified that

$$\inf\{J(u): u\in\mathfrak{U}\}\equiv\alpha>-\infty. \tag{3.1.64}$$

Let $\{u^k\}$ be a sequence of controls in $\mathfrak{U}$ such that

$$\lim_{k\to\infty} J(u^k)=\alpha. \tag{3.1.65}$$

Since $U(x,t)$ is compact and convex for each $(x,t)\in\overline{Q}$, it follows from Proposition 1.4.1 that the class of admissible controls $\mathfrak{U}$ is a w*-compact subset of $L_\infty(Q, R^m)$. Thus, by taking a subsequence, if necessary, we may assume that $u^k\to u^*\in\mathfrak{U}$ in the weak* topology. Since the first and zeroth order coefficients and the free term of the system (3.1.1) are linear in the control variable, it follows from Assumption (3.1.A3) that

$$\begin{aligned}
&b_i(\cdot,\cdot,u^k(\cdot,\cdot))\overset{w^*}{\to}b_i(\cdot,\cdot,u^*(\cdot,\cdot)), \qquad i=1,\dots,n,\\
&c(\cdot,\cdot,u^k(\cdot,\cdot))\overset{w^*}{\to}c(\cdot,\cdot,u^*(\cdot,\cdot)),\\
&f(\cdot,\cdot,u^k(\cdot,\cdot))\overset{w^*}{\to}f(\cdot,\cdot,u^*(\cdot,\cdot))
\end{aligned} \tag{3.1.66}$$

in $L_\infty(Q)$ as $k\to\infty$. Thus, it follows from Theorem 3.1.4 that

$$\begin{aligned}
&\phi(u^k)\overset{u}{\to}\phi(u^*),\\
&(\phi(u^k))_{x_i}\overset{u}{\to}(\phi(u^*))_{x_i}, \qquad i=1,\dots,n,
\end{aligned} \tag{3.1.67}$$

on $\overline{Q}$ as $k\to\infty$, and hence

$$\begin{aligned}
&\phi(u^k)(\cdot,T)\overset{u}{\to}\phi(u^*)(\cdot,T),\\
&(\phi(u^k))_{x_i}(\cdot,T)\overset{u}{\to}(\phi(u^*))_{x_i}(\cdot,T), \qquad i=1,\dots,n,
\end{aligned} \tag{3.1.68}$$

on $\overline{\Omega}$ as $k\to\infty$. Since Ω is bounded, it follows from the above relation that, as $k\to\infty$,

$$\begin{aligned}
&\phi(u^k)(\cdot,T)\overset{s}{\to}\phi(u^*)(\cdot,T),\\
&(\phi(u^k))_{x_i}(\cdot,T)\overset{s}{\to}(\phi(u^*))_{x_i}(\cdot,T), \qquad i=1,\dots,n,
\end{aligned} \tag{3.1.69}$$

in (the norm topology of) $L_q(\Omega)$ for any $q\geq 1$.

Now let J_1 denote the first component of the cost functional $J\ (=J_1+J_2)$ given in (3.1.5) and consider the set B defined by

$$B \equiv \{\gamma_0(\cdot, \phi(u)(\cdot, T), (\phi(u))_x(\cdot, T)) : u \in \mathfrak{U}\}. \tag{3.1.70}$$

On the basis of the inequality appearing in Assumption (3.1.A5) and the estimate (3.1.4), we can find a constant N_0 such that

$$\int_\Omega \{|y(x)|\}\, dx \le N_0 \tag{3.1.71}$$

for all $y \in B$.

By virtue of the estimate (3.1.4), the inequality (3.1.71), the fact that Ω is bounded and the definition of the set B, we can easily show that γ_0 defines an operator mapping $L_q(\Omega, R^{n+1})$ into $L_1(\Omega)$, where $q \ge 1$. Thus, it follows from Theorem 1.2.13 that γ_0 defines a continuous operator from $L_q(\Omega, R^{n+1})$ into $L_1(\Omega)$. Therefore, by using this fact, including (3.1.69), we conclude that

$$\lim_{k\to\infty} \int_\Omega \{\gamma_0(x, \phi(u^k)(x, T), (\phi(u^k))_x(x, T)\}\, dx$$

$$= \int_\Omega \{\gamma_0(x, \phi(u^*)(x, T), (\phi(u^*))_x(x, T))\}\, dx. \tag{3.1.72}$$

This implies that

$$\lim_{k\to\infty} J_1(u^k) = J_1(u^*). \tag{3.1.73}$$

For the second component of the cost functional J, we have

$$J_2(u^k) = \iint_Q \{\gamma_1(x, t, \phi(u^k)(x, t), (\phi(u^k))_x(x, t), u^k(x, t))\}\, dx\, dt$$

$$= \iint_Q \{\gamma_1(x, t, \phi(u^k)(x, t), (\phi(u^k))_x(x, t), u^k(x, t))$$

$$-\gamma_1(x, t, \phi(u^*)(x, t), (\phi(u^*))_x(x, t), u^k(x, t))\}\, dx\, dt$$

$$+ \iint_Q \{\gamma_1(x, t, \phi(u^*)(x, t), (\phi(u^*))_x(x, t), u^k(x, t))$$

$$-\gamma_1(x, t, \phi(u^*)(x, t), (\phi(u^*))_x(x, t), u^*(x, t))\}\, dx\, dt$$

$$+ \iint_Q \{\gamma_1(x, t, \phi(u^*)(x, t), (\phi(u^*))_x(x, t), u^*(x, t))\}\, dx\, dt. \tag{3.1.74}$$

Since Q is bounded, it follows from the first inequality appearing in Assumption (3.1.A6) and the relation (3.1.69) that

$$\lim_{k\to\infty}\iint_Q\{\gamma_1(x,t,\phi(u^k)(x,t),(\phi(u^k))_x(x,t),u^k(x,t))$$
$$-\gamma_1(x,t,\phi(u^*)(x,t),(\phi(u^*))_x(x,t),u^k(x,t))\}\,dx\,dt=0. \tag{3.1.75}$$

Letting $v^1=u^k(x,t)$, $v^2=u^*(x,t)$, $\theta_0=\phi(u^*)(x,t)$, and $\theta_i=(\phi(u^*))_{x_i}(x,t)$, $i=1,\dots,n$, in the second inequality of Assumption (3.1.A6), it follows that

$$-\sum_{i=1}^m\iint_Q\{G_i(x,t)(u_i^k(x,t)-u_i^*(x,t))\}\,dx\,dt$$
$$\le\iint_Q\{\gamma_1(x,t,\phi(u^*)(x,t),(\phi(u^*))_x(x,t),u^k(x,t))$$
$$-\gamma_1(x,t,\phi(u^*)(x,t),(\phi(u^*))_x(x,t),u^*(x,t))\}\,dx\,dt$$
$$\le\sum_{i=1}^m\iint_Q\{G_i(x,t)(u_i^k(x,t)-u_i^*(x,t))\}\,dx\,dt. \tag{3.1.76}$$

Since $u^k\overset{w^*}{\to}u^*$ in $L_\infty(Q,R^m)$, we conclude from (3.1.76) that

$$\lim_{k\to\infty}\iint_Q\{\gamma_1(x,t,\phi(u^*)(x,t),(\phi(u^*))_x(x,t),u^k(x,t))$$
$$-\gamma_1(x,t,\phi(u^*)(x,t),(\phi(u^*))_x(x,t),u^*(x,t))\}\,dx\,dt=0. \tag{3.1.77}$$

Using the equality (3.1.75) and the above equality, it is clear from the expression (3.1.74) that

$$\lim_{k\to\infty}J_2(u^k)=\iint_Q\{\gamma_1(x,t,\phi(u^*)(x,t),(\phi(u^*))_x(x,t),u^*(x,t))\}\,dx\,dt$$
$$=J_2(u^*). \tag{3.1.78}$$

Combining (3.1.73) and (3.1.78), we have

$$\lim_{k\to\infty}J(u^k)=J(u^*). \tag{3.1.79}$$

Thus, the conclusion of the theorem follows from (3.1.64), (3.1.65), and (3.1.79). □

Note that the above theorem is proved under the assumption that the first- and zeroth-order coefficients and the free term of system (3.1.1) are linear in the control variable. In the next theorem, we remove the linearity assumption of the free term. However, the cost functional J is assumed to have the form

$$J(u)=\int_{\Omega}\{\tilde{\gamma}_0(x,\phi(u)(x,T))\}\,dx+\iint_{Q}\{\tilde{\gamma}_1(x,t,\phi(u)(x,t))\}\,dx\,dt. \tag{3.1.80}$$

For convenience, Problem (3.1.P1) with its cost functional J replaced by (3.1.80) will be referred to as Problem (3.1.P1)$'$.

For Problem (3.1.P1)$'$, we assume that the real-valued functions $\tilde{\gamma}_0$ and $\tilde{\gamma}_1$ satisfy the following conditions:

Assumption (3.1.A10)

i. $\tilde{\gamma}_0$ is a Carathéodory function defined on $\bar{\Omega}\times R$.
ii. There exists a nonnegative constant δ_4 and a nonnegative function $g_3\in L_1(Q)$ such that

$$|\tilde{\gamma}_0(x,\theta)|\leq g_3(x)+\delta_4|\theta| \text{ a.e. on } \Omega \text{ for all } \theta\in R.$$

iii. For each $x\in\Omega$, $\tilde{\gamma}_0(x,\cdot)$ is a monotonically increasing function defined on R.

Assumption (3.1.A11)

i. $\tilde{\gamma}_1$ is a Carathéodory function defined on $\bar{Q}\times R$.
ii. There exists a nonnegative constant δ_5 and a nonnegative function $g_4\in L_1(Q)$ such that

$$|\tilde{\gamma}_1(x,t,\theta)|\leq g_4(x,t)+\delta_5|\theta|$$

a.e. on Q for all $\theta\in R$.
iii. For each $(x,t)\in Q$, $\tilde{\gamma}_1(x,t,\cdot)$ is a monotonically increasing function defined on R.

Theorem 3.1.7. *Suppose that the following conditions hold*:

i. *Assumptions* (2.1.A1), (3.1.A1)–(3.1.A4), (3.1.A10), *and* (3.1.A11) *are satisfied.*
ii. *The multifunction U satisfies the corresponding hypotheses of Lemma* 3.1.4.
iii. *The first- and zeroth-order coefficients of system* (3.1.1) *are linear in the control variable.*
iv. *The function $v\to f(x,t,v)$ is convex for each $(x,t)\in\bar{Q}$.*

Then the problem (3.1.P1)$'$ *has a solution.*

PROOF. Since the sets Q and V are bounded, it follows from Assumptions (3.1.A10)–(3.1.A11) and the estimate (3.1.4) that

$$\inf\{J(u): u \in \mathcal{U}\} \equiv \alpha > -\infty. \tag{3.1.81}$$

Let $\{u^k\}$ be a sequence of controls in $\mathcal{U}$ such that

$$\lim_{k\to\infty} J(u^k) = \alpha. \tag{3.1.82}$$

Since, for each $(x,t) \in \bar{Q}$, $U(x,t)$ is a compact and convex subset of R^m, it follows from Proposition 1.4.1 that the class of admissible controls $\mathcal{U}$ is compact in the weak* topology of $L_\infty(Q, R^m)$. Thus, by taking a subsequence, if necessary, we may assume that $u^k \overset{w^*}{\to} u^* \in \mathcal{U}$ in $L_\infty(Q, R^m)$. From the hypothesis, we recall that the first- and zeroth-order coefficients of the system (3.1.1) are linear in the control variable. Thus, it follows from Assumption (3.1.A3) that

$$\begin{aligned} &b_i(\cdot,\cdot,u^k(\cdot,\cdot)) \overset{w^*}{\to} b_i(\cdot,\cdot,u^*(\cdot,\cdot)), \qquad i=1,\dots,n, \\ &c(\cdot,\cdot,u^k(\cdot,\cdot)) \overset{w^*}{\to} c(\cdot,\cdot,u^*(\cdot,\cdot)) \end{aligned} \tag{3.1.83}$$

in $L_\infty(Q)$.

Let h be any nonnegative continuous function on Q. Then by Theorem 1.2.14, we have

$$\begin{aligned} &\iint_Q \{h(x,t) f(x,t,u^*(x,t))\}\, dx\, dt \\ &\qquad \le \liminf_{k\to\infty} \iint_Q \{h(x,t) f(x,t,u^k(x,t))\}\, dx\, dt. \end{aligned} \tag{3.1.84}$$

From the condition imposed on F [see Assumption (3.1.A3)], we may, by taking a subsequence, if necessary, assume that $f(\cdot,\cdot,u^k(\cdot,\cdot)) \overset{w^*}{\to} f^*$ in $L_\infty(Q)$. Thus, the inequality (3.1.84) reduces to

$$\iint_Q \{h(x,t) f(x,t,u^*(x,t))\}\, dx\, dt \le \iint_Q \{h(x,t) f^*(x,t)\}\, dx\, dt. \tag{3.1.85}$$

Since this is true for each nonnegative continuous function h, it follows that

$$f(x,t,u^*(x,t)) \le f^*(x,t) \qquad \text{a.e. on } Q. \tag{3.1.86}$$

Now, let us consider the following first boundary value problems:

$$\begin{aligned} L(u^k)\phi(x,t) &= f(x,t,u^k(x,t)), && (x,t)\in Q,\\ \phi(x,t) &= 0, && (x,t)\in\partial\Omega\times[0,T],\\ \phi(x,0) &= \phi_0(x), && x\in\Omega, \\ k &= 1,2,\ldots, \end{aligned} \tag{3.1.87}$$

and

$$\begin{aligned} L(u^*)\phi(x,t) &= f^*(x,t), && (x,t)\in Q,\\ \phi(x,t) &= 0, && (x,t)\in\partial\Omega\times[0,T],\\ \phi(x,0) &= \phi_0(x), && x\in\Omega. \end{aligned} \tag{3.1.88}$$

By Theorem 2.1.3, system (3.1.88) has a unique solution ϕ^* (from $W_p^{2,1}(Q)$, $3/2<p<\infty$), and by Theorem 3.1.4, we conclude that

$$\begin{aligned} \phi(u^k) &\overset{u}{\to} \phi^*,\\ (\phi(u^k))_{x_i} &\overset{u}{\to} \phi^*_{x_i}, \qquad (i=1,\ldots,n), \end{aligned} \tag{3.1.89}$$

on $\bar{Q}$ as $k\to\infty$. Thus, by virtue of inequality (3.1.86), it can be shown from Theorem 2.1.6 that

$$\phi(u^*)(x,t)\le\phi^*(x,t) \tag{3.1.90}$$

for all $(x,t)\in\bar{Q}$.

In view of the definition of the cost functional J given in (3.1.80), we have

$$\begin{aligned} J(u^k) &= \int_\Omega\{\tilde{\gamma}_0(x,\phi(u^k)(x,T))\}\,dx + \iint_Q\{\tilde{\gamma}_1(x,t,\phi(u^k)(x,t))\}\,dx\,dt\\ &= \int_\Omega\{\tilde{\gamma}_0(x,\phi(u^k)(x,T)) - \tilde{\gamma}_0(x,\phi^*(x,T))\}\,dx\\ &\quad + \iint_Q\{\tilde{\gamma}_1(x,t,\phi(u^k)(x,t)) - \tilde{\gamma}_1(x,t,\phi^*(x,t))\}\,dx\,dt\\ &\quad + \int_\Omega\{\tilde{\gamma}_0(x,\phi^*(x,T))\}\,dx + \iint_Q\{\tilde{\gamma}_1(x,t,\phi^*(x,t))\}\,dx\,dt. \end{aligned} \tag{3.1.91}$$

Note that Q is bounded. Thus, by virtue of the inequalities appearing in Assumptions (3.1.A10)–(3.1.A11) and the relation (3.1.89), it follows from arguments similar to that given for the expression (3.1.72) that the equality

(3.1.91), in the limit with respect to k, reduces to

$$\lim_{k\to\infty} J(u^k)=\int_\Omega\{\tilde{\gamma}_0(x,\phi^*(x,T))\}\,dx+\iint_Q\{\tilde{\gamma}_1(x,t,\phi^*(x,t))\}\,dx\,dt. \tag{3.1.92}$$

Recall that $\tilde{\gamma}_0(x,\cdot)$ and $\tilde{\gamma}_1(x,t,\cdot)$ are monotonically increasing functions [see Assumptions (3.1.A10) and (3.1.A11)]. Thus, it is clear from the inequality (3.1.90) that

$$\begin{aligned}\int_\Omega\{\tilde{\gamma}_0(x,\phi^*(x,T))\}\,dx&+\iint_Q\{\tilde{\gamma}_1(x,t,\phi^*(x,t))\}\,dx\,dt\\ &\geq\int_\Omega\{\tilde{\gamma}_0(x,\phi(u^*)(x,T))\}\,dx+\iint_Q\{\tilde{\gamma}_1(x,t,\phi(u^*)(x,t))\}\,dx\,dt.\end{aligned} \tag{3.1.93}$$

Combining (3.1.92) and the above inequality, we have

$$\lim_{k\to\infty} J(u^k)\geq J(u^*). \tag{3.1.94}$$

Thus, the conclusion of the theorem follows from (3.1.81), (3.1.82), and (3.1.94). This completes the proof. □

In Theorems 3.1.6 and 3.1.7, the first- and zeroth-order coefficients of system (3.1.1) are assumed to be linear in the control variable. This assumption is dropped in Theorem 3.1.8 below at the expense of others. In addition, the form of the cost functional to be considered in Theorem 3.1.8 is less general than that of (3.1.5) but is more general than that of (3.1.80). The precise definition of the cost functional is given by

$$\begin{aligned}J(u)=&\int_\Omega\{\hat{\gamma}_0(x,\phi(u)(x,T),(\phi(u))_x(x,T))\}\,dx\\ &+\iint_Q\{\hat{\gamma}_1(x,t,\phi(u)(x,t),(\phi(u))_x(x,t))\}\,dx\,dt.\end{aligned} \tag{3.1.95}$$

Corresponding to the above cost functional, we impose the following conditions on the real-valued functions $\hat{\gamma}_0$ and $\hat{\gamma}_1$.

Assumption (3.1.A12). $\hat{\gamma}_0$ satisfies Assumption (3.1.A5).

Assumption (3.1.A13). $\hat{\gamma}_1$ is a Carathéodory function defined on $\bar{Q}\times R^{n+1}$; furthermore, there exists a nonnegative constant δ_6 and a nonnegative function $g_5\in L_1(Q)$ such that

$$|\hat{\gamma}_1(x,t,\theta_0,\theta)|\leq g_5(x,t)+\delta_6\sum_{j=0}^{n}|\theta_j|$$

for almost all $(x,t)\in Q$ and for all $\theta_0\in R$ and $\theta\in R^n$.

For convenience, Problem (3.1.P1), with its cost functional J replaced by (3.1.95), will be referred to as Problem (3.1.P1)$''$.

In the next theorem, we present an existence theorem for Problem (3.1.P1)$''$ without the linearity assumption (on b_i, $i=1,\ldots,n$, c, f).

Theorem 3.1.8. *Suppose that Assumptions* (2.1.A1), (3.1.A1)–(3.1.A4), (3.1.A12), *and* (3.1.A13) *are satisfied. Let the multifunctions U and F satisfy the corresponding hypotheses as given in Lemmas* 3.1.4 *and* 3.1.5 *respectively. Then Problem* (3.1.P1)$''$ *has a solution.*

PROOF. Since Q and V are bounded, it follows from Assumptions (3.1.A12) –(3.1.A13) and the estimate (3.1.4) that

$$\inf\{J(u): u\in\mathfrak{U}\}\equiv\alpha>-\infty. \tag{3.1.96}$$

Let $\{u^k\}$ be a sequence of controls in $\mathfrak{U}$ such that

$$\lim_{k\to\infty} J(u^k)=\alpha. \tag{3.1.97}$$

From Theorem 3.1.5, there exist a subsequence $(u^{k(l)}\}$ of the sequence $\{u^k\}$ and a control $u^*\in\mathfrak{U}$ such that

$$\begin{aligned} &\phi(u^{k(l)})\overset{\mathrm{u}}{\to}\phi^*,\\ &\phi(u^{k(l)})_{x_i}\overset{\mathrm{u}}{\to}\phi^*_{x_i}, \qquad i=1,\ldots,n, \end{aligned} \tag{3.1.98}$$

on $\overline{Q}$ as $l\to\infty$ and that $\phi^*=\phi(u^*)$. Note that u^* is not necessarily the w*-limit of $u^{k(l)}$.

Let the first component of the cost functional J [given in (3.1.95)] be denoted by J_1. Then, by using precisely the same arguments as that given for the relation (3.1.73), we deduce that

$$\lim_{l\to\infty} J_1(u^{k(l)})=J_1(u^*). \tag{3.1.99}$$

Let the second component of the cost functional J be denoted by J_2. Note that Q is bounded. Thus, by virtue of the inequality of Assumption (3.1.A13) and the relation (3.1.98), it follows from an argument similar to that given for the expression (3.1.73) that

$$\lim_{l\to\infty} J_2(u^{k(l)})=J_2(u^*). \tag{3.1.100}$$

Combining (3.1.99), and (3.1.100), (3.1.96), and (3.1.97), the proof is complete. □

3.2. First Boundary Value Problems in Divergence Form

3.2.1. Introduction

In this section, we consider a class of systems governed by parabolic partial differential equations in divergence form with first boundary conditions. Here the controls appear in all the coefficients and free terms in contrast to the case considered in the previous section, where the second-order coefficients were assumed to be independent of controls.

In Section 3.2.2, we describe the system and impose certain basic assumptions. Result concerning the existence and uniqueness of solutions of the system and certain important properties of the solutions are given in Section 3.2.3. In Sections 3.2.4 and 3.2.5, an optimal control problem is solved giving the necessary conditions for optimality. The question of existence of optimal controls is discussed in Section 3.2.6.

3.2.2. *System Description*

Let Ω be a bounded domain in R^n with boundary $\partial\Omega$, T a positive constant, $Q \equiv \Omega \times (0, T)$, and $\mathcal{U}$ the class of admissible controls (to be defined later). It is assumed throughout this section that the domain Ω has piecewise smooth boundary (see Definition 1.2.1).

We consider the system

$$\begin{aligned} L(u)\phi(x,t) &= \sum_{j=1}^{n} \left(F_j(x,t,u(x,t))\right)_{x_j} + f(x,t,u(x,t)), \qquad (x,t) \in Q, \\ \phi(x,t) &= \phi_0(x), \qquad x \in \Omega, \\ \phi(x,t) &= 0, \qquad (x,t) \in \partial\Omega \times [0,T], \end{aligned} \tag{3.2.1}$$

where $u \in \mathcal{U}$ and, for each $u \in \mathcal{U}$, the second-order parabolic partial differential operator $L(u)$ is given by

$$\begin{aligned} L(u)\psi(x,t) \equiv \psi_t(x,t) &- \sum_{j=1}^{n} \left(\sum_{i=1}^{n} a_{ij}(x,t,u(x,t))\psi_{x_i}(x,t) \right. \\ &\left. + a_j(x,t,u(x,t))\psi(x,t) \right)_{x_j} \\ &- \sum_{j=1}^{n} b_j(x,t,u(x,t))\psi_{x_j}(x,t) - c(x,t,u(x,t))\psi(x,t). \end{aligned} \tag{3.2.2}$$

Let V be a given nonempty compact subset of R^m and $U: \bar{Q} \to R^m$ a measurable multifunction (see Section 1.4.1) with values from the class of nonempty closed convex subsets of V. A measurable function $u: \bar{Q} \to V$ is called an *admissible control* if $u(x,t) \in U(x,t)$ a.e. on $\bar{Q}$. Let $\mathcal{U}$ denote *the class of admissible controls*.

We recall that the definitions of Banach spaces $L_p(\Omega)$, $1 \le p \le \infty$, $L_{p,r}(Q)$, $W_2^{1,0}(Q)$, $\overset{\circ}{W}{}_2^{1,0}(Q)$, $W_2^{1,1}(Q)$, $\overset{\circ}{W}{}_2^{1,1}(Q)$, $W_p^{2,1}(Q)$, $1 \le p \le \infty$, $V_2(Q)$, $\overset{\circ}{V}_2(Q)$, $V_2^{1,0}(Q)$, and $\overset{\circ}{V}{}_2^{1,0}(Q)$ are given in Section 1.2.3. Further, we introduce the following notation:

$$\tilde{\mathfrak{L}}(u)(\psi,\zeta)(t)$$

$$\equiv \int_\Omega \left\{ -\psi(x,t)\zeta_t(x,t) \right.$$

$$+ \sum_{j=1}^n \left(\sum_{i=1}^n a_{ij}(x,t,u(x,t))\psi_{x_i}(x,t) + a_j(x,t,u(x,t))\psi(x,t) \right) \zeta_{x_j}(x,t)$$

$$\left. - \sum_{j=1}^n b_j(x,t,u(x,t))\psi_{x_j}(x,t)\zeta(x,t) - c(x,t,u(x,t))\psi(x,t)\zeta(x,t) \right\} dx \tag{3.2.3}$$

for any pair of functions $\psi \in \mathring{W}_2^{1,0}(Q)$ and $\zeta \in \mathring{W}_2^{1,1}(Q)$;

$$\tilde{F}(u)(x,t) \equiv \sum_{j=1}^n \left(F_j(x,t,u(x,t)) \right)_{x_j} + f(x,t,u(x,t)); \tag{3.2.4}$$

and

$$(\tilde{F}(u),\zeta)(t) \equiv -\int_\Omega \left\{ \sum_{j=1}^n F_j(x,t,u(x,t))\zeta_{x_j}(x,t) \right.$$

$$\left. - f(x,t,u(x,t)\zeta(x,t) \right\} dx\,dt \tag{3.2.5}$$

for any $\zeta \in \mathring{W}_2^{1,0}(Q)$.

For system (3.2.1), we introduce the following definition.

Definition 3.2.1. For each $u \in \mathfrak{U}$, a function $\phi(u)$: $\bar{Q} \to R$ is said to be a *weak solution from* $V_2^{1,0}(Q)$ *of the system* (3.2.1) if

i. $\phi(u) \in \mathring{V}_2^{1,0}(Q)$ and
ii. for any $\tau \in [0,T]$,

$$\int_\Omega \{\phi(u)(x,\tau)\eta(x,\tau)\}\,dx + \int_0^\tau \{\tilde{\mathfrak{L}}(u)(\phi(u),\eta)(t) - (\tilde{F}(u),\eta)(t)\}\,dt$$

$$= \int_\Omega \{\phi_0(x)\eta(x,0)\}\,dx$$

for any $\eta \in \mathring{W}_2^{1,1}(Q)$.

Throughout this section, we need the following assumptions.

Assumption (3.2.A1). a_{ij}, $i,j=1,\ldots,n$, a_j, b_j, $j=1,\ldots,n$, c, F_j, $j=1,\ldots,n$, and f are real-valued Carathéodory functions defined on $\bar{Q}\times U$, where U is an open subset of R^m containing V.

Assumption (3.2.A2). There exist constants $\alpha_l, \alpha_u > 0$ such that

$$\alpha_l \sum_{i=1}^{n} (\xi_i)^2 \le \sum_{i,j=1}^{n} a_{ij}(x,t,v)\xi_i\xi_j \le \alpha_u \sum_{i=1}^{n} (\xi_i)^2$$

for almost all $(x,t)\in Q$ and for all $v\in V$.

Assumption (3.2.A3). There exists a constant K_1 such that $|\beta(x,t,v| \le K_1$ for all $(x,t,v)\in \bar{Q}\times V$, where β denotes any of the functions a_{ij}, $i,j=1,\ldots,n$, a_j, b_j, $j=1,\ldots,n$, and c.

Assumption (3.2.A4). $F_j(\cdot,\cdot,u(\cdot,\cdot))\in L_2(Q)$, $u\in\mathfrak{U}$, and $j\in\{1,\ldots,n\}$.

Assumption (3.2.A5). $f(\ ,\ ,u(\cdot,\cdot))\in L_{2,s}(Q)$, $u\in\mathfrak{U}$, where $s\in(1,\infty)$.

Assumption (3.2.A6). $\phi_0\in L_2(\Omega)$.

3.2.3. *Existence and Uniqueness of Solutions*

In this subsection, we wish to show the existence and uniqueness of solutions of system (3.2.1) and investigate certain important properties of these solutions.

REMARK 3.2.1. If the quantity q in condition (2.2.7) is taken as ∞, then the corresponding quantity r is equal to 1. However, since T is finite, it is clear from Assumption (3.2.A3) that

$$\left\| \sum_{j=1}^{n} \left(a_j(\cdot,\cdot,u(\cdot,\cdot))\right)^2 \right\|_{\infty,1,Q} \le K_2,$$

$$\left\| \sum_{j=1}^{n} \left(b_j(\cdot,\cdot,u(\cdot,\cdot))\right)^2 \right\|_{\infty,1,Q} \le K_2,$$

and

$$\|c(\cdot,\cdot,u(\cdot,\cdot))\|_{\infty,1,Q} \le K_2$$

for all $u\in\mathfrak{U}$, where $\|\cdot\|_{p,r,Q}$ is the norm in the Banach space $L_{p,r}(Q)$ and $K_2 \equiv \max\{K_1 T, n(K_1)^2 T\}$.

REMARK 3.2.2. If the quantity q_1 in condition (2.2.8) is taken as 2, then the corresponding quantity r_1 is equal to 1. However, since T is finite, it follows from Assumption (3.2.A5) and Hölder's inequality that, for any $u\in\mathfrak{U}$,

$$\|f(\cdot,\cdot,u(\cdot,\cdot))\|_{2,1,Q} \le (T)^{1/s'} \|f(\cdot,\cdot,u(\cdot,\cdot))\|_{2,s,Q},$$

where $(1/s)+(1/s')=1$ and $s\in(1,\infty)$. This, in turn, implies that $f(\cdot,\cdot,u(\cdot,\cdot))\in L_{2,1}(Q)$, $u\in\mathcal{U}$.

In view of the above two remarks, we observe that Assumptions (3.2.A1)–(3.2.A6) imply Assumptions (2.2.A1)–(2.2.A5). Thus, all results up to Remark 2.2.8 given in Section 2.2 remain valid here. Further, the estimate (2.2.83) can be reduced to

$$\|\phi(u)\|_Q \le K_3\left\{\|\phi_0\|_{2,\Omega}+\left(\iint_Q\left\{\sum_{j=1}^n (F_j(x,t,u(x,t)))^2\right\}dx\,dt\right)^{1/2}\right.$$
$$\left.+\|f(\cdot,\cdot,u(\cdot,\cdot))\|_{2,1,Q}\right\}$$
$$\le K_4\left\{\|\phi_0\|_{2,\Omega}+\left(\iint_Q\left\{\sum_{j=1}^n (F_j(x,t,u(x,t)))^2\right\}dx\,dt\right)^{1/2}\right.$$
$$\left.+\|f(\cdot,\cdot,u(\cdot,\cdot))\|_{2,s,Q}\right\},\tag{3.2.6}$$

where $\|\cdot\|_Q$ and $\|\cdot\|_{2,\Omega}$ denote, respectively, the norms in the Banach spaces $V_2^{1,0}(Q)$ and $L_2(\Omega)$ (see Section 1.2.3), $K_3\equiv K_3(n,\alpha_l,\alpha_u,K_2)$, $K_2=\max\{K_1T,n(K_1)^2T_2\}$, K_1 is as given in Assumption (3.2.A3), $K_4\equiv\max\{K_3,(T)^{1/s'}K_3\}$, $(1/s)+(1/s')=1$, and $s\in(1,\infty)$.

REMARK 3.2.3. Since, under Assumptions (3.2.A1)–(3.2.A6), all results up to Remark 2.2.8 presented in Section 2.2 are valid, it follows from Remark 2.2.3 that condition (ii) of Definition 3.2.1 is equivalent to condition (ii)′:

$$\int_0^T\{\tilde{\mathfrak{L}}(u)(\phi(u),\eta)(t)-(\tilde{F}(u),\eta)(t)\}\,dt=\int_\Omega\{\phi_0(x)\eta(x,0)\}\,dx$$

for any $\eta\in\mathring{W}_2^{1,1}(Q)$ that is equal to zero for $t=T$.

From Theorems 2.2.1 and 2.2.4 and the estimate (3.2.6), we have the following result.

Theorem 3.2.1. *Consider system* (3.2.1). *Suppose that Assumptions* (3.2.A1)–(3.2.A6) *are satisfied. Then, for each* $u\in\mathcal{U}$, *the system has a unique weak solution* $\phi(u)$ *from* $V_2^{1,0}(Q)$. *Further,*

$$\|\phi(u)\|_Q \le K_4\left\{\|\phi_0\|_{2,\Omega}+\left(\iint_0\left\{\sum_{j=1}^n (F_j(x,t,u(x,t)))^2\right\}dx\,dt\right)^{1/2}\right.$$
$$\left.+\|f(\cdot,\cdot,u(\cdot,\cdot))\|_{2,s,Q}\right\},\tag{3.2.7}$$

where $s\in(1,\infty)$ *and the constant* K_4 *is as given for the estimate* (3.2.6).

3.2.4. Formulation of Some Control Problems

In this subsection, we shall formulate some control problems subject to system (3.2.1). For this, several cost functionals will be introduced, the first of which is the following:

$$J(u)=\int_{\Omega}\{\gamma(x,\phi(u)(x,T))\}\,dx, \tag{3.2.8}$$

where $\phi(u)$ is the weak solution [from $V_2^{1,0}(Q)$] of the system (3.2.1) corresponding to the control $u\in\mathcal{U}$. In what follows, we shall impose certain conditions on the real-valued function γ.

Assumption (3.2.A7)

i. γ is a Carathéodory function defined on $\Omega\times R$.
ii. There exists a Carathéodory function $g\colon\Omega\times R\to R$ such that

$$\gamma(x,\theta^1)\le\gamma(x,\theta^2)+g(x,\theta^2)(\theta^1-\theta^2)$$

a.e. on Ω, for each θ^1, $\theta^2\in R$;
iii. There exists a constant K_5 and an element $g_1\in L_2(\Omega)$ such that

$$|g(x,\theta)|\le g_1(x)+K_5|\theta|\quad\text{a.e. on}\quad\Omega,$$

for each $\theta\in R$.

Assumption (3.2.A8). $\gamma(\cdot,\bar{\theta}(\cdot))\in L_1(\Omega)$ for *some* $\bar{\theta}\in L_2(\Omega)$. We now consider the following optimal control problem.

Problem (3.2.P1). Subject to the system (3.2.1), find a control $u\in\mathcal{U}$ such that the cost functional (3.2.8) is minimized.

Note that the cost functional (3.2.8) is of a very special form. It depends only on the (weak) solutions of the system (3.2.1) at the terminal time.

The next (optimal control) problem is more general than Problem (3.2.P1).

Problem (3.2.P2). Subject to the system (3.2.1), find a control $u\in\mathcal{U}$ such that the cost functional

$$J(u)=\int_{\Omega}\{\gamma_0(x,\phi(u)(x,T))\}\,dx+\iint_{Q}\{\gamma_1(x,t,\phi(u)(x,t))\}\,dx\,dt \tag{3.2.9}$$

is minimized.

We now impose certain conditions on the real-valued functions γ_0 and γ_1.

Assumption (3.2.A9). γ_0 satisfies Assumptions (3.2.A7)–(3.2.A8).

Assumption (3.2.A10)

i. γ_1 is a continuous function defined on $Q\times R$.
ii. For each $(x,t)\in Q, \gamma_1(x,t,\cdot)$ is convex on R.
iii. $\gamma_1(\cdot,\cdot,0)\in L_1(Q)$.
iv. There exists a constant K_6, an element g_2 in $L_2(Q)$, and a real-valued Carathéodory function g_3 defined on $Q\times R$ such that

$$|g_3(x,t,\theta)|\leq|g_2(x,t)|+K_6|\theta|$$

for all $(x,t,\theta)\in Q\times R$ and

$$g_3(x,t,\theta^2)(\theta^2-\theta^1)\leq\gamma_1(x,t,\theta^1)-\gamma_1(x,t,\theta^2)$$
$$\leq g_3(x,t,\theta^2)(\theta^1-\theta^2)$$

for all θ^1, $\theta^2\in R$ and for all $(x,t)\in Q$.

In what follows, we consider an optimal control problem in which the cost functional depends also on the control variable.

Problem (3.2.P3). Subject to the system (3.2.1), find a control $u\in\mathcal{U}$ such that the cost functional

$$J(u)=\iint_Q\{\gamma(x,t,\phi(u)(x,t),u(x,t))\}\,dx\,dt \tag{3.2.10}$$

is minimized.

We assume that the real-valued function γ satisfies the following conditions:

Assumption (3.2.A11)

i. γ is a Carathéodory function defined on $Q\times\{R\times V\}$.
ii. For each $(x,t,v)\in Q\times V$, $\gamma(x,t,\cdot,v)$ is convex on R.
iii. $\gamma(\cdot,\cdot,0,0)\in L_2(Q)$.
iv. There exists an element g_4 in $L_2(Q)$ such that

$$-g_4(x,t)(\theta^1-\theta^2)\leq\gamma(x,t,\theta^1,v)-\gamma(x,t,\theta^2,v)\leq g_4(x,t)(\theta^1-\theta^2)$$

a.e. on Q, for all θ^1, $\theta^2\in R$ and for all $v\in V$.
v. There exists an element G in $L_1(Q,R^m)\equiv\{\xi:Q\to R^m,\ \xi$ measurable and $(\sum_{i=1}^m\|\xi_i\|_{1,Q}^2)^{1/2}<\infty\}$ such that

$$-\sum_{i=1}^m G_i(x,t)(v_i^1-v_i^2)\leq\gamma(x,t,\theta,v^1)-\gamma(x,t,\theta,v^2)$$
$$\leq\sum_{i=1}^m G_i(x,t)(v_i^1-v_i^2)$$

a.e. on Q, for all $\theta\in R$ and for all v^1, $v^2\in V$.

3.2.5. *Necessary Conditions for Optimality*

The aim of this subsection is to derive necessary conditions for optimality for the terminal control problem (3.2.P1).

With this end in view, we first show that, for each $u \in \mathfrak{U}$, the adjoint system, to be introduced later, has a (weak) unique solution. Then, assuming the existence of an optimal control for the problem, we derive the necessary conditions. The question concerning the existence of optimal controls will be considered in the later subsection.

In view of Theorem 3.2.1, we observe that $\phi(u)$ is the weak solution [from $V_2^{1,0}(Q)$] of system (3.2.1) corresponding to the control $u \in \mathfrak{U}$ and, further, that it satisfies the estimate (3.2.7). Thus, for each $u \in \mathfrak{U}$, $\phi(u)(\cdot, T)$ is a known function and belongs to $L_2(\Omega)$. On this basis, it follows from the second inequality of Assumption (3.2.A7) that

$$g(\cdot, \phi(u)(\cdot, T)) \in L_2(\Omega), \qquad u \in \mathfrak{U}. \tag{3.2.11}$$

Consider the following system, which is called the *adjoint system* of the system (3.2.1):

$$\begin{aligned} L^*(u)z(x,t) &= 0, \qquad (x,t) \in Q, \\ z(x,T) &= g(x, \phi(u)(x,T)), \qquad x \in \Omega, \\ z(x,t) &= 0, \qquad (x,t) \in \partial\Omega \times [0,T], \end{aligned} \tag{3.2.12}$$

where, for each $u \in \mathfrak{U}$, the operator $L^*(u)$ is given by

$$\begin{aligned} L^*(u)\psi(x,t) \equiv -\psi_t(x,t) - \sum_{i=1}^{n} \Bigg(&\sum_{j=1}^{n} a_{ij}(x,t,u(x,t))\psi_{x_j}(x,t) \\ &- b_i(x,t,u(x,t))\psi(x,t) \Bigg)_{x_i} \\ &+ \sum_{i=1}^{n} a_i(x,t,u(x,t))\psi_{x_i}(x,t) - c(x,t,u(x,t))\psi(x,t). \end{aligned} \tag{3.2.13}$$

This operator is the formal adjoint of the operator $L(u)$.

We are required to show that, for each $u \in \mathfrak{U}$, the adjoint system (3.2.12) has a unique (weak) solution in the sense of the following definition.

Definition 3.2.2. For each $u \in \mathfrak{U}$, a function $z(u): \bar{Q} \to R$ is said to be a *weak solution from* $V_2^{1,0}(Q)$ *of system* (3.2.12) if

i. $z(u) \in \overset{\circ}{V}{}_2^{1,0}(Q)$ and
ii. for all $\tau \in [0, T]$,

$$\int_\Omega \{z(u)(x,\tau)\eta(x,\tau)\}\, dx + \int_\tau^T \{\tilde{\mathfrak{L}}^*(z(u),\eta)(t)\}\, dt$$
$$= \int_\Omega \{g(x, \phi(u)(x,T))\eta(x,T)\}\, dx$$

for any $\eta \in \overset{\circ}{W}{}_2^{1,1}(Q)$, where

$$\begin{aligned}\tilde{\mathfrak{L}}^*(\psi,\zeta)(t) = \int_\Omega \Bigg\{ &\psi(x,t)\zeta_t(x,t) + \sum_{i=1}^n \left(\sum_{j=1}^n a_{ij}(x,t,u(x,t))\psi_{x_j}(x,t) \right.\\ &\left. - b_i(x,t,u(x,t))\psi(x,t) \right) \zeta_{x_i}(x,t)\\ &+ \sum_{i=1}^n a_i(x,t,u(x,t))\psi_{x_i}(x,t)\zeta(x,t)\\ &- c(x,t,u(x,t))\psi(x,t)\zeta(x,t) \Bigg\}\, dx\end{aligned}$$

for any $\psi \in \overset{\circ}{W}{}_2^{1,0}(Q)$ and $\zeta \in \overset{\circ}{W}{}_2^{1,1}(Q)$.

The next theorem contains a result concerning the existence of a unique (weak) solution of system (3.2.12).

Theorem 3.2.2. *Consider the adjoint system* (3.2.12). *Suppose that Assumptions* (3.2.*A*1)–(3.2.*A*7) *are satisfied. Then, for each* $u \in \mathfrak{U}$, *the adjoint system admits a unique weak solution* $z(u)$ *from* $V_2^{1,0}(Q)$. *Further, there exists a constant* $K_7 \equiv K_7(n, \alpha_l, \alpha_u, T, s, K_1)$ *such that*

$$\|z(u)\|_Q \le K_7 \|g(\cdot, \phi(u)(\cdot, T))\|_{2,\Omega}, \tag{3.2.14}$$

where $\phi(u)$ *is the weak solution from* $V_2^{1,0}(Q)$ *of the system* (3.2.1).

PROOF. The proof is based on Theorem 2.2.1. Letting $t = T - t'$ and setting $z(x, T-t) \equiv \hat{z}(x,t)$, the adjoint system (3.2.12) can be written as

$$\begin{aligned}\hat{L}^*(u)\hat{z}(x,t) &= 0, \qquad (x,t) \in Q,\\ \hat{z}(x,0) &= g(x,\phi(u)(x,T)), \qquad x \in \Omega,\\ \hat{z}(x,t) &= 0, \qquad (x,t) \in \partial\Omega \times [0,T],\end{aligned} \tag{3.2.15}$$

where, for each $u \in \mathfrak{U}$, the operator $\hat{L}^*(u)$ is given by

$$\begin{aligned}\hat{L}^*(u)\psi(x,t) &\equiv \psi_t(x,t) - \sum_{i=1}^n \left(\sum_{j=1}^n a_{ij}(x,T-t,u(x,T-t))\psi_{x_j}(x,t) \right.\\ &\left. - b_i(x,T-t,u(x,T-t))\psi(x,t) \right)_{x_i}\\ &+ \sum_{i=1}^n a_i(x,T-t,u(x,T-t))\psi_{x_i}(x,t) - c(x,T-t,u(x,T-t))\psi(x,t).\end{aligned} \tag{3.2.16}$$

By virtue of Assumptions (3.2.A1)–(3.2.A6) and the relation (3.2.11), we can easily verify that all the hypotheses of Theorem 2.2.1 are satisfied. Thus, it follows that, for each $u \in \mathcal{U}$, the system (3.2.15) has a unique weak solution $\hat{z}(u)$ [from $V_2^{1,0}(Q)$]; furthermore, $\hat{z}(u)$ satisfies the estimate (3.2.14). This, in turn, implies that $z(u)(x,t) \equiv \hat{z}(u)(x,T-t)$ is the unique weak solution [from $V_2^{1,0}(Q)$] of the adjoint system (3.2.12) and satisfies the estimate (3.2.14). Thus, the proof is complete. □

In our later analysis, we need to take integral averages [for definition, see (1.2.4)] of the coefficients and data of the system (3.2.1) and the adjoint system (3.2.12). Thus, for each $u \in \mathcal{U}$, we shall adopt the following convention:

$$L(u)\psi(x,t) \equiv \psi_t(x,t) - \sum_{i=1}^{n} \psi_{x_i x_i}(x,t) \quad \text{for all} \quad (x,t) \in R^{n+1} \backslash Q, \tag{3.2.17}$$

$$F_j(x,t,u(x,t)) \equiv 0, \qquad j=1,\dots,n, \quad \text{for all} \quad (x,t) \in R^{n+1} \backslash Q, \tag{3.2.18}$$

$$f(x,t,u(x,t)) \equiv 0, \quad \text{for all} \quad (x,t) \in R^{n+1} \backslash Q, \tag{3.2.19}$$

and

$$\phi_0(x) = 0 \quad \text{for all} \quad x \in R^n \backslash \Omega. \tag{3.2.20}$$

For each $u \in \mathcal{U}$ and for each integer $\sigma \geq 1$, let $\beta^\sigma(\cdot,\cdot,u(\cdot,\cdot))$ denote the integral average of $\beta(\cdot,\cdot,u(\cdot,\cdot))$, where β stands for any of the functions a_{ij}, $i,j=1,\dots,n$, a_j, b_j, $j=1,\dots,n$, c, F_j, $j=1,\dots,n$, and f (on R^{n+1}). Further, for each integer $\sigma \geq 1$, let ϕ_0^σ denote the integral average of ϕ_0 (on R^n).

Let $\{\Omega^\sigma\}$ be a sequence of open domains with sufficiently smooth boundaries such that $\overline{\Omega^\sigma} \subset \Omega^{\sigma+1} \subset \overline{\Omega^{\sigma+1}} \subset \Omega$ for all integers $\sigma \geq 1$ and $\lim_{\sigma\to\infty} \Omega^\sigma = \Omega$. For each positive integer σ, let d_σ be an element in $C_0^\infty(\Omega)$ such that $d_\sigma(x)=1$ on Ω^σ and $0 \leq d_\sigma(x) \leq 1$ elsewhere.

We now consider the following sequence of first boundary value problems:

$$\begin{aligned} L^\sigma(u)\phi(x,t) &= \sum_{j=1}^{n} \left(F_j^\sigma(x,t,u(x,t))\right)_{x_j} + f^\sigma(x,t,u(x,t)), \qquad (x,t) \in Q, \\ \phi(x,0) &= d_\sigma(x)\phi_0^\sigma(x), \qquad x \in \Omega, \\ \phi(x,t) &= 0, \qquad (x,t) \in \partial\Omega \times [0,T], \end{aligned} \tag{3.2.21}$$

where, for each $u \in \mathcal{U}$ and for each integer $\sigma \geq 1$, the operator $L^\sigma(u)$ is as defined by the operator $L(u)$ with the coefficients in the operator $L(u)$ replaced by the corresponding integral averages.

Definition 3.2.3. For each $u \in \mathcal{U}$ and for each positive integer σ, a function $\phi^\sigma(u): \bar{Q} \to R$ is said to be a *solution of system* (3.2.21) if $\phi^\sigma(u) \in W_p^{2,1}(Q)$, $3/2 < p < \infty$, and it satisfies the differential equation a.e. on Q and the boundary conditions everywhere in their domains of definition.

REMARK 3.2.4. Clearly, for each $u \in \mathcal{U}$ and for each integer $\sigma \geq 1$, a solution of the system (3.2.21) is also a weak solution from $V_2^{1,0}(Q)$.

To present the next result, we recall that the complete expressions of the norms $\|\cdot\|_{p,Q}^{(2,1)}$, $\|\cdot\|_{p,\Omega}^{(2)}$ and $|\cdot|_Q^{(1+\mu,(1+\mu)/2)}$ are given in (1.2.6), (1.2.5), and (1.2.3), respectively.

Theorem 3.2.3. *Consider system* (3.2.21). *Suppose that Assumptions* (2.1.A1) *and* (3.2.A1)–(3.2.A6) *are satisfied. Then, for each $u \in \mathcal{U}$ and for each integer $\sigma \geq 1$, the system admits a unique solution $\phi^\sigma(u)$. Moreover,*

$$\|\phi^\sigma(u)\|_{p,Q}^{(2,1)} \leq K_8 \tag{3.2.22}$$

for all $p \geq 3/2$, where, for given Q and the second-order coefficients, the constant K_8 depends only on n, $\|d_\sigma\phi^\sigma\|_{\infty,\Omega}^{(2)}$, and the bounds for the functions $a_j^\sigma(\cdot,\cdot,u(\cdot,\cdot))$, $(a_j^\sigma(\cdot,\cdot,u(\cdot,\cdot)))_{x_j}$, $b_j^\sigma(\cdot,\cdot,u(\cdot,\cdot))$, $j=1,\ldots,n$, $c^\sigma(\cdot,\cdot,u(\cdot,\cdot))$, $(F_j^\sigma(\cdot,\cdot,u(\cdot,\cdot)))_{x_j}$, $j=1,\ldots,n$, and $f^\sigma(\cdot,\cdot,u(\cdot,\cdot))$ (on $\bar{Q}$). In addition, $\phi^\sigma(u)$ satisfies also the estimate

$$|\phi^\sigma(u)|_Q^{(1+\mu,(1+\mu)/2)} \leq K_9 \tag{3.2.23}$$

for all $\mu \in (0,1)$, where the constant K_9 depends only on n, Q, μ, and K_8.

PROOF. Using the hypotheses of the theorem, the properties of the integral averages (see Theorem 1.2.1), and the definition of d_σ, we can easily verify that, for each $u \in \mathcal{U}$ and for each integer $\sigma \geq 1$, Assumptions (2.1.A1)–(2.1.A5) are satisfied. Thus, the conclusion of the theorem follows from Remark 2.1.1. □

Since, for each $u \in \mathcal{U}$ and for each integer $\sigma \geq 1$, the coefficients and data of the system (3.2.21) satisfy the corresponding Assumptions (3.2.A1)–(3.2.A6), it follows from Theorem 3.2.1 that the system has a unique weak solution $\phi^\sigma(u)$ from $V_2^{1,0}(Q)$; furthermore, $\phi^\sigma(u)$ satisfies the corresponding version of estimate (3.2.7). Thus, by virtue of this fact, Theorem 3.2.3, and Remark 3.2.4, we conclude that, for each $u \in \mathcal{U}$ and for each integer $\sigma \geq 1$, the system (3.2.21) has a unique solution $\phi^\sigma(u)$ that satisfies the corresponding version of the estimate (3.2.7), and is also the unique weak solution [from $V_2^{1,0}(Q)$].

We now consider the following sequence of first boundary value problems:

$$\begin{aligned} L^{*\sigma}(u)z(x,t)&=0, & (x,t)&\in Q,\\ z(x,T)&=d_\sigma(x)g^\sigma(x,\phi(u)(x,T)) & x&\in\Omega, \qquad (3.2.24)\\ z(x,t)&=0, & (x,t)&\in\partial\Omega\times[0,T], \end{aligned}$$

where, for each $u\in\mathfrak{U}$ and for each integer $\sigma\geq 1$, $g^\sigma(\cdot,\phi(u)(\cdot,T))$ denotes the integral average of $g(\cdot,\phi(u)(\cdot,T))$ and the operator $L^{*\sigma}(u)$ is as defined by the operator $L^*(u)$ with the coefficients in the operator $L^*(u)$ replaced by the corresponding integral averages.

For the above system, we introduce the following definition.

Definition 3.2.4. For each $u\in\mathfrak{U}$ and for each integer $\sigma\geq 1$, a function $z^\sigma(u):\overline{Q}\to R$ is said to be a *solution of system* (3.2.24) if $z^\sigma(u)\in W_p^{2,1}(Q)$, $3/2<p<\infty$, and it satisfies the differential equation a.e. on Q and the boundary conditions everywhere in their domains of definition.

Theorem 3.2.4. *Consider system (3.2.24). Suppose that Assumptions (2.1.A1) and (3.2.A1)–(3.2.A7) are satisfied. Then, for each $u\in\mathfrak{U}$ and for each integer $\sigma\geq 1$, the system has a unique solution $z^\sigma(u)$. Moreover,*

$$\|z^\sigma(u)\|_{p,Q}^{(2,1)}\leq K_{10} \qquad (3.2.25)$$

for any $p\geq 3/2$, where, for given Q and the second-order coefficients, the constant K_{10} depends only on n, $\|d_\sigma(\cdot)g^\sigma(\cdot,\phi(u)(\cdot,T))\|_{\infty,\Omega}^{(2)}$, and the bounds for the functions $a_i^\sigma(\cdot,\cdot,u(\cdot,\cdot))$, $b_i^\sigma(\cdot,\cdot,u(\cdot,\cdot))$, $(b_i^\sigma(\cdot,\cdot,u(\cdot,\cdot)))_{x_i}$, $i=1,\ldots,n$, and $c^\sigma(\cdot,\cdot,u(\cdot,\cdot))$ (on $\overline{Q}$). In addition, $z^\sigma(u)$ satisfies also the estimate

$$|z^\sigma(u)|_{\overline{Q}}^{(1+\mu,(1+\mu)/2)}\leq K_{11} \qquad (3.2.26)$$

for all $\mu\in(0,1)$, where the constant K_{11} depends only on Q, n, μ and the constant K_{10}.

PROOF. Letting $t=T-t'$ and then setting $z(x,T-t)\equiv\hat{z}(x,t)$, the system (3.2.24) is reduced to the form of system (3.2.21). Further, in view of the relation (3.2.11), we recall that $g(\cdot,\phi(u)(\cdot,T))\in L_2(\Omega)$, $u\in\mathfrak{U}$. Thus, it follows from Theorem 3.2.3 that, for each $u\in\mathfrak{U}$ and each integer $\sigma\geq 1$, the reduced system has a unique solution $\hat{z}^\sigma(u)$. Further,

$$\|\hat{z}(u)\|_{p,Q}^{(2,1)}\leq K_{10} \qquad (3.2.27)$$

for all $p\geq 3/2$, where, for given Q and the second-order coefficients, the constant K_{10} depends only on n, $\|d_\sigma(\cdot)g^\sigma(\cdot,\phi(u)(\cdot,T))\|_{\infty,\Omega}^{(2)}$ and the bounds for the functions $a_i^\sigma(\cdot,\cdot,u(\cdot,\cdot))$, $b_i^\sigma(\cdot,\cdot,u(\cdot,\cdot))$, $(b_i^\sigma(\cdot,\cdot,u(\cdot,\cdot)))_{x_i}$, $i=1,\ldots,n$, and $c^\sigma(\cdot,\cdot,u(\cdot,\cdot))$ (on $\overline{Q}$). In addition, $\hat{z}^\sigma(u)$ satisfies also the estimate

$$|\hat{z}^\sigma(u)|_Q^{(1+\mu,(1+\mu)/2)} \le K_{11} \tag{3.2.28}$$

for all $\mu \in (0,1)$, where the constant K_{11} depends only on Q, n, μ, and K_{10}. Consequently, for each $u \in \mathcal{U}$ and for each integer $\sigma \ge 1$, $z^\sigma(u)(x,t) \equiv \hat{z}^\sigma(u)(x, T-t)$ is the unique solution of the system (3.2.24), and it satisfies also the estimates (3.2.25) and (3.2.26). This completes the proof. □

In the next theorem, we shall show that the sequence of solutions of system (3.2.24) converges strongly in $V_2^{1,0}(Q)$ to the weak solution of system (3.2.1).

Theorem 3.2.5. *Suppose that Assumptions* (2.1.A1) *and* (3.2.A1)–(3.2.A6) *are satisfied. Then, for each* $u \in \mathcal{U}$, $\phi^\sigma(u) \xrightarrow{s} \phi(u)$ *in* $V_2^{1,0}(Q)$ *as* $\sigma \to \infty$, *where, for each* $u \in \mathcal{U}$ *and for each integer* $\sigma \ge 1$, $\phi^\sigma(u)$ *is the solution of the system* (3.2.21), *while, for each* $u \in \mathcal{U}$, $\phi(u)$ *is the weak solution from* $V_2^{1,0}(Q)$ *of the system* (3.2.1).

PROOF. The proof is based on Theorem 2.2.5. To apply this theorem, we first establish certain results below.

i. Since Q is bounded, it is clear from Assumptions (3.2.A2)–(3.2.A3) that all the coefficients of the operator $L(u)$ belong to $L_\lambda(Q)$ for any $\lambda \in [1,\infty]$. Let $\beta(u) \equiv \beta(\cdot,\cdot,u(\cdot,\cdot))$ stand for any of these coefficients. Then, by virtue of Theorem 1.2.1(iii), we obtain

$$\beta^\sigma(u) \equiv \beta^\sigma(\cdot,\cdot,u(\cdot,\cdot)) \xrightarrow{s} \beta(u) \tag{3.2.29}$$

in $L_\lambda(Q)$, where $\lambda \in [1,\infty]$, and hence, there exists a subsequence of the sequence $\{\beta^\sigma(u)\}$ that converges to $\beta(u)$ a.e. on Q.

ii. By Assumption (3.2.A4), it follows from Theorem 1.2.1(iii) that

$$F_j^\sigma(u) \xrightarrow{s} F_j(u) \tag{3.2.30}$$

in $L_2(Q)$.

iii. Since T is finite and $s \in (1,\infty)$, it follows from Hölder's inequality and Assumption (3.2.A5) that, for each $u \in \mathcal{U}$, $f(u) \in L_{2,1}(Q)$. Thus, it follows from Theorem 1.2.1(iii) that

$$f^\sigma(u) \xrightarrow{s} f(u) \tag{3.2.31}$$

in $L_{2,1}(Q)$.

iv. Recall that $|d_\sigma(x)| \le 1$ on Ω for all integers $\sigma \ge 1$ and $\lim_{\sigma\to\infty} d_\sigma(x) = 1$ for every $x \in \Omega$. Further, by Assumption (3.2.A6), $\phi_0 \in L_2(\Omega)$ and, by the relation (3.2.11), $g(\cdot, \phi(u)(\cdot, T)) \in L_2(\Omega)$ for any

$u \in \mathfrak{U}$. Thus, we obtain

$$d_\sigma(\cdot)\phi_0^\sigma(\cdot) \xrightarrow{s} \phi_0(\cdot) \tag{3.2.32}$$

and

$$d^\sigma(\cdot)g^\sigma(\cdot, \phi(u)(\cdot, T)) \xrightarrow{s} g(\cdot, \phi(u)(\cdot, T)) \tag{3.2.33}$$

in $L_2(\Omega)$.

On the basis of the results given in (i)–(iv) above, it follows from Theorem 2.2.5 that there exists a subsequence of the solutions $\{\phi^\sigma(u)\}$ of the system (3.2.21) that converges to the weak solution $\phi(u)$ [from $V_2^{1,0}(Q)$] of the system (3.2.1). However, $\phi(u)$ is unique. Thus, it is easy to show from the results given in (i)–(iv) and Theorem 2.2.5 that every subsequence of the sequence $\{\phi^\sigma(u)\}$ contains a convergent subsequence. Each of these convergent subsequences converges to the same limit, $\phi(u)$, strongly in $V_2^{1,0}(Q)$. Therefore, the whole sequence converges to $\phi(u)$ in the same topology. Indeed, if this conclusion were false, then there would exist an $\varepsilon > 0$ and a subsequence $\{\phi^{\sigma(k)}(u)\}$ of the sequence $\{\phi^\sigma(u)\}$ such that

$$\|\phi^{\sigma(k)}(u) - \phi(u)\|_Q > \varepsilon$$

for all integers $k \geq 1$. This is clearly a contradiction. Thus, the proof is complete. □

Letting $t = T - t'$ and setting $z(x, T-t) \equiv \hat{z}(x, t)$, systems (3.2.24) and (3.2.12) are reduced to the form of systems (3.2.21) and (3.2.1), respectively. Thus, by using a similar argument as that given for Theorem 3.2.5, we obtain the following result.

Theorem 3.2.6. *Suppose that Assumptions* (2.1.A1) *and* (3.2.A1)–(3.2.A6) *are satisfied. Then, for each* $u \in \mathfrak{U}$, $z^\sigma(u) \xrightarrow{s} z(u)$ *in* $V_2^{1,0}(Q)$ *as* $\sigma \to \infty$, *where, for each* $u \in \mathfrak{U}$ *and for each integer* $\sigma \geq 1$, $z^\sigma(u)$ *is the solution of the system* (3.2.24), *while, for each* $u \in \mathfrak{U}$, $z(u)$ *is the weak solution from* $V_2^{1,0}(Q)$ *of the system* (3.2.12).

To proceed further, we introduce the following notation:

$$\begin{aligned}
&\mathfrak{L}(u)(\psi, \zeta)(t) \\
&\equiv \int_\Omega \left\{ \sum_{j=1}^n \left(\sum_{i=1}^n a_{ij}(x, t, u(x, t))\psi_{x_i}(x, t) + a_j(x, t, u(x, t))\psi(x, t) \right) \right. \\
&\quad \times \zeta_{x_j}(x, t) - \sum_{j=1}^n b_j(x, t, u(x, t))\psi_{x_j}(x, t)\zeta(x, t) \\
&\quad \left. - c(x, t, u(x, t))\psi(x, t)\zeta(x, t) \right\} dx
\end{aligned} \tag{3.2.34}$$

for any $\psi, \zeta \in \mathring{W}_2^{1,0}(Q)$;

$$\begin{aligned}
&\mathfrak{L}^*(u)(\psi,\zeta)(t) \\
&\equiv \int_\Omega \left\{ \sum_{i=1}^n \left(\sum_{j=1}^n a_{ij}(x,t,u(x,t))\psi_{x_j}(x,t) - b_i(x,t,u(x,t))\psi(x,t) \right) \right. \\
&\quad \times \zeta_{x_i}(x,t) + \sum_{i=1}^n a_i(x,t,u(x,t))\psi_{x_i}(x,t)\zeta(x,t) \\
&\quad \left. - c(x,t,u(x,t))\psi(x,t)\zeta(x,t) \right\} dx \qquad (3.2.35)
\end{aligned}$$

for any $\psi, \zeta \in \mathring{W}_2^{1,0}(Q)$; and

$$\langle f, g \rangle \equiv \iint_Q \{ f(x,t) g(x,t) \} \, dx \, dt \qquad (3.2.36)$$

for any $f, g \in L_2(Q)$.

Clearly,

$$\mathfrak{L}(u)(\psi,\zeta)(t) = \tilde{\mathfrak{L}}(u)(\psi,\zeta)(t) + \int_\Omega \{ \psi(x,t)\zeta_t(x,t) \} \, dx \qquad (3.2.34)'$$

for any $\psi \in \mathring{W}_2^{1,0}(Q)$ and $\zeta \in \mathring{W}_2^{1,1}(Q)$, and

$$\mathfrak{L}^*(u)(\psi,\zeta)(t) = \tilde{\mathfrak{L}}^*(u)(\psi,\zeta)(t) - \int_\Omega \{ \psi(x,t)\zeta_t(x,t) \} \, dx \qquad (3.2.35)'$$

for any $\psi \in \mathring{W}_2^{1,0}(Q)$ and $\zeta \in \mathring{W}_2^{1,1}(Q)$.

REMARK 3.2.5. Let $\mathfrak{L}^\sigma(u)(\psi,\zeta)(t)$ and $\mathfrak{L}^{*\sigma}(u)(\psi,\zeta)(t)$ be as defined by Equations (3.2.34) and (3.2.35), respectively, with the coefficients in these equations replaced by their corresponding integral averages.

Lemma 3.2.1. *Suppose that Assumptions* (2.1.A1) *and* (3.2.A1)–(3.2.A6) *are satisfied. Then,*

$$\lim_{\sigma\to\infty} \lim_{\rho\to\infty} \int_0 \{ \tilde{\mathfrak{L}}(u^0)(\phi^\rho(u^0) - \phi^\rho(u), z^\sigma(u^0))(t) \} \, dt = 0 \qquad (3.2.37)$$

for any pair of controls $u^0, u \in \mathfrak{U}$, where, for each positive integer ρ, $\phi^\rho(u^0)$ $(\phi^\rho(u))$ is the solution of the system (3.2.21) *corresponding to $u^0 \in \mathfrak{U}$ $(u \in \mathfrak{U})$, while, for each positive integer σ, $z^\sigma(u^0)$ is the solution of the system* (3.2.24) *corresponding to $u^0 \in \mathfrak{U}$.*

PROOF. In view of Theorem 3.2.3, we observe that, for any $u \in \mathfrak{U}$ and for each integer $\rho \geq 1$, $\phi^\rho(u) \in \mathring{W}_p^{2,1}(Q)$, $3/2 < p < \infty$, and satisfies the estimate (3.2.23). Similarly, it is clear from Theorem 3.2.4 that, for each integer $\sigma \geq 1$, $z^\sigma(u) \in \mathring{W}_p^{2,1}(Q)$, $3/2 < p < \infty$, and satisfies the estimate (3.2.26).

Thus, by virtue of the equalities (3.2.34)′ and (3.2.36), we have

$$\int_0^T \{\tilde{\mathcal{L}}(u^0)(\phi^\rho(u^0)-\phi^\rho(u), z^\sigma(u^0))(t)\}\,dt$$

$$= -\langle \phi^\rho(u^0)-\phi^\rho(u), (z^\sigma(u^0))_t \rangle + \int_0^T \{\mathcal{L}(u^0)(\phi^\rho(u^0)-\phi^\rho(u), z^\sigma(u^0))(t)\}\,dt. \tag{3.2.38}$$

Since, for each integer $\sigma \geq 1$, $z^\sigma(u^0)$ is the solution of the system (3.2.24) corresponding to $u^0 \in \mathfrak{U}$, it follows from the second condition of Definition 3.2.4 that

$$\langle L^{*\sigma}(u^0)z^\sigma(u^0), \phi^\rho(u^0)-\phi^\rho(u)\rangle = 0. \tag{3.2.39}$$

Next, in view of Theorem 3.2.3, we observe that, for each integer $\rho \geq 1$,

i. $\phi^\rho(u^0), \phi^\rho(u) \in C(\bar{Q})$;

ii. $\phi^\rho(u^0)(\cdot,t), \phi^\rho(u)(\cdot,t) \in C^1(\bar{\Omega})$ for each $t \in [0,T]$; (3.2.40)

iii. $\phi^\rho(u^0)(x,t) = \phi^\rho(u)(x,t) = 0$ on $\partial\Omega \times [0,T]$.

Thus, by performing integration by parts to the terms with coefficients appearing under x-differential in (3.2.39), it follows from the definition of $\mathcal{L}^{*\sigma}(u^0)(\psi,\zeta)(t)$ (see Remark 3.2.5) that

$$\langle (z^\sigma(u^0))_t, \phi^\rho(u^0)-\phi^\rho(u)\rangle = \int_0^T \{\mathcal{L}^{*\sigma}(u^0)(z^\sigma(u^0), \phi^\rho(u^0)-\phi^\rho(u))(t)\}\,dt. \tag{3.2.41}$$

Rearranging the indices of summation, we deduce that

$$\int_0^T \{\mathcal{L}^{*\sigma}(u^0)(z^\sigma(u^0), \phi^\rho(u^0)-\phi^\rho(u))(t)\}\,dt = \int_0^T \{\mathcal{L}^\sigma(u^0)(\phi^\rho(u^0)-\phi^\rho(u), z^\sigma(u^0))(t)\}\,dt. \tag{3.2.42}$$

From the equality (3.2.38) and the above two equalities, we obtain

$$\int_0^T \{\tilde{\mathcal{L}}(u^0)(\phi^\rho(u^0)-\phi^\rho(u), z^\sigma(u^0))(t)\}\,dt$$

$$= \int_0^T \{\mathcal{L}(u^0)(\phi^\rho(u^0)-\phi^\rho(u), z^\sigma(u^0))(t) - \mathcal{L}^\sigma(u^0)(\phi^\rho(u^0)-\phi^\rho(u), z^\sigma(u^0))(t)\}\,dt. \tag{3.2.43}$$

Since T is finite, it follows from Theorem 3.2.5 that $\phi^\rho(u^0)$ ($\phi^\rho(u)$) converges to $\phi(u^0)$ ($\phi(u)$) strongly in $W_2^{1,0}(Q)$ as $\rho \to \infty$. Further, we recall that the coefficients of the operators $\mathfrak{L}(u^0)$ and $\mathfrak{L}^\sigma(u^0)$ are bounded on $\overline{Q}$ and that $z^\sigma(u^0)$ satisfies the estimate (3.2.26). Thus, it follows that the equality (3.2.43), in the limit with respect to ρ, reduces to

$$\lim_{\rho\to\infty}\int_0^T\{\tilde{\mathfrak{L}}(u^0)(\phi^\rho(u^0)-\phi^\rho(u),z^\sigma(u^0))(t)\}\,dt$$
$$=\int_0^T\{\mathfrak{L}(u^0)(\phi(u^0)-\phi(u),z^\sigma(u^0))(t)$$
$$-\mathfrak{L}^\sigma(u^0)(\phi(u^0)-\phi(u),z^\sigma(u^0))(t)\}\,dt. \tag{3.2.44}$$

The right-hand side of the above equality can be written as

$$\int_0^T\{\mathfrak{L}(u^0)(\phi(u^0)-\phi(u),z^\sigma(u^0))(t)$$
$$-\mathfrak{L}^\sigma(u^0)(\phi(u^0)-\phi(u),z^\sigma(u^0))(t)\}\,dt$$
$$=\int_0^T\{\mathfrak{L}(u^0)(\phi(u^0)-\phi(u),z^\sigma(u^0))(t)$$
$$-\mathfrak{L}(u^0)(\phi(u^0)-\phi(u),z(u^0))(t)\}\,dt$$
$$+\int_0^T\{\mathfrak{L}(u^0)(\phi(u^0)-\phi(u),z(u^0))(t)$$
$$-\mathfrak{L}^\sigma(u^0)(\phi(u^0)-\phi(u),z(u^0))(t)\}\,dt$$
$$+\int_0^T\{\mathfrak{L}^\sigma(u^0)(\phi(u^0)-\phi(u),z(u^0))(t)$$
$$-\mathfrak{L}^\sigma(u^0)(\phi(u^0)-\phi(u),z^\sigma(u^0))(t)\}\,dt. \tag{3.2.45}$$

To show that the above integral converges to zero as $\sigma\to\infty$, we first establish the following results.

i. Let β stand for any of the coefficients of the operator $L(u)$ throughout the proof. Then, by the definition of the integral averages and assumptions (3.2.A2)–(3.2.A3), we can easily verify that $|\beta^\sigma(x,t,v)|\leq K_1$ for all $(x,t,v)\in\overline{Q}\times V$ uniformly with respect to integers $\sigma\geq 1$, where β^σ denotes the integral average of β.
ii. There exists a subsequence of the sequence $\{\beta^\sigma(\cdot,\cdot,u(\cdot,\cdot))\}$ that converges to $\beta(\cdot,\cdot,u(\cdot,\cdot))$ a.e. on $\overline{Q}$ as $\sigma\to\infty$ (see the proof of Theorem 3.2.5).
iii. Since T is finite, it follows from Theorem 3.2.6 that $z^\sigma(u^0)\xrightarrow{s} z(u^0)$ in $W_2^{1,0}(Q)$ as $\sigma\to\infty$.
iv. Since T is finite, it follows from Theorems 3.2.1 and 3.2.2 that $\phi(u)\in W_2^{1,0}(Q)$, $\phi(u^0)\in W_2^{1,0}(Q)$, and $z(u^0)\in W_2^{1,0}(Q)$.

Let the first, second, and third integrals on the right-hand side of the equality (3.2.45) be denoted by γ_1^σ, γ_2^σ, and γ_3^σ, respectively. Then, by using the results given in (i), (iii), and (iv) above, and Hölder's inequality, we deduce that $\gamma_1^\sigma \to 0$ and $\gamma_3^\sigma \to 0$ as $\sigma \to \infty$.

It remains to show that $\gamma_2^\sigma \to 0$ as $\sigma \to \infty$. By virtue of the results given in (i), (ii), and (iv) above, the Lebesgue's dominated convergence theorem, it follows that there exists a subsequence $\{\gamma_2^{\sigma(k)}\}$ of the sequence $\{\gamma_2^\sigma\}$ such that $\gamma_2^{\sigma(k)} \to 0$ as $k \to \infty$. In fact, it can be shown easily that every subsequence of the sequence $\{\gamma_2^\sigma\}$ contains a further subsequence that converges to zero. Thus, we conclude that the whole sequence converges to zero, that is, $\gamma_2^\sigma \to 0$ as $\sigma \to \infty$. Therefore, the proof is complete. □

Lemma 3.2.2. *Suppose that all the hypotheses of Lemma* 3.2.1 *are satisfied. Then*

$$\lim_{\rho \to \infty} \int_0^T \{\tilde{\mathfrak{L}}(u^0)(\phi(u) - \phi^\rho(u), z^\sigma(u^0))(t)\}\, dt = 0 \qquad (3.2.46)$$

for any pair of controls $u, u^0 \in \mathfrak{U}$ *and any integer* $\sigma \geq 1$.

PROOF. In view of the equalities (3.2.34)′, and (3.2.36), we have

$$\int_0^T \{\tilde{\mathfrak{L}}(u^0)(\phi(u) - \phi^\rho(u), z^\sigma(u^0))(t)\}\, dt$$

$$= -\langle \phi(u) - \phi^\rho(u), (z^\sigma(u^0))_t \rangle$$

$$+ \int_0^T \{\mathfrak{L}(u^0)(\phi(u) - \phi^\rho(u), z^\sigma(u^0))(t)\}\, dt. \qquad (3.2.47)$$

Recall that (i) $z^\sigma(u^0) \in W_p^{2,1}(Q), 3/2 < p < \infty$, and satisfies the estimate (3.2.26); (ii) $\phi^\rho(u) \overset{s}{\to} \phi(u)$ in $W_2^{1,0}(Q)$ as $\rho \to \infty$; and (iii) the coefficients of the system (3.2.1) are bounded uniformly on $\bar{Q}$. By virtue of these facts, it follows that the above equality, in the limit with respect to ρ, reduces to (3.2.46). This completes the proof. □

In the next lemma, we shall compute the variation in cost due to the variation in control.

Lemma 3.2.3. *Consider Problem* (3.2.P1). *Suppose that Assumptions* (2.1.A1) *and* (3.2.A1)–(3.2.A8) *are satisfied. Then*

$$J(u) - J(u^0) \leq \int_0^T \{\mathfrak{L}(u^0)(\phi(u), z(u^0))(t) - (\tilde{F}(u^0), z(u^0))(t)$$

$$- \mathfrak{L}(u)(\phi(u), z(u^0))(t) + (\tilde{F}(u), z(u^0))(t)\}\, dt$$

$$(3.2.48)$$

for any pair of controls $u, u^0 \in \mathfrak{U}$.

PROOF. Clearly, for any integer $\sigma \geq 1$ and for any pair of controls u, $u^0 \in \mathfrak{U}$, we can write

$$\int_0^T \{\tilde{\mathfrak{L}}(u^0)(\phi(u^0), z^\sigma(u^0))(t) - \tilde{\mathfrak{L}}(u)(\phi(u), z^\sigma(u^0))(t)\} \, dt$$

$$= \int_0^T \{\tilde{\mathfrak{L}}(u^0)(\phi(u), z^\sigma(u^0))(t) - \tilde{\mathfrak{L}}(u)(\phi(u), z^\sigma(u^0))(t)\} \, dt$$

$$+ \int_0^T \{\tilde{\mathfrak{L}}(u^0)(\phi^\rho(u^0) - \phi^\rho(u), z^\sigma(u^0))(t)\} \, dt$$

$$+ \int_0^T \{\tilde{\mathfrak{L}}(u^0)(\phi(u^0) - \phi^\rho(u^0), z^\sigma(u^0))(t)\} \, dt$$

$$+ \int_0^T \{\tilde{\mathfrak{L}}(u^0)(\phi^\rho(u) - \phi(u), z^\sigma(u^0))(t)\} \, dt, \tag{3.2.49}$$

where ρ is any positive integer.

Note that, the left-hand side of the above equality is independent of integers $\rho \geq 1$. Thus, by letting $\rho \to \infty$, it follows from Lemma 3.2.2 that

$$\int_0^T \{\tilde{\mathfrak{L}}(u^0)(\phi(u^0), z^\sigma(u^0))(t) - \tilde{\mathfrak{L}}(u)(\phi(u), z^\sigma(u^0))(t)\} \, dt$$

$$= \int_0^T \{\tilde{\mathfrak{L}}(u^0)(\phi(u), z^\sigma(u^0))(t) - \tilde{\mathfrak{L}}(u)(\phi(u), z^\sigma(u^0))(t)\} \, dt$$

$$+ \lim_{\rho \to \infty} \int_0^T \{\tilde{\mathfrak{L}}(u^0)(\phi^\rho(u^0) - \phi^\rho(u), z^\sigma(u^0))(t)\} \, dt. \tag{3.2.50}$$

For each integer $\sigma \geq 1$, $z^\sigma(u^0)$ is the solution of the system (3.2.24) corresponding to $u^0 \in \mathfrak{U}$. Further, $\phi(u)$ $[\phi(u^0)]$ is the weak solution from $V_2^{1,0}(Q)$ of the system (3.2.1) corresponding to $u \in \mathfrak{U}$ $[u^0 \in \mathfrak{U}]$. Thus, it follows from condition (ii) of Definition 3.2.1 and the last part of Definition 3.2.4 that

$$\int_0^T \{\tilde{\mathfrak{L}}(u^0)(\phi(u^0), z^\sigma(u^0))(t) - \tilde{\mathfrak{L}}(u)(\phi(u), z^\sigma(u^0))(t)\} \, dt$$

$$= \int_0^T \{-(\tilde{F}(u), z^\sigma(u^0))(t) + (\tilde{F}(u^0), z^\sigma(u^0))(t)\} \, dt$$

$$- \int_\Omega \{d_\sigma(x) g^\sigma(x, \phi(u^0)(x,T))(\phi(u^0)(x,T) - \phi(u)(x,T))\} \, dx. \tag{3.2.51}$$

Combining the above two equalities, we have

$$\int_0^T \{\tilde{\mathfrak{L}}(u^0)(\phi(u), z^\sigma(u^0))(t) - \tilde{\mathfrak{L}}(u)(\phi(u), z^\sigma(u^0))(t)\} \, dt$$

$$+ \lim_{\rho \to \infty} \int_0^T \{\tilde{\mathfrak{L}}(u^0)(\phi^\rho(u^0) - \phi^\rho(u), z^\sigma(u^0))(t)\} \, dt$$

$$= \int_0^T \{ -(\tilde{F}(u), z^\sigma(u^0))(t) + (\tilde{F}(u^0), z^\sigma(u^0))(t) \} \, dt$$

$$- \int_\Omega \{ d_\sigma(x) g^\sigma(x, \phi(u^0)(x,T)) (\phi(u^0)(x,T) - \phi(u)(x,T)) \} \, dx. \tag{3.2.52}$$

To show convergence of the above equality with respect to σ, we recall that

i. $\phi(u)$ belongs to $\overset{\circ}{V}{}_2^{1,0}(Q)$ and hence $\overset{\circ}{W}{}_2^{1,0}(Q)$ (because T is finite) and that

ii. the coefficients of the system (3.2.1) are bounded on $\bar{Q} \times V$.

In addition, since $z^\sigma(u^0) \overset{s}{\to} z(u^0)$ in $V_2^{1,0}(Q)$ as $\sigma \to \infty$ and T is finite, it is easy to verify that, as $\sigma \to \infty$,

iii. $z^\sigma(u^0) \overset{s}{\to} z(u^0)$ in $W_2^{1,0}(Q)$ and

iv. $z^\sigma(u^0) \overset{s}{\to} z(u^0)$ in $L_{2,s/(s-1)}(Q)$, where $s \in (1, \infty)$.

By using the results given in (i)–(iii) above, we obtain

$$\lim_{\sigma \to \infty} \int_0^T \{ \tilde{\mathfrak{L}}(u^0)(\phi(u), z^\sigma(u^0))(t) - \tilde{\mathfrak{L}}(u)(\phi(u), z^\sigma(u^0))(t) \} \, dt$$

$$= \lim_{\sigma \to \infty} \int_0^T \{ \mathfrak{L}(u^0)(\phi(u), z^\sigma(u^0))(t) - \mathfrak{L}(u)(\phi(u), z^\sigma(u^0))(t) \} \, dt$$

$$= \int_0^T \{ \mathfrak{L}(u^0)(\phi(u), z(u^0))(t) - \mathfrak{L}(u)(\phi(u), z(u^0))(t) \} \, dt. \tag{3.2.53}$$

Next, by virtue of the results given in (iii) and (iv) above and Assumptions (3.2.A4)–(3.2.A5), we can easily verify that

$$\lim_{\sigma \to \infty} \int_0^T \{ -(\tilde{F}(u), z^\sigma(u^0))(t) + (\tilde{F}(u^0), z^\sigma(u^0))(t) \} \, dt$$

$$= \int_0^T \{ -(\tilde{F}(u), z(u^0))(t) + (\tilde{F}(u^0), z(u^0))(t) \} \, dt. \tag{3.2.54}$$

Furthermore, since $\phi(u^0)(\cdot, T) \in L_2(\Omega)$, $\phi(u)(\cdot, T) \in L_2(\Omega)$, and $d_\sigma(\cdot) g^\sigma(\cdot, \phi(u^0)(\cdot, T)) \overset{s}{\to} g(\cdot, \phi(u^0)(\cdot, T))$ in $L_2(\Omega)$ as $\sigma \to \infty$, it follows that

$$\lim_{\sigma \to \infty} \int_\Omega \{ d_\sigma(x) g^\sigma(x, \phi(u^0)(x,T)) (\phi(u^0)(x,T) - \phi(u)(x,T)) \} \, dx$$

$$= \int_\Omega \{ g(x, \phi(u^0)(x,T)) (\phi(u^0)(x,T) - \phi(u)(x,T)) \} \, dx. \tag{3.2.55}$$

Using the relations (3.2.53)–(3.2.55) and Lemma 3.2.1, we observe that the equality (3.2.52), in the limit with respect to σ, reduces to

$$\int_0^T \{\mathfrak{L}(u^0)(\phi(u), z(u^0))(t) - \mathfrak{L}(u)(\phi(u), z(u^0))(t)\}\, dt$$

$$= \int_0^T \{-(\tilde{F}(u), z(u^0))(t) + (\tilde{F}(u^0), z(u^0))(t)\}\, dt$$

$$-\int_\Omega \{g(x, \phi(u^0)(x,T))(\phi(u^0)(x,T) - \phi(u)(x,T))\}\, dx. \quad (3.2.56)$$

By virtue of Assumptions (3.2.A8) and (3.2.A7), it is easy to show that $\gamma(\cdot, \phi(\cdot)) \in L_1(\Omega)$ for any $\phi \in L_2(\Omega)$, and hence, $\gamma(\cdot, \phi(\mathrm{u})(\cdot, T)) \in L_1(\Omega)$ for any $u \in \mathfrak{U}$. Thus, it follows from the first inequality of Assumption (3.2.A7) that

$$\int_\Omega \{\gamma(x, \phi(u)(x,T)) - \gamma(x, \phi(u^0)(x,T))\}\, dx$$

$$\leq \int_\Omega \{g(x, \phi(u^0)(x,T))(\phi(u)(x,T) - \phi(u^0)(x,T))\}\, dx. \quad (3.2.57)$$

Combining the equality (3.2.56) and the above inequality, and using the definition of the cost functional J [given in (3.2.8)], we obtain the inequality (3.2.48). This completes the proof. □

Replacing u^0 by an optimal control u^* in the above lemma, we obtain the following result.

Lemma 3.2.4. *Suppose that all the hypotheses of Lemma* 3.2.3 *are satisfied and let $u^* \in \mathfrak{U}$ be an optimal control. Then*

$$\int_0^T \{\mathfrak{L}(u^*)(\phi(u), z(u^*))(t) - (\tilde{F}(u^*), z(u^*))(t)\}\, dt$$

$$-\mathfrak{L}(u)(\phi(u), z(u^*))(t) + (\tilde{F}(u), z(u^*))(t)\}\, dt \geq 0 \quad (3.2.58)$$

for all $u \in \mathfrak{U}$.

Let E be a measurable subset of R^{n+1}, G an open subset of R^m, and ξ a real-valued function defined on $E \times G$. Define

$$\xi_{y_i}(x,t,v) \equiv \partial\xi(x,t,v_1,\ldots,v_{i-1},y_i,v_{i+1},\ldots,v_m)/\partial y_i|_{y_i = v_i}. \quad (3.2.59)$$

In addition, we introduce the following terms:

$$\begin{aligned}
&\mathcal{T}(\psi,\zeta,u^0,u)(t)\\
&=\int_\Omega\Big\{\big[a_{iju_k}(x,t,u^0(x,t))\big(u_k(x,t)-u_k^0(x,t)\big)\psi_{x_i}(x,t)\\
&\quad+a_{ju_k}(x,t,u^0(x,t))\big(u_k(x,t)-u_k^0(x,t)\big)\psi(x,t)\big]\zeta_{x_j}(x,t)\\
&\quad-b_{ju_k}(x,t,u^0(x,t))(u_k(x,t)-u_k^0(x,t))\psi_{x_j}(x,t)\zeta(x,t)\\
&\quad-c_{u_k}(x,t,u^0(x,t))\big(u_k(x,t)-u_k^0(x,t)\big)\psi(x,t)\zeta(x,t)\Big\}\,dx
\end{aligned}\tag{3.2.60}$$

for any u^0, $u\in\mathfrak{U}$, $\psi\in\overset{\circ}{W}{}_2^{1,0}(Q)$, and $\zeta\in\overset{\circ}{W}{}_2^{1,0}(Q)$; and

$$\begin{aligned}
&\mathcal{F}(\tilde{F}(u^0),\zeta,u)(t)\\
&=\int_\Omega\Big\{F_{ju_k}(x,t,u^0(x,t))\big(u_k(x,t)-u_k^0(x,t)\big)\zeta_{x_j}(x,t)\\
&\quad-f_{u_k}(x,t,u^0(x,t))\big(u_k(x,t)-u_k^0(x,t)\big)\zeta(x,t)\Big\}\,dx
\end{aligned}\tag{3.2.61}$$

for any u^0, $u\in\mathfrak{U}$ and $\zeta\in\overset{\circ}{W}{}_2^{1,1}(Q)$, where, for convenience, we make use of the standard convention of taking summation (i, j from 1 to n and k from 1 to m) over repeated indices. This convention is used throughout the rest of this subsection.

We are now in a position to derive a necessary condition for optimality in integral form for the terminal control problem (3.2.P1).

Theorem 3.2.7. *Consider the problem* (3.2.P1). *Suppose that the following conditions hold*:

i. *Assumptions* (2.1.A1) *and* (3.2.A1)–(3.2.A8) *are satisfied*;
ii. *$U(x,t)\equiv V$, a compact and convex subset of R^m; and*
iii. *$\xi(x,t,\cdot)$ belongs to $C^1(U)$ a.e. on Q with gradient bounded in R^m for almost every $(x,t)\in Q$ and for every $v\in U$, where ξ stands for any of the coefficients and free terms of the system* (3.2.1), *and U is an open subset of R^m containing V.*

Then, if $u^*\in\mathfrak{U}$ is an optimal control,

$$\int_0^T\big\{\mathcal{T}(\phi(u^*),z(u^*),u^*,u)(t)-\mathcal{F}(\tilde{F}(u^*),z(u^*),u)(t)\big\}\,dt\le 0\tag{3.2.62}$$

for all $u\in\mathfrak{U}$.

PROOF. For any $u \in \mathfrak{U}$, let $\varepsilon \in [0,1]$ and $\tilde{u} \equiv u - u^*$. Since V is convex, it is clear that $u^* + \varepsilon\tilde{u} \in \mathfrak{U}$. Dividing the inequality (3.2.58) by $(-\varepsilon)$, $0 < \varepsilon < 1$, and replacing u by $u^* + \varepsilon\tilde{u}$, we obtain

$$\frac{1}{\varepsilon}\int_0^T \{\mathfrak{L}(u^* + \varepsilon\tilde{u})(\phi(u^* + \varepsilon\tilde{u}), z(u^*))(t) - \mathfrak{L}(u^*)(\phi(u^* + \varepsilon\tilde{u}), z(u^*))(t)$$

$$-(\tilde{F}(u^* + \varepsilon\tilde{u}), z(u^*))(t) + (\tilde{F}(u^*), z(u^*))(t)\}\, dt \leq 0. \tag{3.2.63}$$

By condition (iii) of the hypotheses, $\xi(x,t,\cdot) \in C^1(U)$ a.e. on Q with gradients bounded in R^m for almost every $(x,t) \in Q$ and every $v \in U$. Thus, it follows that

$$\lim_{\varepsilon \downarrow 0} \frac{\xi(x,t,u^*(x,t) + \varepsilon\tilde{u}(x,t)) - \xi(x,t,u^*(x,t))}{\varepsilon}$$

$$= \xi_{u_k}(x,t,u^*(x,t))\tilde{u}_k(x,t) \tag{3.2.64}$$

for almost every $(x,t) \in Q$, where ξ_{u_k} are as defined in (3.2.59) and $\xi_{u_k}(x,t,u^*(x,t))\tilde{u}_k(x,t)$ are bounded for almost all $(x,t) \in Q$. Therefore, there exist constants ε_0, $0 < \varepsilon_0 \leq 1$, and $N_0 > 0$ such that

$$\left| \frac{\xi(x,t,u^*(x,t) + \varepsilon\tilde{u}(x,t)) - \xi(x,t,u^*(x,t))}{\varepsilon} \right| \leq N_0 \tag{3.2.65}$$

a.e. on Q, for all ε, $0 < \varepsilon < \varepsilon_0$.

In view of Assumption (3.2.A1), we observe that

$$\xi(x,t,u^*(x,t) + \varepsilon\tilde{u}(x,t)) \to \xi(x,t,u^*(x,t)) \tag{3.2.66}$$

a.e. on Q as $\varepsilon \to 0$. Further, by virtue of Assumptions (3.2.A4)–(3.2.A5) and the convexity of V, it can be easily shown, by contradiction, that there exist constants N_1 and N_2 such that

$$\sup_{0 \leq \varepsilon \leq 1} \|F_j(\cdot,\cdot,u^*(\cdot,\cdot) + \varepsilon\tilde{u}(\cdot,\cdot))\|_{2,Q} \leq N_1 \tag{3.2.67}$$

for all $j = 1, \ldots, n$ and

$$\sup_{0 \leq \varepsilon \leq 1} \|f(\cdot,\cdot,u^*(\cdot,\cdot) + \varepsilon\tilde{u}(\cdot,\cdot))\|_{2,s,Q} \leq N_2. \tag{3.2.68}$$

Using Assumptions (3.2.A1)–(3.2.A6), the relation (3.2.66), and the inequalities (3.2.67)–(3.2.68), it follows from Theorem 2.2.6 that there exists a sequence $\{\phi(u^* + \varepsilon_\kappa\tilde{u})\}_{\kappa=1}^{\infty}$ of the set $\{\phi(u^* + \varepsilon\tilde{u}) : 0 \leq \varepsilon \leq 1\}$ such that

$$\phi(u^* + \varepsilon_\kappa\tilde{u}) \xrightarrow{w} \phi(u^*)$$

$$(\phi(u^* + \varepsilon_\kappa\tilde{u}))_{x_i} \xrightarrow{w} (\phi(u^*))_{x_i}, \qquad i = 1, \ldots, n, \tag{3.2.69}$$

in $L_2(Q)$ as $\kappa \to \infty$, where $\varepsilon_\kappa \downarrow 0$ as $\kappa \uparrow \infty$. Further, with reference to the sequence $\{\phi(u^*+\varepsilon_\kappa\tilde{u}\}_{\kappa=1}^{\infty}$, the inequality (3.2.63) reduces to

$$\frac{1}{\varepsilon_\kappa}\int_0^T\{\mathfrak{L}(u^*+\varepsilon_\kappa\tilde{u})(\phi(u^*+\varepsilon_\kappa\tilde{u}),z(u^*))(t)-\mathfrak{L}(u^*)(\phi(u^*+\varepsilon_\kappa\tilde{u}),z(u^*))(t)$$
$$-(\tilde{F}(u^*+\varepsilon_\kappa\tilde{u}),z(u^*))(t)+(\tilde{F}(u^*),z(u^*))(t)\}\,dt\leq 0 \tag{3.2.70}$$

for all integers $\kappa\geq 1$.

We shall show that the above inequality, in the limit with respect to κ, reduces to the inequality (3.2.62). By virtue of the estimate (3.2.7), Assumption (3.2.A6), and the inequalities (3.2.67)–(3.2.68), it follows that there exists a constant N_3, independent of ε, such that

$$\|\phi(u^*+\varepsilon\tilde{u})\|_Q\leq N_3 \tag{3.2.71}$$

for all $\varepsilon\in[0,1]$. Since $z(u^*)\in V_2^{1,0}(Q)$, we obtain

$$\|z(u^*)\|_{2,(s/s-1),Q}\leq(T)^{1-(1/s)}\|z(u^*)\|_Q, \tag{3.2.72}$$

and hence,

$$z(u^*)\in L_{2,(s/s-1)}(Q), \tag{3.2.73}$$

where $s\in(1,\infty)$.

By virtue of (3.2.64), (3.2.69), (3.2.65), (3.2.71), (3.2.73), and the fact that Q is bounded, it follows from Theorems 1.2.9 and 1.2.10 that the inequality (3.2.70), in the limit with respect to κ, reduces to the inequality (3.2.62). This completes the proof. □

Using the above theorem and Theorem 1.1.19, specialized to the finite-dimensional case, we can derive a pointwise necessary condition for optimality for Problem (3.2.P1).

Theorem 3.2.8. *Consider the control problem* (3.2.P1). *Suppose that all the hypotheses of Theorem* 3.2.7 *are satisfied. Then*

$$\Big[\big(a_{iju_k}(x,t,u^*(x,t))(\phi(u^*))_{x_i}(x,t)$$
$$+a_{ju_k}(x,t,u^*(x,t))\phi(u^*)(x,t)\big)(z(u^*))_{x_j}(x,t)$$
$$-b_{ju_k}(x,t,u^*(x,t))(\phi(u^*))_{x_j}(x,t)z(u^*)(x,t)$$
$$-c_{u_k}(x,t,u^*(x,t))\phi(u^*)(x,t)z(u^*)(x,t)$$
$$-F_{ju_k}(x,t,u^*(x,t))(z(u^*))_{x_j}(x,t)$$
$$+f_{u_k}(x,t,u^*(x,t))z(u^*)(x,t)\Big](v_k-u_k^*(x,t))\leq 0 \tag{3.2.74}$$

for almost all $(x,t)\in Q$ *and for all* $v\in V$.

PROOF. Let $v \in V$ be arbitrary and let $(x^0, t^0) \in Q$ be a regular point for all the functions involved on the left-hand side of (3.2.74). Further, for a positive integer l, let

$$\mathfrak{N}_l \equiv \{(x,t) \in Q : |(x,t) - (x^0, t^0)| \leq 1/l\}. \tag{3.2.75}$$

Define

$$u^l(x,t) = \begin{cases} v & \text{for} \quad (x,t) \in \mathfrak{N}_l \\ u^*(x,t) & \text{for} \quad (x,t) \in Q \setminus \mathfrak{N}_l. \end{cases} \tag{3.2.76}$$

Clearly, $u^l \in \mathfrak{U}$ for every positive integer l.

Dividing the inequality (3.2.62) by (the Lebesgue measure) $|\mathfrak{N}_l|$, replacing the control u by u^l, and then letting $l \to \infty$, we obtain from Theorem 1.1.19, specialized to the finite-dimensional case, the inequality (3.2.74) for $(x,t) = (x^0, t^0)$. However, almost all $(x,t) \in Q$ are regular points and V is separable; the proof follows from condition (iii) of Theorem 3.2.7. □

3.2.6. *Existence of Optimal Controls*

In the previous subsection, we presented necessary conditions for optimality. In the derivation of these necessary conditions, it was assumed that an optimal control exists. Here, we wish to present sufficient conditions ensuring the existence of optimal controls. However, due to the limitation of the approach used in this subsection, the second- and the first-order coefficients of the system are not allowed to depend on the control variable. More precisely, we consider the following system:

$$\begin{aligned} L^\#(u)\phi(x,t) &= \sum_{j=1}^{n} \left(F_j(x,t,u(x,t))\right)_{x_j} + f(x,t,u(x,t)), \qquad (x,t) \in Q, \\ \phi(x,0) &= \phi_0(x), \qquad x \in \Omega, \\ \phi(x,t) &= 0, \qquad (x,t) \in \partial\Omega \times [0,T], \end{aligned} \tag{3.2.77}$$

where, for each $u \in \mathfrak{U}$, the operator $L^\#(u)$ is given by

$$\begin{aligned} L^\#(u)\psi(x,t) &= \psi_t(x,t) - \sum_{j=1}^{n} \left(\sum_{i=1}^{n} a_{ij}(x,t)\psi_{x_i}(x,t) + a_j(x,t,u(x,t))\psi(x,t) \right)_{x_j} \\ &\quad - \sum_{j=1}^{n} b_j(x,t)\psi_{x_j}(x,t) - c(x,t,u(x,t))\psi(x,t). \end{aligned} \tag{3.2.78}$$

For convenience, Problem (3.2.P2) [Problem (3.2.P3)] with system (3.2.1) replaced by system (3.2.77) will be referred to as Problem (3.2.P2)′ [Problem (3.2.P3)′].

We now assume that the free terms of the system (3.2.77) satisfy the following conditions.

Assumption (3.2.A12). There exists a constant K_{12} such that

$$|F_j(x,t,v)| \le K_{12}, \qquad j=1,\dots,n, \quad \text{and} \quad |f(x,t,v)| \le K_{12}$$

for all $(x,t,v) \in \bar{Q} \times V$.

REMARK 3.2.6. Since Q is bounded, Assumptions (3.2.A2)–(3.2.A3) and (3.2.A12) imply Assumption (2.2.A6).

Lemma 3.2.5. *Consider system* (3.2.77). *Suppose that Assumptions* (3.2.A1)–(3.2.A3) *and* (3.2.A12) *are satisfied and that* ϕ_0 *is a real-valued bounded measurable function defined on* $\bar{\Omega}$. *Then,* $\{\phi(u) : u \in \mathcal{U}\}$ *is equicontinuous and uniformly bounded on any compact subset of* $\Omega \times (0,T]$, *where, for each* $u \in \mathcal{U}$, $\phi(u)$ *is the weak solution from* $V_2^{1,0}(Q)$ *of the system* (3.2.77).

PROOF. Since Q is bounded, we can easily verify that the hypotheses of the lemma imply the hypotheses of Theorem 3.2.1. Thus, it follows that, for each $u \in \mathcal{U}$, the system (3.2.77) has a unique weak solution $\phi(u)$ from $V_2^{1,0}(Q)$. Let $\hat{Q}$ be a compact subset of $\Omega \times (0,T]$. Then, it is clear from Remark 2.2.10 that $\phi(u)$ is also a weak solution from $V_2^{1,0}(Q)$ of the differential equation of the system (3.2.77) defined on $\hat{Q}$ (in the sense of Definition 2.2.2). Further, by virtue of Remark 3.2.6 and the fact that ϕ_0 is bounded measurable on $\bar{\Omega}$, we can easily verify that all the hypotheses of Theorem 2.2.7 are satisfied. Thus, it follows from Theorem 2.2.7 that $|\phi(u)(x,t)| \le N$ for almost all $(x,t) \in \Omega \times (0,T]$, where the constant N is independent of the controls $u \in \mathcal{U}$. On this basis, we obtain from Theorem 2.2.9 that $\{\phi(u) : u \in \mathcal{U}\}$ is equicontinuous on $\hat{Q}$. Since $\hat{Q}$ is arbitrary, the proof is complete. □

Using the above lemma and the Ascoli–Arzelà theorem, we obtain the following result.

Lemma 3.2.6. *Suppose all the hypotheses of Lemma* 3.2.5 *are satisfied and let* $\{u^k\}$ *be a sequence of controls from* $\mathcal{U}$. *Then there is a subsequence of the sequence* $\{u^k\}$, *again denoted by* $\{u^k\}$, *such that* $\phi(u^k)$ *converges pointwise on* $\Omega \times (0,T]$ *and uniformly on any compact subset of* $\Omega \times (0,T]$.

For the proof of existence of optimal controls, it is often convenient to formulate the problem as a sort of contingent problem. Define a multi-function F on Q with values

$$F(x,t) \equiv \{\xi \in R^{2n+2} : \xi = g(x,t,v), v \in U(x,t)\}, \qquad (x,t) \in Q, \quad (3.2.79)$$

where

$$g(x,t,v) \equiv (a_1(x,t,v),\dots,a_n(x,t,v),$$
$$c(x,t,v), F_1(x,t,v),\dots,F_n(x,t,v), f(x,t,v)) \qquad (3.2.80)$$

and U is a measurable multifunction with values nonempty closed convex subsets of the compact set $V(\subset R^m)$.

We now impose the following assumption on the multifunction F.

Assumption 3.2.A13). The multifunction F is measurable and has convex values $F(x,t)$ for each $(x,t)\in Q$.

Lemma 3.2.7. *Consider system* (3.2.77). *Suppose that all the hypotheses of Lemma* 3.2.5 *and Assumption* (3.2.A13) *are satisfied. If* $\{u^k\}$ *is a sequence of controls from* $\mathfrak{U}$, *then there exists a subsequence* $\{u^{k(l)}\}$ *of the sequence* $\{u^k\}$ *and a control* u^0 *in* $\mathfrak{U}$ *such that, as* $l\to\infty$,

$$a_j(\cdot,\cdot,u^{k(l)}(\cdot,\cdot))\overset{w^*}{\to}a_j(\cdot,\cdot,u^0(\cdot,\cdot)),\qquad j=1,\dots,n,$$

$$c(\cdot,\cdot,u^{k(l)}(\cdot,\cdot))\overset{w^*}{\to}c(\cdot,\cdot,u^0(\cdot,\cdot)),$$

$$F_j(\cdot,\cdot,u^{k(l)}(\cdot,\cdot))\overset{w^*}{\to}F_j(\cdot,\cdot,u^0(\cdot,\cdot)),\qquad j=1,\dots,n,$$

$$f(\cdot,\cdot,u^{k(l)}(\cdot,\cdot))\overset{w^*}{\to}f(\cdot,\cdot,u^0(\cdot,\cdot))$$

in $L_\infty(Q)$.

PROOF. By virtue of Assumptions (3.2.A2)–(3.2.A3) and (3.2.A12), it is clear that $g^k(\cdot,\cdot)\equiv g(\cdot,\cdot,u^k(\cdot,\cdot))$, as given in (3.2.80), is bounded on $\bar{Q}$ uniformly with respect to $u^k\in\mathfrak{U}$. Thus, there exists a subsequence of the sequence $\{u^k\}$, again denoted by $\{u^k\}$, and a function $g^*\in[L_\infty(Q)]^{2n+2}$ such that $g^k\overset{w^*}{\to}g^*$ in $[L_\infty(Q)]^{2n+2}$, where $[L_\infty(Q)]^{2n+2}$ denotes the $(2n+2)$-copies of $L_\infty(Q)$. Further, since $U(x,t)$ is compact for each $(x,t)\in Q$, we can verify from Assumption (3.2.A1) that $F(x,t)$ is compact for each $(x,t)\in Q$. Thus, by using this fact, including the fact that $g^k(x,t)\in F(x,t)$, $(x,t)\in Q$, for all integers $k\geq 1$, and the convexity assumption (3.2.A13), we observe from Proposition 1.4.1 that

$$g^*(x,t)\in F(x,t),\qquad (x,t)\in Q. \tag{3.2.81}$$

By virtue of the above relation and Assumption (3.2.A1) and the first part of Assumption (3.2.A13), it follows from Theorem 1.4.6 that there exists a control u^0 in $\mathfrak{U}$ such that $g^*(x,t)=g(x,t,u^0(x,t))$ for all $(x,t)\in Q$. This completes the proof.

Theorem 3.2.9. *Consider the system* (3.2.77). *Suppose all the hypotheses of Lemma* 3.2.7 *are satisfied and let* $\{u^k\}$ *be a sequence of controls from* $\mathfrak{U}$. *Then, there exists a subsequence* $\{u^{k(l)}\}$ *of the sequence* $\{u^k\}$ *and a control*

$u^0 \in \mathfrak{U}$ *such that, as* $l \to \infty$,

i. $\phi(u^{k(l)}) \to \phi(u^0)$ *weakly in* $L_2(Q)$, *pointwise on* $\Omega \times (0, T]$, *and uniformly on any compact subset of* $\Omega \times (0, T]$; *and*

ii. $(\phi(u^{k(l)}))_{x_i} \xrightarrow{w} (\phi(u^0))_{x_i}$, $i = 1, \ldots, n$, *in* $L_2(Q)$.

PROOF. By virtue of Lemma 3.2.7, there exists a subsequence of the sequence $\{u^k\}$, again denoted by $\{u^k\}$, and a control $u^0 \in \mathfrak{U}$ such that, as $k \to \infty$,

$$\begin{aligned} a_j(\cdot, \cdot, u^k(\cdot, \cdot)) &\xrightarrow{w^*} a_j(\cdot, \cdot, u^0(\cdot, \cdot)), \qquad j = 1, \ldots, n, \\ c(\cdot, \cdot, u^k(\cdot, \cdot)) &\xrightarrow{w^*} c(\cdot, \cdot, u^0(\cdot, \cdot)), \\ F_j(\cdot, \cdot, u^k(\cdot, \cdot)) &\xrightarrow{w^*} F_j(\cdot, \cdot, u^0(\cdot, \cdot)), \qquad j = 1, \ldots, n, \\ f(\cdot, \cdot, u^k(\cdot, \cdot)) &\xrightarrow{w^*} f(\cdot, \cdot, u^0(\cdot, \cdot)) \end{aligned} \tag{3.2.82}$$

in $L_\infty(Q)$.

Using the estimate (3.2.7), the hypotheses on the coefficients and data of system (3.2.77), and the fact that Q is bounded, we obtain

$$\|\phi(u^k)\|_Q \le N_0 \tag{3.2.83}$$

for all integers $k \ge 1$, where the constant N_0 is independent of k. By virtue of the above inequality and the fact that T is finite, it follows that

$$\|\phi(u^k)\|_{2,Q} \le (T)^{1/2} \|\phi(u^k)\|_Q \le (T)^{1/2} N_0 \equiv N_1. \tag{3.2.84}$$

On the basis of the above two estimates, we can show, by using an argument similar to that given in the proof of Theorem 2.2.6, that there exists a subsequence of the sequence $\{\phi(u^k)\}$, again denoted by $\{\phi(u^k)\}$, and an element ϕ in $\overset{\circ}{V}_2(Q)$ such that, as $k \to \infty$,

$$\begin{aligned} \phi(u^k) &\xrightarrow{w} \phi, \\ (\phi(u^k))_{x_i} &\xrightarrow{w} \phi_{x_i}, \qquad i = 1, \ldots, n, \end{aligned} \tag{3.2.85}$$

in $L_2(Q)$. Thus, ϕ is a weak solution from $V_2(Q)$ of the system (3.2.77) if it satisfies also condition (ii) of Definition 2.2.3. From Lemma 3.2.6, there is a subsequence $\{u^{k(l)}\}$ of the sequence $\{u^k\}$ such that $\phi(u^{k(l)}) \to \hat{\phi}$ pointwise in $\Omega \times (0, T]$, and uniformly on any compact subset of $\Omega \times (0, T]$. Since $\{\phi(u^{k(l)})\}$ is a subsequence of $\{\phi(u^k)$, it is clear that $\phi(u^{k(l)})$ and $(\phi(u^{k(l)}))_{x_i}$, $i = 1, \ldots, n$, also converge, respectively, to ϕ and ϕ_{x_i} weakly in $L_2(Q)$ as $l \to \infty$. Thus, we conclude that $\phi(x, t) = \hat{\phi}(x, t)$ for almost all $(x, t) \in Q$. Indeed, if this conclusion were false, then there would exist a set

$G_0(\subset Q)$ with positive measure such that $\phi(x,t)\neq\hat{\phi}(x,t)$ for all $(x,t)\in G_0$. Clearly, $\phi(u^{k(l)})\to\hat{\phi}$ pointwise in G_0 and $\phi(u^{k(l)})\overset{w}{\to}\phi$ in $L_2(G_0)$. However, by virtue of the estimates (3.2.84) and the fact that $\phi(u^{k(l)})\to\hat{\phi}$ pointwise in G_0, it follows from Theorem 1.2.10 that $\phi(u^{k(l)})\overset{w}{\to}\hat{\phi}$ in $L_2(G_0)$ (through a subsequence, if necessary). By the uniqueness of the weak limit, $\hat{\phi}(x,t)=\phi(x,t)$ for almost all $(x,t)\in G_0$. This is a contradiction. Thus, $\hat{\phi}(x,t)=\phi(x,t)$ for almost all $(x,t)\in Q$. Changing the values of the functions on a set of measure zero (if necessary), we conclude that $\hat{\phi}(x,t)=\phi(x,t)$ on Q. With this result, we can now verify that ϕ satisfies condition (ii) of Definition 2.2.3. Let η be an arbitrary element of $\overset{\circ}{W}{}_2^{1,1}(Q)$ that is equal to zero for $t=T$. Since $\phi(u^{k(l)})$ is the weak solution from $V_2^{1,0}(Q)$ of the system (3.2.77) corresponding to $u^{k(l)}\in\mathfrak{U}$, it follows from Remark 2.2.3 that, for each integer $l\geq 1$,

$$\int_0^T\left\{\tilde{\mathfrak{L}}^{\#}(u^{k(l)})(\phi(u^{k(l)}),\eta)(t)-(\tilde{F}(u^{k(l)}),\eta)(t)\right\}dt=\int_\Omega\{\phi_0(x)\eta(x)\}\,dx,\tag{3.2.86}$$

where $\tilde{\mathfrak{L}}^{\#}(u)(\psi,\zeta)(t)$ is as defined by Equation (3.2.3) with the coefficients a_{ij}, $i,j=1,\ldots,n$, and b_j, $j=1,\ldots,n$, in this equation being independent of the control variable.

Now, by virtue of Assumptions (3.2.A2)–(3.2.A3) and (3.2.A13), and using the relations (3.2.82) and (3.2.85), and the facts that (i) $\phi(u^{k(l)})\to\phi$ pointwise in $\Omega\times(0,T]$ and hence in Q, (ii) $\phi(u^{k(l)})$ is essentially bounded uniformly with respect to l (see the proof of Lemma 3.2.5), (iii) Q is bounded, and (iv) the Lebesgue dominated convergence theorem, we can show that the expression (3.2.86), in the limit with respect to l, reduces to

$$\int_0^T\left\{\tilde{\mathfrak{L}}^{\#}(u^0)(\phi,\eta)(t)-(\tilde{F}(u^0),\eta)(t)\right\}dt=\int_\Omega\{\phi_0(x)\eta(x,0)\}\,dx.\tag{3.2.87}$$

Since the above equality holds for any $\eta\in\overset{\circ}{W}{}_2^{1,1}(Q)$ equal to zero for $t=T$, it is clear from Lemma 2.2.3 that condition (ii) of Definition 2.2.3 is satisfied. Thus, ϕ is a weak solution from $V_2(Q)$ of the system (3.2.81) corresponding to $u^0\in\mathfrak{U}$, and hence, it is written as $\phi(u^0)$. On this basis, it follows from Theorem 2.2.2 that $\phi(u^0)\in\overset{\circ}{V}{}_2^{1,0}(Q)$. Therefore, we conclude from Theorem 2.2.4 that $\phi(u^0)$ is the unique weak solution from $V_2^{1,0}(Q)$ of system (3.2.77) corresponding to $u^0\in\mathfrak{U}$. This completes the proof. □

As a special case, the above theorem holds if the functions a_j, $j=1,\ldots,n$, c, F_j, $j=1,\ldots,n$, and f are linear in the control variable rather than satisfying the convexity assumption (3.2.A13). However, in this particular case, stronger results can be obtained. More precisely, we have the following result.

Theorem 3.2.10. *Consider system* (3.2.77). *Suppose that all the hypotheses of Lemma* 3.2.7 *are satisfied. Further, let the functions* a_j, $j=1,\ldots,n$, c, F_j, $j=1,\ldots,n$, *and* f *be linear in the control variable, and let* $\{u^k\}$ *be a sequence of controls from* $\mathcal{U}$. *Then there exists a subsequence* $\{u^{k(l)}\}$ *of the sequence* $\{u^k\}$ *and a control* $u^0 \in \mathcal{U}$ *such that, as* $l \to \infty$,

i. $u^{k(l)} \overset{w^*}{\to} u^0$ *in* $L_\infty(Q, R^m)$,

ii. $\phi(u^{k(l)}) \to \phi(u^0)$ *weakly in* $L_2(Q)$, *pointwise in* $\Omega \times (0,T]$, *and uniformly on any compact subset of* $\Omega \times (0,T]$, *and*

iii. $(\phi(u^{k(l)}))_{x_i} \overset{w}{\to} (\phi(u^0))_{x_i}$, $i=1,\ldots,n$, *in* $L_2(Q)$.

PROOF. Since $\{u^k\}$ is a sequence of $\mathcal{U}$, there is a subsequence of the sequence $\{u^k\}$, again denoted by $\{u^k\}$, and a control $u^0 \in \mathcal{U}$ such that $u^k \overset{w^*}{\to} u^0$ in $L_\infty(Q, R^m) \equiv \{\xi : Q \to R^m : \xi \text{ measurable and } \sum_{i=1}^m \|\xi_i\|_{\infty,Q} < \infty\}$. From Lemma 3.2.6, there is a further subsequence of the subsequence $\{u^k\}$, again denoted by $\{u^k\}$, such that $\phi(u^k)$ converges to a function ϕ pointwise in $\Omega \times (0,T]$ and uniformly on any compact subset of $\Omega \times (0,T]$. By hypotheses, the functions a_j, $j=1,\ldots,n$, c, F_j, $j=1,\ldots,n$, and f are linear in the control variable and satisfy Assumptions (3.2.A3) and (3.2.A12). Thus, $\beta(\cdot,\cdot,u^k(\cdot,\cdot)) \overset{w^*}{\to} \beta(\cdot,\cdot,u^0(\cdot,\cdot))$ in $L_\infty(Q)$ as $k \to \infty$, where β denotes any of the functions a_j, $j=1,\ldots,n$, c, F_j, $j=1,\ldots,n$, and f. The rest of the proof is exactly the same as that given in the proof of Theorem 3.2.9. This completes the proof. □

By virtue of Assumptions (3.2.A8) and (3.2.A7), Minkowski's inequality, Hölder's inequality, and the estimate (3.2.7), we can show that

$$\left| \int_\Omega \{\gamma_0(x, \phi(u)(x,T))\}\, dx \right| \le K_{13} \tag{3.2.88}$$

for all $u \in \mathcal{U}$, where $\phi(u)$ is the weak solution from $V_2^{1,0}(Q)$ of the system (3.2.77) and the constant K_{13} is independent of the controls $u \in \mathcal{U}$.

Similarly, we can verify from Assumption (3.2.A10), Minkowski's inequality, Hölder's inequality, and the estimate (3.2.7) that

$$\left| \int_Q \{\gamma_1(x,t,\phi(u)(x,t))\}\, dx\, dt \right| \le K_{14} \tag{3.2.89}$$

for all $u \in \mathcal{U}$, where the constant K_{14} is independent of $u \in \mathcal{U}$.

We are now in a position to present a result on the existence of optimal controls for Problem (3.2.P2)′.

Theorem 3.2.11. *Suppose Assumptions* (3.2.A1)–(3.2.A3), (3.2.A9)–(3.2.A10), *and* (3.2.A12)–(3.2.A13) *are satisfied, and let* ϕ_0 *be a bounded measurable function on* $\bar{\Omega}$. *Then, Problem* (3.2.P2)′ *has a solution.*

PROOF. From the inequalities (3.2.88) and (3.2.89), we obtain

$$\inf\{J(u): u\in\mathfrak{U}\}\equiv\alpha>-\infty. \tag{3.2.90}$$

Let $\{u^k\}$ be a sequence of controls in $\mathfrak{U}$ such that

$$\lim_{k\to\infty} J(u^k)=\alpha. \tag{3.2.91}$$

By virtue of Theorem 3.2.9, there exists a subsequence $\{u^{k(l)}\}$ of the sequence $\{u^k\}$ and a control $u^0\in\mathfrak{U}$ such that, as $l\to\infty$,

i. $\phi(u^{k(l)})\to\phi(u^0)$ weakly in $L_2(Q)$, pointwise in $\Omega\times(0,T]$, and uniformly on any compact subset of $\Omega\times(0,T]$; and
ii. $(\phi(u^{k(l)}))_{x_i}\overset{w}{\to}(\phi(u^0))_{x_i}$, $i=1,\ldots,n$, in $L_2(Q)$.

Let J_1 denote the first component of J $(=J_1+J_2)$ [given in (3.2.9)]. We shall show that $\lim_{l\to\infty} J_1(u^{k(l)})=J_1(u^0)$.

From the estimate (3.2.7), the inequality (3.2.88), and the fact that T is finite and Ω bounded, we can verify that γ_0 is an operator mapping $L_2(\Omega)$ into $L_1(\Omega)$. Thus, it follows from Theorem 1.2.13 that γ_0 defines a continuous operator from $L_2(\Omega)$ into $L_1(\Omega)$.

For each $u\in\mathfrak{U}$, $\phi(u)$ is the corresponding weak solution from $V_2^{1,0}(Q)$ of the system (3.2.77). Further, by hypothesis, ϕ_0 is a bounded measurable function defined on $\bar{\Omega}$. These together with Assumptions (3.2.A2)–(3.2.A3) and (3.2.A12) imply the hypotheses of Theorem 2.2.7. Thus, it follows from Theorem 2.2.7 that

$$|\phi(u)(x,t)|\le N_0 \tag{3.2.92}$$

for almost all $(x,t)\in Q$ and for all $u\in\mathfrak{U}$. Recall that $\phi(u^{k(l)})\to\phi(u^0)$ pointwise in $\Omega\times(0,T]$. Thus, in particular, $\phi(u^{k(l)})(\cdot,T)\to\phi(u^0)(\cdot,T)$ pointwise in Ω. Therefore, by using the inequality (3.2.92) and the fact that Ω is bounded, it is easy to verify that $\phi(u^{k(l)})\overset{s}{\to}\phi(u^0)$ in $L_2(\Omega)$. By virtue of this fact, including the fact that γ_0 is a continuous operator mapping $L_2(\Omega)$ into $L_1(\Omega)$, and the definition of J_1, we obtain

$$\lim_{l\to\infty} J_1(u^{k(l)})=J_1(u^0). \tag{3.2.93}$$

Following an argument similar to that given for (3.2.93) with some obvious modifications, we can show that

$$\lim_{l\to\infty} J_2(u^{k(l)})\to J_2(u^0). \tag{3.2.94}$$

Combining (3.2.93), (3.2.94), (3.2.91), and (3.2.90), we obtain the proof. □

The next theorem contains a result on the existence of optimal controls for Problem (3.2.P3)′. Note that the cost functional considered in Problem (3.2.P3)′ depends also on the control variable.

Theorem 3.2.12. *Suppose that Assumptions* (3.2.A1)–(3.2.A3), (3.2.A6), *and* (3.2.A11)–(3.2.A12) *are satisfied. Further, let the coefficients* a_j, $j = 1,\dots,n$, c, F_j, $j=1,\dots,n$, *and* f *be linear in the control variable. Then, Problem* (3.2.P3)′ *has a solution.*

PROOF. Using Assumption (3.2.A11), the definition of the cost functional $J(u)$ [given in (3.2.10)], Minkowski's inequality, Hölder's inequality, the estimate (3.2.7), and the fact that T is finite, we can verify that

$$|J(u)| \leq \|\gamma(\cdot,\cdot,0,0)\|_{1,Q} + (\|g_4\|_{2,Q})(\|\phi(u)\|_{2,Q}) \leq N_0 < \infty, \quad (3.2.95)$$

where the constant N_0 is independent of the controls $u \in \mathcal{U}$. Thus, it follows that

$$\inf\{J(u): u \in \mathcal{U}\} \equiv \alpha > -\infty. \quad (3.2.96)$$

Let $\{u^k\}$ be a sequence of controls from $\mathcal{U}$ such that

$$\lim_{k\to\infty} J(u^k) = \alpha. \quad (3.2.97)$$

From Theorem 3.2.10, there is a subsequence of the sequence $\{u^k\}$, again denoted by $\{u^k\}$, and a control $u^0 \in \mathcal{U}$ such that, as $k \to \infty$,

(i) $$u^k \xrightarrow{w^*} u^0 \quad \text{in } L_\infty(Q, R^m);$$

and,

(ii) $$\phi(u^k) \xrightarrow{w} \phi(u^0) \quad \text{in } L_2(Q).$$

For each positive integer ν, let

$$\phi^\nu \equiv \frac{1}{\nu} \sum_{k=1}^{\nu} \phi(u^k). \quad (3.2.98)$$

Clearly

$$\phi^\nu \xrightarrow{w} \phi(u^0) \quad (3.2.99)$$

in $L_2(Q)$ as $\nu \to \infty$. Thus, it is easy to verify that

$$\phi^\nu - \phi(u^\nu) \xrightarrow{w} 0 \quad (3.2.100)$$

in $L_2(Q)$ as $\nu \to \infty$.

Let $k \in [1,\nu]$ be an arbitrary integer. Then it follows from the first inequality of Assumption (3.2.A11) that

$$\begin{aligned} &-g_4(x,t)(\phi(u^k)(x,t) - \phi(u^\nu)(x,t)) \\ &\quad \leq \gamma(x,t,\phi(u^k)(x,t),u^\nu(x,t)) - \gamma(x,t,\phi(u^\nu)(x,t),u^\nu(x,t)) \\ &\quad \leq g_4(x,t)(\phi(u^k)(x,t) - \phi(u^\nu)(x,t)) \end{aligned} \quad (3.2.101)$$

a.e. on Q.

Summing the above inequalities over k from 1 to ν, dividing the obtained inequalities by ν, integrating over Q, and using the definition of J, we obtain

$$-\iint_Q \{g_4(x,t)(\phi^\nu(x,t)-\phi(u^\nu)(x,t))\}\,dx\,dt$$

$$\leq \iint_Q \left\{\frac{1}{\nu}\sum_{k=1}^{\nu} \gamma\big(x,t,\phi(u^k)(x,t),u^\nu(x,t)\big)\right\} dx\,dt - J(u^\nu)$$

$$\leq \iint_Q \{g_4(x,t)(\phi^\nu(x,t)-\phi(u^\nu)(x,t))\}\,dx\,dt. \tag{3.2.102}$$

Since $g_4 \in L_2(Q)$, it follows from the relation (3.2.100) that

$$\lim_{\nu\to\infty} \iint_Q \{g_4(x,t)(\phi^\nu(x,t)-\phi(u^\nu)(x,t))\}\,dx\,dt = 0. \tag{3.2.103}$$

Thus, by taking the limit with respect to ν in (3.2.102), we obtain

$$\lim_{\nu\to\infty} \iint_Q \left\{\frac{1}{\nu}\sum_{k=1}^{\nu} \gamma\big(x,t,\phi(u^k)(x,t),u^\nu(x,t)\big)\right\} dx\,dt = \lim_{\nu\to\infty} J(u^\nu) \tag{3.2.104}$$

Since, for each $(x,t,v)\in Q\times V$, $\gamma(x,t,\cdot,v)$ is convex in R [see Assumption (3.2.A11)], it follows that

$$\frac{1}{\nu}\sum_{k=1}^{\nu} \gamma\big(x,t,\phi(u^k)(x,t),:u^\nu(x,t)\big) \geq \gamma\big(x,t,\phi^\nu(x,t),u^\nu(x,t)\big). \tag{3.2.105}$$

Thus, from the relation (3.2.104) and the above inequality, we have

$$\lim_{\nu\to\infty} \iint_Q \{\gamma(x,t,\phi^\nu(x,t),u^\nu(x,t))\}\,dx\,dt \leq \lim_{\nu\to\infty} J(u^\nu). \tag{3.2.106}$$

We now show that the left-hand side of the above inequality is equal to $J(u^0)$. From the first inequality of Assumption (3.2.A11) with $v=u^\nu(x,t)$, $\theta^1=\phi^\nu(x,t)$, and $\theta^2=\phi(u^0)(x,t)$, we obtain

$$-\iint_Q \{g_4(x,t)(\phi^\nu(x,t)-\phi(u^0)(x,t))\}\,dx\,dt$$

$$\leq \iint_Q \{\gamma(x,t,\phi^\nu(x,t),u^\nu(x,t))\}\,dx\,dt$$

$$-\iint_Q \{\gamma(x,t,\phi(u^0)(x,t),u^\nu(x,t))\}\,dx\,dt$$

$$\leq \iint_Q \{g_4(x,t)(\phi^\nu(x,t)-\phi(u^0)(x,t))\}\,dx\,dt. \tag{3.2.107}$$

Since $g_4 \in L_2(Q)$, it follows from the relation (3.2.99) that the above inequalities, in the limit with respect to ν, reduces to

$$\lim_{\nu\to\infty} \iint_Q \{\gamma(x,t,\phi^\nu(x,t),u^\nu(x,t))\}\,dx\,dt$$

$$= \lim_{\nu\to\infty} \iint_Q \{\gamma(x,t,\phi(u^0)(x,t),u^\nu(x,t))\}\,dx\,dt. \quad (3.2.108)$$

We now apply the second inequality of Assumption (3.2.A11) to the right-hand side of the above relation. For this, let $v^1 = u^\nu(x,t)$, $v^2 = u^0(x,t)$, and $\theta = \phi(u^0)(x,t)$ in that inequality. This gives rise to

$$-\iint_Q \left\{ \sum_{i=1}^m G_i(x,t)\big(u_i^\nu(x,t) - u_i^0(x,t)\big) \right\} dx\,dt$$

$$\le \iint \{\gamma(x,t,\phi(u^0)(x,t),u^\nu(x,t))\}\,dx\,dt - J(u^0)$$

$$\le \iint_Q \left\{ \sum_{i=1}^m G_i(x,t)\big(u_i^\nu(x,t) - u_i^0(x,t)\big) \right\} dx\,dt. \quad (3.2.109)$$

Since $G \in L_1(Q, R^m)$ and $u^\nu \overset{w^*}{\to} u^0$ in $L_\infty(Q, R^m)$, we have

$$\lim_{\nu\to\infty} \iint_Q \left\{ \sum_{i=1}^m G_i(x,t)\big(u_i^\nu(x,t) - u_i^0(x,t)\big) \right\} dx\,dt = 0. \quad (3.2.110)$$

Thus, by taking the limit with respect to ν in the inequalities (3.2.109), we obtain

$$\lim_{\nu\to\infty} \iint_Q \{\gamma(x,t,\phi(u^0)(x,t),u^\nu(x,t))\}\,dx\,dt = J(u^0). \quad (3.2.111)$$

Combining (3.2.106), (3.2.108), and (3.2.111), we have

$$J(u^0) \le \lim_{\nu\to\infty} J(u^\nu). \quad (3.2.112)$$

Thus, the conclusion of the theorem follows from (3.2.112), (3.2.97), and (3.2.96). □

3.3. Computational Methods

3.3.1. Introduction

The aim of this section is to devise methods for solving the optimal control problems considered in the previous two sections.

The computational method developed in Section 3.3.2 is based on a strong variational technique, while that developed in Section 3.3.3 is based on a gradient technique.

3.3.2. Strong Variational Technique

In this subsection, we shall use a strong variational technique to obtain a computational method for solving Problem (3.1.P2), which consists of the parabolic system in general form and a terminal cost functional.

Since any nontrivial problem can only be solved by a computer, it appears to be more practical if admissible controls are taken as piecewise continuous functions rather than as bounded measurable functions as considered in Section 3.1. This (smaller) class of admissible controls is denoted by

$$\mathcal{U}_0 \equiv \{u: \textit{piecewise continuous on } \bar{Q},\ u(x,t)\in V \textit{ for all } (x,t)\in\bar{Q}\},$$

where V is a compact subset of R^m.

Since $\mathcal{U}_0 \subset \mathcal{U}$, all results presented in Sections 3.1.3–3.1.5 remain valid with $\mathcal{U}$ replaced by $\mathcal{U}_0$.

The basic idea behind the computational method presented here is this: If an admissible control u^0 does not satisfy the necessary condition for optimality, then a new admissible control u^1 can be constructed so that $J(u^1)<J(u^0)$. Repeating this process, we obtain a sequence of controls with strictly decreasing cost.

For convenience, Problem (3.1.P2) with $\mathcal{U}$ replaced by $\mathcal{U}_0$ will be referred to as Problem (3.1.P2)′.

Let $\phi(u)$ and $z(u)$ denote the solutions of the system (3.1.1) and the adjoint system (3.1.10), both corresponding to the control $u\in\mathcal{U}_0$.

We introduce the following definition.

Definition 3.3.1. For each $u\in\mathcal{U}_0$, a function $H(u):\bar{Q}\times V\to R$ given by

$$H(u)(x,t,v) \equiv \left[\sum_{i=1}^{n} b_i(x,t,v)(\phi(u))_{x_i}(x,t) + c(x,t,v)\phi(u)(x,t)+f(x,t,v)\right] z(u)(x,t) \quad (3.3.1)$$

is called the *Hamiltonian function*.

To proceed further, we need the following assumptions.

Assumption (3.3.A1)

i. For each $u\in\mathcal{U}_0$, there is a partition of $\bar{Q}$, $\{Q_i(u)\}_{i=1}^{n(u)}$, such that u is continuous in each of $Q_i^0(u)$, where $Q_i^0(u)$ denotes the interior of $Q_i(u)$.
ii. For each $u\in\mathcal{U}_0$, $H(u)(\cdot,\cdot,\cdot)$ is continuous on $\{\bigcup_{i=1}^{n(u)}Q_i^0(u)\}\times V$.

Assumption (3.3.A2). For each $u\in\mathcal{U}_0$, $H(u)(x,t,\cdot)$ has a unique minimum in V for each $(x,t)\in\bigcup_{i=1}^{n(u)}Q_i^0(u)$.

For each $u \in \mathfrak{U}_0$, let $\omega(u)$ denote the function mapping $\bar{Q}$ into V such that for each $(x, t) \in \bigcup_{i=1}^{n(u)} Q_i^0(u)$, $\omega(u)$ minimizes $H(u)(x, t, \cdot)$, and for all other $(x, t) \in \bar{Q}$, $\omega(u)(x, t) = u(x, t)$.

Corresponding to each $u \in \mathfrak{U}_0$, the function $\omega(u)$ has the property that

$$H(u)(x, t, \omega(u)(x, t)) \leq H(u)(x, t, v) \tag{3.3.2}$$

for all $(x, t, v) \in \{\bigcup_{i=1}^{n(u)} Q_i^0(u)\} \times V$ and, for each $(x, t) \in \bigcup_{i=1}^{n(u)} Q_i^0(u)$, the equality holds if and only if $v = \omega(u)(x, t)$. Thus, it follows from the definition of the function $\omega(u)$ that, if $\omega(u)(x, t) \neq u(x, t)$, then

$$H(u)(x, t, \omega(u)(x, t)) < H(u)(x, t, u(x, t)). \tag{3.3.3}$$

The main reason for Assumptions (3.3.A1)–(3.3.A2) lies in the following result.

Theorem 3.3.1. *If Assumptions* (3.3.A1)–(3.3.A2) *are satisfied, then* $\omega(u) \in \mathfrak{U}_0$ *for each* $u \in \mathfrak{U}_0$.

PROOF. Let $u \in \mathfrak{U}_0$ and let (x^*, t^*) be an arbitrary but fixed point in $\bigcup_{i=1}^{n(u)} Q_i^0(u)$. Let $\{(x^k, t^k)\} \subset \bigcup_{i=1}^{n(u)} Q_i^0(u)$ be such that $(x^k, t^k) \to (x^*, t^*)$. Since $\{\omega(u)(x^k, t^k)\}$ is a sequence in V which is a compact subset of R^m, we can extract a subsequence $\{\omega(u)(x^{k(l)}, t^{k(l)})\}$ such that

$$\lim_{l \to \infty} \omega(u)(x^{k(l)}, t^{k(l)}) = \omega^*(u). \tag{3.3.4}$$

Further, it is clear that

$$(x^{k(l)}, t^{k(l)}) \to (x^*, t^*). \tag{3.3.5}$$

In view of (3.3.2), we have

$$H(u)\left(x^{k(l)}, t^{k(l)}, \omega(u)(x^{k(l)}, t^{k(l)})\right) \leq H(u)\left(x^{k(l)}, t^{k(l)}, \omega(u)(x^*, t^*)\right). \tag{3.3.6}$$

Using the second part of Assumption (3.3.A1) and the relations (3.3.4) and (3.3.5), we observe that the above inequality, in the limit with respect to l, reduces to

$$H(u)(x^*, t^*, \omega^*(u)) \leq H(u)(x^*, t^*, \omega(u)(x^*, t^*)). \tag{3.3.7}$$

Therefore, it follows from the inequality (3.3.3) and the uniqueness assumption (3.3.A2) that $\omega^*(u) = \omega(u)(x^*, t^*)$, and hence, $\omega(u)$ is continuous on $\bigcup_{i=1}^{n(u)} Q_i^0(u)$. Thus, we conclude from the definition of $\omega(u)$ that it is a piecewise continuous function mapping $\bar{Q}$ into V. This completes the proof. □

The following lemma gives an expression for the difference in the cost corresponding to two different controls.

Lemma 3.3.1. *Consider Problem* (3.1.P2)′. *Suppose that Assumptions* (2.1.A1), (3.1.A1)–(3.1.A4), *and* (3.1.A7)–(3.1.A9) *are satisfied. Let* $\tilde{Q}$ *be a mea-*

surable subset of Q *with positive (Lebesgue) measure. If* $u^1,\, u^2 \in \mathfrak{U}_0$ *and* $u^1(x,t)=u^2(x,t)$ *for all* $(x,t)\notin \tilde{Q}$, *then*

$$J(u^1)-J(u^2) \geq \iint_{\tilde{Q}} \{H(u^1)(x,t,u^1(x,t))-H(u^1)(x,t,u^2(x,t))\}\,dx\,dt$$

$$-\iint_{\tilde{Q}}\Bigg\{\Bigg[\sum_{i=1}^{n}\big(b_i(x,t,u^1(x,t))-b_i(x,t,u^2(x,t))\big)$$

$$\times\big((\phi(u^1))_{x_i}-(\phi(u^2))_{x_i}(x,t)\big)$$

$$+\big(c(x,t,u^1(x,t))-c(x,t,u^2(x,t))\big)\big(\phi(u^1)(x,t)$$

$$-\phi(u^2)(x,t)\big)\Bigg]z(u^1)(x,t)\Bigg\}\,dx\,dt. \tag{3.3.8}$$

PROOF. Let $\Phi=\phi(u^1)-\phi(u^2)$. Then, for almost all $(x,t)\in Q$,

$$\begin{aligned}
&L(u^1)\Phi(x,t)\\
&\quad=L(u^1)\phi(u^1)(x,t)-L(u^2)\phi(u^2)(x,t)\\
&\qquad+\big(L(u^1)-L(u^2)\big)\big(\phi(u^1)(x,t)-\phi(u^2)(x,t)\big)\\
&\qquad+\big(L(u^2)-L(u^1)\big)\phi(u^1)(x,t)\\
&\quad=f(x,t,u^1(x,t))-f(x,t,u^2(x,t))\\
&\qquad-\sum_{i=1}^{n}\big(b_i(x,t,u^1(x,t))-b_i(x,t,u^2(x,t))\big)\\
&\qquad\times\big((\phi(u^1))_{x_i}(x,t)-(\phi(u^2))_{x_i}(x,t)\big)\\
&\qquad-\big(c(x,t,u^1(x,t))-c(x,t,u^2(x,t))\big)\\
&\qquad\times\big(\phi(u^1)(x,t)-\phi(u^2)(x,t)\big)\\
&\qquad-\sum_{i=1}^{n}\big(b_i(x,t,u^2(x,t))-b_i(x,t,u^1(x,t))\big)\big(\phi(u^1)\big)_{x_i}(x,t)\\
&\qquad-\big(c(x,t,u^2(x,t))-c(x,t,u^1(x,t))\big)\phi(u^1)(x,t)\\
&\quad\equiv Y(x,t).
\end{aligned} \tag{3.3.9}$$

We also have

$$\Phi(x,t)=0, \qquad (x,t)\in\partial\Omega\times[0,T], \tag{3.3.10}$$

and

$$\Phi(x,0)=0, \qquad x\in\Omega. \tag{3.3.11}$$

From Lemma 3.1.1, we recall that

$$\langle L(u^1)\Phi, z(u^1)\rangle = \int_\Omega \{\Phi(x,T)g(x,\phi(u^1)(x,T))\}\,dx, \tag{3.3.12}$$

where $\langle\cdot,\cdot\rangle$ denotes the scalar product in $L_2(Q)$.

By virtue of the definition of the cost functional J, given in (3.1.6), and the inequality of Assumption (3.1.A7), we obtain

$$J(u^1)-J(u^2) = -\int_\Omega \{\gamma(x,\phi(u^2)(x,T))-\gamma(x,\phi(u^1)(x,T))\}\,dx$$

$$\geq \int_\Omega \{\Phi(x,T)g(x,\phi(u^1)(x,T))\}\,dx. \tag{3.3.13}$$

Combining the equality (3.3.12) and the above inequality, we have

$$J(u^1)-J(u^2)\geq \langle L(u^1)\Phi, z(u^1)\rangle. \tag{3.3.14}$$

Substituting the expression (3.3.9) into the right-hand side of the above inequality, noting that, for all $(x,t)\in\tilde{Q}$,

$$b_i(x,t,u^1(x,t))=b_i(x,t,u^2(x,t)), \qquad i=1,\dots,n,$$

$$c(x,t,u^1(x,t))=c(x,t,u^2(x,t)),$$

$$f(x,t,u^1(x,t))=f(x,t,u^2(x,t)),$$

and using the definition of the Hamiltonian function $H(u)$, given in (3.3.1), we obtain inequality (3.3.8). This completes the proof. □

Definition 3.3.2. A control $u\in\mathcal{U}_0$ is said to be an *extremal control* if

$$u(x,t)=\omega(u)(x,t)$$

for all $(x,t)\in\bigcup_{i=1}^{n(u)}Q_i^0(u)$.

REMARK 3.3.1. Clearly, if $u\in\mathcal{U}_0$ is not an extremal control, then there *must* exist a point $(x',t')\in\bigcup_{i=1}^{n(u)}Q_i^0(u)$ such that

$$u(x',t')\neq\omega(u)(x',t').$$

Theorem 3.3.2. *Consider the control Problem* (3.1.P2)′. *Suppose that Assumptions* (2.1.A1), (3.1.A1)–(3.1.A4), (3.1.A7)–(3.1.A9) *and* (3.3.A1), (3.3.A2) *are satisfied. Let* $(x',t')\in\bigcup_{i=1}^{n(u)}Q_i^0(u)$ *be such that* $u(x',t')\neq\omega(u)(x',t')$. *Further, for each integer* $k\geq1$, *let* $\Omega_k\equiv\{x\in\Omega:|x-x'|<1/k\}$, $L_k\equiv\{t\in(0,T):|t-t'|<1/k\}$, *and* $Q_k\equiv\Omega_k\times L_k$. *Then, there exists an integer* $k_0\geq1$ *such that*

$$J(u^k)<J(u) \tag{3.3.15}$$

for all $k > k_0$, *where* u^k *is defined by*

$$u^k(x,t) \equiv (1-\chi_{Q_k}(x,t))u(x,t) + \chi_{Q_k}(x,t)\omega(u)(x,t) \quad (3.3.16)$$

and χ_{Q_k} *is the characteristic function of the set* Q_k.

PROOF. Define

$$\alpha_k \equiv \frac{1}{|Q_k|} \iint_Q \left\{ \sum_{i=1}^{n} (b_i(u) - b_i(u^k))\big((\phi(u))_{x_i} - (\phi(u^k))_{x_i}\big)z(u) \right\} dx\,dt, \quad (3.3.17)$$

$$\beta_k \equiv \frac{1}{|Q_k|} \iint_Q \{ c(u) - c(u^k))(\phi(u) - \phi(u^k))z(u) \} \,dx\,dt. \quad (3.3.18)$$

and

$$\gamma_k \equiv \frac{1}{|Q_k|} \iint_Q \{ (H(u)(u) - H(u)(u^k)) \} \,dx\,dt, \quad (3.3.19)$$

where, for convenience, the variable (x,t) is suppressed, $h(u)$ is used to denote $h(x,t,u(x,t))$, and $|Q_k|$ denotes the Lebesgue measure of the set Q_k.

Since $u^k(x,t) = u(x,t)$ for all $(x,t) \notin Q_k$, it follows from Lemma 3.3.1 that

$$\frac{1}{|Q_k|}\{J(u) - J(u^k)\} \geq \gamma_k - \alpha_k - \beta_k. \quad (3.3.20)$$

From Lemma 3.1.3, we have

$$\alpha_k \to 0 \quad \text{and} \quad \beta_k \to 0 \quad \text{as } k \to \infty. \quad (3.3.21)$$

Next, by examining the proof of Theorem 3.3.1, we observe that (x', t') is not only a continuity point of u but also a continuity point of $\omega(u)$. Thus, it follows from the second part of Assumption (3.3.A1) that $H(u)(\cdot,\cdot,u(\cdot,\cdot))$ and $H(u)(\cdot,\cdot,\omega(u)(\cdot,\cdot))$ are continuous at (x', t'). Since $\omega(u)(x',t') \neq u(x',t')$, it is clear from the inequality (3.3.3) that

$$H(u)(x',t',\omega(u)(x',t')) < H(u)(x',t',u(x',t')). \quad (3.3.22)$$

Therefore, we conclude that

$$\lim_{k\to\infty} \gamma_k > 0. \quad (3.3.23)$$

Combining (3.3.20), (3.3.21), and (3.3.23), we have

$$\lim_{k\to\infty} \frac{1}{|Q_k|}\{J(u) - J(u^k)\} > 0, \quad (3.3.24)$$

and consequently, there exists an integer $k_0 > 0$ such that

$$J(u) - J(u^k) > 0$$

for all $k > k_0$. This completes the proof. □

REMARK 3.3.2. Using expressions (3.3.17)–(3.3.19), the relation (3.3.21), the inequality (3.3.20), and Definition 3.3.2, we obtain a necessary condition for optimality entirely similar to the necessary condition given in Theorem 3.1.3 [see inequality (3.1.54)].

REMARK 3.3.3. If $\mathcal{U}_0$ is replaced by the larger class $\mathcal{U}$, which consists of measurable functions mapping $\bar{Q}$ into V, then results presented in Theorem 3.3.2 can be derived without requiring the first part of Assumption (3.3.A1).

For a given nonextremal control u, it follows from Theorem 3.3.2 that if u is replaced by $\omega(u)$ on a sufficiently small set and kept the same elsewhere, then this new control will give an improvement in the cost. Thus, a computational method may be devised as follows:

Algorithm 3.3.I

(*Initialization*)

1. Set $k=0$ and guess a control $u^{(k)} \in \mathcal{U}_0$.
2. Determine $\phi(u^{(k)})$, $J(u^{(k)})$, $z(u^{(k)})$ and $\omega(u^{(k)})$, where, corresponding to the control $u^{(k)}$, $\phi(u^{(k)})$ is the solution of the system (3.1.1) (in the sense of Definition 3.1.1) and $z(u^{(k)})$ is the weak solution of the adjoint system (3.1.10) (in the sense of Definition 3.1.2).

(*Control Variation*)

3. Calculate

$$\Delta \tilde{H}(u^{(k)}) = \tilde{H}(u^{(k)})(x,t,u^{(k)}(x,t)) - \tilde{H}(u^{(k)})(x,t,\omega(u^{(k)})(x,t)).$$

4. Find a point $(x',t') \in Q$ such that $\Delta\tilde{H}(u^{(k)})$ is maximum (see Remark 3.3.4).
5. Let δ_1 and δ_2 be positive numbers such that the cube $\Omega(x';\delta_1) \equiv \{x \in R^n : |x-x'| < \delta_1\}$ is contained in Ω and $(t'-\delta_2, t'+\delta_2) \subset (0,T)$. Let $l=0$.
6. Let

$$Q^{(l)} \equiv \Omega\left(x'; \frac{\delta_1}{(2)^l}\right) \times \left(t' - \frac{\delta_2}{(2)^l}, t' + \frac{\delta_2}{(2)^l}\right).$$

7. Determine

$$u^{(k,l)}(x,t) \equiv \left(1-\chi_{Q^{(l)}}(x,t)\right)u^{(k)}(x,t) + \chi_{Q^{(l)}}(x,t)\,\omega(u^{(k)})(x,t).$$

8. Determine $\phi(u^{(k,l)})$ and then $J(u^{(k,l)})$.

(*Acceptance Test*)

9. If $J(u^{(k,l)}) < J(u^{(k)})$ then go to Step 10; otherwise set $l=l+1$ and go to Step 6 (see Remark 3.3.5).

(*Reinitialization*)

10. Set $k=k+1$, $u^{(k)} = u^{(k-1,l)}$, $\phi(u^{(k)}) = \phi(u^{(k-1,l)})$, $z(u^{(k)}) = z(u^{(k-1,l)})$, and $J(u^{(k)}) = J(u^{(k-1,l)})$.

11. Determine $\omega(u^{(k)})$.

(*Termination Test*)
12. If $u^{(k)} = \omega(u^{(k)})$ stop, otherwise go to Step 3 (see Remark 3.3.6).

REMARK 3.3.4. Intuitively, the maximum improvement in the cost may be achieved if the point (x', t') in Step 4 is chosen such that $\Delta \tilde{H}(u^k)$ is maximum. However, it is not always true, and also it is not necessary. Thus, any point in Q for which $\omega(u^{(k)})(x,t) \neq u^k(x,t)$ may be chosen for (x', t') in Step 4.

REMARK 3.3.5. From Theorem 3.3.2, it is clear that the test for the improvement of the cost functional will be met for some positive integer l. Thus, the loop between Step 6 and Step 9 is finite.

REMARK 3.3.6. In practice, the algorithm terminates if any of the following conditions is met:

i. The improvement in the cost functional per iteration is less than some number $\varepsilon > 0$.
ii. $\int\int_Q \{ |u^{(k)}(x,t) - \omega(u^{(k)})(x,t)| \} \, dx \, dt$ is less than some number $\varepsilon > 0$.
iii. $\int\int_Q \{ \Delta H(u^{(k)})(x,t) \} \, dx \, dt$ is less than some number $\varepsilon > 0$.

In Algorithm 3.3.I, it is assumed that, corresponding to each $u^{(k)}$, we can solve the system (3.1.1) (in the sense of Definition 3.1.1) and the adjoint system (3.1.10) (in the sense of Definition 3.1.2). In general, direct calculation of the weak solution of the adjoint system fails to be accessible in practice. However, it is possible to calculate approximate solutions of the adjoint system. For details, we refer the reader to the discussion following the algorithm to be given in the next subsection.

In view of Theorem 2.2.7, we observe that $z(u)(x,t) \geq 0$ for all $u \in \mathcal{U}_0$ and for all $(x,t) \in Q$ if the function g appearing in the system (3.1.10) is such that $g(x, \phi(u)(x,T)) \geq 0$ for all $x \in \Omega$ and for all $u \in \mathcal{U}_0$. In fact, if Assumption (3.3.A1) and (3.3.A2) are also satisfied, then it can be shown, by contradiction, that $z(u)(x,t) > 0$ for all $u \in \mathcal{U}_0$ and for all $(x,t) \in Q$. In this case, we do not need to compute the (weak) solutions of the adjoint system (3.1.10) in Algorithm 3.3.I. More precisely, Algorithm 3.3.I with $z(u)$ deleted and $H(u)$ replaced by $\tilde{H}(u)$ remains valid, where

$$\tilde{H}(u)(x,t,v) \equiv \sum_{i=1}^{n} b^i(x,t,v)(\phi(u))_{x_i}(x,t) + c(x,t,v)\phi(u)(x,t) + f(x,t,v). \tag{3.3.25}$$

This version of the algorithm is referred to as Algorithm 3.3.II.

Computationally, the most difficult part in Algorithm 3.3.II is the calculation of the solutions of the system (3.1.1). In general, this is done numerically. At each iteration of the algorithm, we have a solution of the system (3.1.1) for one control, and require a solution to be calculated for a better control. Frequently, the new control is only slightly different from the old one. Thus, it is logical to make use of the solution already calculated to speed up the computation of the new solution. Therefore, we propose to use a fully implicit discretization scheme that is solved by relaxation [Sau.1], with the previous solution as an initial approximation.

An Example. For illustration, we consider the following stochastic nonlinear regulator problem

$$\min_{u\in\mathcal{U}_0} \mathcal{E}\left[\int_0^{\tau}\left\{c_1(\xi_1(t)-c_2)^2+c_3(u(\xi(t),t))^2-c_4\right\}dt\right] \tag{3.3.26}$$

subject to

$$\begin{aligned} d\xi_1(t)&=\big(c_5u(\xi(t),t)\xi_1(t)-c_6(\xi_1(t))^2+c_7\xi_2(t)\big)\,dt+\sigma_1\,dW_1(t),\\ d\xi_2(t)&=\big(c_8\xi_1(t)-c_9\xi_2(t)\big)\,dt+\sigma_2\,dW_2(t) \end{aligned} \tag{3.3.27}$$

for $t\in[0,T]$,

$$\xi(0)=\xi_0, \quad \textit{a random variable with density } q_0\equiv\chi_{[0,1]\times[0,1]}, \tag{3.3.28}$$

where $\mathcal{E}[\cdot]$ denotes the mathematical expectation of the random variable within the square brackets, τ is the stopping time of the process $\xi(t)$ defined by

$$\tau=\inf\left[\{t\in[0,T]:\xi(t)\notin\Omega\}\cup\{T\}\right],$$

$\xi(t)\equiv(\xi_1(t),\xi_2(t))$, $W\equiv(W_1,W_2)$ is the standard 2-dimensional Wiener process, ξ_0 is independent of $W(t)$, $\Omega\equiv(0,1)\times(0,1)$, $Q\equiv\Omega\times(0,T)$, and $\mathcal{U}_0\equiv\{u:Q\to[c_{10},c_{11}]: u \text{ piecewise continuous}\}$.

By [TA.1], the above stochastic optimal control problem is reduced to the following optimal control problem of distributed parameter system:

$$\min_{u\in\mathcal{U}_0}\int_\Omega\{\phi(u)(x,T)\}\,dx \tag{3.3.29}$$

subject to

$$\begin{aligned} (\phi(u))_t(x,t)&=\tfrac{1}{2}(\sigma_1)^2(\phi(u))_{x_1x_1}(x,t)+\tfrac{1}{2}(\sigma_2)^2(\phi(u))_{x_2x_2}(x,t)\\ &\quad+\big(c_5u(x,t)x_1-c_6(x_1)^2+c_7x_2\big)(\phi(u))_{x_1}(x,t)\\ &\quad+(c_8x_1-c_9x_2)(\phi(u))_{x_2}(x,t)\\ &\quad+c_1(x_1-c_2)^2+c_3(u(x,t))^2-c_4, \qquad (x,t)\in Q,\\ \phi(u)(x,0)&=0, \qquad x\in\Omega,\\ \phi(u)(x,t)&=0, \qquad (x,t)\in\partial\Omega\times[0,T]. \end{aligned} \tag{3.3.30}$$

The Hamiltonian function for this problem is given by

$$\tilde{H}(u)(x,t,v) \equiv \left(c_5 v x_1 - c_6 (x_1)^2 + c_7 x_2\right)(\phi(u))_{x_1}(x,t) + (c_8 x_1 - c_9 x_2)(\phi(u))_{x_2}(x,t) + c_1(x_1 - c_2)^2 + c_3(v)^2 - c_4. \tag{3.3.31}$$

This problem with the following constraints was successfully solved in [TRB.1] by using a computer program developed on the basis of Algorithm 3.3.II:

$$\begin{array}{lllll} T=2, & \sigma_1 = 0.5, & \sigma_2 = 0.6, & c_1 = 1, & c_2 = 0.6, \\ c_3 = 0.4, & c_4 = 10, & c_5 = 1.0, & c_6 = 0.05, & c_7 = 0.2, \\ c_8 = 1, & c_9 = 1, & c_{10} = -0.5, & c_{11} = 0.5. & \end{array}$$

Interested readers are referred to the original literature.

3.3.3. *Gradient Technique*

In this subsection, we shall use a gradient technique to obtain a computational method for solving Problem (3.2.P1), which consists of the parabolic system in divergence form and a terminal cost functional.

As in Section 3.3.1, we consider the following class of admissible controls:

$$\mathcal{U}_0 \equiv \left\{u : u \textit{ piecewise continuous on } \bar{Q}, u(x,t) \in V \textit{ for all } (x,t) \in \bar{Q}\right\},$$

where V is a compact and convex subset of R^m.

For convenience, Problem (3.2.P1) with $\mathcal{U}$ replaced by $\mathcal{U}_0$ will be referred to as Problem (3.2.P1)′.

REMARK 3.3.7. Examining the proof of Theorem 2.2.6, we note that every subsequence of the sequence $\{\phi^k\}$ contains a convergent subsequence. Each of these convergent subsequences converges to the same limit, ϕ, in the sense of Theorem 2.2.6. Thus, we conclude that the whole sequence $\{\phi^k\}$ converges to ϕ in the same topology.

For each $v \in V$, define

$$H(v,\theta,z,x,t) = \sum_{j=1}^{n}\left(\sum_{i=1}^{n} a_{ij}(x,t,v)\theta_{x_i}(x,t) + a_j(x,t,v)\theta(x,t)\right) z_{x_j}(x,t) - \sum_{j=1}^{n} b_j(x,t,v)\theta_{x_j}(x,t)z(x,t) - c(x,t,v)\theta(x,t)z(x,t) + \sum_{j=1}^{n} F_j(x,t,v)z_{x_j}(x,t) - f(x,t,v)z(x,t), \tag{3.3.32}$$

where θ, $z \in W_2^{1,0}(Q)$. Similarly, for each $v \in V$, let $H_{u_k}(v, \theta, z, x, t)$ be as defined by Equation (3.3.32) with the functions a_{ij}, $i, j = 1, \ldots, n$, a_j, b_j, $j = 1, \ldots, n$, c, F_j, $j = 1, \ldots, n$, and f in this equation replaced, respectively, by a_{iju_k}, $j = 1, \ldots, n$, a_{ju_k}, b_{ju_k}, $j = 1, \ldots, n$, c_{u_k}, F_{ju_k}, $j = 1, \ldots, n$, and f_{u_k}, where ξ_{u_k} is the partial derivative of ξ with respect to u_k as defined in (3.2.59).

Let $\phi(u)$ and $z(u)$ be the (weak) solutions [from $V_2^{1,0}(Q)$] of the system (3.2.1) and the adjoint system (3.2.12), both corresponding to the control $u \in \mathcal{U}_0$. Then, we define

$$M_k(u)(x,t) \equiv H_{u_k}(u(x,t), \phi(u)(x,t), z(u)(x,t), x, t). \quad (3.3.33)$$

The main result of this subsection is contained in the following theorem.

Theorem 3.3.3. *Consider Problem* (3.2.P1)′. *Suppose that all the hypotheses of Theorem* 3.2.7 *are satisfied. Let u^0 be an arbitrary control in $\mathcal{U}_0$. Then*

$$\begin{aligned}\overline{\lim_{\varepsilon \downarrow 0}}\, & \frac{J(u^0 + \varepsilon(u - u^0)) - J(u^0)}{\varepsilon} \\ & = J'(u^0)(u - u^0) \\ & \leq -\int_0^T \{\mathcal{T}(\phi(u^0), z(u^0), u^0, u)(t) \\ & \quad - \mathcal{F}(\tilde{F}(u^0), z(u^0), u)(t)\}\, dt \end{aligned} \quad (3.3.34)$$

for any $u \in \mathcal{U}_0$, where $\mathcal{T}$ and $\mathcal{F}$ are as defined in (3.2.60) *and* (3.2.61), *respectively.*

PROOF. Let $\varepsilon \in (0, 1]$ and $u \in \mathcal{U}_0$ arbitrary. Since $u^0 \in \mathcal{U}_0$ and V is convex, it follows that $u^0 + \varepsilon(u - u^0) \in \mathcal{U}_0$ for all ε, $0 \leq \varepsilon \leq 1$. Thus, dividing the inequality (3.2.48) by ε, $0 < \varepsilon \leq 1$, and replacing u by $u^0 + \varepsilon(u - u^0)$, we obtain

$$\begin{aligned} & \frac{J(u^0 + \varepsilon(u - u^0)) - J(u^0)}{\varepsilon} \\ & \leq -\frac{1}{\varepsilon} \int_0^T \{\mathcal{L}(u^0 + \varepsilon(u - u^0))(\phi(u^0 + \varepsilon(u - u^0)), z(u^0))(t) \\ & \quad - \mathcal{L}(u^0)(\phi(u^0 + \varepsilon(u - u^0)), z(u^0))(t) \\ & \quad - (\tilde{F}(u^0 + \varepsilon(u - u^0)), z(u^0))(t) + (\tilde{F}(u^0), z(u^0))(t)\}\, dt. \end{aligned}$$
(3.3.35)

By following an argument similar to that given in the proof for Theorem 3.2.7 and noting the previous remark, we obtain

$$\lim_{\varepsilon\downarrow 0}\Big(\frac{1}{\varepsilon}\int_0^T\{\mathfrak{L}(u^0+\varepsilon(u-u^0))(\phi(u^0+\varepsilon(u-u^0)),z(u^0))(t)$$

$$-\mathfrak{L}(u^0)(\phi(u^0+\varepsilon(u-u^0)),z(u^0))(t)$$

$$-(\tilde{F}(u^0+\varepsilon(u-u^0)),z(u^0))(t)+(\tilde{F}(u^0),z(u^0))(t)\}\,dt\Big)$$

$$=\int_0^T\{\mathfrak{T}(\phi(u^0),z(u^0),u^0,u)(t)-\mathfrak{F}(\tilde{F}(u^0),z(u^0),u)(t)\}\,dt. \tag{3.3.36}$$

Thus, by using the above relation and the inequality (3.3.35), the result of the theorem follows.

To proceed further, we introduce the following assumption. □

Assumption 3.3.A3. For each $u\in\mathfrak{U}_0$, $\Sigma_{k=1}^m M_k(u)(\cdot,\cdot)$ is a piecewise continuous function on $\bar{Q}$.

Definition 3.3.3. A control $u\in\mathfrak{U}_0$ is said to be an *extremal control* if it satisfies the necessary condition (3.2.74) of Theorem 3.2.8 for almost all $(x,t)\in Q$.

On the basis of Theorem 3.3.3 *and Definition* 3.3.3, *we have*

Theorem 3.3.4. *Consider the control problem* (3.2.P1)′. *Suppose that all the hypotheses of Theorem* 3.2.7 *and Assumption* (3.3.A3) *are satisfied. If* $u^0\in\mathfrak{U}_0$ *is not an extremal control, then there exists an* $\varepsilon_0>0$ *such that, for all* ε, $0<\varepsilon<\varepsilon_0$,

$$J(u^0+\varepsilon(\hat{u}^0-u^0))<J(u^0), \tag{3.3.37}$$

where $\hat{u}^0$ *is any element from* $\mathfrak{U}_0$ *such that*

$$\sum_{k=1}^m M_k(u^0)(x,t)\hat{u}_k^0(x,t)>\sum_{k=1}^m M_k(u^0)(x,t)u_k^0(x,t) \tag{3.3.38}$$

for almost all (x,t) *in a subset of* Q *with positive (Lebesgue) measure, and* $\hat{u}^0=u^0$ *elsewhere.*

PROOF. Since $u^0\in\mathfrak{U}_0$ is not an extremal control, it follows that

$$\iint_Q\Big\{\sum_{k=1}^m M_k(u^0)(x,t)(\hat{u}_k^0(x,t)-u_k^0(x,t))\Big\}\,dx\,dt>0. \tag{3.3.39}$$

Thus, by Theorem 3.3.3, we have

$$\overline{\lim_{\varepsilon\downarrow 0}} \frac{J(u^0+\varepsilon(\hat{u}^0-u^0))-J(u^0)}{\varepsilon}<0. \tag{3.3.40}$$

This, in turn, implies that the relation (3.3.37) holds true for $\varepsilon>0$ sufficiently small. Thus, the proof is complete. □

On the basis of Theorem 3.3.4, we can devise a conceptual algorithm for solving the control problem (3.2.P1)′. This algorithm guarantees a sequence of controls with strictly decreasing cost.

The algorithm may be stated as follows:

Algorithm 3.3.III

1. Set $k=0$; guess a control $u^{(k)}\in\mathfrak{U}_0$.
2. Compute $\phi(u^{(k)})$, $J(u^{(k)})$, and $z(u^{(k)})$.
3. Determine $v^{(k)}\in\mathfrak{U}_0$ such that

$$\sum_{\kappa=1}^{m} M_\kappa(u^{(k)})(x,t)v_\kappa^{(k)}(x,t)=\max\left\{\sum_{\kappa=1}^{m} M_\kappa(u^{(k)})(x,t)v_\kappa: v\in V\right\}$$

4. Determine

$$\hat{u}_\kappa^{(k)}(x,t)=\begin{cases} u_\kappa^{(k)}(x,t) & \text{if } M_\kappa(u^{(k)})(x,t)=0\\ v_\kappa^{(k)}(x,t) & \text{otherwise.}\end{cases}$$

5. Solve for $0\le\lambda_k\le 1$ (by a line search) so that

$$J(u^{(k)}+\lambda_k(\hat{u}^{(k)}-u^{(k)}))=\min\{J(u^{(k)}+\lambda(\hat{u}^{(k)}-u^{(k)})):0\le\lambda\le 1\}$$

6. Set $u^{(k+1)}=u^{(k)}+\lambda_k(\hat{u}^{(k)}-u^{(k)})$.
7. If $\lambda_k=0$, go to Step 8; otherwise go back to Step 1 with $u^{(k+1)}=u^{(k)}$.
8. Calculate $J(u^{(k)})$.
9. Stop.

In the above algorithm, there are two objections. The first of these is the termination test given in Step 7. The other objection is that we are required to compute the weak solution of the system (3.2.1) and the system (3.2.12) in every iteration; in particular, the system (3.2.1) has to be solved a number of times in each iteration (see Step 5 of the algorithm). Clearly, these fail to be accessible in practice. Thus, the algorithm requires some modifications. For the first objection, we replace Step 7 by Step 7′: Go to Step 8 if any of the following conditions is met.

i. the improvement in the cost functional per iteration is less than some number $\varepsilon>0$;
ii. λ_k is less than some number $\varepsilon>0$; and

iii.

$$\iint_Q \left\{ \sum_{\kappa=1}^{m} M_\kappa(u^{(k)})(x,t)\left(v_\kappa^{(k)}(x,t)-u_\kappa^{(k)}(x,t)\right) \right\} dx\,dt$$

is less than some number $\varepsilon>0$; otherwise go back to Step 1 with $u^{(k+1)}=u^{(k)}$.

The second objection to the above algorithm may be handled as follows: Consider the system (3.2.1). For each $u\in\mathfrak{U}_0$ and for each integer $N\geq 1$, let $\phi^N(u)$ be defined by

$$\phi^N(u)(x,t)\equiv \sum_{\kappa=1}^{N} \gamma_\kappa^N(u)(t)\psi_\kappa(x),$$

where $\{\psi_\kappa\}$ is a fundamental system of functions in $\overset{\circ}{W}{}_2^1(\Omega)$ (for definition, see Section 1.2.2) such that

$$\langle \psi_\kappa, \psi_l\rangle_\Omega \equiv \int_\Omega \{\psi_\kappa(x)\psi_l(x)\}\,dx = \begin{cases} 1 & \text{if } \kappa=l \\ 0 & \text{if } \kappa\neq l, \end{cases}$$

while the functions $\gamma_\kappa^N(u)$ are determined by the corresponding version of the ordinary differential equation (2.2.115) and the initial condition (2.2.114). By examining the proof of Theorem 2.2.3, we note that every subsequence $\{\phi^{N(k)}(u)\}$ of the sequence $\{\phi^N\}$ contains a convergent subsequence, again denoted by $\{\phi^{N(k)}(u)\}$, such that

$$\phi^{N(k)}(u) \overset{w}{\to} \phi(u),$$

$$(\phi^{N(k)}(u))_{x_i} \overset{w}{\to} (\phi(u))_{x_i}, \qquad i=1,\dots,n,$$

in $L_2(Q)$ as $k\to\infty$, where $\phi(u)$ is the unique weak solution of the system (3.2.1). Thus, we conclude that the whole sequence converges to $\phi(u)$ in the same topology. On this basis, $\phi(u)$ is, in practice, approximated by $\phi^{N_0}(u)$, where N_0 is a sufficiently large positive integer.

Similarly, for each $u\in\mathfrak{U}_0$, the weak solution $\hat{z}(u)$ of the system (3.2.15) is, in practice, approximated by $\hat{z}^{N_1}(u)$, where N_1 is a sufficiently large positive integer. Thus, for each $u\in\mathfrak{U}_0$, the function $z(u)$ defined by

$$z^{N_1}(u)(x,t)\equiv\hat{z}(u)(x,T-t)$$

is an approximate solution of the system (3.2.12).

3.4. Notes

Optimal control problems involving distributed parameter systems arise in many physical problems. In most of these problems, the controls act on the boundary and/or the initial conditions, though in some cases the control also appears in the free term of the system. The basic theory and methods dealing with many such problems may be found in the books of Lions [Li.2] and Butkovskiy [Bu.1].

It is known [F1.2, TA.1] that a class of (stochastic) optimal control problems involving an Ito stochastic differential equation with Markov terminal time can be reduced to a class of (deterministic) optimal control problems of the form considered in this chapter. In the reduced problem, the controls appear in the coefficients of the differential operator. Thus, the results presented in [Li.2, Bu.1] are not applicable here. However, there are many other results available in the literature for dealing with this class of problems. Interested readers are referred to [AT.4, AT.5, Fl.1, Fl.2, Fl.3, N.1, NT.1, SC.1, Te.1, Te.2, TA.2, Z.2] (for necessary conditions for optimality), [AT.1, Fl.2, NT.2, NNT.1, RT.2, Te.1, Te.3, TA.3, Tri.1, Z.1] (for existence of optimal controls), [AT.3, RT.1, Te.1, TAW.1] (for optimal parameter selection), and [Boy.1, Fl.1, R.1, RT.3, TRB.1] (for computational methods).

Theorem 3.1.3 contains a necessary condition for optimality for Problem (3.1.P2) (which consists of a parabolic system in general form and a terminal cost functional). Its proof is based on Lemma 3.1.3. However, it is not known whether Lemma 3.1.3 can be extended to cases with more general cost functionals. Thus, results similar to Theorem 3.1.3 are not available for Problem (3.1.P1). Variants of Theorem 3.1.3 may be found in [Fl.2, Li.2, Te.1]. The extension of Theorem 3.1.3 to the case with the time-delayed arguments in the coefficients of the differential operator is given in [Te.2]. For the quasilinear case, see [TA.2].

In reference [AW.1], Ahmed and Wong derive a necessary condition for an optimal control problem for a class of parabolic partial differential equations arising naturally from the optimal control problem involving a stochastic differential equation with jump parameter.

In Section 3.1.6, three theorems for the existence of optimal control (Theorems 3.1.6, 3.1.7, and 3.1.8) are proved. Theorems 3.1.7 and 3.1.8 are essentially due to Fleming [Fl.2] and Teo [Te.3], respectively. The first- and zeroth-order coefficients and the free term of the system considered in Theorem 3.1.6 are linear in the control variable; in Theorem 3.1.7, the free term of the system is only required to be convex in the control variable, while in Theorem 3.1.8, the first- and zeroth-order coefficients and the free term of the system are allowed to be nonlinear in the control variable. However, the cost functional considered in Theorem 3.1.8 is more general than that of Theorem 3.1.7 but less general than that of Theorem 3.1.6. The results of these three theorems are related. For example, if the cost functional of Theorem 3.1.6 is reduced to that of Theorem 3.1.8, then the result of Theorem 3.1.6 is contained in that of Theorem 3.1.8 as a special case. On the other hand, Theorem 3.1.8 does not include Theorem 3.1.7 as a special case even with the same cost functional. The reader is referred to Chapter 5 for a more general approach. For results on the existence of optimal controls for linear parabolic time-lag systems (in general form), see [Te.3, TA.3, Wa.5].

Questions concerning necessary conditions for optimality and existence of optimal controls for the optimal control problems of the form considered in Section 3.1. with controls appearing in the highest-order coefficients remain open. However, there are some results available in the literature [Fl.3, FlR.1] for the case of Lipschitzian rather than bounded measurable controls.

In Section 3.2, optimal control problems involving parabolic systems in divergence form together with a first boundary condition are considered. Since the system is in divergence form, the highest-order coefficients are not required to be smooth. Thus, we are allowed to include the control variable in these highest-order coefficients. A necessary condition for optimality in integral form and its pointwise version for Problem (3.2.P1) are derived, respectively, in Theorems 3.2.7 and 3.2.8. Theorem 3.2.8 is essentially the same as Theorem 1 of Zolezzi [Z.2]. Its proof is based on Lemma 3.2.3. As for Lemma 3.1.3, we do not know whether Lemma 3.2.3 can be extended to cases with more general cost functionals. Thus, similar results to those given in Theorems 3.2.7 and 3.2.8 are not available for Problem (3.2.P2) or (3.2.P3).

In Section 3.2.6, two existence theorems for optimal controls (Theorems 3.2.11 and 3.2.12) are proved. Unlike the results on the necessary conditions of optimality, the proofs of results concerning existence of optimal controls do not depend on Lemma 3.2.3. Thus, we can handle a more general cost functional than that of Problem (3.2.P1). However, the first- and second-order coefficients are not allowed to include the control variable. Without this assumption, it is not possible to prove the existence of optimal controls using the approach of Section 3.2.6. We now discuss the main difficulty involved. Let $\{u^n\}$ be a minimizing sequence of admissible controls and $\{\phi^n\}$ the corresponding sequence of weak solutions of the system (3.2.1). Then, from the assumptions on the class of admissible controls, the coefficients, and the estimates of solutions, we can find functions u^0 and ϕ^0 such that $u^n \overset{w^*}{\to} u^0$ in L_∞ and $\phi^n \overset{w}{\to} \phi^0$ in $W_2^{1,1}(Q)$ as $n \to \infty$. However, it is not possible to show that ϕ^0 is the weak solution of the system (3.2.1) corresponding to u^0. Without this result, the approach of Section 3.2.6 breaks down.

For parabolic time-lag systems in divergence form, the reader is referred to [N.1, NT.1, NT.4] (for necessary conditions) and [Li.2, NT.3, NNT.1, RT.2] (for existence of optimal controls).

In Section 3.3, we present computational methods for solving the optimal control problems considered in Sections 3.1 and 3.2, more precisely, for Problem (3.1.P2) and Problem (3.2.P1).

To solve Problem (3.1.P2), a direct method is to use the necessary condition (3.1.54) to obtain the form of the extremal policy. The resulting expression, as a functional of the state ϕ and the adjoint state z, is then substituted into the system equations (3.1.1) and the adjoint system equations (3.1.10). This gives rise to a two-point boundary value problem involving a coupled system of nonlinear parabolic partial differential equations associated with the boundary conditions as given in the system equations (3.1.1) and (3.1.10). A similar argument also applies to the optimal control problem (3.2.P1).

In [Boy.1, Fl.1, R.1, RT.3, TRB.1, WA.1], various iterative methods are presented. The basic idea behind these iterative methods is to construct a sequence of control $\{u^k\}$ from a nonextremal control u^0 such that $J(u^{k+1}) < J(u^k)$, $k=0,1,2,\ldots$. In each of these iterative methods, we are required to solve a sequence of linear partial differential equations rather than a coupled system of highly nonlinear differential equations arising from using the two-point boundary value problem approach mentioned above.

The iterative method presented in Section 3.3.2 is due to Teo, Reid, and Boyd [TRB.1] and is developed on the basis of a strong variational technique. Given a nonextremal control u^0, this iterative method generates a sequence of controls $\{u^k\}$ such that $J(u^{k+1}) < J(u^k)$, $k=0,1,2,\ldots$. However, we have not shown that the limit controls of the sequence $\{u^k\}$ (if they exist) satisfy the necessary conditions. This is the main subject matter in references [R.1, RT.3]. To achieve this task, we require a more subtle way of constructing the sequence of subsets on which the control is perturbed. More precisely, instead of being any sequence of rectangles that converges to a point, it is necessary to specify the Lebesgue measure of the set as part of the construction. Interested readers are referred to references [R.1, TR.3] for details. Several numerical examples solved by using Algorithm II of Section 3.3.2 may be found in [Boy.1, R.1, TRB.1].

The iterative method developed in Section 3.3.3 is based on a gradient technique. A variant of this result may be found in [R.1, WA.1]. Several numerical examples may also be found in these articles. As for the strong variational method given in Section 3.3.2, the gradient method presented in Section 3.3.3 constructs a sequence of controls $\{u^k\}$ from a nonextremal control u^0 such that $J(u^{k+1}) < J(u^k)$, $k=0,1,2,\ldots$. However, reference [R.1] goes beyond this; it contains a result on the convergence of the algorithm similar to that of reference [RT.1].

Instead of assuming that the sequence of controls converges to a limit point as considered in references [R.1, RT.1], it would be more interesting and useful if one could show that the sequence converges. This question remains open.

Note that all the optimal control problems considered in this chapter are of an open loop nature. Results concerning feedback control for these problems are not known.

In [Te.1, AT.4, AT.5], an optimal control problem for linear second-order parabolic partial differential equation in divergence form with Cauchy condition is considered. This class of problems arises naturally from a class of optimal control problems involving an Ito stochastic differential equation with fixed terminal time [TA.1]. Recall the optimal control problem involving parabolic system in divergence form with Cauchy condition:

Subject to the dynamical system

$$L(u)\phi(x,t)=f(x,t,u(x,t)), \qquad (x,t)\in R^n\times(0,T),$$
$$\phi(x,0)=\phi_0(x), \qquad x\in R^n, \tag{1}$$

find a control $u\in\mathfrak{U}$ such that the cost functional

$$J(u)=\int_\Omega\{g(x,\phi(u)(x,T))\}\,dx \tag{2}$$

is minimized.

Since the Cauchy problem has unbounded domain, it is more involved than the first boundary value problem considered in Section 3.2. One approach to tackle this problem is now briefly indicated.

Let $\Omega_k\equiv\{x\in R^n:|x|<k\}$. Then, we consider the following sequence of first boundary value problems:

$$L(u)\phi(x,t)=f(x,t,u(x,t)), \qquad (x,t)\in\Omega_k\times(0,T),$$
$$\phi(x,0)=\phi_0(x), \qquad x\in\Omega_k, \tag{3}$$
$$\phi(x,t)=0, \qquad (x,t)\in\partial\Omega_k\times[0,T].$$

Note that the estimate (2.2.83) of Theorem 2.2.1 does not depend on the bound of the spatial domain Ω. Thus, by applying this estimate to the first boundary value problems (3), we can show that

i. the same estimate remains valid for the Cauchy problem;
ii. for each $u\in\mathfrak{U}$, the Cauchy problem (1) has a unique weak solution $\phi(u)$,

$$\phi^k(u)\xrightarrow{\text{w}}\phi(u)$$

$$(\phi^k(u))_{x_i}\xrightarrow{\text{w}}(\phi(u))_{x_i}, \qquad i=1,\ldots,n,$$

in $L_2(R^n\times(0,T))$ as $k\to\infty$, where

$$\phi^k(u)(x,t)\equiv\begin{cases}\tilde{\phi}^k(u)(x,t) & \text{on } \Omega_k\times(0,T)\\ 0 & \text{elsewhere,}\end{cases}$$

and for each $u\in\mathfrak{U}$ and for each positive integer k, $\tilde{\phi}^k(u)$ is the unique weak solution of the first boundary value problem (3); and
iii. the results stated in (i) and (ii) above are valid for the adjoint system of the Cauchy problem (1).

On the basis of the facts mentioned in (i)–(iii) above, we can prove, by using arguments similar to those given for the necessary conditions for optimality presented in Theorems 3.2.7 and 3.2.8, that these two theorems are valid for the optimal control problem involving the Cauchy problem (1) and the cost functional (2). For details, the reader is referred to [Te.1, AT.4, AT.5].

Following similar arguments as those given for the results concerning the existence of optimal controls presented in Theorems 3.2.11 and 3.2.12, we can again show that these two theorems remain valid for the Cauchy problem. The details may be found in [RT.2].

The necessary conditions for optimality for the optimal control problem involving the system (1) and the cost functional (2) have been extended to the case in which the time-delayed arguments appear in the coefficients of the differential operator of the system. Interested readers are referred to [N.1, NT.1] for details.

CHAPTER FOUR

OPTIMAL CONTROL OF HYPERBOLIC PARTIAL DIFFERENTIAL SYSTEMS

The optimal control theory of hyperbolic partial differential equations has been studied in many works, including [Ah.1, Az.1, Bu.1, Eg.1, Fa.9, Li.2, Pu.1, PS.1, Ru.1, Ru.3, Ru.8, Sa.1, Sch.1, Su.1, Su.2, Su.3, Su.4, Wa.4]. Certainly, its potential applications are numerous. For instance, the problems considered in [Ah.1, Ru.3, Ru.8, Wa.4] arise in studying the optimal control of power flow in electrical networks and of counterflow processes in chemical engineering, vibration, and plasma confinement.

This chapter is divided into four sections. In Section 4.1, we consider a class of optimal control problems involving first-order hyperbolic systems with one spatial variable and with boundary controls. The cost functional to be optimized is quadratic in both the state and control variables. This class of problems arises in studying counterflow processes in chemical engineering. The main reference of this section is the work of Russell [Ru.8]. Note that much work along similar lines is available in the literature on parabolic equations; interested readers are referred to [AM.1, D.2, KE.1, KG.1, LR.1]. In Section 4.2, we consider a class of optimal control problems involving second-order hyperbolic systems in one spatial variable with data from the space of distributions. This class of problems arises in studying power flow in electrical engineering. The main reference of this section is the work of Ahmed [Ah.1]. In Section 4.3, we consider a class of optimal control problems involving a general second-order nonlinear hyperbolic system with the control appearing in the coefficients of the differential equations. The boundary condition is of Darboux type. The main reference of this section is the work of Suranarayana [Su.1–Su.4]. Finally, Section 4.4 briefly discusses certain related results and poses several open problems.

4.1. First-Order Hyperbolic Systems in One Spatial Variable with Boundary Control

4.1.1. Introduction

Transmission lines in electrical engineering, problems of vibration, and counterflow processes in chemical engineering [Ru.3, Ru.8] can be described, under simplifying assumptions, by linear hyperbolic systems. In this section, we consider a class of symmetric linear hyperbolic systems with controls acting on the boundary of the spatial domain. We assume that the performance of the controlled system is measured through a cost functional that is quadratic in both the state and the control variable.

In Section 4.1.2, we describe the system and impose certain basic assumptions. Results concerning the existence and uniqueness of solutions of the system and certain important properties of the solutions are given in Section 4.1.3. In Sections 4.1.4 and 4.1.5., an optimal control problem is solved giving the necessary and sufficient conditions for optimality. The question of existence of optimal controls is discussed in a remark.

4.1.2. System Description

As in Chapter 3, we shall denote time by t and spatial variable by x. Let $\Omega \equiv (0,1)$ and $I \equiv (0,T)$ with $T<\infty$. For a measurable subset G of R^s, let $\bar{G}$ denote the closure of G. We consider the class of linear hyperbolic partial differential equations of the form

$$\frac{\partial\tilde{\phi}(x,t)}{\partial t}=\tilde{A}(x)\frac{\partial\tilde{\phi}(x,t)}{\partial x}+\tilde{B}(x)\tilde{\phi}(x,t), \tag{4.1.1}$$

$(x,t)\in\Omega\times I\equiv Q$, where $x\to\tilde{A}(x)$ and $x\to\tilde{B}(x)$ are $n\times n$ matrix-valued functions on $\bar{\Omega}$ and $(x,t)\to\tilde{\phi}(t,x)$ is an n vector-valued function on $\bar{Q}$ representing the state of the process. Depending on the problem, the boundary conditions of the process $\tilde{\phi}$ may be specified accordingly. We assume that these conditions are described as follows:

$$\begin{aligned} &\tilde{\phi}(x,0)=\tilde{\phi}_0(x), && x\in\bar{\Omega},\\ &\tilde{C}_0\tilde{\phi}(0,t)=0, && t\in\bar{I}, \\ &\tilde{C}_1\tilde{\phi}(1,t)=Cu(t), && t\in\bar{I}, \end{aligned} \tag{4.1.2}$$

where u is an r-control vector and $\tilde{C}_0, \tilde{C}_1, C$ are matrices of compatible dimensions.

Let G be any connected subset of R^s and l a nonnegative integer. Denote by $C^l(G, R^r)$ the class of all those functions $\{z\}$ mapping G into R^r such that $z_i\in C^l(G)$, $i=1,\ldots,r$. Similarly, let $C^l(G, R^{r_1\times r_2})$ denote the class of all those $r_1 \times r_2$ matrix-valued functions $\{z\}$ defined on G such that $z_{ij} \in C^l(G)$, $i, j=1,\ldots,n$.

Let $C(\bar{I}, L_2(\Omega, R^n))$ be the class of all those functions $\{z\}$ mapping $\bar{Q}$ into R^n such that $\|z(\cdot,t)\|_{2,\Omega}$ are continuous functions on $\bar{I}$. Similarly, let $C(\bar{\Omega}, L_2(I, R^n))$ be the class of all those functions $\{z\}$ mapping $\bar{Q}$ into R^n such that $\|z(x,\cdot)\|_{2,I}$ are continuous functions on $\bar{\Omega}$.

If $\tilde{A}$ belongs to $C^1(\bar{\Omega}, R^{n\times n})$ and, for each $x \in \bar{\Omega}$, $\tilde{A}(x)$ has distinct characteristic roots $(\lambda_1(x),\ldots,\lambda_n(x))$, then [CH.2] there exists an $n\times n$ matrix-valued function Γ, which is also in $C^1(\bar{\Omega}, R^{n\times n})$ and is invertible for each $x \in \bar{\Omega}$, such that

$$(\Gamma(x))^{-1}\tilde{A}(x)\Gamma(x)=A(x);$$

the superscript -1 denotes the inversion and $A(x)=\text{diag}(\lambda_1(x),\ldots,\lambda_n(x))$ with $\lambda_i \in C^1(\bar{\Omega})$, $i=1,\ldots,n$.

If $\tilde{A}$ satisfies the above properties, then the system (4.1.1) is called *strictly hyperbolic.*

Under the above assumptions, the differential equation (4.1.1.) can be written in the normal form

$$\frac{\partial\phi(x,t)}{\partial t}=A(x)\frac{\partial\phi(x,t)}{\partial x}+B(x)\phi(x,t), \qquad (x,t)\in Q, \quad (4.1.3)$$

where

$$\begin{aligned}
A(x)&=(\Gamma(x))^{-1}\tilde{A}(x)\Gamma(x)=\text{diag}(\lambda_1(x),\ldots,\lambda_n(x)),\\
B(x)&=(\Gamma(x))^{-1}\tilde{A}(x)\Gamma_x(x)+(\Gamma(x))^{-1}\tilde{B}(x)\Gamma(x),\\
\tilde{\phi}&=\Gamma(x)\phi(x,t).
\end{aligned}$$

The initial and boundary conditions are given by

$$\begin{aligned}
\phi(x,0)&=\phi_0(x)\equiv(\Gamma(x))^{-1}\tilde{\phi}_0(x),\\
C_0\phi(0,t)&=0, \qquad (4.1.4)\\
C_1\phi(1,t)&=Cu(t),
\end{aligned}$$

where $C_0=\tilde{C}_0\Gamma(0)$ and $C_1=\tilde{C}_1\Gamma(1)$.

For convenience, we may assume that the eigenvalues λ_i have been rearranged, if necessary, in increasing order, that is,

$$\lambda_1(x)<\lambda_2(x)<\cdots<\lambda_p(x)<0<\lambda_{p+1}(x)<\cdots<\lambda_n(x),$$

for $x\in\bar{\Omega}$. Then we write

$$A(x)=\begin{bmatrix} A^-(x) & 0\\ 0 & A^+(x)\end{bmatrix}, \qquad (4.1.5)$$

where $A^-(x)=\text{diag}(\lambda_1(x),\ldots,\lambda_p(x))$ and $A^+(x)=\text{diag}(\lambda_{p+1}(x),\ldots,\lambda_{p+q}(x))$ with $p+q=n$. Accordingly, the state ϕ can be represented as $\phi=(\phi^-,\phi^+)$ with $\phi^-\in R^p$, $\phi^+\in R^q$, and the matrices C_0 and C_1 as

$C_0 = [C_0^- \mid C_0^+]$ and $C_1 = [C_1^- \mid C_1^+]$, respectively. We assume that C_0 and C_1 are $p \times n$ and $q \times n$ matrices, respectively, and that the $p \times p$ matrix C_0^- and $q \times q$ matrix C_1^+ are invertible.

Under these hypotheses (assumed throughout this section) we can rewrite the boundary conditions as

$$\begin{aligned} \phi^-(0,t) &= D_0 \phi^+(0,t), \\ \phi^+(1,t) &= D_1 \phi^-(1,t) + Du(t), \end{aligned} \tag{4.1.6}$$

where $D_0 = -(C_0^-)^{-1}(C_0^+)$, $D_1 = -(C_1^+)^{-1}(C_1^-)$, and $D = (C_1^+)^{-1}C$.

Thus, the mixed initial boundary value problem (4.1.1)–(4.1.2) is reduced to its normal characteristic form:

$$\begin{aligned} \frac{\partial \phi(x,t)}{\partial t} &= A(x)\frac{\partial \phi(x,t)}{\partial x} + B(x)\phi(x,t), && (x,t) \in Q, \\ \phi(x,0) &= \phi_0(x), && x \in \Omega, \\ \phi^-(0,t) &= D_0\phi^+(0,t), && t \in I, \\ \phi^+(1,t) &= D_1\phi^-(1,t) + Du(t), && t \in I. \end{aligned} \tag{4.1.7}$$

REMARK 4.1.1. The above system is precisely in the form in which the characteristic method, in conjunction with Volterra integral equation, can be used to uniquely solve for ϕ. In fact, for each $\phi_0 \in C^1(\bar{\Omega}, R^n)$ and $u \in C^1(\bar{I}, R^r)$, the system (4.1.7) has a unique classical solution $\phi(u) \in C^1(\bar{Q}, R^n)$. This result is contained in [CH.2]. By a classical solution of the system (4.1.7), we mean that all the equations of the system are satisfied everywhere in their domains of definition.

4.1.3. Existence and Uniqueness of Solutions

With the help of an a priori estimate for the solutions, we can prove that, if $\phi_0 \in L_2(\Omega, R^n)$, then system (4.1.7) admits, for each $u \in L_2(I, R^r)$, a unique (weak) solution $\phi(u) \in L_2(Q, R^n)$ in the following sense.

Definition 4.1.1. For each $u \in L_2(I, R^r)$, a function $\phi(u)\colon \bar{Q} \to R^n$ is said to be a *weak solution of system* (4.1.7) if

i. $\phi(u)$ satisfies the differential equations in the generalized sense on Q;
ii. $\phi(u)(x,0) = \phi_0(x)$ a.e. on Ω; and
iii. $(\phi(u))^-(0,t) = D_0(\phi(u))^+(0,t)$ a.e. on I, and $(\phi(u))^+(1,t) = D_1(\phi(u))^-(1,t) + Du(t)$ a.e. on I.

Lemma 4.1.1. *Consider system* (4.1.7). *Suppose that* $A \in C^1(\bar{\Omega}, R^{n \times n})$, $B \in C(\bar{\Omega}, R^{n \times n})$, $\phi_0 \in C^1(\bar{\Omega}, R^n)$, *and* $u \in C^1(\bar{I}, R^r)$. *Then the system admits a unique classical solution* $\phi \in C^1(\bar{Q}, R^n)$. *Further, there exists a constant* K

(>0), *independent of* ϕ_0 *and* u, *such that*

$$\|\phi(\cdot,t)\|_{2,\Omega}^2 \le K\left[\|\phi_0\|_{2,\Omega}^2 + \int_0^t |u(\theta)|^2\,d\theta\right] \tag{4.1.8}$$

for all $t\in\bar{I}$.

PROOF. The existence and uniqueness of the classical solution of system (4.1.7) is given in Remark 4.1.1. We need only to give a proof of the inequality. Let E be a diagonal $n\times n$ matrix-valued function on $\bar{\Omega}$ with nonzero diagonal entries in $C^1(\bar{\Omega})$ and let $\phi\in C^1(\bar{Q}, R^n)$ denote the classical solution of the system. Let $Q_\tau=\{(x,t): x\in\bar{\Omega},\ 0\le t\le\tau\}$ and consider the expression

$$\int_{Q_\tau}\!\!\int\left\{\frac{\partial}{\partial t}(\phi(x,t), E(x)\phi(x,t)) - \frac{\partial}{\partial x}(\phi(x,t), E(x)A(x)\phi(x,t))\right\}dx\,dt, \tag{4.1.9}$$

where $(\cdot,\cdot)$ denotes the scalar product in any finite-dimensional real euclidean space.

Performing the differentiations within the braces and noting that

$$E(x)\frac{\partial\phi(x,t)}{\partial t} = E(x)A(x)\frac{\partial\phi(x,t)}{\partial x} + E(x)B(x)\phi(x,t)$$

on Q, we obtain

$$\begin{aligned}\int_{Q_\tau}\!\!\int&\left\{\frac{\partial}{\partial t}(\phi(x,t), E(x)\phi(x,t)) - \frac{\partial}{\partial x}(\phi(x,t), E(x)A(x)\phi(x,t))\right\}dx\,dt\\ &= \int_{Q_\tau}\!\!\int\left\{\phi(x,t), \left[E(x)B(x)+(B(x))'E(x)\right.\right.\\ &\qquad \left.\left. -\frac{\partial}{\partial x}(E(x)A(x))\right]\phi(x,t))\right\}dx\,dt,\end{aligned} \tag{4.1.10}$$

where the prime denotes transposition.

For any $t\in\bar{I}$, let us define

$$\langle\phi(\cdot,t), E(\cdot)\phi(\cdot,t)\rangle_\Omega = \int_\Omega\{(\phi(x,t), E(x)\phi(x,t))\}dx. \tag{4.1.11}$$

Then, performing integration by parts in (4.1.10), we obtain

$$\begin{aligned}\langle\phi(\cdot,\tau), E(\cdot)\phi(\cdot,\tau)\rangle_\Omega &= \langle\phi_0, E\phi_0\rangle_\Omega + \int_0^\tau\{(\phi(1,t), E(1)A(1)\phi(1,t))\\ &\qquad -(\phi(0,t), E(0)A(0)\phi(0,t))\}dt\\ &\qquad + \int_{Q_\tau}\!\!\int\left\{\left(\phi(x,t), \left[E(x)B(x)+(B(x))'E(x)\right.\right.\right.\\ &\qquad \left.\left.\left. -\frac{\partial}{\partial x}(E(x)A(x)\right]\phi(x,t)\right)\right\}dx\,dt,\end{aligned} \tag{4.1.12}$$

where all the expressions are well defined since the functions involved are all smooth. From the definition of the matrix E, we can take it to be

$$E(x)=\begin{bmatrix} e^-(x)I_p & 0 \\ 0 & e^+(x)I_q \end{bmatrix}, \tag{4.1.13}$$

where I_p and I_q are identity matrices of dimensions $p\times p$ and $q\times q$, respectively, while e^- and e^+ are C^1-functions on $\overline{\Omega}$ satisfying these additional properties:

$$\varepsilon_0=e^-(0)\le e^-(x)\le e^-(1)=1, \qquad \varepsilon_0>0, \tag{4.1.14}$$

$$\varepsilon_1=e^+(1)\le e^+(x)\le e^+(0)=1, \qquad \varepsilon_1>0. \tag{4.1.15}$$

Then

$$(\phi(1,t),E(1)A(1)\phi(1,t))$$
$$=(\phi^-(1,t),A^-(1)\phi^-(1,t))+\varepsilon_1(\phi^+(1,t),A^+(1)\phi^+(1,t)) \tag{4.1.16}$$

and

$$-(\phi(0,t),E(0)A(0)\phi(0,t))$$
$$=-\{\varepsilon_0(\phi^-(0,t),A^-(0)\phi^-(0,t))+(\phi^+(0,t),A^+(0)\phi^+(0,t))\}. \tag{4.1.17}$$

Substituting the boundary conditions of system (4.1.7) into (4.1.16) and (4.1.17), we obtain

$$(\phi(1,t),E(1)A(1)\phi(1,t))$$
$$\le(\phi^-(1,t),[A^-(1)+2\varepsilon_1(D_1)'A^+(1)D_1]\phi^-(1,t))$$
$$+2\varepsilon_1(u(t),[D'A^+(1)D]u(t)) \tag{4.1.18}$$

and

$$-(\phi(0,t),E(0)A(0)\phi(0,t))$$
$$=(\phi^+(0,t),-[A^+(0)+\varepsilon_0D_0'A^-(0)D_0]\phi^+(0,t)). \tag{4.1.19}$$

In the estimate (4.1.18), we have used the fact that, for a positive definite symmetric matrix M,

$$(y+z,M[y+z])\le2\{(y,My)+(z,Mz)\}. \tag{4.1.20}$$

Since $A^-<0$ and $A^+>0$, we can choose ε_1 and ε_0 sufficiently small so that

$$A^-(1)+2\varepsilon_1D_1'A^+(1)D_1\le0, \tag{4.1.21}$$

$$-[A^+(0)+\varepsilon_0D_0'A^-(0)D_0]\le0. \tag{4.1.22}$$

For such a choice, inequalities (4.1.18) and (4.1.19) reduce, respectively, to

$$(\phi(1,t), E(1)A(1)\phi(1,t)) \leq 2\varepsilon_1(u(t), [D'A^+(1)D]u(t)) \tag{4.1.23}$$

and

$$-(\phi(0,t), E(0)A(0)\phi(0,t)) \leq 0. \tag{4.1.24}$$

Substituting inequalities (4.1.23) and (4.1.24) into (4.1.12), it follows that

$$\begin{aligned}\langle \phi(\cdot,\tau), E(\cdot)\phi(\cdot,\tau)\rangle_\Omega &\\ \leq \langle \phi_0, E\phi_0\rangle_\Omega &+ 2\varepsilon_1 \int_0^\tau \{(u(\theta), [D'A^+(1)D]u(\theta)\} d\theta \\ &+ \int_{Q_\tau}\int \Big\{\Big(\phi(x,t), \Big[E(x)B(x) + (B(x))'E(x) \\ &- \frac{\partial}{\partial x}(E(x)A(x))\Big]\phi(x,t)\Big)\Big\} dx\, dt. \end{aligned} \tag{4.1.25}$$

Since the matrices $D, A^+(1)$ are constant and E, A belong to $C^1(\bar{\Omega}, R^{n\times n})$, there exists a constant $N>0$ such that

$$\max\Big\{|D'A^+(1)D|, \max_{x\in\bar{\Omega}} |E(x)B(x) + (B(x))'E(x) - \frac{\partial}{\partial x}(E(x)A(x))|\Big\} \leq N. \tag{4.1.26}$$

Further, due to our choice of the matrix E, we have

$$\varepsilon |z|^2 \leq (z, E(x)z) \leq |z|^2 \tag{4.1.27}$$

for all $z \in R^n$ uniformly on $\bar{\Omega}$ with $\varepsilon = \min\{\varepsilon_0, \varepsilon_1\}$. Using these estimates in (4.1.25), we obtain

$$\begin{aligned}\|\phi(\cdot,\tau)\|_{2,\Omega}^2 \leq \frac{1}{\varepsilon}\|\phi_0\|_{2,\Omega}^2 &+ \frac{2N\varepsilon_1}{\varepsilon}\int_0^\tau \{|u(t)|^2\} dt \\ &+ \frac{N}{\varepsilon}\int_0^\tau \{\|\phi(\cdot,t)\|_{2,\Omega}^2\} dt \end{aligned} \tag{4.1.28}$$

for all $\tau \in \bar{I}$. Defining $K_1 = \max\{1/\varepsilon, 2N\varepsilon_1/\varepsilon\}$, $K_2 = N/\varepsilon$ and using Gronwall's lemma, it follows from (4.1.28) that, for $t \in \bar{I}$,

$$\|\phi(\cdot,t)\|_{2,\Omega}^2 \leq K\Big[\|\phi_0\|_{2,\Omega}^2 + \int_0^t |u(\theta)|^2 d\theta\Big], \tag{4.1.29}$$

where $K \equiv K(N, \varepsilon_0, \varepsilon_1, T)$.

This completes the proof of the lemma. □

With the help of the above lemma, we prove the following result.

Theorem 4.1.1. *Consider system* (4.1.7). *Suppose that the matrices A and B satisfy the hypotheses of Lemma* 4.1.1. *Then, for every* $\phi_0 \in L_2(\Omega, R^n)$ *and*

$u \in L_2(I, R^r)$, the system has a unique weak solution $\phi \in L_2(Q, R^n)$. Further,

$$\phi \in L_2(Q, R^n) \cap L_{2,\infty}(Q, R^n) \cap C\big(\bar{I}, L_2(\Omega, R^n)\big) \cap C\big(\bar{\Omega}, L_2(I, R^n)\big).$$

PROOF. Let $\{\phi_0^k\} \subset C^1(\bar{\Omega}, R^n)$ and $\{u^k\} \subset C^1(\bar{I}, R^r)$ be two sequences such that $\phi_0^k \xrightarrow{s} \phi$ (strongly) in $L_2(\Omega, R^n)$ and $u^k \xrightarrow{s} u$ (strongly) in $L_2(I, R^r)$. Now, for each positive integer k, let us consider the following system:

$$\begin{aligned} &\frac{\partial \phi(x,t)}{\partial t} = A(x)\frac{\partial \phi(x,t)}{\partial x} + B(x)\phi(x,t), \qquad (x,t) \in Q, \\ &\phi(x,0) = \phi_0^k(x), \qquad x \in \Omega, \\ &\phi^-(0,t) = D_0\phi^+(0,t), \qquad t \in I, \\ &\phi^+(1,t) = D_1\phi^-(1,t) + Du^k(t), \qquad t \in I. \end{aligned} \tag{4.1.30}$$

According to Lemma 4.1.1, the above system has a unique classical solution $\phi^k \in C^1(\bar{Q}, R^n)$.

Using the estimate (4.1.8) and noting that $\bar{I}$ is bounded, it is easily shown that there exists a constant $N_0 > 0$, independent of integers $k \geq 1$, such that

$$\sup_{t \in \bar{I}} \|\phi^k(\cdot, t)\|_{2,\Omega}^2 \leq N_0, \tag{4.1.31}$$

and further,

$$\|\phi^k(\cdot,t) - \phi^l(\cdot,t)\|_{2,\Omega}^2 \leq K\left[\|\phi_0^k - \phi_0^l\|_{2,\Omega}^2 + \int_0^t \{|u^k(\theta) - u^l(\theta)|^2 \, d\theta \right] \tag{4.1.32}$$

for each $t \in \bar{I}$. Thus, $\{\phi^k\}$ is a Cauchy sequence in $L_2(Q, R^n) \cap L_{2,\infty}(Q, R^n)$ $\cap$ $C(\bar{I}, L_2(\Omega, R^n))$. Consequently, there exists an element $\phi \in L_2(Q, R^n) \cap$ $L_{2,\infty}(Q, R^n) \cap C(\bar{I}, L_2(\Omega, R^n))$ such that $\phi^k \to \phi$ strongly. We shall show that ϕ is a weak solution of system (4.1.7). For this, we recall that $\phi^k \in C^1(\bar{Q}, R^n)$ is the classical solution of system (4.1.30). Thus, multiplying the differential equation of system (4.1.30) by $\eta \in C_0^1(Q, R^n)$ and then performing integration by parts, we obtain

$$\iint_Q \{(\phi^k(x,t), L\eta(x,t))\} \, dx\, dt = 0, \tag{4.1.33}$$

where

$$L\psi(x,t) = \frac{\partial \psi(x,t)}{\partial t} - \frac{\partial}{\partial x}((A(x))'\psi(x,t)) + (B(x))'\psi(x,t).$$

Note that identity (4.1.33) holds true for all $\eta \in C_0^1(Q, R^n)$.

Since $\phi^k \to \phi$ strongly in $L_2(Q, R^n)$, we have, on taking the limit in (4.1.33),

$$\int_Q\int \{(\phi(x,t), L\eta(x,t))\} \, dx\, dt = 0 \tag{4.1.34}$$

for all $\eta \in C_0^1(Q, R^n)$. Thus,

$$\frac{\partial \phi(x,t)}{\partial t} = A(x)\frac{\partial \phi(x,t)}{\partial x} + B(x)\phi(x,t) \tag{4.1.35}$$

in the generalized sense on Q.

Multiplying the differential equations of systems (4.1.30) and (4.1.35) by $\eta \in C^1(\bar{Q}, R^n)$ with $\eta(x,T)=0$ for all $x\in\bar{\Omega}$ and $\eta(1,t)=\eta(0,t)=0$ for $t\in\bar{I}$, and then performing integration by parts, we obtain, respectively,

$$0 = \int_Q\int \{(\phi^k(x,t), L\eta(x,t))\}\,dx\,dt + \int_\Omega \{(\phi_o^k(x), \eta(x,0))\}\,dx \tag{4.1.36}$$

and

$$0 = \int_Q\int \{(\phi(x,t), L\eta(x,t))\}\,dx\,dt + \int_\Omega \{(\phi(x,0), \eta(x,0))\}\,dx. \tag{4.1.37}$$

Since $\phi^k \overset{s}{\rightarrow} \phi$ in $L_2(Q, R^n) \cap C(\bar{I}, L_2(\Omega, R^n))$, it is clear that, for each $t\in\bar{I}, \phi(\cdot,t)$ is well defined. Thus, it follows from (4.1.36) and (4.1.37) that

$$\int_\Omega \{(\phi_0(x) - \phi(x,0), \eta(x,0))\}\,dx = 0 \tag{4.1.38}$$

for all $\eta\in C^1(\bar{Q}, R^n)$, where $\eta(x,T)=0$ for all $x\in\bar{\Omega}$ and $\eta(1,t)=\eta(0,t)=0$ for all $t\in\bar{I}$. Thus,

$$\int_\Omega \{(\phi_0(x) - \phi(x,0), \xi(x))\}\,dx = 0 \tag{4.1.39}$$

for all $\xi\in C_0^1(\Omega, R^n)$. Consequently,

$$\phi_0(x) = \phi(x,0) \quad \text{a.e. on } \Omega. \tag{4.1.40}$$

It remains to prove that ϕ satisfies the boundary conditions of system (4.1.7). Since the differential equation of system (4.1.7) is of the first order both in t and x variables and A is invertible, we can conclude that $\phi^k(x,\cdot) \to \phi(x,\cdot)$ strongly in $L_2(I, R^n)$ uniformly in $x\in\bar{\Omega}$. Thus, ϕ belongs to $C(\bar{\Omega}, L_2(I, R^n))$. Therefore, it follows that, for any $\eta\in C^1(\bar{Q}, R^n)$ such that $\eta(x,0)=\eta(x,T)=0$ on $\bar{\Omega}$,

$$\lim_{k\to\infty} \int_I \{(\phi^k(\alpha,t), A(\alpha)\eta(\alpha,t))\}\,dt = \int_I \{(\phi(\alpha,t), A(\alpha)\eta(\alpha,t))\}\,dt \tag{4.1.41}$$

for $\alpha = 0, 1$.

By virtue of the decompositions of ϕ^k into $(\phi^k)^-$ and $(\phi^k)^+$ and ϕ into ϕ^- and ϕ^+, we have, for $\alpha=1$,

$$\lim_{k\to\infty}\int\Big\{\big((\phi^k)^-(1,t),A^-(1)\eta^-(1,t)\big)$$
$$+\big(D_1(\phi^k)^-(1,t)+Du^k(t),A^+(1)\eta^+(1,t)\big)\Big\}dt$$
$$=\int_I\{(\phi^-(1,t),A^-(1)\eta^-(1,t))+(\phi^+(1,t),A^+(1)\eta^+(1,t))\}dt. \tag{4.1.42}$$

Here, the left-hand side of the above relation is obtained by using the fact that, for each integer $k\geq 1$, ϕ^k satisfies the boundary conditions of system (4.1.30).

Since $\phi^k(\alpha,\cdot)\overset{s}{\to}\phi(\alpha,\cdot)$ in $L_2(I,R^n)$ for $\alpha=0,1$ and $u^k\overset{s}{\to}u$ in $L_2(I,R^n)$, we get

$$\int_I\{(\phi^-(1,t)A^-(1)\eta^-(1,t))+(D_1\phi^-(1,t)+Du(t),A^+(1)\eta^+(1,t))\}dt$$
$$=\int_I\{\phi^-(1,t),A^-(1)\eta^-(1,t))+(\phi^+(1,t),A^+(1)\eta^+(1,t))\}dt \tag{4.1.43}$$

for all $\eta\in C^1(\bar{Q},R^n)$ satisfying the properties stated above. On this basis, it follows that

$$\phi^+(1,t)=D_1\phi^-(1,t)+Du(t). \tag{4.1.44}$$

Similarly, we can show that

$$\phi^-(0,t)=D_0\phi^+(0,t) \tag{4.1.45}$$

Note that these two equalities hold in the generalized sense. However, since $\phi(\alpha,\cdot)\in L_2(I,R^n)$ for $\alpha=1$ or 0 and $u\in L_2(I,R^r)$, the equalities hold a.e. on I. This completes the proof. □

4.1.4. Formulation of the Control Problem

We consider system (4.1.7) with the quadratic cost functional

$$J(u)=\int_Q\int\{(\phi(u)(x,t),M(x,t)\phi(u)(x,t))\}dx\,dt$$
$$+\int_I\{(u(t),N(t)u(t))\}dt+\int_\Omega\{(\phi(u)(x,T),R(x)\phi(u)(x,T))\}dx, \tag{4.1.46}$$

where $\phi(u)$ denotes the weak solution of system (4.1.7) corresponding to the control $u \in L_2(I, R^n)$, and hence, $\phi(u)(\cdot, T) \in L_2(\Omega, R^n)$. It is assumed that $M \in L_\infty(Q, R^{n \times n})$, $N \in L_\infty(I, R^{r \times r})$, and $R \in L_\infty(\Omega, R^{n \times n})$. Further, we assume that $M(x,t)$, $R(x)$, and $N(t)$ are symmetric positive definite.

By Theorem 4.1.1, system (4.1.7) admits, for each $u \in L_2(I, R^r)$, a unique weak solution $\phi(u) \in L_2(Q, R^n) \cap C(\bar{I}, L_2(\Omega, R^n)) \cap C(\bar{\Omega}, L_2(I, R^n))$. Thus, under the given assumptions, $J(u)$ is well defined for such controls. We take the class of admissible controls as a closed convex subset $\mathcal{U}_a \subset L_2(I, R^r)$.

For convenience, the optimal control problem defined above is referred to as Problem (4.1.P1).

4.1.5. Necessary and Sufficient Conditions for Optimality

In this subsection, our aim is to derive necessary and sufficient conditions for optimality for Problem (4.1.P1). For this, let us, first of all, introduce the following system, called the *adjoint system* [to system (4.1.7)]:

$$\begin{aligned} \frac{\partial z(x,t)}{\partial t} &= A(x)\frac{\partial z(x,t)}{\partial x} - \left((B(x))' - \frac{\partial A(x)}{\partial x}\right) z(x,t) \\ &\quad - M(x,t)\phi(u)(x,t), \qquad (x,t) \in Q, \\ z(x,T) &= R(x)\phi(u)(x,T), \qquad x \in \Omega, \\ M_0^* z(0,t) &= 0, \qquad t \in I, \\ M_1^* z(1,t) &= 0, \qquad t \in I, \end{aligned} \tag{4.1.47}$$

where

$$M_0^* \psi(0,t) \equiv (A^+(0))^{-1}(D_0)' A^-(0)\psi^-(0,t) + \psi^+(0,t), \qquad t \in I,$$

and

$$M_1^* \psi(1,t) \equiv \psi^-(1,t) + (A^-(1))^{-1}(D_1)' A^+(1)\psi^+(1,t), \qquad t \in I.$$

REMARK 4.1.2. With the reversal of time, $t \to T - t$, the adjoint system (4.1.47) reduces to

$$\begin{aligned} -\frac{\partial \hat{z}(x,t)}{\partial t} &= A(x)\frac{\partial \hat{z}(x,t)}{\partial x} - \left((B(x))' - \frac{\partial A(x)}{\partial x}\right) \hat{z}(x,t) \\ &\quad - M(x,T-t)\phi(u)(x,T-t), \qquad (x,t) \in Q \\ \hat{z}(x,0) &= R(x)\phi(u)(x,T), \qquad x \in \Omega \\ M_0^* \hat{z}(0,t) &= 0, \qquad t \in I \\ M_1^* \hat{z}(1,t) &= 0, \qquad t \in I. \end{aligned} \tag{4.1.48}$$

If $\phi_0 \in C^1(\bar{\Omega}, R^n)$ and $u \in C^1(\bar{I}, R^r)$, then it is clear from Remark 4.1.2 that the system (4.1.7) has a unique classical solution $\phi \in C^1(\bar{Q}, R^n)$. Now, in addition, let us assume that $R \in C^1(\bar{\Omega}, R^{n\times n})$. Then, by using, again, the characteristic method in conjunction with Volterra integral equations, it can be shown that system (4.1.48) has a classical solution $\hat{z} \in C^1(\bar{Q}, R^n)$. Equivalently, the adjoint system (4.1.47) has a unique classical solution $z \in C^1(\bar{Q}, R^n)$. Following an approach similar to that given in Lemma 4.1.1, we obtain an a priori estimate of the type given in (4.1.8). Thus, by an approach similar to that of Theorem 4.1.1, we can conclude that system (4.1.48) admits a unique weak solution $\hat{z} \in L_2(Q, R^n) \cap C(\bar{I}, L_2(\Omega, R^n)) \cap C(\bar{\Omega}, L_2(I, R^n))$ under the assumptions that $M \in L_\infty(Q, R^{n\times n})$, $N \in L_\infty(I, R^{r\times r})$, $R \in L_\infty(\Omega, R^{n\times n})$, and $u \in L_2(I, R^r)$. Equivalently, the adjoint system (4.1.47) has a unique weak solution $z \in L_2(Q, R^n) \cap C(\bar{I}, L_2(\Omega, R^n)) \cap C(\bar{\Omega}, L_2(I, R^n))$.

The following result gives the necessary and sufficient conditions for optimality for Problem (4.1.P1).

Theorem 4.1.2. *Consider Problem* (4.1.P1). *Suppose that* $\mathfrak{U}_a$ *is a closed convex subset of* $L_2(I, R^r)$ *and that* $u^* \in \mathfrak{U}_a$ *is an optimal control. Then, it is necessary and sufficient that*

$$\int_I \left\{ \left(D'A^+(1)(z(u^*))^+(1,t) + N(t)u^*(t),\, u(t) - u^*(t) \right) \right\} dt \geq 0 \tag{4.1.49}$$

for all $u \in \mathfrak{U}_a$, *where* $z(u^*)$ *is the (weak) solution of the adjoint system* (4.1.47) *corresponding to the optimal control* $u^* \in \mathfrak{U}_a$.

PROOF. The cost functional J is of positive quadratic form and is defined on a closed convex subset of a Hilbert space. Thus, it is known that the necessary and sufficient conditions for u^* to be an optimal control is that

$$J'_{u^*}(u - u^*) \geq 0 \tag{4.1.50}$$

for all $u \in \mathfrak{U}_a$, where $J'_v(u - u^*)$ denotes the Gateaux derivative of J at v in the direction $(u - u^*)$. Computing the Gateaux gradient at u^*, we obtain

$$\begin{aligned} J'_{u^*}(u - u^*) = & \int_Q \int \left\{ \left(M(x,t)\phi(u^*)(x,t), \hat{\phi}(x,t) \right) \right\} dx\, dt \\ & + \int_I \{ (N(t)u^*(t), u(t) - u^*(t)) \} dt \\ & + \int_\Omega \left\{ \left(R(x)\phi(u^*)(x,T), \hat{\phi}(x,T) \right) \right\} dx \geq 0 \end{aligned} \tag{4.1.51}$$

for all $u\in\mathcal{U}_a$, where $\hat{\phi}$ is the (weak) solution of the system

$$\begin{aligned}
&\frac{\partial\hat{\phi}(x,t)}{\partial t}=A(x)\frac{\partial\hat{\phi}(x,t)}{\partial x}+B(x)\hat{\phi}(x,t), && (x,t)\in Q,\\
&\hat{\phi}(x,0)=0, && x\in\Omega,\\
&M_0\hat{\phi}(0,t)=0, && t\in I,\\
&M_1\hat{\phi}(1,t)=D(u(t)-u^*(t)), && t\in I,
\end{aligned}\tag{4.1.52}$$

and where the boundary operators are defined by

$$M_0\psi(0,t)=\psi^-(0,t)-D_0\psi^+(0,t)\quad \text{on } I$$

and

$$M_1\psi(1,t)=-D_1\psi^-(1,t)+\psi^+(1,t)\quad \text{on } I.$$

In view of the above system, we observe that it belongs to the same class as that of system (4.1.7). Thus, this system has a unique weak solution $\hat{\phi}\in L_2(Q,R^n)\cap C(\bar{I},L_2(\Omega,R^n))\cap C(\bar{\Omega},L_2(I,R^n))$.

Multiplying the first equation of (4.1.47) by $\hat{\phi}$ and then integrating by parts, it follows that

$$\begin{aligned}
&\int_\Omega\big\{\big(z(u^*)(x,T),\hat{\phi}(x,T)\big)\big\}dx\\
&\quad=\int_I\big\{\big(z(u^*)(1,t),A(1)\hat{\phi}(1,t)\big)-\big(z(u^*)(0,t),A(0)\hat{\phi}(0,t)\big)\big\}dt\\
&\qquad-\int_Q\!\!\int\big\{\big(M(x,t)\phi(u^*)(x,t),\hat{\phi}(x,t)\big)\big\}dx\,dt.
\end{aligned}\tag{4.1.53}$$

Using the final conditions of system (4.1.47) in the above equation, we have

$$\begin{aligned}
&\int_\Omega\big\{\big(R(x)\phi(u^*)(x,T),\hat{\phi}(x,T)\big)\big\}dx\\
&\qquad+\int_Q\big\{\big(M(x,t)\phi(u^*)(x,t),\hat{\phi}(x,t)\big)\big\}dx\,dt\\
&\quad=\int_I\big\{\big(z(u^*)(1,t),A(1)\hat{\phi}(1,t)\big)-\big(z(u^*)(0,t),A(0)\hat{\phi}(0,t)\big)\big\}dt.
\end{aligned}\tag{4.1.54}$$

Combining (4.1.51) and (4.1.54), we obtain

$$\begin{aligned}
J'_{u^*}(u-u^*)=\int_I\big\{&\big(z(u^*)(1,t),A(1)\hat{\phi}(1,t)\big)-\big(z(u^*)(0,t),A(0)\hat{\phi}(0,t)\big)\\
&+\big(N(t)u^*(t),u(t)-u^*(t)\big)\big\}dt\geq 0
\end{aligned}\tag{4.1.55}$$

for all $u \in \mathcal{U}_a$. Now, using the boundary conditions of (4.1.47) and (4.1.52) in the above relation, it follows that

$$J'_{u^*}(u-u^*)$$
$$= \int_I \{(D)'A^+(1)(z(u^*))^+(1,t)+N(t)u^*(t), u(t)-u^*(t))\}dt \geq 0 \tag{4.1.56}$$

for all $u \in \mathcal{U}_a$. This proves the theorem. □

REMARK 4.1.3. If there are no control constraints other than $\mathcal{U}_a \equiv L_2(I, R^r)$, then (4.1.56) yields

$$u^*(t) = -(N(t))^{-1}(D)'A^+(1)(z(u^*))^+(1,t). \tag{4.1.57}$$

REMARK 4.1.42. In summary, the optimality conditions for Problem (4.1.P1) are given by

$$\begin{aligned}
&\frac{\partial\phi(x,t)}{\partial t} = A(x)\frac{\partial\phi(x,t)}{\partial x} + B(x)\phi(x,t), && (x,t)\in Q,\\
&\phi(x,0) = \phi_0(x), && x\in\Omega,\\
&\phi^-(0,t) = D_0\phi^+(0,t), && t\in I,\\
&\phi^+(1,t) = D_1\phi^-(1,t) + Du(t), && t\in I,
\end{aligned} \tag{4.1.58}$$

$$\begin{aligned}
&\frac{\partial z(x,t)}{\partial t} = A(x)\frac{\partial z(x,t)}{\partial x} - \left((B(x))' - \frac{\partial A(x)}{\partial x}\right)z(x,t)\\
&\qquad - M(x,t)\phi(u)(x,t), && (x,t)\in Q\\
&z(x,T) = R(T)\phi(u)(x,T), && x\in\Omega\\
&M_0^* z(0,t) = 0, && t\in I\\
&M_1^* z(1,t) = 0, && t\in I,
\end{aligned} \tag{4.1.59}$$

and the inequality

$$\int_I \{((D)'A^+(1)(z(u))^+(1,t) + N(t)u(t), v(t)-u(t))\}dt \geq 0 \tag{4.1.60}$$

for all $v \in \mathcal{U}_a$, where u is the optimal control.

REMARK 4.1.5. By virtue of Lemma 4.1.1 and the linearity of the system (4.1.7), it is clear that $u \to \phi$ is an affine continuous map from $\mathcal{U}_a$ $[\subset L_2(I, R^r)]$ into $L_2(I, L_2(\Omega, R^n))$. Thus the cost functional $J(\cdot)$ is a quadratic functional of u in $L_2(I, R^r)$ and, since N is strictly positive, it is also strictly convex and $J(u) \to +\infty$ as $\|u\|_{2,I} \to \infty$. As a consequence of this, the optimal control problem (4.1.P1) has a unique solution.

Since similar results are given in Chapter 5 under much more general conditions, we omit the proof of this fact.

4.2. Second-Order Hyperbolic Systems in One Spatial Variable with Data from the Space of Distributions

4.2.1. Introduction

In this section, we consider a class of problems of optimal control of systems governed by a second-order hyperbolic partial differential equation with Dirichlet boundary conditions. This class of optimal control problems arises naturally in the study of optimal generation and transmission of power over large networks. For more details, the reader is referred to the original paper [Ah.1].

In Section 2.2, we describe the system. In Section 2.3, we use the principle of transposition to show that, corresponding to each admissible control, the system admits a unique generalized solution. In Section 2.4, we define a cost functional and then formulate a class of optimal control problems. In Section 2.5, necessary and sufficient conditions for optimality are derived. The question of existence and uniqueness of an optimal control is dealt with in Section 2.6. For computation of optimal controls, iterative techniques based on the necessary and sufficient conditions for optimality are developed in Section 2.7.

4.2.2. System Description

Let l_0 be a positive constant, $\Omega \equiv (0, l_0)$, $I \equiv (0, T)$ be the time interval of interest, and $Q \equiv \Omega \times I$. Let $\partial\Omega \equiv \{0, l_0\}$ denote the end points of Ω and let $\Sigma \equiv \partial\Omega \times I$. With this notation, we can describe the system considered in this section as follows:

$$
\begin{aligned}
&\phi_{tt}(x,t) + a\phi_t(x,t) + b\phi(x,t) \\
&\qquad - c\phi_{xx}(x,t) = \alpha f_x(x,t), && (x,t) \in Q, \\
&\qquad \phi(x,0) = \phi_0(x), && x \in \Omega, \\
&\qquad \phi_t(x,0) = \phi_1(x), && x \in \Omega, \\
&\qquad \phi(0,t) = u_1(t), && t \in I, \\
&\qquad \phi(l_0,t) = u_2(t), && t \in I,
\end{aligned}
\tag{4.2.1}
$$

where $u \equiv (u_1, u_2) \in \mathcal{U}_a$ ($\mathcal{U}_a$ is the class of admissible controls, to be defined later) and a, b, c, α are positive constants.

4.2.3. *Existence and Uniqueness of Solutions*

We shall use the method of transposition to prove the existence and uniqueness of solutions of the system (4.2.1). For this, let us, first of all, define the map F as

$$\psi \to \psi_{tt} + a\psi_t + b\psi - c\psi_{xx}.$$

Clearly, the formal adjoint of F, which is denoted by F^*, is given by

$$\psi \to \psi_{tt} - a\psi_t + b\psi - c\psi_{xx}.$$

For the solution of the nonhomogeneous boundary value problem (4.2.1), we consider the following homogeneous boundary value problem,

$$\begin{aligned} F^*(\psi)(x,t) &= h(x,t), && (x,t)\in Q,\\ \psi(x,T) &= 0, && x\in\Omega,\\ \psi_t(x,T) &= 0, && x\in\Omega,\\ \psi(x,t) &= 0, && (x,t)\in\Sigma, \end{aligned} \tag{4.2.2}$$

and construct an isomorphism, which we transpose to obtain the solution of the original problem (4.2.1).

To proceed further, we recall the notation of Sections 1.2.1–1.2.3.

Definition 4.2.1. For each $h\in L_2(Q)$, a function $\psi\in W_2^{1,1}(Q)$ is said to be a *generalized solution of system* (4.2.2) *corresponding to* $h\in L_2(Q)$ if

i. the differential equation of the system is satisfied in the generalized (distributional) sense,
ii. $\psi(x,t)=0$ for all $(x,t)\in\Sigma$, and
iii. $\psi(x,T)=\psi_t(x,T)=0$ for all $x\in\Omega$.

Lemma 4.2.1. *Consider system* (4.2.2). *Suppose that a, b, c are positive constants. Then, for each $h\in L_2(Q)$, the system admits a unique generalized solution $\psi\in W_2^{1,1}(Q)$.*

PROOF. We shall follow Galerkin's procedure. To start with, let $\{\zeta_i\}$ be a fundamental system of functions in $\mathring{W}_2^1(\Omega)$ such that

$$\langle \zeta_k, \zeta_l\rangle_\Omega \equiv \int \{\zeta_k(x)\zeta_l(x)\}dx = \delta_{kl}, \tag{4.2.3}$$

where

$$\delta_{kl} = \begin{cases} 1 & \text{if} \quad k=l \\ 0 & \text{if} \quad k\neq l. \end{cases} \tag{4.2.4}$$

We shall look for an approximate solution ψ^N of system (4.2.2) in the form

$$\psi^{(N)}(x,t) = \sum_{j=1}^{N} \gamma_j^N(t)\zeta_j(x), \tag{4.2.5}$$

where the functions γ_j^N, $j=1,\dots,N$, are determined by the following system of equations:

$$\sum_{j=1}^{N} \ddot{\gamma}_j^{(N)}(t)\langle\zeta_j,\zeta_i\rangle_\Omega - a\sum_{j=1}^{N}\dot{\gamma}_j^{(N)}(t)\langle\zeta_j,\zeta_i\rangle_\Omega$$

$$+b\sum_{j=1}^{N}\gamma_j^{(N)}(t)\langle\zeta_j,\zeta_i\rangle_\Omega + c\sum_{j=1}^{N}\gamma_j^{(N)}(t)\langle(\zeta_j)_x,(\zeta_i)_x\rangle_\Omega$$

$$=\langle h(\cdot,t),\zeta_i(\cdot)\rangle_\Omega,$$

$$\gamma_i^{(N)}(T)=0,$$

$$\dot{\gamma}_i^{(N)}(T)=0,$$

$$i=1,\dots,N, \qquad t\in I. \tag{4.2.6}$$

Here, $\dot{\gamma}_j^{(N)}$ and $\ddot{\gamma}_j^{(N)}$ denote, respectively, the first- and second-order differentials of the function $\gamma_j^{(N)}$ with respect to t.

Since, for each positive integer N, the system (4.2.6) is finite dimensional and linear, it has a unique solution $\gamma_i^{(N)}$, $i=1,\dots,N$. Thus ψ^N is uniquely determined by Equation (4.2.5).

Multiplying the first equation of system (4.2.6) by $\dot{\gamma}_i^{(N)}$, summing the resulting equality over i from 1 to N, and then using equation (4.2.5), we obtain

$$\int_\Omega\{\psi_{tt}^{(N)}(x,t)\psi_t^{(N)}(x,t)\}dx - a\int_\Omega\{\psi_t^{(N)}(x,t)\psi_t^{(N)}(x,t)\}dx$$

$$+b\int_\Omega\{\psi^{(N)}(x,t)\psi_t^{(N)}(x,t)\}dx + c\int_\Omega\{\psi_x^{(N)}(x,t)\psi_{xt}^{(N)}(x,t)\}dx$$

$$=\int_\Omega\{h(x,t)\psi_t^{(N)}(x,t)\}dx. \tag{4.2.7}$$

Denoting $\left(\int_\Omega|\xi(x,t)|^2\,dx\right)^{1/2}$ by $\|\xi(\cdot,t)\|_{2,\Omega}$, we define

$$\|Y^{(N)}(\cdot,t)\|_{2,\Omega}^2=\|\psi_t^{(N)}(\cdot,t)\|_{2,\Omega}^2+b\|\psi^{(N)}(\cdot,t)\|_{2,\Omega}^2+c\|\psi_x^{(N)}(\cdot,t)\|_{2,\Omega}^2. \tag{4.2.8}$$

Then

$$\left(\frac{d}{dt}\|Y^{(N)}(\cdot,t)\|_{2,\Omega}^2\right)=2\int_\Omega\{\psi_t^{(N)}(x,t)\psi_{tt}^{(N)}(x,t)\}dx$$

$$+2b\int_\Omega\{\psi^{(N)}(x,t)\psi_t^{(N)}(x,t)\}dx$$

$$+2c\int_\Omega\{\psi_x^{(N)}(x,t)\psi_{xt}^{(N)}(x,t)\}dx. \tag{4.2.9}$$

Combining (4.2.9) and (4.2.7), we have

$$\frac{d}{dt}\left(\|Y^{(N)}(\cdot,t)\|_{2,\Omega}^2\right)-2a\|\psi_t^{(N)}(\cdot,t)\|_{2,\Omega}^2=2\int_\Omega\{h(x,t)\psi_t^{(N)}(x,t)\}dx. \tag{4.2.10}$$

Integrating the above equality with respect to t, we obtain

$$\begin{aligned}\|Y^{(N)}(\cdot,t)\|_{2,\Omega}^2-2a\int_0^t\int_\Omega\{[\psi_t^{(N)}(x,\tau)]^2\}dx\,d\tau\\ =\|Y^{(N)}(\cdot,0)\|_{2,\Omega}^2+2\int_0^t\int_\Omega\{h(x,\tau)\psi_t^{(N)}(x,\tau)\}dx\,d\tau\end{aligned} \tag{4.2.11}$$

for all $t\in[0,T]$. In particular, it is true for $t=T$:

$$\begin{aligned}\|Y^{(N)}(\cdot,T)\|_{2,\Omega}^2-2a\int_0^T\int_\Omega\{[\psi_t^{(n)}(x,\tau)]^2\}dx\,d\tau\\ =\|Y^{(N)}(\cdot,0)\|_{2,\Omega}^2+2\int_0^T\int_\Omega\{h(x,\tau)\psi_t^{(N)}(x,\tau)\}dx\,d\tau.\end{aligned} \tag{4.2.12}$$

Subtracting (4.2.12) from (4.2.11), we have

$$\begin{aligned}\|Y^{(N)}(\cdot,t)\|_{2,\Omega}^2+2a\int_t^T\int_\Omega\{[\psi_t^{(N)}(x,\tau)]^2\}dx\,d\tau\\ =\|Y^{(N)}(\cdot,T)\|_{2,\Omega}^2-2\int_t^T\int_\Omega\{h(x,\tau)\psi_t^{(N)}(x,\tau)\}dx\,d\tau.\end{aligned} \tag{4.2.13}$$

Since $\gamma_i^{(N)}(T)=\dot{\gamma}_i^{(N)}(T)=0$, it follows from the definitions of $\|Y^{(N)}(\cdot,t)\|_{2,\Omega}^2$ and $\psi^{(N)}$ that

$$\|Y^{(N)}(\cdot,T)\|_{2,\Omega}^2=0. \tag{4.2.14}$$

From the definition of $\psi^{(N)}$, it is easily observed that $\psi^{(N)},\psi_t^{(N)}\in\mathring{W}_2^{1,1}(Q)$. Further, $\psi^{(N)}$ and $\psi_t^{(N)}$ are equal to zero for $t=T$.

Using Hölder's inequality and recalling (4.2.14), it follows from (4.2.13) that

$$\begin{aligned}\|Y^{(N)}(\cdot,t)\|_{2,\Omega}^2+2a\int_t^T\int_\Omega\{[\psi_t^{(N)}(x,\tau)]^2\}dx\,d\tau\\ \le 2\left(\int_t^T\int_\Omega\{[h(x,\tau)]^2\}dx\,d\tau\right)^{1/2}\left(\int_t^T\int_\Omega\{[\psi_t^{(N)}(x,\tau)]^2\}dx\,d\tau\right)^{1/2}\end{aligned} \tag{4.2.15}$$

Now, by virtue of the elementary inequality

$$\alpha\beta = \frac{1}{2\varepsilon}\alpha^2 + \frac{\varepsilon}{2}\beta^2, \qquad \varepsilon > 0,$$

the above inequality reduces to

$$\begin{aligned}\|Y^{(N)}(\cdot, t)\|_{2,\Omega}^2 &+ 2a\int_t^T\int_\Omega\{[\psi_t^{(N)}(x,\tau)]^2\}dx\,d\tau \\ &\leq \frac{1}{\varepsilon}\int_t^T\int_\Omega\{[h(x,\tau)]^2\}dx\,d\tau + \varepsilon\int_t^T\int_\Omega\{[\psi_t^{(N)}(x,\tau)]^2\}dx\,d\tau\end{aligned} \tag{4.2.16}$$

for all $t\in[0, T]$ and for all $\varepsilon > 0$.

Thus,

$$\begin{aligned}\|Y^{(N)}(\cdot, t)\|_{2,\Omega}^2 &+ (2a-\varepsilon)\int_t^T\int_\Omega\{[\psi_t^{(N)}(x,\tau)]^2\}dx\,d\tau \\ &\leq \frac{1}{\varepsilon}\int_t^T\int_\Omega\{[h(x,\tau)]^2\}dx\,d\tau.\end{aligned} \tag{4.2.17}$$

Since $a > 0$, we can choose an $\varepsilon > 0$ such that $(2a-\varepsilon) > 0$. Thus, the inequality (4.2.17) reduces to

$$\|Y^{(N)}(\cdot, t)\|_{2,\Omega}^2 \leq \frac{1}{\varepsilon}\int_t^T\int \quad \{[h(x,\tau)]^2\}dx\,d\tau \tag{4.2.18}$$

for all $t\in[0, T]$.

Integrating the above inequality with respect to t and then using the definition of $\|Y^{(N)}(\cdot, t)\|_{2,\Omega}^2$, we obtain

$$\begin{aligned}\int_0^T\int_\Omega\{[\psi_t^{(N)}(x,t)]^2\}dx\,dt &+ b\int_0^T\int_\Omega\{[\psi^{(N)}(x,t)]^2\}dx\,dt \\ + c\int_0^T\int_\Omega\{[\psi_x^{(N)}(x,t)]^2\}dx\,dt &\leq \frac{T}{\varepsilon}\int_0^T\int_\Omega\{[h(x,t)]^2\}dx\,dt \equiv K.\end{aligned} \tag{4.2.19}$$

Thus, we can choose a subsequence of $\{\psi^{(N)}\}$, which is denoted by the original sequence, and a function $\psi\in W_2^{1,1}(Q)$ such that

$$\psi^{(N)} \xrightarrow{s} \psi \quad \text{(strongly) in } L_2(Q),$$

$$\left.\begin{aligned}\psi_t^{(N)} &\xrightarrow{w} \psi_t \\ \psi_x^{(N)} &\xrightarrow{w} \psi_x\end{aligned}\right\} \quad \text{(weakly) in } L_2(Q),$$

as $N\to\infty$, where ψ_t and ψ_x denote, respectively, the generalized derivatives of ψ with respect to t and x. The final and boundary conditions of system

(4.2.2) for ψ are satisfied because of the facts stated in the paragraph following Equation (4.2.14) and Theorem 1.2.8.

We shall now show that ψ satisfies the differential equation of system (4.2.2) in the generalized sense. For this, by virtue of Equation (4.2.5), the differential equation of system (4.2.6) can be written as

$$\langle \psi_{tt}^{(N)}(\cdot,t), \zeta_i(\cdot)\rangle_\Omega - a\langle \psi_t^{(N)}(\cdot,t), \zeta_i(\cdot)\rangle_\Omega + b\langle \psi^{(N)}(\cdot,t), \zeta_i(\cdot)\rangle_\Omega$$
$$+ c\langle \psi_x^{(N)}(\cdot,t), (\zeta_i)_x(\cdot)\rangle_\Omega = \langle h(\cdot,t), \zeta_i(\cdot)\rangle_\Omega. \tag{4.2.20}$$

Let Φ be an arbitrary function of $\mathring{W}_2^{1,1}(Q)$ that is equal to zero for $t=0$. Then, there is a sequence $\Phi^M(x,t)=\sum_{i=1}^M d_i^M(t)\zeta_i(x)$ such that Φ^M converges to Φ in the norm of $W_2^{1,1}(Q)$, where $d_i^M(t)$ are smooth functions that are equal to zero for $t=0$. Multiplying (4.2.20) by $d_i^M(t)$, summing the resulting equality over i from 1 to M, we obtain

$$+\langle \psi_{tt}^{(N)}(\cdot,t), \Phi^M(\cdot,t)\rangle_\Omega - a\langle \psi_t^{(N)}(\cdot,t), \Phi^M(\cdot,t)\rangle_\Omega$$
$$+ b\langle \psi^{(N)}(\cdot,t), \Phi^M(\cdot,t)\rangle_\Omega + c\langle \psi_x^{(N)}(\cdot,t), \Phi_x^M(\cdot,t)\rangle_\Omega$$
$$= \langle h(\cdot,t), \Phi^M(\cdot,t)\rangle_\Omega. \tag{4.2.21}$$

Integrating the above equality with respect to t, we have

$$-\int_Q\int \{\psi_t^{(N)}(x,t)\Phi_t^M(x,t)\}\,dx\,dt - a\int_Q\int \{\psi_t^{(N)}(x,t)\Phi^M(x,t)\}\,dx\,dt$$
$$+ b\int_Q\int \{\psi^{(N)}(x,t)\Phi^M(x,t)\}\,dx\,dt + c\int_Q\int \{\psi_x^{(N)}(x,t)\Phi_x^M(x,t)\}\,dx\,dt$$
$$= \int_Q\int \{h(x,t)\Phi^M(x,t)\}\,dx\,dt. \tag{4.2.22}$$

From this relation, if we fix M and let N tend to infinity through an appropriate subsequence (if necessary), we obtain

$$-\int_Q\int \{\psi_t(x,t)\Phi_t^M(x,t)\}\,dx\,dt - a\int_Q\int \{\psi_t(x,t)\Phi^M(x,t)\}\,dx\,dt$$
$$+ b\int_Q\int \{\psi(x,t)\Phi^M(x,t)\}\,dx\,dt + c\int_Q\int \{\psi_x(x,t)\Phi_x^M(x,t)\}\,dx\,dt$$
$$= \int_Q\int \{h(x,t)\Phi^M(x,t)\}\,dx\,dt. \tag{4.2.23}$$

Letting $M \to \infty$ in the above equation, we have

$$-\int_Q\int \{\psi_t(x,t)\Phi_t(x,t)\}\,dx\,dt - a\int_Q\int \{\psi_t(x,t)\Phi(x,t)\}\,dx\,dt$$

$$+b\int_Q\int \{\psi(x,t)\Phi(x,t)\}\,dx\,dt + c\int_Q\int \{\psi_x(x,t)\Phi_x(x,t)\}\,dx\,dt$$

$$=\int_Q\int \{h(x,t)\Phi(x,t)\}\,dx\,dt. \tag{4.2.24}$$

This relation is true for all $\Phi \in \mathring{W}_2^{1,1}(Q)$ that are equal to zero for $t=0$. Choosing Φ with compact support in Q, it follows that ψ satisfies the differential equation of system (4.2.2) in the generalized sense. Thus, ψ is a generalized solution of system (4.2.2).

Next we shall prove the uniqueness of the solution. For this, let ψ be a generalized solution of system (4.2.2). We multiply the first equation of system (4.2.2) by ψ_t and integrate with respect to x. This gives rise to Equation (4.2.7) with exponent N deleted. Thus, by following an argument similar to that given for inequality (4.2.19), we obtain the same inequality with the exponent N deleted. This inequality is true for any generalized solution of the system (4.2.2). By virtue of this fact, we observe that $\psi(x,t)=0$ on Q if $\|h\|_{2,Q}=0$. Let ψ_1 and ψ_2 be two generalized solutions of the system (4.2.2). It is clear that $\psi_1-\psi_2$ is also a generalized solution of system (4.2.2) with $h(x,t)=0$ a.e. on Q. Thus, $\psi_1(x,t)-\psi_2(x,t)=0$ on Q. This shows that system (4.2.2) admits a unique generalized solution. Thus the proof is complete. □

REMARK 4.2.1. The strong convergence of $\psi^{(N)} \to \psi$ [in $L_2(Q)$] follows from the facts that $\phi^{(N)} \overset{w}{\to} \psi$ in $W_2^{1,1}(Q)$ and that the injection map $W_2^{1,1}(Q) \hookrightarrow L_2(Q)$ is compact.

Using Lemma 4.2.1 and the principle of transposition, to be described shortly, we can show that, for each $u \in \mathcal{U}_a$, system (4.2.1) has a unique solution in the sense of the following definition.

Definition 4.2.2. For each $u \in \mathcal{U}_a$, a function $\phi(u) \in L_2(Q)$ is said to be a *solution of system* (4.2.1) if

i. the differential equation of the system is satisfied in the generalized sense, and
ii. $\phi(x,0)=\phi_0(x)$ a.e. on Ω, $\phi_t(x,0)=\phi_1(x)$ a.e. on Ω, $\phi(0,t)=u_1(t)$ a.e. on I, $\phi(l_0,t)=u_2(t)$ a.e. on I.

By use of Lemma 4.2.1, we construct an isomorphism that we transpose to solve the nonhomogeneous boundary value problem (4.2.1). For this purpose, we introduce the set X given by

$$X \equiv \left\{ \begin{array}{l} \psi \in W_2^{1,1}(Q): \psi(x,T) = \psi_t(x,T) = 0 \text{ for } x \in \Omega, \psi(x,t) = 0 \\ \text{for } (x,t) \in \Sigma \text{ and } F^*(\psi) \in L_2(Q) \end{array} \right\}.$$

The set X equipped with the norm topology $|||\psi|||_X \equiv \| F^*(\psi) \|_{2,Q}$ is a Hilbert space. We observe that, by virtue of Lemma 4.2.1, F^* is an isomorphism of X onto $L_2(Q)$. This is the isomorphism that we transpose to prove the existence and uniqueness of solution of the nonhomogeneous system (4.2.1).

Theorem 4.2.1. *Consider the system* (4.2.1) *with* $a, b, c > 0$ *and constant and* α *a real number. Suppose* $\phi_0 \in W_2^1(\Omega)$, $\phi_1 \in L_2(\Omega)$, $f \in L_2(Q)$, *and* $\mathfrak{U}_a$ *is a closed convex subset of* $L_2(I) \times L_2(I)$ *such that for each* $u \in \mathfrak{U}_a$,

$$\int_I \{u_1(t)\psi_x(0,t) - u_2(t)\psi_x(l_0,t)\}\,dt$$

is defined for all $\psi \in X$. *Then the system* (4.2.1) *admits a unique generalized solution* $\phi(u) \in L_2(Q)$, *for each* $u \in \mathfrak{U}_a$. *Further, the map* $u \to \phi(u)$ *is affine continuous.*

PROOF. For a fixed but arbitrary $u \in \mathfrak{U}_a$, we introduce the functional l given by

$$l(\psi) \equiv -\alpha \int_Q \int \{f(x,t)\psi_x(x,t)\}\,dx\,dt + \int_\Omega \{\phi_0(x)[a\psi(x,0) - \psi_t(x,0)]$$

$$+ \phi_1(x)\psi(x,0)\}\,dx + c\int_I \{u_1(t)\psi_x(0,t) - u_2(t)\psi_x(l_0,t)\}\,dt. \tag{4.2.25}$$

Under the hypotheses of the theorem, we can verify that l is well defined on X and that it is a continuous linear functional thereon. First we show that there exists a unique $\phi(u) \in L_2(Q)$ such that

$$\int_Q \int \{\phi(u)(x,t) F^*(\psi)(x,t)\}\,dx\,dt = l(\psi) \tag{4.2.26}$$

for all $\psi \in X$. Since F^* is an isomorphism of X onto $L_2(Q)$ (Lemma 4.2.1), F^{*-1} exists and it is a continuous linear operator from $L_2(Q)$ into X. On the other hand, $l \in X^*$, and consequently, the composition (lF^{*-1}) is a continuous linear functional on $L_2(Q)$. Thus, there exists a unique $\phi(u) \in L_2(Q)$ such that

$$(lF^{*-1})(g) = \int_Q \{\phi(u)(x,t) g(x,t)\}\,dx\,dt \quad \text{for all } g \in L_2(Q). \tag{4.2.27}$$

Again, due to Lemma 4.2.1 this is equivalent to the statement that

$$l(\psi)=\int_Q\{\phi(u)(x,t)F^*(\psi)(x,t)\}dx\,dt$$

for all $\psi\in\text{range}(F^{*-1})=X$. The system (4.2.26) is equivalent to (4.2.1). Indeed, for every $\psi\in X$ with compact support in Q, it follows from (4.2.25) and (4.2.26) that

$$\int_Q\{(F\phi)(x,t)\psi(x,t)\}dx\,dt=\alpha\int_Q f_x(x,t)\psi(x,t)dx\,dt. \quad (4.2.28)$$

Since (4.2.28) holds for arbitrary $\psi\in X$ with compact support in Q, we conclude that $\phi(u)$ satisfies the differential equation

$$F\phi(u)=\alpha f_x \quad \text{in } Q \quad (4.2.29)$$

in the sense of distribution. Multiplying (4.2.29) by an arbitrary ψ from X and integrating by parts over Q, we obtain

$$\int_Q\int\{\phi(u)(x,t)F^*(\psi)(x,t)\}dx\,dt=-\alpha\int_Q\int\{f(x,t)\psi_x(x,t)\}dx\,dt$$

$$+\int_\Omega\{\phi(u)(x,0)[a\psi(x,0)-\psi_t(x,0)]\}dx$$

$$+\int_\Omega\{\phi_t(u)(x,0)\psi(x,0)\}dx+c\int_I\{\phi(u)(0,t)\psi_x(0,t)$$

$$-\phi(u)(l_0,t)\psi_x(l_0,t)\}dt. \quad (4.2.30)$$

Comparing (4.2.26) with (4.2.30), we obtain $\phi(u)(x,0)=\phi_0(x)$ on Ω, $\phi_t(u)(x,0)=\phi_1(x)$ on Ω, and $\phi(u)(0,t)=u_1(t)$ and $\phi(u)(l_0,t)=u_2(t)$ on I. Thus, we have proved that, for each $u\in\mathfrak{U}_a$, the system (4.2.1) has a unique generalized solution in the sense of (4.2.26). It remains to show that $u\to\phi(u)$ is affine continuous. Let $\{l^n\}$ be a sequence from X^* such that $l^n(\psi)\to 0$ for each $\psi\in X$, and denote by $\{\phi^n\}$ the corresponding sequence of solutions of the problem given by (4.2.26) and (4.2.25). Clearly the sequence $\{\phi^n\}$ is contained in a bounded subset of $L_2(Q)$, and hence, there exists a subsequence $\{n(k)\}$ of the sequence (n) and a $\phi^*\in L_2(Q)$ such that $\phi^{n(k)}\overset{w}{\to}\phi^*$. Therefore, for each $\psi\in X$,

$$\lim_{k\to\infty}\int_Q\int\phi^{n(k)}F^*(\psi)dx\,dt=\lim_k l^{n(k)}(\psi)=0,$$

and hence,

$$\int_Q\int\phi^*F^*(\psi)dx\,dt=0 \quad \text{for each } \psi\in X.$$

Since F^* is an isomorphism of X onto $L_2(Q)$, the above equality implies that $\phi^*=0$. Thus, $l\to\phi$ is a continuous linear map from X^* into $L_2(Q)$ and hence, in particular, $u\to\phi(u)$ is affine continuous. This completes the proof. □

In the sequel, we shall need the following definition.

Definition 4.2.3. A function ϕ mapping $\mathcal{U}_a$ into $L_2(Q)$ is said to have a *weak Gateaux differential at* $u_0\in\mathcal{U}_a$ *in the direction* $v\in\mathcal{U}_a$, if there exists an element $\hat{\phi}(u_0,v)\in L_2(Q)$ such that

$$\int_Q\int \hat{\phi}(u_0,v)(x,t)g(x,t)\,dx\,dt$$

$$=\lim_{\varepsilon\to 0}\int_Q\int\left\{\frac{\phi(u_0+\varepsilon v)(x,t)-\phi(u_0)(x,t)}{\varepsilon}\right\}g(x,t)\,dx\,dt \tag{4.2.31}$$

for all $g\in L_2(Q)$.

Lemma 4.2.2. *Under the hypotheses of Theorem* 4.2.1, *the generalized solution* $\phi(u)$ *of the system* (4.2.1) *corresponding to* $u\in\mathcal{U}_a$ *has a unique weak linear Gateaux differential at every point* $u\in\mathcal{U}_a$ *in any direction* $w\equiv v-u$ *with* $v\in\mathcal{U}_a$. *Further, this differential is the generalized solution* $\hat{\phi}(w)\in L_2(Q)$ *of the following system*:

$$\begin{aligned}
&\hat{\phi}_{tt}(x,t)+a\hat{\phi}_t(x,t)+b\hat{\phi}(x,t)\\
&\qquad -c\hat{\phi}_{xx}(x,t)=0, && (x,t)\in Q,\\
&\qquad\quad \hat{\phi}(x,0)=\hat{\phi}_t(x,0)=0, && x\in\Omega,\\
&\qquad\quad \hat{\phi}(x,t)=w && \textit{for } (x,t)\in\Sigma,
\end{aligned}\tag{4.2.32}$$

where the map $w\to\hat{\phi}(w)$ *is linear.*

PROOF. Let $u,v\in\mathcal{U}_a$. Since $\mathcal{U}_a$ is closed and convex, $u+\varepsilon(v-u)\in\mathcal{U}_a$ for $0\le\varepsilon\le 1$. Let $\phi(u)$ and $\phi(u+\varepsilon(v-u))$ denote the solutions in the variational sense (4.2.26) of the nonhomogeneous boundary value problem (4.2.1) corresponding to the controls u and $u+\varepsilon(v-u)$, respectively. Then, by the use of (4.2.26), it is easily verified that

$$\frac{1}{\varepsilon}\int_Q\int(\phi(u+\varepsilon(v-u))-\phi(u))F^*(\psi)\,dx\,dt=\tilde{l}(\psi)\quad\text{for all }\psi\in X,$$

where

$$\tilde{l}(\psi)=c\int_I\{(v_1(t)-u_1(t))\psi_x(0,t)-(v_2(t)-u_2(t))\psi_x(l_0,t)\}\,dt.$$

Letting $\varepsilon \to 0$, we obtain

$$\int_Q \int \hat{\phi}(w) F^*(\psi)\, dx\, dt = \tilde{l}(\psi) \quad \text{for all } \psi \in X,$$

where $w \equiv (v-u)$. Hence, the result follows from arguments identical to that given for Theorem 4.2.1. □

4.2.4. *Formulation of the Control Problem*

Consider the system (4.2.1) and suppose ϕ_0, ϕ_1, and f are given (fixed) elements of $W_2^1(\Omega)$, $L_2(\Omega)$, and $L_2(Q)$, respectively. Let $\phi(u)$ denote the response of the system (4.2.1) corresponding to the control $u \in \mathcal{U}_a$, and let

$$J(u) \equiv \frac{\lambda_0}{2} \int_Q \int |\phi(u) - \phi_d|^2\, dx\, dt + \frac{1}{2} \int_I \{\lambda_1 |u_1|^2 + \lambda_2 |u_2|^2\}\, dt$$
$$+ \frac{\lambda_3}{2} \int_Q \int |\phi(u)|^2\, dx\, dt, \tag{4.2.33}$$

define the cost functional for given positive numbers $\lambda_0, \lambda_1, \lambda_2, \lambda_3$, and a given element $\phi_d \in L_2(Q)$. It is required to find a control $u^* \in \mathcal{U}_a$ that minimizes the functional J on $\mathcal{U}_a$ [subject to the dynamic constraint (4.2.1)].

For future reference, this control problem will be referred to as Problem (4.2.P1).

4.2.5. *Necessary and Sufficient Conditions for Optimality*

In this subsection, we shall derive necessary and sufficient conditions for optimality for Problem (4.2.P1). For this, we assume that an optimal control exists (the existence proof will be given in Section 4.2.6); it will be denoted by u^*.

For the derivation of the necessary and sufficient conditions for optimality, we shall make use of the adjoint system corresponding to the system (4.2.1). This is given by the following homogeneous boundary value problem:

$$\begin{aligned}
&z_{tt}(x,t) - a z_t(x,t) + b z(x,t) - c z_{xx}(x,t) \\
&\qquad = \lambda_0(\phi(u)(x,t) - \phi_d(x,t)) + \lambda_3 \phi(u)(x,t), && (x,t) \in Q, \\
&z(x,T) = z_t(x,T) = 0, && x \in \Omega, \\
&z(x,t) = 0, && (x,t) \in \Sigma,
\end{aligned} \tag{4.2.34}$$

where $\phi(u)$ is the (generalized) solution of (4.2.1) corresponding to the control $u \in \mathcal{U}_a$.

REMARK 4.2.2. For the given data $\phi_0 \in W_2^1$, $\phi_1 \in L_2(\Omega)$, $f \in L_2(Q)$, and the control $u \in \mathcal{U}_a$, it follows from Theorem 4.2.1 that the system (4.2.1) has a unique (generalized) solution $\phi(u) \in L_2(Q)$. Thus, for $\phi_d \in L_2(Q)$, $\lambda_3\phi(u) + \lambda_0(\phi(u)-\phi_d) \in L_2(Q)$, and consequently it follows from Lemma 4.2.1 that the adjoint system (4.2.34) has a unique solution $z(u) \in X$.

With the help of the above remark, we can present a result on the necessary and sufficient conditions for optimality for Problem (4.2.P1).

Theorem 4.2.2. *Consider Problem* (4.2.P1). *Suppose that the hypotheses of Theorem* 4.2.1 *are satisfied. Further, let* $\phi_d \in L_2(Q)$ *and let* $\lambda_i > 0$, $i = 0, 1, 2, 3$. *Then, the necessary and sufficient condition for* $u^* \in \mathcal{U}_a$ *to be an optimal control is that*

$$\int_I \{(\lambda_1 u_1^*(t) + c(z(u^*))_x(0,t))(u_1(t) - u_1^*(t)) + (\lambda_2 u_2^*(t) - c(z(u^*))_x(l_0,t))(u_2(t) - u_2^*(t))\}\, dt \geq 0 \quad (4.2.35)$$

for all $u = (u_1, u_2) \in \mathcal{U}_a$.

PROOF. In view of Lemma 4.2.2, we note that, for arbitrary $w \in L_2(I, R^2)$, the system (4.2.32) has a unique generalized solution $\hat{\phi}(w) \in L^2(Q)$. Thus, by multiplying the first equation of (4.2.34) corresponding to $u = u^*$ by $\hat{\phi}(w)$ and then integrating over Q, we obtain

$$\int_Q \int \{\hat{\phi}(w)(x,t) F^*(z(u^*))(x,t)\}\, dx\, dt = \lambda_0 \int_Q \int \left\{ \hat{\phi}(w)(x,t) \left(\left(1 + \frac{\lambda_3}{\lambda_0} \right) \phi(u^*)(x,t) - \phi_d(x,t) \right) \right\} dx\, dt. \quad (4.2.36)$$

Examining the proof of Lemma 4.2.2, we observe that

$$\int_Q \int \{\hat{\phi}(w)(x,t) F^*(z(u^*))(x,t)\}\, dx\, dt = c \int_I \{(w_1(t)(z(u^*))_x(0,t) - w_2(t)(z(u^*))_x(l_0,t))\}\, dt, \quad (4.2.37)$$

where $w = (w_1, w_2)$.

Since $u \to \phi(u)$ has a weak Gateaux differential on $\mathcal{U}_a$ (Lemma 4.2.2) and J is defined by (4.2.33), it follows that J has a Gateaux differential. Further, J is convex in u. In addition, we recall that $\mathcal{U}_a$ is closed and convex. Thus, it follows from Theorem 1.1.23 that the necessary and sufficient condition for u^* to be an optimal control is that the Gateaux differential

$$J'_{u^*}(u-u^*)\geq 0 \tag{4.2.38}$$

for all $u\in\mathcal{U}_a$.

Computing the Gateaux differential of J at u^* in the direction $(u-u^*)$ as permitted by Lemma 4.2.2, it follows from (4.2.38) that

$$\begin{aligned} J'_{u^*}(u-u^*)=\lambda_0\int_Q\int\left\{\hat{\phi}(u-u^*)\left[\left(1+\frac{\lambda_3}{\lambda_0}\right)\phi(u^*)-\phi_d\right]\right\}dx\,dt \\ +\int_I\{\lambda_1u_1^*(u_1-u_1^*)+\lambda_2u_2^*(u_2-u_2^*)\}dt\geq 0 \end{aligned} \tag{4.2.39}$$

for all $u\in\mathcal{U}_a$, where $\hat{\phi}(u-u^*)$ is the solution of (4.2.32) corresponding to $w=u-u^*$.

Letting $w=(u-u^*)$ in (4.2.36) and (4.2.37) and using these relations in (4.2.39), we obtain the inequality (4.2.35). This completes the proof. □

REMARK 4.2.3. For $\mathcal{U}_a\equiv L_2(I,R^2)$ (i.e., no control constraint), the optimal control $u^*\equiv(u_1^*,u_2^*)$ takes the form

$$\begin{aligned} u_1^*(t)&=-\frac{c}{\lambda_1}(z(u^*))_x(0,t),\\ u_2^*(t)&=\frac{c}{\lambda_2}(z(u^*))_x(l_0,t). \end{aligned} \tag{4.2.40}$$

In this case, the optimal control is determined by the generalized solution of the following two-point boundary value problem:

$$\begin{aligned} &\phi_{tt}(x,t)+a\phi_t(x,t)+b\phi(x,t)\\ &\qquad -c\phi_{xx}(x,t)=\alpha f_x(x,t), && (x,t)\in Q,\\ &\phi(x,0)=\phi_0(x), && x\in\Omega,\\ &\phi_t(x,0)=\phi_1(x), && x\in\Omega,\\ &\phi(0,t)=-\frac{c}{\lambda_1}z_x(0,t), && t\in I,\\ &\phi(1,t)=\frac{c}{\lambda_2}z_x(l_0,t), && t\in I, \end{aligned} \tag{4.2.41a}$$

$$\begin{aligned} &z_{tt}(x,t)-az_t(x,t)+bz(x,t)-cz_{xx}(x,t)\\ &\qquad=\lambda_0(\phi(x,t)-\phi_d(x,t))+\lambda_3\phi(x,t), && (x,t)\in Q,\\ &z(x,T)=z_t(x,T)=0, && x\in\Omega,\\ &z(x,t)=0, && (x,t)\in\Sigma. \end{aligned} \tag{4.2.41b}$$

Next, we present a pointwise version of the necessary and sufficient condition given in Theorem 4.2.2. This result is given in the following corollary.

Corollary 4.2.1. *Consider Problem* (4.2.P1). *Suppose that the hypotheses of Theorem* 4.2.2 *are satisfied and that the class of admissible controls* $\mathcal{U}_a$ *is given by*

$$\mathcal{U}_a = \{u : u \text{ measurable on } I \text{ with values in } V\},$$

where V is a compact and convex subset of R^2. Then, the necessary and sufficient condition for $u^* \in \mathcal{U}_a$ to be an optimal control is that the inequality

$$\{(\lambda_1 u_1^*(t) + c(z(u^*))_x(0,t))(v_1 - u_1^*(t)) \\ + (\lambda_2 u_2^*(t) - c(z(u^*))_x(l_0,t))(v_2 - u_2^*(t))\} \geq 0 \quad (4.2.42)$$

holds for almost all $t \in I$ and for all $v = (v_1, v_2) \in V$.

PROOF. Let $t^* \in (0,T)$ be a regular point for all the functions appearing in the integrand of (4.2.35). Let E be any measurable set containing the point t^* and contracting to the one-point set $\{t^*\}$. Further, let $v \in V$ be arbitrary and define

$$\tilde{u}(t) = \begin{cases} v & \text{for } t \in E \\ u^*(t) & \text{for } t \in I \setminus E. \end{cases}$$

Substituting $\tilde{u}$ for u in (4.2.35) and then dividing the resulting expression by $|E|$ (the Lebesgue measure of E), we obtain

$$\frac{1}{|E|}\int_E \{(\lambda_1 u_1^*(t) + c(z(u^*))_x(0,t))(v_1 - u_1^*(t)) \\ + (\lambda_2 u_2^*(t) - c(z(u^*))_x(l_0,t)(v_2 - u_2^*(t))\} dt \geq 0. \quad (4.2.43)$$

Letting $|E| \to 0$, we obtain

$$(\lambda_1 u_1^*(t^*) + c(z(u^*))_x(0,t^*))(v_1 - u_1^*(t^*)) \\ + (\lambda_2 u_2^*(t^*) - c(z(u^*))_x(l_0,t^*))(v_2 - u_2^*(t^*)) \geq 0. \quad (4.2.44)$$

Since $u^*(\cdot)$, $(z(u^*))_x(0,\cdot)$, and $(z(u^*))_x(l_0,\cdot)$ are measurable functions, and consequently, almost all points of I are regular points with respect to these functions, we obtain the inequality (4.2.42). This completes the proof. □

REMARK 4.2.4. If the control restraint set V is given by

$$V = \{v \in R^2 : |v_j| \leq \beta_j, \beta_j > 0, j = 1,2\},$$

then it follows from the inequality (4.2.42) that the optimal control $u^* \equiv (u_1^*, u_2^*)$ takes the form

$$u_1^*(t) = \gamma_1\left(-\frac{c}{\lambda_1}(z(u^*))_x(0,t)\right),$$

$$u_2^*(t) = \gamma_2\left(\frac{c}{\lambda_2}(z(u^*)_x(l,t)\right),$$

where

$$\gamma_j(y)=\begin{cases}\beta_j & \text{for } y\geq\beta_j\\ y & \text{for } |y|<\beta_j\\ -\beta_j & \text{for } y\leq-\beta_j,\end{cases}$$

and $j=1,2$.

4.2.6. *Existence of Optimal Controls*

In the derivation of the necessary and sufficient conditions for optimality for Problem (4.2.P1), it was assumed that an optimal control exists. In this subsection, we give a proof of this fact. For this, we need the following lemma.

Lemma 4.2.3. *Consider Problem* (4.2.P1). *Suppose that all the hypotheses of Theorem* 4.2.1 *are satisfied and that* $\lambda_i>0$, $i=0,1,2,3$. *Further, let* $\mathcal{U}_a$ *be a closed and convex subset of* $L_2(I, R^2)$. *Then the cost functional J is weakly lower semicontinuous on* $\mathcal{U}_a$.

PROOF. Let $\{u^k\}\subset\mathcal{U}_a$ be arbitrary, so that $u^k\to u^0$ weakly in $L_2(I, R^2)$. It is required to show that

$$J(u^0)\leq \varliminf_{k\to\infty} J(u^k). \tag{4.2.45}$$

From Theorem 4.2.1, we note that there exists a sequence of generalized solutions $\{\phi(u^k)\}\subset L_2(Q)$ for the system (4.2.1), so that, for each integer $k\geq 1$,

$$\int_Q\int\{\phi(u^k)(x,t)F^*(\psi)(x,t)\}dx\,dt=l^k(\psi) \tag{4.2.46}$$

for all $\psi\in X$, where $l^k(\psi)$ is as defined by (4.2.25) with u replaced by u^k. Similarly, for the control u^0, there exists a unique $\phi(u^0)\in L_2(Q)$ that satisfies the equality

$$\int_Q\int\{\phi(u^0)(x,t)F^*(\psi)(x,t)\}dx\,dt=l^0(\psi) \tag{4.2.47}$$

for all $\psi\in X$, where $l^0(\psi)$ corresponds to u^0.

Since $u^k\to u^0$ weakly in $L_2(I, R^2)$ and $\psi\in X$, it is clear from the expressions for l^k and l^0 that $l^k\to l^0$ in the weak* topology of X^* (dual of X). Therefore,

$$\int_Q\int\{((\phi(u^k)(x,t)-\phi(u^0)(x,t))F^*(\psi)(x,t))\}dx\,dt\to 0 \tag{4.2.48}$$

for all $\psi\in X$. Since F^* is an isomorphism of X onto $L_2(Q)$, it follows that

$\phi(u^k) \to \phi(u^0)$ weakly in $L_2(Q)$. Further,

$$J(u^k) - J(u^0)$$
$$= \lambda_0 \int_Q \int \tfrac{1}{2}\left\{ \left(\phi(u^k)(x,t) - \phi_d(x,t) \right)^2 - \left(\phi(u^0)(x,t) - \phi_d(x,t) \right)^2 \right\} dx\, dt$$
$$+ \lambda_3 \int_Q \int \tfrac{1}{2}\left\{ \left(\phi(u^k)(x,t) \right)^2 - \left(\phi(u^0)(x,t) \right)^2 \right\} dx\, dt$$
$$+ \lambda_1 \int_I \tfrac{1}{2}\left\{ \left(u_1^k(t) \right)^2 - \left(u_1^0(t) \right)^2 \right\} dt + \lambda_2 \int_I \tfrac{1}{2}\left\{ \left(u_2^k(t) \right)^2 - \left(u_2^0(t) \right)^2 \right\} dt. \tag{4.2.49}$$

Using the following elementary inequality,

$$\tfrac{1}{2}(\theta^2 - \Delta^2) \geq (\theta - \Delta)\Delta,$$

in (4.2.49), we obtain

$$J(u^k) \geq J(u^0) + \lambda_0 \int_Q \int \left\{ \left(\phi(u^k)(x,t) - \phi(u^0)(x,t) \right) \left[\left(1 + \frac{\lambda_3}{\lambda_0} \right) \phi(u^0)(x,t) \right.\right.$$
$$\left.\left. - \phi_d(x,t) \right] \right\} dx\, dt + \int_I \left\{ \sum_{j=1}^{2} \lambda_j \left(u_j^k(t) - u_j^0(t) \right) u_j^0(t) \right\} dt. \tag{4.2.50}$$

Since $u^k \to u^0$ weakly in $L_2(I, R^2)$ and $\phi(u^k) \to \phi(u^0)$ weakly in $L_2(Q)$, it follows from taking limit inferior on both sides of (4.2.50) that

$$J(u^0) \leq \varliminf_{k \to \infty} J(u^k). \tag{4.2.51}$$

This completes the proof. □

Theorem 4.2.3. *Consider Problem* (4.2.P1). *Suppose that all the hypotheses of Lemma* 4.2.3 *are satisfied. Then there exists a unique optimal control* $u^* \in \mathcal{U}_a$.

PROOF. Define

$$\inf\{ J(u) : u \in \mathcal{U}_a \} = \gamma. \tag{4.2.52}$$

Since, by definition, $J \geq 0$ and $J(u) < \infty$ for each $u \in \mathcal{U}_a$, it follows that $0 \leq \gamma < \infty$. Let $\{u^k\} \subset \mathcal{U}_a$ be a minimizing sequence such that

$$\lim_{k \to \infty} J(u^k) = \gamma. \tag{4.2.53}$$

Since $0 \leq \gamma < \infty$ and $J(u) \to \infty$ as $\|u\|_{2,I} \to \infty$, it follows that $\{u^k\}$ is a bounded subset of $\mathcal{U}_a \subset L_2(I, R^2)$. Thus, there exists a subsequence of $\{u^k\}$, again denoted by $\{u^k\}$, and a $u^* \in L_2(I, R^2)$ such that $u^k \to u^*$ weakly in $L_2(I, R^2)$. Since $\mathcal{U}_a$ is closed and convex, it is weakly closed, and therefore, $u^* \in \mathfrak{U}_a$. Now, in view of (4.2.52), we have

$$\gamma \leq J(u^*). \tag{4.2.54}$$

On the other hand, it follows from Lemma 4.2.3 that

$$J(u^*) \leq \varliminf_{k\to\infty} J(u^k). \tag{4.2.55}$$

Combining (4.2.53), (4.2.54), and (4.2.55), we obtain

$$\gamma \leq J(u^*) \leq \varliminf_{k\to\infty} J(u^k) \leq \lim_{k\to\infty} J(u^k) = \gamma. \tag{4.2.56}$$

This shows that u^* is an optimal control. However, J is strictly convex in u, and therefore, u^* is unique. This completes the proof. □

4.2.7. *Computational Considerations*

Based on Theorem 4.2.2, we can devise simple iterative techniques for computing optimal controls.

Following the proof of Theorem 4.2.2, it is easily observed that

$$\begin{aligned} J'_{u^0}(u-u^0) = \int_I \Big\{ &\big(\lambda_1 u_1^0(t) + c\big(z(u^0)\big)_x(0,t)\big)\big(u_1(t)-u_1^0(t)\big) \\ &+ \big(\lambda_2 u_2^0(t) - c\big(z(u^0)\big)_x(l_0,t)\big)\big(u_2(t)-u_2^0(t)\big)\Big\}\,dt \end{aligned} \tag{4.2.57}$$

for any pair of controls $u, u^0 \in \mathfrak{U}_a$. By the definition of $J'_{u^0}(u-u^0)$, we have

$$\begin{aligned} \lim_{\varepsilon\to 0} &\frac{J(u^0+\varepsilon(u-u^0)) - J(u^0)}{\varepsilon} \\ &\equiv J'_{u^0}(u-u^0) \\ &= \int_I \Big\{\big(\lambda_1 u_1^0(t) + c\big(z(u^0)\big)_x(0,t)\big)\big(u_1(t)-u_1^0(t)\big) \\ &\quad + \big(\lambda_2 u_2^0(t) - c\big(z(u^0)\big)_x(l_0,t)\big)\big(u_2(t)-u_2^0(t)\big)\Big\}\,dt \end{aligned} \tag{4.2.58}$$

for any pair of controls $u, u^0 \in \mathfrak{U}_a$.

On the basis of (4.2.58), we can devise a computational method for each of the following two cases:

i. $\mathfrak{U}_a \equiv L_2(I, R^2)$.
ii. $\mathfrak{U}_a$ as defined in Corollary 4.2.1.

Let us consider case (i). Suppose that $u^0 \in \mathcal{U}_a$ is not an optimal control and that $\nu \in \mathcal{U}_a$ is defined by

$$\nu_1(t) = -\frac{c}{\lambda_1}\big(z(u^0)\big)_x(0,t), \qquad t \in I$$

$$\nu_2(t) = \frac{c}{\lambda_2}\big(z(u^0)\big)_x(l_0,t), \qquad t \in I. \tag{4.2.59}$$

Then we observe that

$$\begin{aligned}
&\int_I \Big\{ \big(\lambda_1 u_1^0(t) + c\big(z(u^0)\big)_x(0,t)\big)\big(\nu_1(t) - u_1^0(t)\big) \\
&\qquad + \big(\lambda_2 u_2^0(t) - c\big(z(u^0)\big)_x(l_0,t)\big)\big(\nu_2(t) - u_2^0(t)\big)\Big\}\,dt \\
&\quad = \int_I \Big\{ \big(\lambda_1 u_1^0(t) + c\big(z(u^0)\big)_x(0,t)\big)\Big(-\frac{c}{\lambda_1}\big(z(u^0)\big)_x(0,t) - u_1^0(t)\Big) \\
&\qquad + \big(\lambda_2 u_2^0(t) - c\big(z(u^0)\big)_x(l_0,t)\big)\Big(\frac{c}{\lambda_2}\big(z(u^0)\big)_x(l_0,t) - u_2^0(t)\Big)\Big\}\,dt \\
&\quad = \int_I \Big\{ -\frac{1}{\lambda_1}\big(\lambda_1 u_1^0(t) + c\big(z(u^0)\big)_x(0,t)\big)^2 \\
&\qquad - \frac{1}{\lambda_2}\big(\lambda_2 u_2^0(t) - c\big(z(u^0)\big)_x(l_0,t)\big)^2\Big\}\,dt < 0.
\end{aligned} \tag{4.2.60}$$

Thus, it follows from (4.2.58) with $u = \nu$ and (4.2.60) that

$$\lim_{\varepsilon \to 0} \frac{J\big(u^0 + \varepsilon(\nu - u^0)\big) - J(u^0)}{\varepsilon} < 0. \tag{4.2.61}$$

This, in turn, implies that there exists an $\varepsilon_0 > 0$ such that

$$J\big(u^0 + \varepsilon(\nu - u^0)\big) < J(u^0) \tag{4.2.62}$$

for all ε, $0 < \varepsilon < \varepsilon_0$.

The above discussion leads us to the following algorithm [case (i)]:

1. Guess $u^0 \in \mathcal{U}_a$.
2. Compute $\phi(u^0)$, $z(u^0)$, and $J(u^0)$.
3. Find ν according to (4.2.59).
4. Define $u^1 = u^0 + \varepsilon(\nu - u^0)$ with $\varepsilon \in (0,1)$.
5. Compute $J(u^1)$. If $J(u^1) < J(u^0)$, go to Step 2 with u^1 replacing u^0. Otherwise, reduce ε by a suitable factor and go to Step 4.

For convenience, the above algorithm will be referred to as Algorithm 4.2.I.

REMARK 4.2.5. In practice, the algorithm is terminated if any of the following two conditions is satisfied:

i. $|J(u^{n+1})-J(u^n)|$ is less than some prespecified tolerance.
ii. $\|u^{n+1}-u^n\|_{2,I}$ is less than some prespecified tolerance.

Now let us consider case (ii). Suppose that $u^0\in\mathfrak{U}_a$ is not an optimal control and that $\nu\in\mathfrak{U}_a$ is a control defined by

$$\nu_1(t)=\gamma_1\Big(\frac{c}{\lambda_1}(z(u^0))_x(0,t)\Big),$$

$$\nu_2(t)=\gamma_2\Big(\frac{c}{\lambda_2}(z(u^0))_x(l,t)\Big), \tag{4.2.63}$$

where

$$\gamma_j(y)=\begin{cases}\beta_j & \text{for } y\geq\beta_j\\ y & \text{for } |y|<\beta_j\\ -\beta_j & \text{for } y\leq-\beta_j,\end{cases} \tag{4.2.64}$$

and $j=1,2$. Then, it can be easily shown that

$$\int_I\{(\lambda_1u_1^0(t)+c(z(u^0))_x(0,t))(\nu_1(t)-u_1^0(t))$$

$$+(\lambda_2u_2^0(t)-c(z(u^0))_x(l,t))(\nu_2(t)-u_2^0(t))\}\,dt<0. \tag{4.2.65}$$

Thus, it follows from (4.2.58) with $u=\nu$ and (4.2.65) that

$$\lim_{\varepsilon\to0}\frac{J(u^0+\varepsilon(\nu-u^0))-J(u^0)}{\varepsilon}<0. \tag{4.2.66}$$

This, in turn, implies that there exists an $\varepsilon_0>0$, so that

$$J(u^0+\varepsilon(\nu-u^0))<J(u^0) \tag{4.2.67}$$

for all ε, $0<\varepsilon<\varepsilon_0$.

The discussion given for case (ii) above leads us to an algorithm for computing an optimal control. For convenience, this algorithm will be referred to as Algorithm 4.2.II. Algorithm 4.2.II is similar to Algorithm 4.2.I with Step 3 replaced by Step 3′: find ν according to (4.2.63). Further, the practical termination test for Algorithm 4.2.II remains the same as that given in Remark 4.2.5.

REMARK 4.2.6. Note that the above algorithms are equivalent to the gradient technique suggested in [Ah.1].

4.3. Second-Order Hyperbolic Systems in One Spatial Variable with Boundary Control

4.3.1. Introduction

In this section, we consider a class of systems governed by hyperbolic partial differential equations with Darboux boundary conditions. Here, the controls appear in the differential equation of the system.

In Section 4.3.2, we describe the system and impose certain basic assumptions. Results concerning the existence and uniqueness of solutions of the system and certain important properties of the solutions are given in Section 4.3.3. In Sections 4.3.4 and 4.3.5, an optimal control problem is solved giving the necessary conditions for optimality. The questions of existence of optimal controls is discussed in Section 4.3.6, and a computational method for finding optimal controls is given in Section 4.3.7.

4.3.2. System Description

Let x_0, l_1, t_0, l_2 be fixed positive constants and define $I_1 \equiv [x_0, x_0 + l_1]$, $I_2 \equiv [t_0, t_0 + l_2]$ and $Q \equiv I_1 \times I_2$.

We consider the following system:

$$\begin{aligned} &\phi_{xt}(x,t) = f(x,t,\phi(x,t),\phi_x(x,t),\phi_t(x,t),u(x,t)), \qquad (x,t)\in Q,\\ &\phi(x,t_0) = \phi^0(x), \qquad x \in I_1,\\ &\phi(x_0,t) = \phi^b(t), \qquad t\in I_2, \end{aligned} \tag{4.3.1}$$

where, for each $(x, t) \in Q$, $\phi(x, t) \equiv (\phi_1(x, t), \ldots, \phi_n(x, t)) \in R^n$, $f \equiv (f_1, \ldots, f_n)$, $u \in \mathscr{U}$, $\mathscr{U}$ is the class of admissible controls to be defined later, $\phi^0 \equiv (\phi_1^0, \ldots, \phi_n^0)$, and $\phi^b \equiv (\phi_1^b, \ldots, \phi_n^b)$.

For $z \in R^n$, we define its norm as $|z| \equiv \sum_{i=1}^n |z_i|$. For a measurable set G, let $L_p(G, R^n)$ be the n-copies of $L_p(G)$ with the norm defined by

$$\|z\|_{p,G} \equiv \sum_{i=1}^{n} \|z_i\|_{p,G},$$

where $z \equiv (z_1, \ldots, z_n)$ and $z_i \in L_p(G)$, $i = 1, \ldots, n$.

Let $W_p^{1,1}(Q)$, $1 \le p \le \infty$, denote the space of all those real-valued functions $\{z\}$ such that z together with all its generalized partial derivatives z_x, z_t are in $L_p(Q)$. $W_p^{1,1}(Q)$, $1 \le p \le \infty$, is a vector space and is a Banach space with respect to the norm

$$\|z\|_{p,Q}^{(1,1)} \equiv \|z\|_{p,Q} + \|z_x\|_{p,Q} + \|z_t\|_{p,Q}.$$

Let $W_p^{1,1}(Q, R^n)$, $1 \le p \le \infty$, denote the n-copies of $W_p^{1,1}(Q)$ with the norm defined by

$$\|z\|_{p,Q}^{(1,1)} \equiv \sum_{i=1}^{n} \|z_i\|_{p,Q}^{(1,1)},$$

where $z \equiv (z_1, \ldots, z_n)$ and $z_i \in W_p^{1,1}(Q)$, $i=1,\ldots,n$. Clearly, $W_p^{1,1}(Q, R^n)$, $1 \le p \le \infty$, equipped with the norm defined above is a Banach space.

Let V be a given nonempty compact subset of R^m and U a measurable multifunction (see Section 1.4.1 on $\bar{Q}$ taking values from the class of nonempty closed convex subsets of V. A measurable function $u: \bar{Q} \to V$ is called an *admissible control* if $u(x,t) \in U(x,t)$ a.e. on $\bar{Q}$. Let $\mathcal{U}$ denote *the class of admissible controls*.

For system (4.3.1), we introduce the following definition.

Definition 4.3.1. For each $u \in \mathcal{U}$, a function $\phi(u) \in W_p^{1,1}(Q, R^n)$, for some $p \in [1, \infty]$, is said to be a *solution of system* (4.3.1) if it satisfies the differential equation a.e. on Q and the boundary conditions everywhere in their corresponding domains of definition.

Throughout this section, we need the following assumptions.

Assumption (4.3.A1)

i. The functions ϕ^0 and ϕ^b are absolutely continuous on I_1 and I_2, respectively.
ii. $\phi_x^0 \in L_p(I_1, R^n)$ and $\phi_t^b \in L_p(I_2, R^n)$ for some $p \in [1, \infty]$.
iii. $\phi^b(t_0) = \phi^0(x_0)$.

Assumption (4.3.A2)

i. The function $(x, t, \theta^1, \theta^2, \theta^3, v) \to f(x, t, \theta^1, \theta^2, \theta^3, v)$ is defined on $Q \times R^{3n} \times V$.
ii. $f(\cdot, \cdot, \theta^1, \theta^2, \theta^3, v)$ is measurable on Q.
iii. $f(x, t, \theta^1, \theta^2, \theta^3, \cdot)$ is continuous on V.

Assumption (4.3.A3). For each $u \in \mathcal{U}$, the function $h(u)(\cdot, \cdot) \equiv f(\cdot, \cdot, 0, 0, 0, u(\cdot, \cdot))$ belongs to $L_p(Q, R^n)$ with p as in Assumption (4.3.A1).

Assumption (4.3.A4). There exists a constant K_1 such that

$$|f(x, t, \theta^1, \theta^2, \theta^3, v) - f(x, t, \tilde{\theta}^1, \tilde{\theta}^2, \tilde{\theta}^3, v)|$$

$$\le K_1 \{ |\theta^1 - \tilde{\theta}^1| + |\theta^2 - \tilde{\theta}^2| + |\theta^3 - \tilde{\theta}^3| \}$$

for all $(x, t) \in Q$, θ^i, $\tilde{\theta}^i \in R^n$, $i = 1, 2, 3$, and $v \in V$.

4.3.3. Existence and Uniqueness of Solutions

In this subsection, we wish to prove the existence and uniqueness of solutions of system (4.3.1) and investigate certain important properties of these solutions.

Theorem 4.3.1. *Consider system* (4.3.1). *Suppose that Assumptions* (4.3.A1)–(4.3.A4) *are satisfied. Then, for each* $u \in \mathcal{U}$, *the system admits a unique solution* $\phi(u) \in W_p^{1,1}(Q, R^n)$. *Further,* $\phi(u)$ *is continuous on* Q *and there exists a constant* $K_2 \equiv K_2(l_1, l_2, p, K_1)$ *such that*

$$\|\phi(u)\|_{p,Q}^{(1,1)} \le K_2\{\|\phi_x^0\|_{p,I_1} + \|\phi^0\|_{p,I_1} + \|\phi_t^b\|_{p,I_2} + \|\phi^b\|_{p,I_2} + \|h(u)\|_{p,Q}\} \tag{4.3.2}$$

for all $u \in \mathcal{U}$, *where the function* $h(u)$ *and the constant* K_1 *are defined in Assumptions* (4.3.A3) *and* (4.3.A4), *respectively.*

PROOF. The proof is based on the existence (and uniqueness) of solutions of the following functional equation

$$\phi(x,t) = \phi^0(x) + \phi^b(t) - \phi^0(x_0) + \int_{x_0}^{x}\int_{t_0}^{t}\{f(\eta,\tau,\phi(\eta,\tau),\phi_x(\eta,\tau),\phi_t(\eta,\tau),u(\eta,\tau))\}\,d\eta\,d\tau. \tag{4.3.3}$$

By Assumptions (4.3.A2) and (4.3.A4), we can easily verify that if ϕ, ϕ_x, and ϕ_t are measurable on Q, then $f(\cdot,\cdot,\phi(\cdot,\cdot),\phi_x(\cdot,\cdot),\phi_t(\cdot,\cdot),u(\cdot,\cdot))$ is measurable on Q for each $u \in \mathcal{U}$. Since $\phi^0(x_0) = \phi^b(t_0)$, it is clear from (4.3.3) that $\phi(x,t_0) = \phi^0(x)$ and $\phi(x_0,t) = \phi^b(t)$. In other words, any solution of the functional equation (4.3.3) satisfies the boundary conditions of system (4.3.1). We shall show that if this functional equation has a solution, then it is also a solution of a system of Volterra integral equations. Let $\phi(u)$ be a solution of the functional equation corresponding to the control $u \in \mathcal{U}$. Further, if $\phi(u) \in W_p^{1,1}(Q, R^n)$, then we observe that the "usual" partial derivatives $(\phi(u))_x$, $(\phi(u))_t$, $(\phi(u))_{xt}$ of $\phi(u)$ exist a.e. on Q and are given by

$$(\phi(u))_x(x,t) = \phi_x^0(x) + \int_{t_0}^{t}\{f(x,\tau,\phi(u)(x,\tau),(\phi(u))_x(x,\tau),(\phi(u))_t(x,\tau),u(x,\tau))\}\,d\tau, \tag{4.3.4}$$

$$(\phi(u))_t(x,t) = \phi_t^b(t) + \int_{x_0}^{x}\{f(\eta,t,\phi(u)(\eta,t),(\phi(u))_x(\eta,t),(\phi(u))_t(\eta,t),u(\eta,t))\}\,d\eta \tag{4.3.5}$$

and

$$(\phi(u))_{xt}(x,t) = f(x,t,\phi(u)(x,t),(\phi(u))_x(x,t),(\phi(u))_t(x,t),u(x,t)) \tag{4.3.6}$$

for almost all $(x,t) \in Q$. By virtue of the hypotheses of the theorem, it follows that these "usual" partial derivatives of $\phi(u)$ belong to $L_p(Q, R^n)$. Thus, they are also the generalized derivatives of $\phi(u)$. Integrating the functions $(\phi(u))_x$, $(\phi(u))_t$, and $(\phi(u))_{xt}$ and using Equations (4.3.4)–(4.3.6), we obtain

$$\phi(u)(x,t)=\frac{1}{2}(\phi^0(x)+\phi^b(t))+\frac{1}{2}\int_{x_0}^{x}\{(\phi(u))_x(\eta,t)\}\,d\eta$$
$$+\frac{1}{2}\int_{t_0}^{t}\{(\phi(u))_t(x,\tau)\}\,d\tau,$$
$$(\phi(u))_x(x,t)=\phi_x^0(x)+\int_{x_0}^{x}\{0\}\,d\eta$$
$$+\int_{t_0}^{t}\{f(x,\tau,\phi(u)(x,\tau),(\phi(u))_x(x,\tau),(\phi(u))_t$$
$$(x,\tau),u(x,\tau)\}\,d\tau,$$
$$(\phi(u))_t(x,t)=\phi_t^b(t)+\int_{x_0}^{x}\{f(\eta,t,\phi(u)(\eta,t),(\phi(u))_x(\eta,t),$$
$$(\phi(u))_t(\eta,t),u(\eta,t))\}\,d\eta+\int_{t_0}^{t}\{0\}\,d\tau. \tag{4.3.7}$$

Thus, $\hat{\phi}(u)\equiv(\phi(u),\ (\phi(u))_x,\ (\phi(u))_t)$ belongs to $L_p(Q, R^{3n})$ and is a solution of the following system of Volterra integral equations:

$$w^1(x,t)=\frac{1}{2}(\phi^0(x)+\phi^b(t))+\frac{1}{2}\int_{x_0}^{x}\{w^2(\eta,t)\}\,d\eta+\frac{1}{2}\int_{t_0}^{t}\{w^3(x,\tau)\}\,d\tau,$$
$$w^2(x,t)=\phi_x^0(x)+\int_{x_0}^{x}\{0\}\,d\eta+\int_{t_0}^{t}\{f(x,\tau,w(x,\tau),u(x,\tau))\}\,d\tau,$$
$$w^3(x,t)=\phi_t^b(t)+\int_{x_0}^{x}\{f(\eta,t,w(\eta,t),u(\eta,t))\}\,d\eta+\int_{t_0}^{t}\{0\}\,d\tau, \tag{4.3.8}$$

where $w\equiv(w^1,w^2,w^3)$. Now, we shall show that if this system of Volterra integral equations has a solution, then it is also the solution of the functional equation (4.3.3). Let $w(u)\in L_p(Q, R^{3n})$ be a solution of system (4.3.8) corresponding to the control $u\in\mathfrak{U}$. Then we can easily verify that $w^2(u)$ and $w^3(u)$ belong to $L_p(Q, R^n)$; further, $w^2(u)=(w^1(u))_x$ and $w^3(u)=(w^1(u))_t$. Thus,

$$w^1(u)(x,t)=\phi^0(x)+\phi^b(t)-\phi^0(x_0)$$
$$+\int_{x_0}^{x}\int_{t_0}^{t}\{f(\eta,\tau,w(\eta,\tau),u(\eta,\tau))\}\,d\eta\,d\tau, \tag{4.3.9}$$

belongs to $W_p^{1,1}(Q, R^n)$ and satisfies the functional equation (4.3.3).

At this stage, we conclude that every solution $[\phi(u)\in W_p^{1,1}(Q, R^n)]$ of the functional equation (4.3.3) corresponds uniquely to a solution $[w(u)\equiv(w^1(u),\ w^2(u),\ w^3(u))\in L_p(Q, R^{3n})]$ of system (4.3.8) and conversely. Now we prove the existence and uniqueness of solutions of the integral equation (4.3.8). We observe that system (4.3.8) is of the form of system (2.3.1). Further, we can easily verify that the hypotheses of the theorem imply the assumptions of Theorem 2.3.1. Thus, it follows from Theorem 2.3.1 that system (4.3.8) has a unique solution $w(u) \in L_p(Q, R^{3n})$, and hence, the functional equation (4.3.3) has a unique solution $\phi(u) \in W_p^{1,1}(Q, R^n)$. In view of Equation (4.3.6), it is clear that $\phi(u)$ satisfies also the differential equation of system (4.3.1) a.e. on Q. Further, we recall that $\phi(u)$ satisfies the boundary conditions of the system. Thus, $\phi(u)$ belongs to $W_p^{1,1}(Q, R^n)$ and is the unique solution of system (4.3.1). It remains to establish the estimate (4.3.2). By applying Theorem 2.3.1 to system (4.3.8), we obtain

$$\| w(u) \|_{p,Q} \leq \gamma_0,$$

where the constant γ_0 depends only on l_1, l_2, p, K_1, and the norms of $h(u)$, ϕ^0, ϕ_x^0, ϕ^b, and ϕ_t^b. Furthermore, by examining the proof of Theorem 2.3.1, we observe that there exists a constant $K_2 \equiv K_2(l_1, l_2, p, K_1)$ such that

$$\gamma_0 \leq K_2 \left\{ \| \phi_x^0 \|_{p,I_1} + \| \phi^0 \|_{p,I_1} + \| \phi_t^b \|_{p,I_2} + \| \phi^b \|_{p,I_2} + \| h(u) \|_{p,Q} \right\}.$$

Since $\| w(u) \|_{p,Q} \equiv \| \phi(u) \|_{p,Q}^{(1,1)}$, we obtain the estimate (4.3.2). This completes the proof. □

The next theorem presents a result concerning certain pointwise estimates for the solutions of system (4.3.1).

Theorem 4.3.2. *Consider system* (4.3.1). *Suppose that all the hypotheses of Theorem* 4.3.1 *are satisfied. Then there exists a constant* $K_3 \equiv K_3(l_1, l_2, K_1, p)$ *such that for all* $(x, t) \in Q$,

$$|\phi(u)(x,t)| \leq K_3 \left\{ \| \phi^0 \|_{I_1}^{(0)} + \| \phi^b \|_{I_2}^{(0)} + \| \phi_x^0 \|_{p,I_1} + \| \phi_t^b \|_{p,I_2} + \int_Q\int \{ |h(u)(x,t)| \} \, dx \, dt \right\}, \tag{4.3.10}$$

and, for almost all $(x, t) \in Q$,

$$|(\phi(u))_x(x,t)| \leq K_3 \left\{ |\phi_x^0(x)| + |\phi^0(x)| + \int_{I_2} |h(u)(x,t)| \, dt + \| \phi_x^0 \|_{p,I_1} + \| \phi_t^b \|_{p,I_2} + \| \phi^0 \|_{I_1}^{(0)} + \| \phi^b \|_{I_2}^{(0)} + \int_Q\int |h(u)(x,t)| \, dx \, dt \right\}, \tag{4.3.11}$$

and

$$|(\phi(u)_t(x,t)| \le K_3\Big\{ |\phi_t^b(t)| + |\phi^b(t)| + \int_{I_1} |h(u)(x,t)|\,dx$$

$$+ \|\phi_x^0\|_{p,I_1} + \|\phi_t^b\|_{p,I_2} + \|\phi^0\|_{I_1}^{(0)} + \|\phi^b\|_{I_2}^{(0)}$$

$$+ \int_Q\int \{h(u)(x,t)\}\,dx\,dt\Big\}, \tag{4.3.12}$$

where $\|z\|_I^{(0)} \equiv \Sigma_{i=1}^n \max\{|z_i(y)| : y \in I\}$ *and the constant* K_1 *and the function* $h(u)$ *are as given in Assumptions* (4.3.A4) *and* (4.3.A3), *respectively.*

PROOF. By virtue of Equation (4.3.4) and Assumptions (4.3.A3)–(4.3.A4), we obtain

$$|(\phi(u)_x(x,t)| \le |\phi_x^0(x)| + \int_{t_0}^t |h(u)(x,\tau)|\,dx$$

$$+ K_1\int_{t_0}^t \{|\phi(u)(x,\tau)| + |(\phi(u)_x(x,\tau)| + |(\phi(u)_t(x,\tau)|\}\,d\tau \tag{4.3.13}$$

for almost all $x \in I_1$ and for all $t \in I_2$. Similarly, we can verify from Equation (4.3.5) and Assumptions (4.3.A3)–(4.3.A4) that

$$|(\phi(u))_t(x,t)| \le |\phi_t^b(t)| + \int_{x_0}^x |h(u)(\eta,t)|\,d\eta$$

$$+ K_1\int_{x_0}^x \{|\phi(u)(\eta,t)| + |(\phi(u))_x(\eta,t)|$$

$$+ |(\phi(u))_t(\eta,t)|\}\,d\eta \tag{4.3.14}$$

for all $x \in I_1$ and for almost all $t \in I_2$.

Let I_1^0 (I_2^0) be the subset of I_1 [I_2] such that the inequality (4.3.13) [(4.3.14)] is satisfied on $I_1^0 \times I_2[I_1 \times I_2^0]$. Clearly, $|I_1^0| = |I_1|$ and $|I_2^0| = |I_2|$, where $|I|$ denotes the Lebesgue measure of I.

Since

$$\phi(u)(x,t) = \phi^b(t) + \int_{x_0}^x \{(\phi(u))_x(\eta,t)\}\,d\eta$$

$$= \phi^0(x) + \int_{t_0}^t \{(\phi(u))_t(x,\tau)\}\,d\tau, \tag{4.3.15}$$

we have

$$\int_{x_0}^x \{|\phi(u)(\eta,t)|\}\,d\eta \le l_1\left[|\phi^b(t)| + \int_{x_0}^x \{|(\phi(u))_x(\eta,t)|\}\,d\eta\right] \tag{4.3.16}$$

and

$$\int_{t_0}^{t}\{|\phi(u)(x,\tau)|\}\,d\tau \le l_2\left[|\phi^0(x)| + \int_{t_0}^{t}\{|(\phi(u))_t(x,\tau)|\}\,d\tau\right]. \tag{4.3.17}$$

From inequalities (4.3.13) and (4.3.17), it follows that

$$|(\phi(u))_x(x,t)| \le \gamma_0(x) + N_0\int_{t_0}^{t}\{|(\phi(u))_t(x,\tau)|\}\,d\tau + K_1\int_{t_0}^{t}\{|(\phi(u))_x(x,\tau)|\}\,d\tau \tag{4.3.18}$$

for all $(x,t)\in I_1^0\times I_2$, where

$$\gamma_0(x) \equiv |\phi_x^0(x)| + K_1 l_2|\phi^0(x)| + \int_{I_2}|h(u)(x,\tau)|\,d\tau$$

and $N_0 \equiv K_1(1+l_2)$.

Applying Gronwall's lemma to the above inequality, we obtain

$$|(\phi(u))_x(x,t)| \le N_1\gamma_0(x) + N_2\int_{t_0}^{t}\{|(\phi(u))_t(x,\tau)|\}\,d\tau \tag{4.3.19}$$

for all $(x,t)\in I_1^0\times I_2$, where $N_1 \equiv 1 + K_1 l_2 \exp\{K_1 l_2\}$ and $N_2 \equiv N_0 N_1$.

Similarly, we can verify from the inequalities (4.3.14) and (4.3.16) and Gronwall's lemma that

$$|(\phi(u))_t(x,t)| \le \tilde{N}_1\tilde{\gamma}_0(t) + \tilde{N}_2\int_{x_0}^{x}\{|(\phi(u))_x(\eta,t)|\}\,d\eta \tag{4.3.20}$$

for all $(x,t)\in I_1\times I_2^0$, where $\tilde{N}_1 \equiv 1 + K_1 l_1 \exp\{K_1 l_1\}$,

$$\tilde{N}_2 \equiv \tilde{N}_1 K_1(1+l_1) \quad \text{and} \quad \tilde{\gamma}_0(t) \equiv |\phi_t^b(t)| + K_1 l_1|\phi^b(t)| + \int_{I_1}|h(u)(\eta,t)|\,d\eta.$$

Substituting the above inequality into the inequality (4.3.19), we have

$$|(\phi(u))_x(x,t)| \le N_1\gamma_0(x) + N_3\int_{t_0}^{t}\{\tilde{\gamma}_0(\tau)\}\,d\tau + N_4\int_{t_0}^{t}\int_{x_0}^{x}\{|(\phi(u))_x(\eta,\tau)|\}\,d\eta\,d\tau, \tag{4.3.21}$$

where $N_3 \equiv N_2\tilde{N}_1$ and $N_4 \equiv N_2\tilde{N}_2$.

Integrating both sides of the above inequality with respect to x from x_0 to $x_0 + l_1$ and defining $w(u)(t) \equiv \int_{I_1}\{|(\phi(u))_x(\eta,t)|\}\,d\eta$, we obtain

$$w(u)(t) \le N_1\int_{I_1}\{\gamma_0(x)\}\,dx + N_3 l_1\int_{I_2}\left\{\tilde{\gamma}_0(t)\,dt + N_4 l_1\int_{t_0}^{t}\{w(u)(\tau)\}d\tau.\right. \tag{4.3.22}$$

Thus, it follows from Gronwall's lemma that

$$w(u)(t) \le N_5 \int_{I_1} \{\gamma_0(x)\}\,dx + N_6 \int_{I_2} \{\tilde{\gamma}_0(t)\}\,dt, \tag{4.3.23}$$

where $N_5 \equiv N_1 N_7$, $N_6 \equiv N_3 l_1 N_7$, and $N_7 \equiv (1 + N_4 l_1 l_2 \exp\{N_4 l_1 l_2\})$.

By virtue of inequality (4.3.21), the definition of $w(u)$, and the above inequality, we can verify that

$$|(\phi(u))_x(x,t)| \le N_1 \gamma_0(x) + N_8 \int_{I_2} \{\tilde{\gamma}_0(t)\}\,dt + N_9 \int_{I_1} \{\gamma_0(x)\}\,dx, \tag{4.3.24}$$

where $N_8 \equiv N_3 + N_4 N_6 l_2$ and $N_9 \equiv N_4 N_5 l_2$. Thus, by using the definitions of γ_0 and $\tilde{\gamma}_0$, it follows from Hölder's inequality that

$$|(\phi(u))_x(x,t)| \le N_{10} \Big\{ |\phi_x^0(x)| + |\phi^0(x)| + \int_{I_2} |h(u)(x,\tau)|\,d\tau + \|\phi_x^0\|_{p,I_1} + \|\phi_t^b\|_{p,I_2} + \|\phi^0\|_{I_1}^{(0)} + \|\phi^b\|_{I_2}^{(0)} + \int_Q \int |h(u)(x,t)|\,dx\,dt \Big\} \tag{4.3.25}$$

for all $(x,t) \in I_1^0 \times I_2$, where $N_{10} \equiv \max\{N_1,\ N_1 K_1 l_2,\ N_8 (l_2)^{1/q},\ N_8 K_1 l_1 l_2,\ N_8 + N_9,\ N_9 (l_1)^{1/q},\ N_9 K_1 l_1 l_2\}$, $(1/p) + (1/q) = 1$, and $\|z\|_I^{(0)} \equiv \sum_{i=1}^n \max\{|z_i(y)| : y \in I\}$.

Similarly, we can show that

$$|(\phi(u))_t(x,t)| \le N_{11} \Big\{ |\phi_t^b(t)| + |\phi^b(t)| + \int_{I_1} |h(u)(\eta,t)|\,d\eta + \|\phi_x^0\|_{p,I_1} + \|\phi_t^b\|_{p,I_2} + \|\phi^0\|_{I_1}^{(0)} + \|\phi^b\|_{I_2}^{(0)} + \int_Q \int |h(u)(x,t)|\,dx\,dt \Big\}, \tag{4.3.26}$$

for all $(x,t) \in I_1 \times I_2^0$, where the constant N_{11} is obtained by interchanging l_1 and l_2 in all the constants involved in N_{10}.

Substituting (4.3.25) and (4.3.26) into the first equation of (4.3.7) and using Hölder's inequality, we obtain

$$|\phi(u)(x,t)| \le N_{12} \Big[\|\phi^0(x)\|_{I_1}^{(0)} + \|\phi^b\|_{I_2}^{(0)} + \|\phi_x^0\|_{p,I_1} + \|\phi_t^b\|_{p,I_2} + \int_Q \int |h(u)(x,t)|\,dx\,dt \Big], \tag{4.3.27}$$

where

$$N_{12} \equiv \tfrac{1}{2}\max\{1+2N_{10}l_1+N_{11}l_2, 1+N_{10}l_1+2N_{11}l_2, N_{10}(l_1)^{1/q}+N_{10}l_1+N_{11}l_2,$$
$$\times N_{10}l_1+N_{11}(l_2)^{1/q}+N_{11}l_2, N_{10}l_1+N_{10}+N_{11}+N_{11}l_2\}.$$

Letting $K_3 \equiv \max\{N_{10}, N_{11}, N_{12}\}$, the conclusion of the theorem follows from the above three inequalities. □

In the next theorem, we present a result concerning the dependence of solutions of system (4.3.1) with respect to the data of the system.

Theorem 4.3.3. *Consider system* (4.3.1). *Suppose that* $\{\phi^{0,1}, \phi^{b,1}\}, \{\phi^{0,2}, \phi^{b,2}\}$ *are any two paris of functions satisfying Assumption* (4.3.A1) *and* $u^1, u^2 \in \mathcal{U}$. *Let the function f in the system* (4.3.1) *satisfy assumptions* (4.3.A2)–(4.3.A4) *and let* ϕ^1 *and* ϕ^2 *be the solutions of the system corresponding to the data* $\{\phi^{0,1}, \phi^{b,1}, u^1\}$ *and* $\{\phi^{0,2}, \phi^{b,2}, u^2\}$, *respectively. Define* $\phi \equiv \phi^1 - \phi^2$, $\phi^0 \equiv \phi^{0,1} - \phi^{0,2}$, $\phi^b \equiv \phi^{b,1} - \phi^{b,2}$, *and*

$$z(u^1)(x,t,u^2(x,t)) \equiv |f(x,t,\phi^1(x,t),\phi_x^1(x,t),\phi_t^1(x,t),u^1(x,t))$$
$$-f(x,t,\phi^1(x,t),\phi_x^1(x,t),\phi_t^1(x,t),u^2(x,t))|. \tag{4.3.28}$$

Then, for all $(x,t)\in Q$,

$$|\phi(x,t)| \le K_3\left\{ \|\phi^0\|_{I_1}^{(0)} + \|\phi^b\|_{I_2}^{(0)} + \|\phi_x^0\|_{p,I_1} + \|\phi_t^b\|_{p,I_2} + \int_Q\int \{z(u^1)(x,t,u^2(x,t))\}\,dx\,dt \right\}, \tag{4.3.29}$$

and, for almost all $(x,t)\in Q$,

$$|\phi_x(x,t)| \le K_3\left\{ |\phi_x^0(x)| + |\phi^0(x)| + \int_{I_2} \{z(u^1)(x,t,u^2(x,t))\}\,dt + \|\phi_x^0\|_{p,I_1} + \|\phi_t^b\|_{p,I_2} + \|\phi^0\|_{I_1}^{(0)} + \|\phi^b\|_{I_2}^{(0)} + \int_Q\int \{z(u^1)(x,t,u^2(x,t))\}\,dx\,dt \right\}, \tag{4.3.30}$$

and

$$|\phi_t(x,t)| \le K_3\left\{ |\phi_t^b(t)| + |\phi^b(t)| + \int_{I_1} \{z(u^1)(x,t,u^2(x,t))\}\,dx + \|\phi_x^0\|_{p,I_1} + \|\phi_t^b\|_{p,I_2} + \|\phi^0\|_{I_1}^{(0)} + \|\phi^b\|_{I_2}^{(0)} + \int_Q\int \{z(u^1)(x,t,u^2(x,t))\}\,dx\,dt \right\}, \tag{4.3.31}$$

where K_3 *and* $\|\cdot\|_I^{(0)}$ *are as given in Theorem* 4.3.1.

PROOF. The proof is similar to that given for Theorem 4.3.1. □

REMARK 4.3.1. Let $\phi^{0,1}=\phi^{0,2}$ on I_1, $\phi^{b,1}=\phi^{b,2}$ on I_2, and let (x^0,t^0) be an interior point of Q. Further, let $\delta>0$ be such that $N_\delta \equiv N_\delta(x^0,t^0) \equiv \{(x,t):|x-x^0|\le\delta, |t-t^0|\le\delta\} \subset Q$. If $u^1(x,t)=u^2(x,t)$ on $Q\backslash N_\delta$, then it is easily observed that the following conclusions are satisfied.

i. $z(u^1)(x,t,u^2(x,t))=0$ for all $(x,t)\in Q\backslash N_\delta$.
ii. $|\phi(x,t)|\le K_3 \int_{N_\delta}\int \{z(u^1)(x,t,u^2(x,t))\}dx\,dt$ for all $(x,t)\in Q$.
iii. $|\phi_x(x,t)|\le K_3[\int_{t^0-\delta}^{t^0+\delta}\{z(u^1)(x,t,u^2(x,t))\}\,dt+\int_{N_\delta}\int\{z(u^1)$ $\times(x,t,u^2(x,t))\}\,dx\,dt]$ for almost all $(x,t)\in Q$.
iv. $|\phi_t(x,t)|\le K_3[\int_{x^0-\delta}^{x^0+\delta}\{z(u^1)(x,t,u^2(x,t))\}\,dx+\int_{N_\delta}\int\{z(u^1)$ $\times(x,t,u^2(x,t))\}\,dx\,dt]$ for almost all $(x,t)\in Q$.

4.3.4. Formulation of Some Control Problems

Given the system (4.3.1), we can formulate an optimal control problem by including a cost functional that gives a measure of performance of the system corresponding to each control. In this subsection, two such cost functionals will be defined. The first of these is

$$J(u)=\alpha\cdot\phi(u)(x_0+l_1,t_0+l_2), \tag{4.3.32}$$

where $\alpha\in R^n$ and, for any $a,b\in R^n$, $a\cdot b\equiv\sum_{i=1}^n a_i b_i$.

With reference to the above cost functional we consider the following optimal control problem.

Problem (4.3.P1). Subject to system (4.3.1), find a control $u\in\mathfrak{U}$ such that the cost functional (4.3.32) is minimized.

The following problem is more general.

Problem (4.3.P2). Subject to system (4.3.1), find a control $u\in\mathfrak{U}$ such that the cost functional

$$J(u)=\gamma(\phi(u)(x_0+l_1,t_0+l_2)) \tag{4.3.33}$$

is minimized, here γ is a continuous function mapping R^n into R.

REMARK 4.3.2. It can be shown easily that Problem (4.3.P2) is equivalent to an optimal control problem with the same dynamical system but with the more general cost functional

$$J(u)=\gamma_0(\phi(u)(x_0+l_1,t_0+l_2))+\int_Q\int\{\gamma_1(x,t,\phi(u)(x,t),u(x,t))\}\,dx\,dt,$$

where γ_0 is a continuous function mapping R^n into R and, for any $u\in\mathfrak{U}$, $\gamma_1(\cdot,\cdot,\phi(u)(\cdot,\cdot),u(\cdot,\cdot))\in L_1(Q)$.

A similar result is well known in the case of optimal control problems involving only ordinary differential equations.

In order to derive necessary conditions for optimality for the terminal control problem (4.3.P1), we need certain preparation.

First of all, the partial differential equations of system (4.3.1) are written in Dieudonné–Rashevsky form. On this basis, we consider a corresponding class of optimal control problems. Then the concept of the conjugate problem for this class of control problems is introduced.

To start with, let Λ be the class of all pairs of functions $\{u, \phi\}$ satisfying the following conditions:

i. $u \in \mathfrak{U}$ and $\phi \in W_p^{1,1}(Q, R^n)$ for some $p \geq 1$.
ii. $\{u, \phi\}$ satisfies the system of $(n_1 + n_2)$ first-order differential equations

$$\phi_{ix}(x,t) = f_i(x,t,\phi(x,t),u(x,t)),\ i \in N_1,$$
$$\phi_{jt}(x,t) = g_j(x,t,\phi(x,t),u(x,t)),\ j \in N_2, \qquad (4.3.34a)$$

a.e., on Q, where f_i and g_j are real-valued Carathéodory functions defined on $Q \times \{R^n \times V\}$, while, for each $k \in \{1,2\}$, N_k consists of n_k elements of $\{1,\dots,n\}$ and $N_1 \cup N_2 = \{1,\dots,n\}$.

Define

$$S_1 \equiv \{(x,t): x \in I_1, t = t_0 + l_2\}, \qquad S_2 \equiv \{(x,t): x \in I_1, t = t_0\},$$
$$S_3 \equiv \{(x,t): x = x_0 + l_1, t \in I_2\}, \qquad S_4 \equiv \{(x,t): x = x_0, t \in I_2\}.$$

Note that S_k, $k = 1,2,3,4$, are the four sides of the rectangle Q. For every $k = 1,2,3,4$ and side S_k of Q, let $\{i\}_k$ denote any collection of indices $i = 1,\dots,n$, that depends on k and may be empty. To proceed further, we impose the following boundary conditions on elements in Λ.

$$\phi_i(x,t) = \beta_{ki}(x,t), \qquad i \in \{i\}_k, \quad k = 1,2,3,4, \quad \text{for all} \quad (x,t) \in S_k, \qquad (4.3.34b)$$

where β_{ki} belong to $L_p(S_k)$.

With reference to system (4.3.34), we consider the following optimal control problem.

Problem (4.3.P3). Subject to system (4.3.34), find an element $\{u, \phi\} \in \Lambda$ such that the cost functional

$$J(u,\phi) = \int_{I_1} \left\{ \sum_{j \in N_2} \theta_j^1(x)\phi_j(x, t_0 + l_2) \right\} dx + \int_{I_1} \left\{ \sum_{j \in N_2} \theta_j^2(x)\phi_j(x, t_0) \right\} dx$$
$$+ \int_{I_2} \left\{ \sum_{i \in N_1} h_i^1(t)\phi_i(x_0 + l_1, t) \right\} dt + \int_{I_2} \left\{ \sum_{i \in N_1} h_i^2(t)\phi_i(x_0, t) \right\} dt \qquad (4.3.35)$$

is minimized, where $\theta_j^1, \theta_j^2, h_i^1, h_i^2$ are given functions satisfying the following assumptions.

Assumption (4.3.A5)

i. $\theta_j^1, \theta_j^2 \in L_q(I_1)$, $h_i^1, h_i^2 \in L_q(I_2)$, and $(1/p)+(1/q)=1$.
ii. if $i \in \{i\}_k$, then the corresponding θ_i^1, θ_i^2, h_i^1 or h_i^2 is identically zero.

For any $\lambda \in R^{n_1}$, $\zeta \in R^{n_2}$, we define a Hamiltonian function on $Q \times R^n \times V \times R^{n_1} \times R^{n_2}$ for the control problem (4.3.P3) as follows:

$$H(x,t,\phi,v,\lambda,\zeta)=\sum_{i\in N_1}\lambda_i f_i(x,t,\phi,v)+\sum_{j\in N_2}\zeta_j g_j(x,t,\phi,v). \tag{4.3.36}$$

The conjugate problem for Problem (4.3.P3) is described by the system of n first-order partial differential equations

$$\begin{aligned}
&\lambda_{ix}(x,t)+\zeta_{it}(x,t)\\
&\quad=-\left\{\sum_{j\in N_1}\lambda_j(x,t)\frac{\partial f_j(x,t,\phi(x,t),u(x,t))}{\partial\phi_i}\right.\\
&\qquad\left.+\sum_{j\in N_2}\zeta_j(x,t)\frac{\partial g_j(x,t,\phi(x,t),u(x,t))}{\partial\phi_i}\right\},\quad i\in N_1\cap N_2,\\
&\lambda_{ix}(x,t)=-\left\{\sum_{j\in N_1}\lambda_j(x,t)\frac{\partial f_j(x,t,\phi(x,t),u(x,t))}{\partial\phi_i}\right.\\
&\qquad\left.+\sum_{j\in N_2}\zeta_j(x,t)\frac{\partial g_j(x,t,\phi(x,t),u(x,t))}{\partial\phi_i}\right\},\quad i\in N_1\backslash N_2,
\end{aligned} \tag{4.3.37a}$$

$$\begin{aligned}
&\zeta_{it}(x,t)=-\left\{\sum_{j\in N_1}\lambda_j(x,t)\frac{\partial f_j(x,t,\phi(x,t),u(x,t))}{\partial\phi_i}\right.\\
&\qquad\left.+\sum_{j\in N_2}\zeta_j(x,t)\frac{\partial g_j(x,t)\phi(x,t)u(x,t))}{\partial\phi_i}\right\},\quad i\in N_2\backslash N_1,
\end{aligned}$$

with boundary conditions

$$\begin{cases}
\zeta_i(x,t_0+l_2)=\theta_i^1(x), & x\in I_1,\quad i\in\{i\}_1^c\cap N_2\\
\zeta_i(x,t_0)=-\theta_i^2(x), & x\in I_1,\quad i\in\{i\}_2^c\cap N_2\\
\lambda_i(x_0+l_1,t)=h_i^1(t), & t\in I_2,\quad i\in\{i\}_3^c\cap N_1\\
\lambda_i(x_0,t)=-h_i^2(t), & t\in I_2,\quad i\in\{i\}_4^c\cap N_1,
\end{cases} \tag{4.3.37b}$$

where $\{i\}_k^c$ denotes $\{1,\ldots,n\}\backslash\{i\}_k$.

We now introduce the following notation:

$$\phi^1 \equiv \phi,$$
$$\phi^2 \equiv \phi_x, \tag{4.3.38}$$
$$\phi^3 \equiv \phi_t.$$

Then system (4.3.1) can be converted into the following equivalent form (known as Dieudonné–Rashevsky form):

$$\begin{aligned}
&\phi_x^1(x,t) = \phi^2(x,t), \qquad (x,t)\in Q,\\
&\phi_x^3(x,t) = f(x,t,\phi^1(x,t),\phi^2(x,t),\phi^3(x,t),u(x,t)), \quad (x,t)\in Q,\\
&\phi_t^1(x,t) = \phi^3(x,t), \qquad (x,t)\in Q, \qquad\qquad (4.3.39a)\\
&\phi_t^2(x,t) = f(x,t,\phi^1(x,t),\phi^2(x,t),\phi^3(x,t),u(x,t)), \qquad (x,t)\in Q,
\end{aligned}$$

with boundary conditions

$$\begin{aligned}
&\phi^1(x,t_0) = \phi^0(x), \qquad x\in I_1,\\
&\phi^2(x,t_0) = \phi_x^0(x), \qquad x\in I_1,\\
&\phi^1(x_0,t) = \phi^b(t), \qquad t\in I_2, \qquad\qquad (4.3.39b)\\
&\phi^3(x_0,t) = \phi_t^b(t), \qquad t\in I_2.
\end{aligned}$$

Note that the system (4.3.39) appears to be overdetermined (four equations in three unknowns). However, it is not actually so, since the second and fourth equations are actually equivalent.

By Theorem 4.3.1, it follows that, for each $u\in\mathcal{U}$, system (4.3.1) [and equivalently, system (4.3.39)] has a unique solution $\phi(u)$. Let $\phi^1(u)\equiv\phi(u)$, $\phi^2(u)\equiv(\phi(u))_x$ and $\phi^3(u)\equiv(\phi(u))_t$. Then by performing integration and using the boundary conditions (4.3.39b), we obtain

$$\begin{aligned}
&\frac{1}{2}\int_{I_1}\{\alpha\cdot\phi^2(u)(x,t_0+l_2)\}\,dx + \frac{1}{2}\int_{I_2}\{\alpha\cdot\phi^3(u)(x_0+l_1,t)\}\,dt\\
&\qquad = \alpha\cdot\phi(u)(x_0+l_1,t_0+l_2) - \frac{1}{2}\alpha\cdot\phi^b(t_0+l_2) - \frac{1}{2}\alpha\cdot\phi^0(x_0+l_1).
\end{aligned} \tag{4.3.40}$$

Thus, the cost functional (4.3.32) is equivalent to

$$\bar{J}(u) = \int_{I_1}\left\{\frac{1}{2}\alpha\cdot\phi^2(u)(x,t_0+l_2)\right\}dx + \int_{I_2}\left\{\frac{1}{2}\alpha\cdot\phi^3(u)(x_0+l_1,t)\right\}dt. \tag{4.3.41}$$

Since the system (4.3.1) and the cost functional (4.3.32) are equivalent to the system (4.3.39) and the cost functional (4.3.41), respectively, it is clear that Problem (4.3.P1) is equivalent to the following (optimal control) problem.

Problem (4.3.P1)′. Subject to system (4.3.39), find a control $u \in \mathfrak{U}$ so that the cost functional (4.3.41) is minimized.

Since the above problem is in the form of Problem (4.3.P3), the corresponding Hamiltonian functional H is given by

$$H(x,t,\hat{\phi},v,\lambda,\zeta)=\lambda^1\cdot\phi^2+\lambda^3\cdot f(x,t,\hat{\phi},v)+\zeta^1\cdot\phi^3+\zeta^2\cdot f(x,t,\hat{\phi},v). \tag{4.3.42}$$

where $\lambda \equiv (\lambda^1, \lambda^3)$, $\zeta \equiv (\zeta^1, \zeta^2)$, $\hat{\phi} \equiv (\phi^1, \phi^2, \phi^3)$, and $\lambda^1, \lambda^3, \zeta^1, \zeta^2, \phi^l$, $l=1,2,3$, belong to R^n.

The conjugate problem of Problem (4.3.P1)′ [and equivalently, Problem (4.3.P1)] can be written as the system of the $3n$ first-order differential equations

$$\begin{aligned}
&\lambda^1_{ix}(x,t)+\zeta^1_{it}(x,t)\\
&\qquad=-(\lambda^3(x,t)+\zeta^2(x,t))\cdot\frac{\partial f(x,t,\phi(u)(x,t),u(x,t))}{\partial\phi^1_i},\\
&\qquad\qquad(x,t)\in Q,\\
&\zeta^2_{it}(x,t)\\
&\qquad=-\lambda^1_i(x,t)-(\lambda^3(x,t)+\zeta^2(x,t))\cdot\frac{\partial f(x,t,\hat{\phi}(u)(x,t),u(x,t))}{\partial\phi^2_i},\\
&\qquad\qquad(x,t)\in Q,\\
&\lambda^3_{ix}(x,t)\\
&\qquad=-\zeta^1_i(x,t)-(\lambda^3(x,t)+\zeta^2(x,t))\cdot\frac{\partial f(x,t,\hat{\phi}(u)(x,t),u(x,t))}{\partial\phi^3_i},\\
&\qquad\qquad(x,t)\in Q,
\end{aligned} \tag{4.3.43a}$$

$i=1,\ldots,n$, with boundary conditions

$$\begin{aligned}
\lambda^1(x_0+l_1,t)&=0, && t\in I_2,\\
\zeta^1(x,t_0+l_2)&=0, && x\in I_1,\\
\zeta^2(x,t_0+l_2)&=\alpha/2, && x\in I_1,\\
\lambda^3(x_0+l_1,t)&=\alpha/2, && t\in I_2,
\end{aligned} \tag{4.3.43b}$$

where α is as given in (4.3.32), $u \in \mathfrak{U}$, and $\phi(u)$ is the corresponding solution of system (4.3.1).

4.3.5. *Necessary Conditions for Optimality*

The aim of this subsection is to derive a necessary condition for optimality for Problem (4.3.P1).

With this end in view, we first show that, for each $u \in \mathcal{U}$, the system (4.3.43) has a solution. Then, assuming the existence of an optimal control for the problem, we derive the necessary condition. The existence of optimal controls is considered in the later subsection. A solution of the system (4.3.43) is to be understood in the following sense.

Definition 4.3.2. For each $u \in \mathcal{U}$, a function $(\lambda^1(u), \lambda^3(u), \zeta^1(u), \zeta^2(u))$ is said to be a *solution of the system* (4.3.43) if it belongs to $L_\infty(Q, R^{4n})$ and satisfies the differential equations (4.3.43a) a.e. on Q and the boundary conditions (4.3.43b) everywhere in their domains of definition.

For the existence of solutions, we introduce the following assumptions.

Assumption (4.3.A6). The functions

$$(x, t, \theta^1, \theta^2, \theta^3, v) \to \frac{\partial f(x, t, \theta^1, \theta^2, \theta^3, v)}{\partial \theta_i^k}, \qquad i = 1, \ldots, n, \quad k = 1, 2, 3,$$

are measurable on $Q \times R^{3n} \times V$; furthermore, for each $(x, t) \in Q$, the functions

$$\frac{\partial f(x, t, \cdot, \cdot, \cdot, \cdot)}{\partial \theta_i^k}, \qquad i = 1, \ldots, n, \quad k = 1, 2, 3,$$

are continuous on $R^{3n} \times V$.

Assumption (4.3.A7). There exists a constant K_4 such that

$$\left| \frac{\partial f(x, t, \theta^1, \theta^2, \theta^3, v)}{\partial \theta_i^k} \right| \leq K_4, \qquad i = 1, \ldots, n, \quad k = 1, 2, 3,$$

for all $(x, t, \theta^1, \theta^2, \theta^3, v) \in Q \times R^{3n} \times V$.

Note that Assumptions (4.3.A2) and (4.3.A7) together imply Assumption (4.3.A4).

In view of the system (4.3.43), we observe that this system is underdetermined. It has only $3n$ (partial differential) equations and $4n$ functions to be determined. Thus, in general, the system has infinitely many solutions $(\lambda^1(u), \lambda^3(u), \zeta^1(u), \zeta^2(u))$. More precisely, we have the following result.

Theorem 4.3.4. *Consider system* (4.3.43). *Suppose that Assumptions* (4.3.A1)–(4.3.A3) *and* (4.3.A6)–(4.3.A7) *are satisfied. Then, for each* $u \in \mathcal{U}$, *the system admits infinitely many solutions* $(\lambda^1(u), \lambda^3(u), \zeta^1(u), \zeta^2(u)) \in L_\infty(Q, R^{4n})$.

PROOF. Let $\phi(u)$ be the solution of the system (4.3.1) corresponding to $u\in\mathfrak{U}$ and let

$$\hat{\phi}(u)\equiv(\phi^1(u),\phi^2(u),\phi^3(u))\equiv(\phi(u),(\phi(u))_x,(\phi(u))_t).$$

Consider the integral equation $w=Tw$, where $Tw=((Tw)_1,\ldots,(Tw)_n)$ is defined by

$$
\begin{aligned}
(Tw)_i(x,t)=\alpha_i&+\int_x^{x_0+l_1}\int_t^{t_0+l_2}\left\{w(\eta,\tau)\cdot\frac{\partial f(\eta,\tau,\hat{\phi}(\eta,\tau),u(\eta,\tau))}{\partial\phi_i^1}\right\}d\eta\,d\tau\\
&+\int_t^{t_0+l_2}\left\{w(x,\tau)\cdot\frac{\partial f(x,\tau,\hat{\phi}(x,\tau),u(x,\tau))}{\partial\phi_i^2}\right\}d\tau\\
&+\int_x^{x_0+l_1}\left\{w(\eta,t)\cdot\frac{\partial f(\eta,t,\hat{\phi}(\eta,t),u(\eta,t))}{\partial\phi_i^3}\right\}d\eta, \qquad (4.3.44)
\end{aligned}
$$

where $w\equiv(w_1,\ldots,w_n)$.

In view of Assumptions (4.3.A6)–(4.3.A7), we observe that, for each $u\in\mathfrak{U}$,

$$\frac{\partial f(\cdot,\cdot,\hat{\phi}(u)(\cdot,\cdot),u(\cdot,\cdot))}{\partial\phi_i^k},\qquad i=1,\ldots,n,\quad k=1,2,3,$$

are elements of $L_\infty(Q,R^n)$. Thus, we can show, as in Theorem 2.3.1, that some power of the operator T is a contraction and hence the integral equation has a unique solution w in $L_\infty(Q,R^n)$. Since w depends on the choice of the control $u\in\mathfrak{U}$, it will be referred to as $w(u)$. We now return to the question concerning the existence of solutions of the system (4.3.43). We assume that λ_x^1 is a given element in $L_\infty(Q,R^n)$.

Define

$$\lambda_i^1(x,t)\equiv-\int_x^{x_0+l_1}\{\lambda_{ix}^1(\eta,\tau)\}\,d\eta, \qquad (4.3.45)$$

$$
\begin{aligned}
\zeta_i^1(x,t)\equiv&\int_t^{t_0+l_2}\{\lambda_{ix}^1(x,\tau)\}\,d\tau\\
&+\int_t^{t_0+l_2}\left\{w(u)(x,\tau)\cdot\frac{\partial f(x,\tau,\hat{\phi}(x,\tau),u(x,\tau))}{\partial\phi_i^1}\right\}d\tau,
\end{aligned}
$$

$$i=1,\ldots,n, \qquad (4.3.46)$$

$$
\begin{aligned}
\lambda_i^3(x,t)\equiv\frac{\alpha_i}{2}&+\int_x^{x_0+l_1}\int_t^{t_0+l_2}\{\lambda_{ix}^1(\eta,\tau)\}\,d\eta\,d\tau\\
&+\int_x^{x_0+l_1}\int_t^{t_0+l_2}\left\{w(u)(\eta,\tau)\cdot\frac{\partial f(\eta,\tau,\hat{\phi}(u)(\eta,\tau),u(\eta,\tau))}{\partial\phi_i^1}\right\}d\eta\,d\tau\\
&+\int_x^{x_0+l_1}\left\{w(u)(\eta,t)\cdot\frac{\partial f(\eta,t,\hat{\phi}(u)(\eta,t),u(\eta,t))}{\partial\phi_i^3}\right\}d\eta,
\end{aligned}
$$

$$i=1,\ldots,n, \qquad (4.3.47)$$

$$\zeta_i^2(x,t) \equiv \frac{\alpha_i}{2} - \int_t^{t_0+l_2} \int_x^{x_0+l_1} \{\lambda_{ix}^1(\eta,\tau)\}\, d\eta\, d\tau$$

$$+ \int_t^{t_0+l_2} \left\{ w(u)(x,\tau) \cdot \frac{\partial f\big(x,\tau,\hat{\phi}(u)(x,\tau),u(x,\tau)\big)}{\partial \phi_i^3} \right\} d\tau,$$

$$i=1,\ldots,n. \tag{4.3.48}$$

We wish to show that $\lambda^1, \zeta^1, \lambda^3, \zeta^2$, as defined in (4.3.45)–(4.3.48), satisfy the system (4.3.43). It is easily observed that $\lambda^1, \zeta^1, \lambda^3, \zeta^2$ belong to $L_\infty(Q, R^n)$. Further, by using Equations (4.3.45) and (4.3.47), Assumption (4.3.A7), and the fact that $w(u) \in L_\infty(Q, R^n)$, we can easily show that the generalized partial derivatives λ_x^1 and λ_x^3 of λ^1 and λ^3 belong to $L_\infty(Q, R^n)$. Similarly, ζ_t^1 and ζ_t^2 also belong to $L_\infty(Q, R^n)$. On this basis, it can be verified that $(\lambda^1, \lambda^3, \zeta^1, \zeta^2)$ satisfies Equations (4.3.43a) a.e. on Q and the boundary conditions (4.3.43b) everywhere in their domains of definition. Thus, $(\lambda^1, \lambda^3, \zeta^1, \zeta^2)$ is a solution of system (4.3.43). However, since λ_x^1 is an arbitrary element of $L_\infty(Q, R^n)$, there are infinitely many such solutions. This completes the proof. □

REMARK 4.3.3. Let $(\lambda^1(u), \lambda^3(u), \zeta^1(u), \zeta^2(u))$ be a solution of system (4.3.43) corresponding to the control $u \in \mathcal{U}$. By examining the proof of Theorem 4.3.4, we observe that, for each $u \in \mathcal{U}$, $w(u) \equiv \lambda^3(u) + \zeta^2(u)$ is the unique solution of the integral equation $w = Tw$ with Tw defined by (4.3.44). Thus, it is clear that $\lambda^3(u) + \zeta^2(u)$ is independent of the choice of λ_x^1 in $L_\infty(Q, R^n)$.

Let u^1, u^2 be any two elements of $\mathcal{U}$ and $\phi(u^1), \phi(u^2)$ the corresponding solutions of system (4.3.1). Further, let

$$\big(\lambda(u^1), \zeta(u^1)\big) \equiv \big(\lambda^1(u^1), \lambda^3(u^1), \zeta^1(u^1), \zeta^2(u^1)\big)$$

and

$$\big(\lambda(u^2), \zeta(u^2)\big) \equiv \big(\lambda^1(u^2), \lambda^3(u^2), \zeta^1(u^2), \zeta^2(u^2)\big)$$

be solutions of system (4.3.43) corresponding to u^1 and u^2, respectively.

Integrating by parts the following terms,

$$\begin{cases} \text{(i)} & \displaystyle\int_Q\!\!\int \big\{\zeta^2(u^1)(x,t) \cdot \big((\phi^2(u^2))_t(x,t) - (\phi^2(u^1))_t(x,t)\big)\big\}\, dx\, dt, \\ \text{(ii)} & \displaystyle\int_Q\!\!\int \big\{\lambda^3(u^1)(x,t) \cdot \big((\phi^3(u^2))_x(x,t) - (\phi^3(u^1))_x(x,t)\big\}\, dx\, dt, \\ \text{(iii)} & \displaystyle\int_Q\!\!\int \big\{\lambda^1(u^1)(x,t) \cdot \big((\phi^1(u^2))_x(x,t) - (\phi^1(u^1))_x(x,t)\big)\big\}\, dx\, dt, \\ \text{(iv)} & \displaystyle\int_Q\!\!\int \big\{\zeta^1(u^1)(x,t) \cdot \big((\phi^1(u^2))_t(x,t) - (\phi^1(u^2))_t(x,t)\big)\big\}\, dx\, dt, \end{cases} \tag{4.3.49}$$

and using appropriate sets of equations of (4.3.39a), (4.3.39b), (4.3.43a), and (4.3.43b), we obtain

$$\int_{I_1}\left\{\frac{\alpha}{2}\cdot\left(\phi^2(u^2)(x,t_0+l_2)-\phi^2(u^1)(x,t_0+l_2)\right)\right\}dx$$

$$=\int_Q\int\left\{\zeta^2(u^1)(x,t)\cdot\left(f\left(x,t,\hat{\phi}(u^2)(x,t),u^2(x,t)\right)\right.\right.$$

$$\left.\left.-f\left(x,t,\hat{\phi}(u^1)(x,t),u^1(x,t)\right)\right)\right\}dx\,dt$$

$$+\int_Q\int\left\{\left(\zeta^2(u^1)\right)_t(x,t)\cdot\left(\phi^2(u^2)(x,t)-\phi^2(u^1)(x,t)\right)\right\}dx\,dt, \tag{4.3.50}$$

$$\int_{I_2}\left\{\frac{\alpha}{2}\cdot\left(\phi^3(u^2)(x_0+l_1,t)-\phi^3(u^1)(x_0+l_1,t)\right)\right\}dt$$

$$=\int_Q\int\left\{\lambda^3(u^1)(x,t)\cdot\left(f\left(x,t,\hat{\phi}(u^2)(x,t),u^2(x,t)\right)\right.\right.$$

$$\left.\left.-f\left(x,t,\hat{\phi}(u^1)(x,t),u^1(x,t)\right)\right)\right\}dx\,dt$$

$$+\int_Q\int\left\{\left(\lambda^3(u^1)\right)_x(x,t)\cdot\left(\phi^3(u^2)(x,t)-\phi^3(u^1)(x,t)\right)\right\}dx\,dt, \tag{4.3.51}$$

$$\int_Q\int\left\{\lambda^1(u^1)(x,t)\cdot\left(\phi^2(u^2)(x,t)-\phi^2(u^1)(x,t)\right)\right\}dx\,dt$$

$$+\int_Q\int\left\{\left(\lambda^1(u^1)\right)_x(x,t)\cdot\left(\phi^1(u^2)(x,t)-\phi^1(u^1)(x,t)\right)\right\}dx\,dt=0, \tag{4.3.52}$$

and

$$\int_Q\int\left\{\zeta^1(u^1)(x,t)\cdot\left(\phi^3(u^2)(x,t)-\phi^3(u^1)(x,t)\right)\right\}dx\,dt$$

$$+\int_Q\int\left\{\left(\zeta^1(u^1)\right)_t(x,t)\cdot\left(\phi^1(u^2)(x,t)-\phi^1(u^1)(x,t)\right)\right\}dx\,dt=0. \tag{4.3.53}$$

Combining the above four equalities, we have

$$\int_{I_1}\left\{\frac{\alpha}{2}\cdot\left(\phi^2(u^2)(x,t_0+l_2)-\phi^2(u^1)(x,t_0+l_2)\right)\right\}dx$$

$$+\int_{I_2}\left\{\frac{\alpha}{2}\cdot\left(\phi^3(u^2)(x_0+l_1,t)-\phi^3(u^1)(x_0+l_1,t)\right\}dt\right.$$

$$=\int_Q\int\Big\{\zeta^2(u^1)(x,t)\cdot\Big(f\big(x,t,\hat{\phi}(u^2)(x,t),u^2(x,t)\big)$$

$$-f\big(x,t,\hat{\phi}(u^1)(x,t),u^1(x,t)\big)\Big)\Big\}\,dx\,dt$$

$$+\int_Q\int\Big\{\lambda^3(u^1)(x,t)\cdot\Big(f\big(x,t,\hat{\phi}(u^2)(x,t),u^2(x,t)\big)$$

$$-f\big(x,t,\hat{\phi}(u^1)(x,t),u^1(x,t)\big)\Big)\Big\}\,dx\,dt$$

$$+\int_Q\int\left\{\zeta^1(u^1)(x,t)\cdot\left(\phi^3(u^2)(x,t)-\phi^3(u^1)(x,t)\right)\right\}dx\,dt$$

$$+\int_Q\int\left\{\lambda^1(u^1)(x,t)\cdot\left(\phi^2(u^2)(x,t)-\phi^2(u^1)(x,t)\right)\right\}dx\,dt$$

$$+\int_Q\int\left\{\left(\lambda^1(u^1)\right)_x(x,t)\cdot\left(\phi^1(u^2)(x,t)-\phi^1(u^1)(x,t)\right)\right\}dx\,dt$$

$$+\int_Q\int\left\{\left(\zeta^1(u^1)\right)_t(x,t)\cdot\left(\phi^1(u^2)(x,t)-\phi^1(u^1)(x,t)\right)\right\}dx\,dt$$

$$+\int_Q\int\left\{\left(\zeta^2(u^1)\right)_t(x,t)\cdot\left(\phi^2(u^2)(x,t)-\phi^2(u^1)(x,t)\right)\right\}dx\,dt$$

$$+\int_Q\int\left\{\left(\lambda^3(u^1)\right)_x(x,t)\cdot\left(\phi^3(u^2)(x,t)-\phi^3(u^1)(x,t)\right)\right\}dx\,dt. \tag{4.3.54}$$

In the following lemma, we prove that $J(u^2)-J(u^1)$ can be expressed in terms of the integral of the difference of the corresponding Hamiltonian functions over Q.

Lemma 4.3.1. *Consider Problem* (4.3.P1) *and system* (4.3.43). *Suppose that Assumptions* (4.3.A1)–(4.3.A3) *and* (4.3.A6)–(4.3.A7) *are satisfied. Let* u^1, u^2 *be any two controls in* $\mathscr{U}$ *and define*

$$H(u)(x,t,v)\equiv H\big(x,t,\hat{\phi}(u)(x,t),v,\lambda(u)(x,t),\zeta(u)(x,t)\big). \tag{4.3.55}$$

Then

$$J(u^2)-J(u^1)\equiv\int_Q\int\{H(u^1)(x,t,u^2(x,t))-H(u^1)(x,t,u^1(x,t))\}\,dx\,dt$$
$$+w(u^1,u^2,v) \tag{4.3.56}$$

where $J(u)$ *is given in* (4.3.32), $w(u^1,u^2,v)$ *is defined by*

$$w(u^1,u^2,v)$$
$$\equiv\sum_{k=1}^{3}\sum_{j=1}^{n}\int_Q\int\left\{\left[\frac{\partial H(x,t,\hat{\Phi}[v](x,t),u^2(x,t),\lambda(u^1)(x,t),\zeta(u^1)(x,t))}{\partial\phi_j^k}\right.\right.$$
$$\left.\left.-\frac{\partial H(u^1)(x,t,u^1(x,t))}{\partial\phi_j^k}\right]\left((\phi^k(u^2))_{,j}(x,t)-(\phi^k(u^1))_{,j}(x,t)\right)\right\}dx\,dt \tag{4.3.57}$$

and

$$\hat{\Phi}[v](x,t)\equiv\hat{\phi}(u^1)(x,t)-v(x,t)\left(\hat{\phi}(u^2)(x,t)-\hat{\phi}(u^1)(x,t)\right)$$

with $0\le v(x,t)\le 1$.

PROOF. Using (4.3.42) and (4.3.55), we obtain

$$H(u^1)(x,t,u(x,t))$$
$$=\lambda^1(u^1)(x,t)\cdot\phi^2(u^1)(x,t)+\lambda^3(u^1)(x,t)\cdot f\left(x,t,\hat{\phi}(u^1)(x,t),u(x,t)\right)$$
$$+\zeta^1(u^1)(x,t)\cdot\phi^3(u^1)(x,t)+\zeta^2(u^1)(x,t)\cdot f\left(x,t,\hat{\phi}(u^1)(x,t),u(x,t)\right). \tag{4.3.58}$$

Next, it is clear that

$$\int_Q\int\{\zeta^2(u^1)\cdot f(\hat{\phi}(u^2),u^2)+\lambda^3(u^1)\cdot f(\hat{\phi}(u^2),u^2)+\zeta^1(u^1)\cdot\phi^3(u^2)$$
$$+\lambda^1(u^1)\cdot\phi^2(u^2)\}\,dx\,dt$$
$$=\int_Q\int\{\zeta^2(u^1)\cdot f(\hat{\phi}(u^1),u^2)+\lambda^3(u^1)\cdot f(\hat{\phi}(u^1),u^2)+\zeta^1(u^1)\cdot\phi^3(u^1)$$
$$+\lambda^1(u^1)\cdot\phi^2(u^1)\}\,dx\,dt$$
$$+\int_Q\int\{\zeta^2(u^1)\cdot\left(f(\hat{\phi}(u^2),u^2)-f(\hat{\phi}(u^1),u^2)\right)+\lambda^3(u^1)\cdot\left(f(\hat{\phi}(u^2),u^2)\right.$$
$$\left.-f(\hat{\phi}(u^1),u^2)\right)$$
$$+\zeta^1(u^1)\cdot\left(\phi^3(u^2)-\phi^3(u^1)\right)+\lambda^1(u^1)\cdot\left(\phi^2(u^2)-\phi^2(u^1)\right)\}\,dx\,dt, \tag{4.3.59}$$

where, for convenience, the variable (x,t) is suppressed and $f(\hat{\phi},u)$ is used to denote $f(x,t,\hat{\phi}(x,t),u(x,t))$.

By virtue of the definition of $J(u)$ given in (4.3.32), it follows from (4.3.54), (4.3.43a), and the above two equalities that

$$J(u^2)-J(u^1)=\int_{I_1}\left\{\frac{\alpha}{2}\cdot(\phi^2(u^2)(x,t_0+l_2)-\phi^2(u^1)(x,t_0+l_2))\right\}dx$$

$$+\int_{I_2}\left\{\frac{\alpha}{2}\cdot(\phi^3(u^2)(x_0+l_1,t)-\phi^3(u^1)(x_0+l_1,t)\right\}dt$$

$$=\int_Q\int\{H(u^1)(x,t,u^2(x,t))-H(u^1)(x,t,u^1(x,t))\}\,dx\,dt$$

$$+\hat{w}(u^1,u^2), \tag{4.3.60}$$

where

$$\hat{w}(u^1,u^2)$$

$$\equiv\int_Q\int\Big\{\zeta^2(u^1)\cdot\big(f(\hat{\phi}(u^2),u^2)-f(\hat{\phi}(u^1),u^2)\big)$$

$$+\lambda^3(u^1)\cdot\big(f(\hat{\phi}(u^2),u^2)-f(\hat{\phi}(u^1),u^2)\big)$$

$$+\zeta^1(u^1)\cdot(\phi^3(u^2)-\phi^3(u^1))+\lambda^1(u^1)\cdot(\phi^2(u^2)-\phi^2(u^1))\Big\}\,dx\,dt$$

$$-\int_Q\int\Bigg\{(\lambda^3(u^1)+\zeta^2(u^1))\cdot\left[\sum_{i=1}^{n}\frac{\partial f(\hat{\phi}(u^1),u^1)}{\partial\phi_i^1}((\phi^1(u^2))_i-(\phi^1(u^1))_i)\right]$$

$$-\lambda^1(u^1)\cdot(\phi^2(u^2)-\phi^2(u^1))$$

$$-(\lambda^3(u^1)+\zeta^2(u^1))\cdot\left[\sum_{i=1}^{n}\frac{\partial f(\hat{\phi}(u^1),u^1)}{\partial\phi_i^2}((\phi^2(u^2))_i-(\phi^2(u^1))_i)\right]$$

$$-\zeta^1(u^1)\cdot(\phi^3(u^2)-\phi^3(u^1))$$

$$-(\lambda^3(u^1)+\zeta^2(u^1))\cdot\left[\sum_{i=1}^{n}\frac{\partial f(\hat{\phi}(u^1),u^1)}{\partial\phi_i^3}((\phi^3(u^2))_i-(\phi^3(u^1))_i)\right\}\,dx\,dt.$$

Using the mean value theorem and the definition of H [given in (4.3.42)] and $H(u^1)$ [given in (4.3.58)], we can easily show that $\hat{w}(u^1,u^2)=w(u^1,u^2,v)$, where $w(u^1,u^2,v)$ is defined by (4.3.57). Thus, the proof is complete. □

Let u^1,u^2 be any two elements of $\mathcal{U}$, and $\phi(u^1),\phi(u^2)$ the corresponding solutions of system (4.3.1). Further, let $z(u^1)(\cdot,\cdot,u^2)(\cdot,\cdot))$ be as given in

(4.3.28). Then, by using the triangle inequality and Assumptions (4.3.A3) and (4.3.A4) it is clear that

$$\begin{aligned}&z(u^1)(x,t,u^2(x,t))\\&\quad\leq 2K_1\{|\phi(u^1)(x,t)|+|(\phi(u^1))_x(x,t)|+|(\phi(u^1))_t(x,t)|\}\\&\qquad+|h(u^1)(x,t)|+|h(u^2)(x,t)|.\end{aligned}\tag{4.3.61}$$

Since, by Theorem 4.3.1, $\hat{\phi}(u^1)\in L_p(Q,R^{3n})$ and, by Assumption (4.3.A3), $h(u^k)\in L_p(Q)$, $k=1,2$, it follows that $z(u^1)(\cdot,\cdot,u^2(\cdot,\cdot))\in L_p(Q)$, for all $u^1,u^2\in\mathfrak{U}$.

For the derivation of the necessary condition, we are required to impose the following conditions on the function f [see system (4.3.1)].

Assumption (4.3.A8). There exists a nonnegative real-valued function $(x,t,v)\rightarrow r_1(x,t,v)$, $(x,t,v)\in Q\times V$, such that $r_1(\cdot,\cdot,u(\cdot,\cdot))\in L_4(Q)$ for any $u\in\mathfrak{U}$. If $p\in[1,\infty)$, there exists a constant K_5 such that

$$|f(x,t,\hat{\theta},v)|\leq r_1(x,t,v)+K_5\{|\theta^1|+|\theta^2|+|\theta^3|\}^{p/4}$$

for all $(x,t,\hat{\theta},v)\in Q\times R^{3n}\times V$. If $p=\infty$, the above inequality is replaced by

$$|f(x,t,\hat{\theta},v)|\leq r_1(x,t,v)+r_2(|\theta^1|+|\theta^2|+|\theta^3|),$$

where r_2 is a real-valued continuous function defined on $[0,\infty)$ such that $0\leq r_2(\xi)\leq K_5'\xi$ for some constant K_5' and for all $\xi\in[0,\infty)$.

Assumption (4.3.A9). There exists a constant K_6 such that, for $k=1,2,3$,

$$\sum_{i=1}^{n}\left|\frac{\partial f(x,t,\hat{\theta},v)}{\partial\phi_i^k}-\frac{\partial f(x,t,\hat{\bar{\theta}},v)}{\partial\phi_i^k}\right|\leq K_6\sum_{l=1}^{3}\|\theta^l-\bar{\theta}^l\|$$

for all $(x,t,v)\in Q\times V$ and all $\hat{\theta},\hat{\bar{\theta}}\in R^{3n}$.

REMARK 4.3.4. Under the additional assumption (4.3.A8), for any pair of controls $u^1,u^2\in\mathfrak{U}$, $z(u^1)(\cdot,\cdot,u^2(\cdot,\cdot))\in L_4(Q)$.

Lemma 4.3.2. *Suppose Assumptions* (4.3.A1)–(4.3.A3) *and* (4.3.A6)–(4.3.A9) *hold. Let $U(x,t)\equiv V$, $(x,t)\in Q$, a nonempty compact subset of R^m, and let u^0 be an element of the corresponding class of admissible controls $\mathfrak{U}$, v an arbitrary element of V, and (x^0,t^0) an interior point of Q at a distance δ_0 from the boundary ∂Q. Define, for $0<\delta<\delta_0$,*

$$u^\delta(x,t)\equiv\begin{cases}u^0(x,t) & \text{for}\quad (x,t)\in Q\backslash N_\delta\\ v & \text{for}\quad (x,t)\in N_\delta,\end{cases}\tag{4.3.62}$$

where

$$N_\delta \equiv N_\delta(x^0, t^0) = \{(x,t) \in Q : |x - x^0| \le \delta,\ |t - t^0| \le \delta\}.$$

Then there exists a constant K_7, *independent of* u^0, v, *and* δ, *such that* $w_0 \equiv w(u^0, u^\delta, v)$, *as defined in* (4.3.57), *satisfies the inequality*

$$|w_0| \le K_7(\delta)^2 \xi(u^0, v, \delta), \tag{4.3.63}$$

where

$$\begin{aligned}\xi(u^0, v, \delta) \equiv & \int_{N_\delta}\int \left\{ \left(z(u^0)(x,t,v)\right)^2 \right\} dx\,dt \\ & + \left(\int_{N_\delta}\int \left\{ \left(z(u^0)(x,t,v)\right)^2 \right\} dx\,dt \right)^{1/2} \\ & + \left(\int_{N_\delta}\int \left\{ \left(z(u^0)(x,t,v)\right)^4 \right\} dx\,dt \right)^{1/2}\end{aligned} \tag{4.3.64}$$

and $\xi(u^0, v, \delta) \to 0$ *as* $\delta \downarrow 0$.

PROOF. Throughout the proof, the variable (x,t) will be suppressed whenever there is no confusion.

In view of the definitions of H and $H(u)$ given in (4.3.42) and (4.3.55), the expression for w_0 can be written as

$$\begin{aligned} w_0 = \int_Q\int \Bigg\{ \sum_{k=1}^{3} \sum_{i,j=1}^{n} \Bigg[& \left(\lambda_j^2(u^0) + \zeta_j^2(u^0)\right) \left[\frac{\partial f_j(\hat\Phi[v], u^\delta)}{\partial \phi_i^k} - \frac{\partial f_j(\hat\phi(u^0), u^\delta)}{\partial \phi_i^k} \right] \\ & + \left(\lambda_j^2(u^0) + \zeta_j^2(u^0)\right) \left[\frac{\partial f_j(\hat\phi(u^0), u^\delta)}{\partial \phi_i^k} - \frac{\partial f_j(\hat\phi(u^0), u^0)}{\partial \phi_i^k} \right] \Bigg] \\ & \times \left[\phi_i^k(u^\delta) - \phi_i^k(u^0) \right] \Bigg\} dx\,dt, \end{aligned} \tag{4.3.65}$$

where $f_j(\hat\psi, u) \equiv f_j(x, t, \hat\psi(x,t), u(x,t))$. Thus, it follows from Assumption (4.3.A9) that

$$\begin{aligned} |w_0| \le K_6 \|\lambda^2(u^0) + \zeta^2(u^0)\|_{\infty, Q} & \\ \times \Bigg[\int_Q\int \Bigg\{ & \sum_{k,l=1}^{3} \left(|\Phi^k[v] - \phi^k(u^0)|\right)\left(|\phi^l(u^\delta) - \phi^l(u^0)|\right) \Bigg\} dx\,dt \\ + \int_Q\int \Bigg\{ & \sum_{k=1}^{3} \left(|\phi^k(u^\delta) - \phi^k(u^0)|\right) \\ & \times \left[\sum_{i=1}^{n} \left| \frac{\partial f(\hat\phi(u^0), u^\delta)}{\partial \phi_i^k} - \frac{\partial f(\hat\phi(u^0), u^0)}{\partial \phi_i^k} \right| \right] \Bigg\} dx\,dt \Bigg]. \end{aligned} \tag{4.3.66}$$

By virtue of the Assumption (4.3.A7) and the definition of u^δ, we obtain

$$\left|\frac{\partial f(\hat{\phi}(u^0),u^\delta)}{\partial \phi_i^k}-\frac{\partial f(\hat{\phi}(u^0),u^0)}{\partial \phi_i^k}\right|\begin{cases} =0 & \text{if } (x,t)\in Q\backslash N_\delta \\ \leq 2K_4 & \text{if } (x,t)\in N_\delta \end{cases} \tag{4.3.67}$$

for all $k=1,2,3$ and $i=1,\ldots,n$. Thus, it is clear from (4.3.66) that

$$|w_0|\leq K_6\|\lambda^2(u^0)+\zeta^2(u^0)\|_{\infty,Q}\big(w_1(u^0,u^\delta,v)-w_2(u^0,u^\delta)\big), \tag{4.3.68}$$

where

$$w_1(u^0,u^\delta,v)\equiv\int_Q\int\left\{\sum_{k,l=1}^{3}\big(|\Phi^k[v]-\phi^k(u^0)|\big)\big(|\phi^l(u^\delta)-\phi^l(u^0)|\big)\right\}dx\,dt \tag{4.3.69}$$

and

$$w_2(u^0,u^\delta)\equiv 2nK_4\int_{N_\delta}\int\left\{\sum_{k=1}^{3}|\phi^k(u^\delta)-\phi^k(u^0)|\right\}dx\,dt. \tag{4.3.70}$$

Since $0\leq v(x,t)\leq 1$ for all $(x,t)\in Q$, it follows that

$$|\Phi^k[v](x,t)-\phi^k(u^0)(x,t)|\leq|\phi^k(u^\delta)(x,t)-\phi^k(u^0)(x,t)| \tag{4.3.71}$$

for each $k=1,2,3$.

In view of Remark 4.3.1, we observe that, for all $k=1,2,3$,

$$\begin{aligned}&|\phi^k(u^\delta)(x,t)-\phi^k(u^0)(x,t)|\\ &\quad\leq K_3\bigg[\int_{N_\delta}\int\{z(u^0)(x,t,v)\}\,dx\,dt+\int_{t^0-\delta}^{t^0+\delta}\{z(u^0)(x,\tau,u^\delta(x,\tau))\}\,d\tau\\ &\qquad+\int_{x^0-\delta}^{x^0+\delta}\{z(u^0)(\eta,t,u^\delta(\eta,t))\}\,d\eta\bigg].\end{aligned} \tag{4.3.72}$$

Using the following elementary inequality,

$$(a+b+c)^2\leq\tfrac{10}{3}(a^2+b^2+c^2), \tag{4.3.73}$$

it follows from (4.3.69), (4.3.71), (4.3.73), and (4.3.72) that

$$\begin{aligned}w_1(u^0,u^\delta,v)&\leq\int_Q\int\left\{\left(\sum_{k=1}^{3}|\phi^k(u^\delta)-\phi^k(u^0)|\right)^2\right\}dx\,dt\\ &\leq 10(K_3)^2\int_Q\int\left\{\left(\int_{N_\delta}\int\{z(u^0)(\eta,\tau,v)\}\,d\eta\,d\tau\right.\right.\end{aligned}$$

$$+\int_{t^0-\delta}^{t^0+\delta}\{z(u^0)(x,\tau,u^\delta(x,\tau))\}\,d\tau$$

$$\left.+\int_{x^0-\delta}^{x^0+\delta}\{z(u^0)(\eta,t,u^\delta(\eta,\tau))\}\,d\eta\Bigg)^2\right\}dx\,dt. \tag{4.3.74}$$

By virtue of (4.3.73), the above inequality reduces to

$$\begin{aligned} w_1(u^0,u^\delta,v) \le \frac{100}{3}(K_3)^2\int_Q\int\Bigg\{&\left(\int_{N_\delta}\int\{z(u^0)(\eta,\tau,v)\}\,d\eta\,d\tau\right)^2\\ &+\left(\int_{t^0-\delta}^{t^0+\delta}\{z(u^0)(x,\tau,u^\delta(x,\tau))\}\,d\tau\right)^2\\ &+\left(\int_{x^0-\sigma}^{x^0+\delta}\{z(u^0)(\eta,t,u^\delta(\eta,t))\}\,d\eta\right)^2\Bigg\}dx\,dt. \end{aligned} \tag{4.3.75}$$

In view of Remark 4.3.4, we observe that $z(u^1)(\cdot,\cdot,v)\in L_4(Q)$. Thus, it follows from Hölder's inequality, (4.3.69) and Remark 4.3.1(i) that

$$\begin{aligned} w_1(u^0,u^\delta,v) \le \beta_0(\delta)^2\Bigg[&\int_{N_\delta}\int\{(z(u^0)(\eta,\tau,v))^2\}\,d\eta\,d\tau\\ &+\left(\int_{N_\delta}\int\{(z(u^0)(\eta,\tau,v))^4\}\,d\eta\,d\tau\right)^{1/2}\Bigg], \end{aligned} \tag{4.3.76}$$

where $\beta_0\equiv(400/3)(K_3)^2(l_1l_2+l_1+l_2)$.

To establish the estimate for $w_2(u^0,u^\delta)$ (see (4.3.70)), we use Schwarz's inequality, and inequality (4.3.72) to obtain

$$\begin{aligned} w_2(u^0,u^\delta) &\le 6nK_3K_4\int_{N_\delta}\int\Bigg\{\int_{N_\delta}\int\{z(u^0)(\eta,\tau,v)\}\,d\eta\,d\tau\\ &\qquad+\int_{t^0-\delta}^{t^0+\delta}\{z(u^0)(x,\tau,v)\}\,d\tau\\ &\qquad+\int_{x^0-\delta}^{x^0+\delta}\{z(u^0)(\eta,t,v)\}\,d\eta\Bigg\}dx\,dt\\ &\le\beta_1(\delta)^2\left(\int_{N_\delta}\int\{(z(u^0)(x,t,v))^2\}\,dx\,dt\right)^{1/2}, \end{aligned} \tag{4.3.77}$$

where $\beta_1\equiv48nK_3K_4(\max\{l_1,l_2\}+1)$.

Combining (4.3.68), (4.3.76), and (4.3.77), we obtain the estimate (4.3.63) with

$$K_7 \equiv K_6 \max\{\beta_0, \beta_1\} \|\lambda^2(u^0) + \zeta^2(u^0)\|_{\infty, Q}$$

and

$$\begin{aligned}\xi(u^0, v, \delta) \equiv & \int_{N_\delta}\int \left\{ \left(z(u^0)(x,t,v) \right)^2 \right\} dx\, dt \\ & + \left(\int_{N_\delta}\int \left\{ z(u^0)(x,t,v) \right)^4 \right\} dx\, dt \right)^{1/2} \\ & + \left(\int_{N_\delta}\int \left\{ z(u^0)(x,t,v) \right)^2 \right\} dx\, dt \right)^{1/2}.\end{aligned}$$

Since $z(u^0)(\cdot, \cdot, v) \in L_4(Q)$ and Q is bounded, it follows that $\xi(u^0, v, \delta) \to 0$ as $\delta \to 0$. This completes the proof. □

With the help of the above lemma, we now can compute the variation in the cost due to the variation in control. This is given in the following theorem.

Theorem 4.3.5. *If all the hypotheses of the above lemma are satisfied, then*

$$\begin{aligned}J(u^\delta) - J(u^0) = w(u^0, u^\delta, v) + \int_Q\int \{ & H(u^0)(x,t,u^\delta(x,t)) \\ & - H(u^0)(x,t,u^0(x,t)) \} \, dx\, dt,\end{aligned} \tag{4.3.78}$$

where $w(u^0, u^\delta, v)$ *satisfies the estimate* (4.3.63).

PROOF. The proof follows from the application of Lemma 4.3.2 to Equation (4.3.56). □

We are now in a position to derive a necessary condition for optimality for the optimal control problem (4.3.P1).

Theorem 4.3.6. *Consider problem* (4.3.P1). *Suppose all the hypotheses of Lemma* 4.3.2 *are satisfied and let* $u^* \in \mathfrak{U}$ *be an optimal control. Then*

$$H(u^*)(x,t,v) \geq H(u^*)(x,t,u^*(x,t)) \tag{4.3.79}$$

for all $v \in V$ *and for almost all* $(x,t) \in Q$, *where* $H(u)$ *is as given in* (4.3.55), *and, corresponding to* $u^* \in \mathfrak{U}$, $\phi(u^*)$ *is the solution of system* (4.3.1) *and* $(\lambda(u^*), \zeta(u^*))$ *is a solution of the adjoint system* (4.3.43).

PROOF. From Theorem 4.3.1, $\phi(u^*)$ is the solution of the system (4.3.1) corresponding to $u^* \in \mathfrak{U}$. In view of Theorem 4.3.4, we recall that the adjoint system (4.3.43) admits infinitely many solutions $(\lambda(u^*), \zeta(u^*))$ corresponding to $u^* \in \mathcal{U}$. Throughout the proof, we assume that $(\lambda(u^*), \zeta(u^*))$ is a solution.

Let $v \in V$ be arbitrary and let (x^0, t^0) be an interior point of Q such that it is also a regular point of $H(u^*)(\cdot, \cdot, u^*(\cdot, \cdot))$ and $H(u^*)(\cdot, \cdot, v)$. Let u^δ be defined as in (4.3.62) with u^0 replaced by u^*. Then by virtue of Theorem 4.3.5 and the fact that u^* is an optimal control, we obtain

$$\begin{aligned}
0 &\leq J(u^\delta) - J(u^*) \\
&= w(u^*, u^\delta, v) + \int_Q\int \{H(u^*)(x, t, u^\delta(x, t)) \\
&\quad - H(u^*)(x, t, u^*(x, t))\}\, dx\, dt \\
&\leq K_7(\delta)^2 \xi(u^*, v, \delta) + \int_{N_\delta}\int \{H(u^*)(x, t, v) \\
&\quad - H(u^*)(x, t, u^*(x, t))\}\, dx\, dt.
\end{aligned} \tag{4.3.80}$$

Dividing the above inequality by $|N_\delta| = 4\delta$ (where $|N_\delta|$ denotes the Lebesgue measure of N_δ), and letting $\delta \downarrow 0$, it follows from Lemma 1.1.19, specialized to the finite-dimensional case, that

$$H(u^*)(x^0, t^0, v) \geq H(u^*)(x^0, t^0, u^*(x^0, t^0)). \tag{4.3.81}$$

Since V is separable, $H(u^*)$ is continuous on V and almost all $(x, t) \in Q$ are regular points, the proof is complete. □

4.3.6. *Existence of Optimal Controls*

In this subsection, we wish to study the question of existence of optimal controls for the system discussed in the previous subsection. To be more precise, we consider the following system:

$$\begin{aligned}
\phi_{xt}(x, t) &= A(x, t) \cdot \phi_x(x, t) + B(x, t)\phi_t(x, t) \\
&\quad + g(x, t, \phi(x, t), u(x, t)), \qquad (x, t) \in Q, \\
\phi(x, t_0) &= \phi^0(x), \qquad x \in I_1, \\
\phi(x_0, t) &= \phi^b(t), \qquad t \in I_2.
\end{aligned} \tag{4.3.82}$$

For convenience, the optimal control problem (4.3.P2) with system (4.3.1) replaced by the above system will be referred to as problem (4.3.P2)′.

In the sequel, we shall impose certain assumptions on the coefficients and data of the system (4.3.82).

Assumption (4.3.A10)

i. The functions ϕ^0 and ϕ^b are absolutely continuous on I_1 and I_2 respectively.
ii. $\phi_x^0 \in L_p(I_1, R^n)$ and $\phi_t^b \in L_p(I_2, R^n)$ for some $p \in [1, \infty]$.
iii. $\phi^0(t_0) = \phi^b(x_0)$.

Assumption (4.3.A11). $A \in L_\infty(Q, R^{n\times n})$ and $B \in L_\infty(Q, R^{n\times n})$.

Assumption (4.3.A12)

i. The function $(x, t, \theta, v) \to g(x, t, \theta, v)$ is measurable on $Q \times R^n \times V$.
ii. For almost all $(x, t) \in Q$, the function $g(x, t, \cdot, \cdot)$ is continuous on $R^n \times V$.
iii. There exists a nonnegative function $\beta \in L_p(Q)$ such that

$$|g(x, t, 0, v)| \leq \beta(x, t)$$

for almost all $(x, t) \in Q$ and for all $v \in V$, where p is as in Assumption (4.3.A10).
iv. There exists a constant K_8 such that

$$|g(x, t, \theta^1, v) - g(x, t, \theta^2, v)| \leq K_8 |\theta^1 - \theta^2|$$

for all $(x, t, v) \in Q \times V$ and for all $\theta^1, \theta^2 \in R^n$.

$$|||z|||_{p,Q} \equiv \|z\|_{p,Q} + \|z_x\|_{p,Q} + \|z_t\|_{p,Q} + \|z_{xt}\|_{p,Q}.$$

For each $p \in [1, \infty]$, let $W_p(Q)$ denote the space of all real-valued functions $\{z\}$ such that z together with all its generalized partial derivatives z_x, z_t, z_{xt} are in $L_p(Q)$. $W_p(Q)$, $1 \leq p \leq \infty$, is a vector space and is a Banach space with respect to the norm

$$|||z|||_{p,Q} \equiv \|z\|_{p,Q} + \|z_x\|_{p,Q} + \|z_t\|_{p,Q} + \|z_{xt}\|_{p,Q}.$$

For each $p \in [1, \infty]$, let $W_p(Q, R^n)$ denote the n-copies of $W_p(Q)$ with the norm defined by

$$|||z|||_{p,Q} \equiv \sum_{i=1}^{n} |||z_i|||_{p,Q},$$

where $z = (z_i, \ldots, z_n)$ and $z_i \in W_p(Q)$, $i = 1, \ldots, n$. Clearly, $W_p(Q, R^n)$ equipped with the norm defined above is a Banach space.

Under Assumptions (4.3.A10)–(4.3.A12), it follows from Theorem 4.3.1 that, for each $u \in \mathfrak{U}$, system (4.3.82) has a unique solution $\phi(u) \in W_p^{1,1}(Q, R^n)$, and further, $\phi(u)$ satisfies the estimate

$$\|\phi(u)\|_{p,Q}^{(1,1)} \leq K_9, \tag{4.3.83}$$

where the constant K_9 is independent of the control $u \in \mathfrak{U}$. Since the matrices A, B [see system (4.3.82)] are in $L_\infty(Q, R^{n\times n})$ and g satisfies Assumption (4.3.A12), it is easy to verify that there exists a constant K_{10}, independent of $u \in \mathfrak{U}$, such that

$$\|(\phi(u))_{xt}\|_{p,Q} \le K_{10}. \tag{4.3.84}$$

At this point, we conclude that under Assumptions (4.3.A10)–(4.3.A12), system (4.3.82) admits, for each $u \in \mathfrak{U}$, a unique solution $\phi(u) \in W_p(Q, R^n)$.

REMARK 4.3.5. Consider system (4.3.82). Suppose that Assumptions (4.3.A10)–(4.3.A12) with $p=1$ are satisfied. Then it follows from Theorem 4.3.2 that, for all $(x,t) \in Q$,

$$|\phi(u)(x,t)| \le K_{11} \tag{4.3.85}$$

and, for almost all $(x,t) \in Q$,

$$|(\phi(u))_x(x,t)| \le \beta_1(x) + K_{11}, \tag{4.3.86}$$

and

$$|(\phi(u))_t(x,t)| \le \beta_2(t) + K_{11}, \tag{4.3.87}$$

where

$$K_{11} \equiv K_3\left\{ \|\phi^0\|_{I_1}^{(0)} + \|\phi^b\|_{I_2}^{(0)} + \|\phi_x^0\|_{1,I_1} + \|\phi_t^b\|_{1,I_2} + \int_Q\int \{\beta(x,t)\}\,dx\,dt \right\}$$

$$\beta_1(x) \equiv K_3\left\{ |\phi_x^0(x)| + |\phi^0(x)| + \int_{I_2} \{\beta(x,t)\}\,dt \right\},$$

and

$$\beta_2(t) \equiv K_3\left\{ |\phi_t^b(t)| + |\phi^b(t)| + \int_{I_1} \{\beta(x,t)\}\,dx \right\}.$$

Clearly, the right-hand sides of the three inequalities above are independent of $u \in \mathfrak{U}$, $\beta_1 \in L_1(I_1)$, and $\beta_2 \in L_1(I_2)$. Further, since Q is bounded, the matrices A, B are in $L_\infty(Q, R^{n\times n})$, and the function g satisfies Assumption (4.3.A12), we can easily verify that there exists an element $r \in L_1(Q)$, independent of $u \in \mathfrak{U}$, such that

$$|(\phi(u))_{xt}(x,t)| \le r(x,t) \tag{4.3.88}$$

for almost all $(x,t) \in Q$.

Lemma 4.3.3. *Consider system* (4.3.82). *Suppose that Assumptions* (4.3.A10)–(4.3.A12) *are satisfied. Let* $\{u^k\} \subset \mathfrak{U}$ *and let* $\{\phi(u^k)\}$ *be the corresponding sequence of solutions from* $W_p(Q, R^n)$. *Then there exists a subsequence of the sequence* $\{u^k\}$, *again denoted by* $\{u^k\}$, *and a function* $\phi^* \in W_p(Q, R^n)$

such that

$$\left.\begin{aligned} (\phi(u^k))_{xt} &\overset{w(w^*)}{\rightarrow} \phi^*_{xt} \\ (\phi(u^k))_{x} &\overset{w(w^*)}{\rightarrow} \phi^*_{x} \\ (\phi(u^k))_{t} &\overset{w(w^*)}{\rightarrow} \phi^*_{t} \end{aligned}\right\} \begin{aligned} &\textit{in } L_p(Q, R^n) \textit{ if } p\in[1,\infty) \\ &(\textit{in } L_\infty(Q, R^n) \textit{ if } p=\infty) \end{aligned}$$

$$(\phi(u^k) \overset{u}{\rightarrow} \phi^* \quad \text{(uniformly) on } Q$$

and that ϕ^* *is the solution of the system*

$$\begin{aligned} \phi_{xt}(x,t) &= A(x,t)\cdot\phi_x(x,t) \\ &\quad + B(x,t)\cdot\phi_t(x,t) + g^*(x,t), \qquad (x,t)\in Q, \\ \phi(x,t_0) &= \phi^0(x), \qquad x\in I_1, \\ \phi(x_0,t) &= \phi^b(t), \qquad t\in I_2, \end{aligned} \tag{4.3.89}$$

where g^* *is the weak (weak*) limit of* $\{g(\cdot,\cdot,\phi(u^k)(\cdot,\cdot),u^k(\cdot,\cdot))\}$ *in* $L_p(Q, R^n)$ *if* $p\in[1,\infty)$ *(in* $L_\infty(Q, R^n)$ *if* $p=\infty$*).*

PROOF. First of all, we consider the case when $p\in(1,\infty)$. The cases when $p=\infty$ and $p=1$ will be considered later. In view of the estimates (4.3.83) and (4.3.84), we recall that

$$\|\phi(u^k)\|^{(1,1)}_{p,Q} \leq K_9 \tag{4.3.90}$$

and

$$\|(\phi(u^k))_{xt}\|_{p,Q} \leq K_{10}, \tag{4.3.91}$$

where the constants K_9 and K_{10} are independent of k.

Since $L_p(Q, R^n)$ is reflexive when $p\in(1,\infty)$, it is clear by virtue of the above estimates that there exist Φ^*, $\tilde{\Phi}^*$, $\Psi^*\in L_p(Q, R^n)$ and a subsequence of the sequence $\{\phi(u^k)\}$, again denoted by $\{\phi(u^k)\}$, such that

$$\begin{aligned} (\phi(u^k))_x &\overset{w}{\rightarrow} \Phi^* \\ (\phi(u^k))_t &\overset{w}{\rightarrow} \tilde{\Phi}^* \\ (\phi(u^k))_{xt} &\overset{w}{\rightarrow} \Psi^* \end{aligned} \tag{4.3.92}$$

in $L_p(Q, R^n)$.

Note that $\phi(u^k)$ is the solution of system (4.3.82) corresponding to u^k. Thus, it is clear from Assumption (4.3.A10) that

$$\phi(u^k)(x,t) = \phi^0(x) + \phi^b(t) - \phi^b(t_0) + \int_{x_0}^{x}\int_{t_0}^{t}\{(\phi(u^k))_{xt}(\eta,\tau)\}\,d\eta\,d\tau. \tag{4.3.93}$$

Define

$$\phi^*(x,t) \equiv \phi^0(x) + \phi^b(t) - \phi^b(t_0) + \int_{x_0}^{x}\int_{t_0}^{t} \{\Psi^*(\eta,\tau)\}\, d\eta\, d\tau. \tag{4.3.94}$$

Clearly, ϕ^* is continuous on Q. Since $(\phi(u^k))_{xt} \overset{w}{\to} \Psi^*$ in $L_p(Q, R^n)$, it follows from the expressions (4.3.93) and (4.3.94) that $\phi(u^k) \overset{u}{\to} \phi^*$ on Q.

Next, for any $z \in C_0^1(Q, R^n)$, we have

$$\int_Q\int \{(\phi(u^k))_x \cdot z\}\, dx\, dt = -\int_Q\int \{\phi(u^k) \cdot z_x\}\, dx\, dt,$$

$$\int_Q\int \{(\phi(u^k))_t \cdot z\}\, dx\, dt = -\int_Q\int \{\phi(u^k) \cdot z_t\}\, dx\, dt,$$

and

$$\int_Q\int \{(\phi(u^k))_{xt} \cdot z\}\, dx\, dt = -\int_Q\int \{(\phi(u^k))_x \cdot z_t\}\, dx\, dt$$

$$= -\int_Q\int \{(\phi(u^k))_t \cdot z_x\}\, dx\, dt,$$

where, for convenience, the variable (x, t) is suppressed.

Thus, it is easy to verify that (i) Φ^* and $\tilde{\Phi}^*$ are, respectively, the generalized derivatives of ϕ^* with respect to x and t (and hence are written as ϕ_x^* and ϕ_t^*), and (ii) Ψ^* is the generalized derivative of ϕ_x^* (or ϕ_t^*) with respect to t (or x) (and hence is written as ϕ_{xt}^*).

At this point, we have already shown that

$$\phi(u^k) \overset{u}{\to} \phi^* \qquad \text{on } Q$$

$$\left.\begin{aligned} (\phi(u^k))_x &\overset{w}{\to} \phi_x^* \\ (\phi(u^k))_t &\overset{w}{\to} \phi_t^* \\ (\phi(u^k))_{xt} &\overset{w}{\to} \phi_{xt}^* \end{aligned}\right\} \quad \text{in } L_p(Q, R^n) \tag{4.3.95}$$

By virtue of Assumption (4.3.A12) and the inequality (4.3.90), we can easily verify that the sequence $\{g^k\} \equiv \{g(\cdot,\cdot,\phi(u^k)(\cdot,\cdot), u^k(\cdot,\cdot))\}$ is contained in a sphere of $L_p(Q, R^n)$. Thus, there exists a $g^* \in L_p(Q, R^n)$ and a subsequence of $\{g^k\}$, relabled as $\{g^k\}$, such that $g^k \overset{w}{\to} g^*$ in $L_p(Q, R^n)$.

We now show that ϕ^* is a solution of system (4.3.89). For this, we recall that $\phi^* \in W_p(Q, R^n)$ and satisfies the boundary conditions of the system. Thus, it remains to show that ϕ^* satisfies also the differential equations a.e. on Q. Since $\phi(u^k)$ is the solution of system (4.3.82) corresponding to u^k, it is clear that, for almost all $(x,t)\in Q$,

$$\begin{aligned}\phi^*_{xt}-A\cdot\phi^*_x-B\cdot\phi^*_t-g^* &= \left(\phi^*_{xt}-(\phi(u^k))_{xt}\right)-\left[A\cdot\left(\phi^*_x-(\phi(u^k))_x\right)\right.\\ &\quad \left.+B\cdot\left(\phi^*_t-(\phi(u^k))_t\right)+(g^*-g^k)\right]. \end{aligned} \tag{4.3.96}$$

Recall that ϕ^*, ϕ^*_x, ϕ^*_t, ϕ^*_{xt}, g^*, $\phi(u^k)$, $(\phi(u^k))_x$, $(\phi(u^k))_t$, $(\phi(u^k))_{xt}$, g^k are in $L_p(Q, R^n)$ and that the matrices A, B are in $L_\infty(Q, R^{n\times n})$. Thus, it can easily be verified that

$$\phi^*_{xt}-A\cdot\phi^*_x-B\cdot\phi^*_t-g^*$$

and

$$\left(\phi^*_{xt}-(\phi(u^k))_{xt}\right)-\left[A\cdot\left(\phi^*_x-(\phi(u^k))_x\right)+B\cdot\left(\phi^*_t-(\phi(u^k))_t\right)+g^*-g^k\right]$$

are in $L_p(Q, R^n)$. Thus, it follows from the expression (4.3.96) that

$$\begin{aligned}&\int_Q\int\left\{\left[\phi^*_{xt}-A\cdot\phi^*_x-B\cdot\phi^*_t-g^*\right]\cdot z\right\}dx\,dt\\ &=\int_Q\int\left\{\left[\left(\phi^*_{xt}-(\phi(u^k))_{xt}\right)-A\cdot\left(\phi^*_x-(\phi(u^k))_x\right)\right.\right.\\ &\quad\left.\left.+B\cdot\left(\phi^*_t-(\phi(u^k))_t\right)+g^*-g^k\right)\right]\cdot z\right\}dx\,dt\end{aligned} \tag{4.3.97}$$

for any $z\in L_q(Q)$, where $(1/p)+(1/q)=1$.

Using (4.3.95), we observe that the relation (4.3.97), in the limit with respect to k, reduces to

$$\int_Q\int\left\{\left[\phi^*_{xt}-A\cdot\phi^*_x-B\cdot\phi^*_t-g^*\right]\cdot z\right\}dx\,dt=0 \tag{4.3.98}$$

for any $z\in L_q(Q, R^n)$. This, in turn, implies that ϕ^* satisfies the differential equations of system (4.3.89) a.e. on Q. Thus, ϕ^* is a solution of the system. The uniqueness follows from Theorem 4.3.1. This completes the proof of the lemma for the case when $p\in(1,\infty)$.

For the case when $p=\infty$, the proof is similar to the one given above with only the modification that the weak topology is replaced by the weak* topology.

Consider the case when $p=1$. It is known that spheres in $L_1(Q, R^n)$ are not necessarily weakly compact. However, these spheres are weakly compact if they are, in addition, equiabsolutely integrable on Q. Thus, it remains to show that $\{(\phi(u^k))_{xt}\}$, $\{(\phi(u^k))_x\}$, and $\{(\phi(u^k))_t\}$ are equiabsolutely integrable on Q. But these results follow readily from the estimates (4.3.86), (4.3.87), and (4.3.88). Therefore, the proof is complete. □

For the proof of existence of optimal controls, it is often convenient to formulate the problem as a sort of contingent problem. Define, for each

$(x, t, z) \in Q \times R^n$,

$$\mathbb{R}(x, t, z) \equiv \{g(x, t, z, v) : v \in U(x, t)\}, \tag{4.3.99}$$

where U is measurable multifunction with values from the class of nonempty closed convex subsets of a compact set $V(\subset R^m)$.

We impose the following assumptions on $\mathbb{R}$.

Assumption (4.3.A13). For each $(x, t, z) \in Q \times R^n$, the set $\mathbb{R}(x, t, z)$ is a convex and closed subset of R^n.

Assumption (4.3.A14). The multifunction $\mathbb{R} : Q \times R^n \to R^n$ is upper semicontinuous with respect to inclusion (u.s.c.i) for all $(x, t, z) \in Q \times R^n$.

Lemma 4.3.4. *Consider system* (4.3.82). *Suppose that Assumptions* (4.3.A10)–(4.3.A14) *are satisfied. Let* $\{u^k\}$ *be a sequence of controls from* $\mathfrak{U}$, $\{\phi(u^k)\}$ *the corresponding solutions of the system, and* $\phi \in W_p(Q, R^n)$, $g^* \in L_p(Q, R^n)$ *such that*

$$\phi(u^k) \overset{u}{\to} \phi \qquad \textit{on } Q$$

$$\left.\begin{array}{l} (\phi(u^k))_x \overset{w(w^*)}{\to} \phi_x \\ (\phi(u^k))_t \overset{w(w^*)}{\to} \phi_t \\ (\phi(u^k))_{xt} \overset{w(w^*)}{\to} \phi_{xt} \\ g^k \overset{w(w^*)}{\to} g^* \end{array}\right\} \quad \begin{array}{l} \textit{in } L_p(Q, R^n) \textit{ if } p \in [1, \infty) \\ (\textit{in } L_\infty(Q, R^n) \textit{ if } p = \infty), \end{array}$$

where $g^k(x, t) \equiv g(x, t, \phi(u^k)(x, t), u^k(x, t))$. Then there exists a control $u^* \in \mathfrak{U}$ such that

$$g^*(x, t) = g(x, t, \phi(x, t), u^*(x, t))$$

a.e. on Q.

PROOF. The proof is based on an extension of Fillipov's implicit function lemma (Theorem 1.4.5). To apply this result, we need, first of all, to show that $g^*(x, t) \in \mathbb{R}(x, t, \phi(x, t))$ a.e. on Q. There are two cases to be considered.

Case (i): $p \in [1, \infty)$. By hypothesis, we see that $g^k \overset{w}{\to} g^*$ in $L_p(Q, R^n)$. Thus, by virtue of Mazur's theorem (Corollary 1.1.1), there exist, for any integer $k > 0$, an integer $m_k > 0$, a set of integers $i = 1, \ldots, \sigma$, where $\sigma \equiv \sigma(k)$, and a set of nonnegative numbers $s_{k,i}$, $i = 1, \ldots, \sigma$, with $\Sigma_{i=1}^{\sigma} s_{k,i} = 1$, such that $m_{k+1} \geq m_k + \sigma$ and

$$\sum_{i=1}^{\sigma} s_{k,i} g^{m_k + i} \overset{s}{\to} g^* \text{ in } L_p(Q, R^n) \text{ as } k \to \infty.$$

Define

$$y^k \equiv \sum_{i=1}^{\sigma} s_{k,i} g^{m_k+i}.$$

Then, there exists a subsequence of the sequence $\{y^k\}$, again indexed by k, such that

$$y^k \to g^* \quad \text{a.e. } on\ Q. \tag{4.3.100}$$

Let Q_0 denote the subset of Q in which y^k does not converge pointwise to g^*. Then the Lebesgue measure $|Q_0|=0$. Let $Q_1 \equiv Q \backslash Q_0$, $\tilde{Q}_k \equiv \{(x,t) \in Q : u^k(x,t) \notin U(x,t)\}$, $\tilde{Q} \equiv \bigcup_{k=1}^{\infty} \tilde{Q}_k$. Then $|\tilde{Q}|=0$ and $Q_2 \equiv Q_1 \backslash \tilde{Q}$ has full measure. For arbitrary $(x^0,t^0) \in Q_2$, we show that

$$g^*(x^0,t^0) \in \mathbb{R}\big(x^0,t^0,\phi(x^0,t^0)\big). \tag{4.3.101}$$

Since $\mathbb{R}$ is u.s.c.i., it follows that, for any given $\varepsilon>0$, there exists a $\delta \equiv \delta(\varepsilon)>0$ such that

$$\mathbb{R}(x,t,z) \subset R^{\varepsilon}\big(x^0,t^0,\phi(x^0,t^0)\big) \tag{4.3.102}$$

whenever $|(x,t,z)-(x^0,t^0,\phi(x^0,t^0))|<\delta$, where $\mathbb{R}^{\varepsilon}(x,t,z)$ denotes the closed ε-neighborhood of $\mathbb{R}(x,t,z)$. Since $\phi(u^k) \overset{u}{\to} \phi$ on Q, there exists an integer $k_0>0$ such that, for all $k>k_0$,

$$|\phi(u^k)(x,t)-\phi(x,t)|<\delta \tag{4.3.103}$$

for all $(x,t) \in Q$. Therefore, for $k>k_0$, it follows from (4.3.102) and (4.3.103) that

$$\mathbb{R}\big(x^0,t^0,\phi(u^k)(x^0,t^0)\big) \subset \mathbb{R}^{\varepsilon}\big(x^0,t^0,\phi(x^0,t^0)\big). \tag{4.3.104}$$

Since $g^k(x^0,t^0) \in \mathbb{R}(x^0,t^0 \phi(u^k)(x^0,t^0))$ and $\mathbb{R}(x,t,z)$ is convex, by hypothesis, we have

$$y^k(x^0,t^0) \in \mathbb{R}^{\varepsilon}\big(x^0,t^0,\phi(x^0,t^0)\big) \tag{4.3.105}$$

for all $k>k_0$. Therefore, it follows from the relation (4.3.100) and the closure property of $\mathbb{R}^{\varepsilon}(x,t,z)$ that $g^*(x^0,t^0) \in \mathbb{R}^{\varepsilon}(x^0,t^0,\phi(x^0,t^0))$. Since $\varepsilon>0$ was chosen arbitrarily and $\mathbb{R}^{\varepsilon}(x,t,z)$ is closed, the relation (4.3.101) holds. By a modification on a set of measure zero, if necessary, we can assume that $g^*(x,t) \in \mathbb{R}(x,t,\phi(x,t))$ for all $(x,t) \in Q$. Thus, it follows from (the selection) Theorem 1.4.6 that there exists a control $u^* \in \mathfrak{U}$ such that

$$g^*(x,t)=g(x,t,\phi(x,t),u^*(x,t)), \qquad (x,t) \in Q. \tag{4.3.106}$$

This completes the proof of the lemma for case (i), $p \in [1,\infty)$.

Case (ii): $p=\infty$. The proof in this case is similar to the one given above with the only modification that the weak topology is replaced by the weak* topology.

Combining Lemmas 4.3.3 and 4.3.4, we have the following result.

Theorem 4.3.7. *Consider system* (4.3.82). *Suppose that all the hypotheses of Lemma* 4.3.4 *are satisfied. Let* $\{u^k\}$ *be a sequence of controls in* $\mathcal{U}$. *Then there exist a subsequence of the sequence* $\{u^k\}$, *again denoted by* $\{u^k\}$, *and a control* $u^* \in \mathcal{U}$ *such that*

$$\phi(u^k) \overset{\text{u}}{\to} \phi(u^*) \qquad \text{on } Q$$

$$\left.\begin{aligned} (\phi(u^k))_x &\overset{\text{w(w*)}}{\to} (\phi(u^*))_x \\ (\phi(u^k))_t &\overset{\text{w(w*)}}{\to} (\phi(u^*))_t \\ (\phi(u^k))_{xt} &\overset{\text{w(w*)}}{\to} (\phi(u^*))_{xt} \\ g^k &\overset{\text{w(w*)}}{\to} g^* \end{aligned}\right\} \quad \begin{aligned} &\text{in } L_p(Q, R^n) \text{ if } p \in [1, \infty) \\ &(\text{in } L_\infty(Q, R^n) \text{ if } p = \infty), \end{aligned}$$

where $\phi(u^k)(\phi(u^*))$ *is the solution of system* (4.3.82) *corresponding to* $u^k \in \mathcal{U}(u^* \in \mathcal{U})$, g^k *is as defined in Lemma* 4.3.4, *and* $g^*(x,t) \equiv g(x,t,\phi(u^*)(x,t),u^*(x,t))$.

On the basis of the above result, we can solve the optimal problem (4.3.P2)′.

Theorem 4.3.8. *Suppose that all the hypotheses of Lemma* 4.3.4 *are satisfied and that* γ, *as given in* (4.3.33), *is a continuous real-valued function defined on* R^n. *Then Problem* (4.3.P2)′ *has a solution.*

PROOF. Define the set X by $X \equiv \{\phi(u) : u \in \mathcal{U}\}$ and the functional ν on $C(Q, R^n)$ by $\nu(\phi) = \gamma(\phi(x_0 + l_1, t_0 + l_2))$. Clearly, the functional ν is a continuous functional on $C(Q, R^n)$. Then, if we can show that X is a compact subset of $C(Q, R^n)$, this will guarantee the existence of a minimum for the functional ν on X. This, in turn, will imply the existence of a control $u^* \in \mathcal{U}$ such that

$$\nu(\phi(u^*)) \leq \nu(\phi(u))$$

for all $u \in \mathcal{U}$.

Let $\{\phi^k\}$ be any sequence in X. Clearly, this implies that there exists a sequence $\{u^k\} \subset \mathcal{U}$ such that, for each positive integer k, $\phi^k = \phi(u^k)$. By virtue of Theorem 4.3.7, it follows that there exists a subsequence of the sequence $\{u^k\}$, again denoted by $\{u^k\}$, and a control $u^* \in \mathcal{U}$ such that $\phi^k \to \phi(u^*)$ uniformly on Q. Consequently, $\phi(u^*) \in X$. Thus, X is closed and sequentially compact and hence a compact subset of $C(Q, R^n)$. Since ν is continuous and X is compact, it follows from Theorem 1.1.22 that there exists a $\phi^0 \in X$ such that $\nu(\phi^0) \leq \nu(\phi)$ for all $\phi \in X$. Hence, there exists a $u^0 (\in \mathcal{U})$ so that $\phi^0 = \phi(u^0)$. This completes the proof. □

In what follows, we present (without proof) an existence theorem for the optimal control problem (4.3.P2)′ without the convexity condition on the set $\{g(x,t,z,v): v \in U(x,t)\}$ but with more restricted assumptions on the coefficients and data of the system and the control set. More precisely, the following conditions are required.

Assumption (4.3.A15)

i. The initial and boundary conditions (ϕ^0, ϕ^b) satisfy Assumption (4.3.A1).
ii. $U(x,t) = V$ a fixed nonempty convex and compact set.
iii. The function g has the form $g(x,t,z,v) = E(x,t)\cdot z + \hat{g}(x,t,v)$.
iv. The entries of E belong to $L_\infty(Q)$, $\hat{g}$ is defined and measurable on $Q \times V$, $\hat{g}(x,t,\cdot)$ is continuous on V for each $(x,t) \in Q$, and $\hat{g}(\cdot,\cdot,u(\cdot,\cdot))$ belongs to $L_\infty(Q, R^n)$ for $u \in \mathfrak{U}$.

Theorem 4.3.9. *Suppose that Assumption* (4.3.A15) *is satisfied and that* γ [*see* (4.3.33)] *is a continuous real-valued function defined on* R^n. *Let* $(x,t) \to \mu(x,t)$ *be a measurable function defined on* Q *with values in* R^σ *such that* $\mu_k(x,t) \geq 0, k = 1, \ldots, \sigma$, *and* $\sum_{k=1}^{\sigma} \mu_k(x,t) = 1$ *for all* $(x,t) \in Q$. *Let* Λ *denote the set of all such measurable functions and define*

$$G^*(x,t) \equiv \left\{ \xi \in R^n : \xi = \sum_{i=1}^{\sigma} \mu_i(x,t)\hat{g}(x,t,v^i), \right.$$

$$\left. \times \mu \in \Lambda, v^i \in V, i = 1, \ldots, \sigma \right\}.$$

Then, if the multifunction G^* *is measurable on* Q, *Problem* (4.3.P2)′ *has a solution.*

The above result is due to Suryanarayana [Su.4]. Its proof is briefly indicated in Section 4.4. For the detailed proof, the interested readers are referred to the original paper. Note that we will also consider, in Section 5.4, the question of existence of optimal controls for a general class of nonlinear evolution equations without the convexity condition.

4.3.7. Computation Methods

In this subsection, we present a computational algorithm for solving the optimal control problem (4.3.P2).

As in Section 3.3, we consider the following class of admissible controls:

$$\mathfrak{U}_0 \equiv \{u : u \textit{ piecewise continuous on } Q, u(x,t) \in V \textit{ for all } (x,t) \in Q\},$$

where V is a compact subset of R^m.

Since $\mathfrak{U}_0 \subset \mathfrak{U}$, all results presented in Sections 4.3.4 and 4.3.5 remain valid with $\mathfrak{U}$ replaced by $\mathfrak{U}_0$. For convenience, Problem (4.3.P1) with $\mathfrak{U}$ replaced by $\mathfrak{U}_0$ will be referred to as Problem (4.3.P1)″.

To proceed further, we need the following assumptions.

Assumption (4.3.A16). For each $u \in \mathfrak{U}_0$, there is partition of $Q, \{Q_i(u)\}_{i=1}^{n(u)}$, such that u is continuous on each of $Q_i^0(u)$, where $Q_i^0(u)$ denotes the interior of $Q_i(u)$. Further, for each $u \in \mathfrak{U}_0$, $H(u)(\cdot,\cdot,\cdot)$, as given in (4.3.58), is continuous on $\{\bigcup_{i=1}^{n(u)} Q_i^0(u)\} \times V$.

Assumption (4.3.A17). For each $(x,t) \in \bigcup_{i=1}^{n(u)} Q_i^0(u)$, $H(u)(x,t,\cdot)$ has a unique minimum in V.

For each $u \in \mathfrak{U}_0$, let $\omega(u)$ denote the function mapping Q into V such that, for each $(x,t) \in \bigcup_{i=1}^{n(u)} Q_i^0(u)$, $\omega(u)(x,t)$ minimizes $H(u)(x,t,\cdot)$, and, for all other $(x,t) \in Q$, $\omega(u)(x,t) = u(x,t)$. Clearly, for each $u \in \mathfrak{U}_0$, the function $\omega(u)$ has the property that

$$H(u)(x,t,v) \geq H(u)(x,t,\omega(u)(x,t)) \tag{4.3.107}$$

for all $(x,t,v) \in \{\bigcup_{i=1}^{n(u)} Q_i^0(u)\} \times V$, and for each $(x,t) \in \bigcup_{i=1}^{n(u)} Q_i^0(u)$, the equality holds if and only if $v = \omega(u)(x,t)$. Thus, it follows from the definition of the function $\omega(u)$ that if $\omega(u)(x,t) \neq u(x,t)$, then

$$H(u)(x,t,u(x,t)) > H(u)(x,t,\omega(u)(x,t)). \tag{4.3.108}$$

Under Assumptions (4.3.A16)–(4.3.A17), we can show, by using precisely the same argument as that given for Theorem 3.3.1, that, for any $u \in \mathfrak{U}_0$, $\omega(u) \in \mathfrak{U}$. More precisely, $\omega(u)$ is continuous on $\bigcup_{i=1}^{n(u)} Q_i^0(u)$.

Definition 4.3.3. A control $u \in \mathfrak{U}_0$ is said to be *an extremal control* if $u(x,t) = \omega(u)(x,t)$ for all continuity points $(x,t) \in Q$ of u.

REMARK 4.3.6. Clearly, if $u \in \mathfrak{U}_0$ is not an extremal control, then there must be a point $(x',t') \in \bigcup_{i=1}^{n(u)} Q_i^0(u)$ such that

$$u(x',t') \neq \omega(u)(x',t'). \tag{4.3.109}$$

Theorem 4.3.10. *Consider Problem* (4.3.P1)″ *with the class of admissible controls* $\mathfrak{U}_0$. *Suppose that all the hypotheses of Lemma* 4.3.2 *and Assumptions* (4.3.A16)–(4.3.A17) *are satisfied. Let* (x',t') *be a point in* $\bigcup_{i=1}^{n(u)} Q_i^0(u)$ *such that* $u(x',t') \neq \omega(u)(x',t')$. *For each integer* $k \geq 1$, *define*

$$I_{1k} \equiv \{x \in I_1 : |x - x'| < 1/k\}, \qquad I_{2k} \equiv \{t \in I_2 : |t - t'| < 1/k\}$$

and $Q_k \equiv I_{1k} \times I_{2k}$.

Then there exists an integer $k^0 \geq 1$ *such that*

$$J(u^k) < J(u) \tag{4.3.110}$$

for all $k > k^0$, where u^k is defined by

$$u^k(x,t) \equiv \left(1 - \chi_{Q_k}(x,t)\right)u(x,t) + \chi_{Q_k}(x,t)\,\omega(u)(x',t') \tag{4.3.111}$$

and χ_{Q_k} is the characteristic function of Q_k.

PROOF. Since $u(x',t') \neq \omega(u)(x',t')$, it follows from (4.3.108) that

$$H(u)(x',t',u(x',t')) > H(u)(x',t',\omega(u)(x',t')). \tag{4.3.112}$$

Further, by hypothesis, (x',t') is a point in $\bigcup_{i=1}^{n(u)} Q_i^0(u)$. Thus, it is a continuity point of u and $\omega(u)$. Therefore, we can verify easily from the second condition of Assumption (4.3.A16) that $H(u)(\cdot,\cdot,u(\cdot,\cdot))$ and $H(u)(\cdot,\cdot,\omega(u)(\cdot,\cdot))$ are continuous at (x',t').

By virtue of Theorem 4.3.5, we have

$$J(u^k) - J(u) \leq K_7\left(\frac{1}{k}\right)^2 \xi\left(u, \omega(u)(x',t'), \frac{1}{k}\right) + \iint_{Q_k} \{H(u)(x,t,\omega(u)(x',t')) - H(u)(x,t,u(x,t))\}\,dx\,dt \tag{4.3.113}$$

for all integers $k \geq 1$, where the constant K_7 is independent of k and $\xi(u,\omega(u)(x',t'),1/k) \to 0$ as $k \to \infty$. Thus, by dividing the above inequality by $|Q_k| = (1/k)^2$, (where $|Q_k|$ denotes the Lebesgue measure of Q_k), we obtain

$$\frac{1}{|Q_k|}\left(J(u^k) - J(u)\right) \leq K_7 \xi\left(u, \omega(u)(x',t'), \frac{1}{k}\right) + \frac{1}{|Q_k|}\iint_{Q_k} \{H(u)(x,t,\omega(u)(x',t')) - H(u)(x,t,u(x,t))\}\,dx\,dt. \tag{4.3.114}$$

Since $H(u)(\cdot,\cdot,\omega(u)(\cdot,\cdot))$ and $H(u)(\cdot,\cdot,u(\cdot,\cdot))$ are continuous at (x',t'), it follows that

$$\frac{1}{|Q_k|}\iint_{Q_k} \{H(u)(x,t,\omega(u)(x',t')) - H(u)(x,t,u(x,t))\}\,dx\,dt \to H(u)(x',t',\omega(u)(x',t')) - H(u)(x',t',u(x',t')) \tag{4.3.115}$$

as $k \to \infty$. Thus, we conclude from (4.3.114) and (4.3.112) that

$$\lim_{k\to\infty} \frac{1}{|Q_k|}\left(J(u^k) - J(u)\right) < 0. \tag{4.3.116}$$

Consequently, there exists an integer $k_0 > 0$ such that

$$J(u^k) < J(u) \tag{4.3.117}$$

for all $k > k_0$. This completes the proof. □

The above theorem is of the same nature as Theorem 3.3.2. Thus, an algorithm similar to Algorithm 3.3.I can be devised. The only significant difference between these two algorithms is this: in the present algorithm, the adjoint system (4.3.43) admits infinitely many solutions $(\lambda^1(u), \lambda^3(u), \zeta^1(u), \zeta^2(u)) \in L_\infty(Q, R^{4n})$ for each $u \in \mathscr{U}_0$ (see Theorem 4.3.4), whereas, in Algorithm 3.3.I the adjoint system (3.1.10) has only one solution for each $u \in \mathscr{U}_0$ (see Theorem 3.1.1).

However, by examining the proof of Theorem 4.3.4, we observe that if $\lambda^1_x \in L_\infty(Q, R^n)$ is chosen and treated as fixed, then, for each $u \in \mathscr{U}_0$, the adjoint system (4.3.43) has only a unique solution. Thus, in practice, we may choose $\lambda^1_x \equiv 1$ in Step 1 of the present algorithm.

4.4. Notes

In Section 4.1, we consider a class of optimal control problems involving first-order hyperbolic systems with one spatial variable and boundary controls. This class of problems arises naturally in the study of counterflow problems in chemical engineering and low frequency-wave propagation problems in electrical engineering. A necessary condition for optimality is derived in Theorem 4.1.2, while the question concerning the existence of optimal controls is discussed in a remark. For general existence theorems, see Chapter 5. The results of Section 4.1 are essentially due to Russell [Ru.8]. These results have been extended to the case involving many spatial variables in a half-space (for details, see [VJ.1]). Choice of the domain puts a serious limitation on the applicability of these results. It appears from the literature that not much work has been done for a bounded domain. Thus, it would be interesting to consider general hyperbolic problems for bounded domain in R^n with controls exercised from the boundary. Such control problems are encountered in the study of plasma confinement, power flow over interconnected system of networks, structural vibration, acoustics, magnetohydrodynamic shock waves, water waves in tanks, lakes, etc.

In Section 4.2, we consider a class of optimal control problems involving second-order hyperbolic systems in one spatial variable with data from the space of distributions. This class of problems arises naturally from power flow problems in electrical engineering. Necessary and sufficient conditions for optimality are derived in Theorem 4.2.2 and Corollary 4.2.1. A result on the existence of optimal controls is proved in Theorem 4.2.3 and a computational method for finding optimal controls is discussed in Section 4.2.7. The results of this section are essentially due to Ahmed

[Ah.1]. In fact, these results can be easily extended to the vector case. We note that a more realistic mathematical model representing a power system consists of a system of hyperbolic equations governing the wave motion over transmission lines, a system of ordinary differential equations representing the dynamics of generators and prime movers that determine the boundary conditions, and a system of algebraic equations representing the (interconnected) topology of the network. Such large-scale distributed parameter systems have not been considered in the literature. Interest in the area is expected to grow in the near future with the development of large-scale computers. At present, the question remains open.

For plasma and vibration control problems, see [Ru.3, Wa.4, Kom.1, Kom.2].

Most of the results of the review paper [Ru.10] mentioned in the Notes of Chapter 3 are given for hyperbolic systems. The reader is also referred to [Cl.1, CW.1] for certain results concerning the controllability of the hyperbolic systems not included in [Ru.10].

In Section 4.3, we consider a class of second-order nonlinear hyperbolic systems with controls appearing in the coefficients of the differential equations. The boundary condition is of Darboux type. A necessary condition for optimality is derived in Theorem 4.3.6. This result is basically due to Suryanarayana [Su.1, Su.2]. It is interesting to observe that this necessary condition is similar to that for the first boundary value problem (of parabolic type) in general form (see Theorem 3.1.3). This type of necessary condition is said to be of the Pontryagin type.

In [Su.4], an existence theorem for optimal controls is proved without assuming the convexity condition on the "velocity field." This result is quoted without proof in Theorem 4.3.9. However, the basic idea behind the proof may be briefly indicated as follows.

For any fixed $(x_1, t_1) \in Q, (\phi(u))_j(x_1, t_1), 1 \leq j \leq n$, can be represented as a functional of the initial boundary data (ϕ^0, ϕ^b), the coefficients A, B, E, $\hat{g}$, and certain functions

$$\lambda^{1,j} \equiv (\lambda_1^{1,j}, \ldots, \lambda_n^{1,j}), \qquad \lambda^{2,j} \equiv (\lambda_1^{2,j}, \ldots, \lambda_n^{2,j}),$$

$$\zeta^{1,j} \equiv (\zeta_1^{1,j}, \ldots, \zeta_n^{1,j}), \qquad \zeta^{2,j} \equiv (\zeta_1^{2,j}, \ldots, \zeta_n^{2,j}),$$

$j = 1, \ldots, n$, where $\phi(u)$ is the solution of the system (4.3.82) [under Assumption (4.3.A15)] corresponding to the control $u \in \mathfrak{U}$, and for each $j \in \{1, \ldots, n\}$, $(\lambda^{1,j}, \lambda^{2,j}, \zeta^{1,j}, \zeta^{2,j})$ is a solution of the system (4.3.43), specialized to the situation of Theorem 4.3.9. Then the system (4.3.82) is "convexified" by replacing $\hat{g}(x, t, u)$ by $\sum_{i=1}^{\nu} \mu_i(x, t)\hat{g}(x, t, u_i)$ on the right-hand side of the differential equation of the system, where, for each $i \in \{1, \ldots, \nu\}, (x, t) \to \mu_i(x, t)$ is a measurable function such that $\mu_i(x, t) \geq 0$ and $\sum_{i=1}^{\nu} \mu_i(x, t) = 1$. A solution of this convexified system is called a *generalized solution of the original system*. By using a result of Cesari (see Theorem 5 of [Ce.7]), one can show that, for each generalized solution, there is a "usual" solution such that they are equal at $(x,t)=(x_0+l_1, t_0+l_2)$. By Theorem 4.3.8, the optimal control problem involving the convexified system has a solution. Thus, Problem (4.3.P2)′ [under Assumption (4.3.A15)] also has a solution.

Theorem 4.3.8 appears to be new. Its proof is based on a standard argument used for optimal control problems involving only ordinary differential equations.

In reference [Su.3], an existence theorem for optimal controls is proved for an optimal control problem involving a general nonlinear hyperbolic partial differential equation with Darboux boundary condition. In the proof of this result, it was assumed that, for each $u\in\mathfrak{U}$, the system has a solution in a particular Sobolev space suitable for later analysis. This assumption is very difficult, or even impossible, to justify by imposing sufficient conditions on the data of the system. Thus, in Section 4.3.6, we use a more natural approach. We first obtain a set of sufficient conditions on the data of the system ensuring the existence of a unique solution in some Sobolev space corresponding to each $u\in\mathfrak{U}$. However, for this reason, we cannot handle systems more general than the one used in (4.3.82).

Note that both the domain and the boundary condition considered in Section 4.3 are rather restrictive even though the system equations are quite general. Thus, it would be very useful if more general domain and boundary conditions could be included.

Another interesting problem that arises in power system voltage regulation is given by the first-order hyperbolic system [Ah.6]

$$\frac{\partial}{\partial t}(A(u)y)+By=f, \qquad (t,x)\in I\times\Omega\equiv Q,$$
$$y(0,x)=y_0(x), \qquad x\in\Omega,$$
$$y(t,x)=y_b(t,x), \qquad (t,x)\in I\times\partial\Omega$$

with fixed initial and boundary conditions. Here B is a matrix of first-order differential operators in Ω with positive coefficients generally defined on Q and A is a positive definite matrix with elements defined on $Q\times U$, where U is a compact convex subset of R^m. The class of admissible controls is given by $\mathfrak{U}=L_\infty(Q,U)$. The problem on existence of optimal controls for this class of systems with suitable cost function (quadratic, for example) remains to be resolved.

For optimal control of the class of hyperbolic systems described by

$$\frac{d^2y}{dt^2}+A(t)y=f+B(t)u,$$
$$y(0)=y_0, \qquad \dot{y}(0)=y_1$$

with $A\in L_\infty(I,\mathfrak{L}(V,V^*))$, $A^*\in L_\infty(I,\mathfrak{L}(V,V^*))$, A coercive and $B\in L_\infty(I,\mathfrak{L}(\mathfrak{U},H))$, and $f\in L_2(I,H)$, where $V\subset H\subset V^*$, we refer the reader to [Li.2]. In this reference, the reader will also find an interesting discussion of boundary control (Dirichlet and Neumann type) of the above class of systems with either distributed or boundary observation.

In Section 4.3.7, we suggest a method to construct an improved control from any nonextremal control. This leads to an iterative scheme for seeking optimal controls. The basic idea behind this computational method is similar to that of Section 4.3.2: from any nonextremal control, we can construct a sequence of controls $\{u^k\}$ such that $J(u^{k+1})<J(u^k)$, $k=0,1,2,\ldots$. If there is an integer $k_0>0$ such that u^{k_0} satisfies the necessary condition, then we set $u^{k_0+1}=u^{k_0}$ and the algorithm is terminated. However, $\{u^k\}$ is, in general, an infinite sequence. Thus, for this general case, it would be interesting to show that this sequence of controls converges to a limit that satisfies the necessary condition. This result is not yet available in the literature. However, the approach suggested in reference [RT.3] appears to be valid for the class of optimal control problems considered in Section 4.3.7. This is yet to be investigated.

CHAPTER FIVE

OPTIMAL CONTROL OF EVOLUTION EQUATIONS ON BANACH SPACES

5.1. Linear Evolution Equations in Coercive Form

5.1.1. *Introduction*

A very large class of distributed parameter systems can be modeled as an abstract differential equation on a suitable Banach space or on a suitable manifold therein. The advantage of an abstract formulation lies not only in its generality but also in the insight that can be gained about the many common unifying properties that tie together apparently diverse problems. In this section we wish to study some typical control problems for a class of systems governed by linear evolution equations on a Hilbert space.

5.1.2. *System Description*

Let H be a Hilbert space and V a subset of H also with the structure of a Hilbert space, so that V is dense in H. Identifying H with its dual H^* and denoting the (topological) dual of V by V^*, we have $V \subset H \subset V^*$. We denote the scalar product in H by $(\cdot, \cdot)$ and the norm by $|\cdot|_H$. The duality pairing between V^* and V is denoted by $\langle \cdot, \cdot \rangle$ or by $(\cdot, \cdot)_{V^*-V}$ and the norms in V and V^* by $\|\cdot\|_V$ and $\|\cdot\|_{V^*}$, respectively. Let $\mathcal{L}(E, F)$ denote the class of linear operators, bounded or unbounded, from a topological space E into a topological space F.

An operator $A \in \mathcal{L}(V, V^*)$ is said to be coercive if there exist numbers α, λ, $\alpha > 0$, $\lambda \geq 0$, such that

$$\langle A\psi, \psi \rangle + \lambda |\psi|_H^2 \geq \alpha \|\psi\|_V^2 \quad \text{for all } \psi \in V.$$

In general we are concerned with a family of operators $\{A(t), t \in I\} \subset \mathcal{L}(V, V^*)$ uniformly coercive on $I = (0, T)$. Associated with this family of operators, there exists a family of bilinear forms (sesquilinear forms in the complex case) $a: I \times V \times V \to R(C)$ such that

$$a(t, \psi, \psi) + \lambda |\psi|_H^2 \geq \alpha \|\psi\|_V^2$$

for all $\psi \in V$, uniformly in t on I, where

$$a(t, \psi, \phi) = \langle A(t)\psi, \phi \rangle.$$

Similarly, given a family of bilinear forms $a(t, \psi, \phi)$ on V, there exists a family of operators $\{A(t), t \in I\} \in \mathcal{L}(V, V^*)$ such that $a(t, \psi, \phi) = \langle A(t)\psi, \phi \rangle$. Throughout this section we shall be concerned only with the real case and bounded bilinear forms. That is, from now on we consider $\mathcal{L}(E, F)$ to be the space of bounded linear operators from E into F. Thus, suppose we are given a family of operators $\{A(t), t \in I\} \subset \mathcal{L}(V, V^*)$. Let $a(t, \psi, \phi) \equiv \langle A(t)\psi, \phi \rangle, t \in I$, denote the corresponding family of bilinear forms. We assume that a satisfies the following properties:

a1. For each pair $\psi, \phi \in V$, the function $t \to a(t, \psi, \phi)$ is Lebesgue measurable on I and there exists a constant $c > 0$, independent of $\{t, \psi, \phi\} \in I \times V \times V$, such that

$$|a(t, \psi, \phi)| \leq c \|\psi\|_V \|\phi\|_V. \tag{5.1.1}$$

a2. There exist two numbers $\alpha > 0$, $\lambda \geq 0$ independent of $\{t, \psi\} \in I \times V$ such that

$$a(t, \psi, \psi) + \lambda |\psi|_H^2 \geq \alpha \|\psi\|_V^2. \tag{5.1.2}$$

Clearly the class of operators A satisfying property (a1) can be denoted by $L_\infty(I, \mathcal{L}(V, V^*))$. We wish to consider the following evolution equation:

$$\frac{dy}{dt} + A(t)y = f, \qquad t \in I = (0, T),$$

$$y(0) = y_0 \tag{5.1.3}$$

in V, where y_0 and f are the given data, $\{A(t), t \in I\}$ is a family of operators with values in $\mathcal{L}(V, V^*)$ satisfying properties (a1) and (a2), and y, with values $y(t) \in V$, is the solution sought. A linear evolution equation given by (5.1.3) is said to be *in the coercive form* if the corresponding linear operator A is coercive.

Before we can discuss the question of existence and uniqueness of a solution for the problem, we must find appropriate (admissible) function spaces for the given data $\{y_0, f\}$ and discover the space in which we can look for the solution y. This problem does not have a universal solution. The function spaces to be chosen for y_0, f, and y must be sufficiently general to cover a wide class of physical problems.

Let $(E,|\cdot|_E)$ be an arbitrary Banach space, and denote by $L_p(I, E)$, $1 \le p < \infty$, the equivalence classes of strongly measurable functions on $I \equiv (0, T)$ taking values in E and endowed with the norm

$$\|y\| = \left(\int_I |y(t)|_E^p \, dt \right)^{1/p}.$$

If E is a reflexive Banach space and $p \in (1, \infty)$, then $L_p(I, E)$ is also a reflexive Banach space with dual $L_q(I, E^*)$ where $(1/p)+(1/q)=1$. For $p=2$ and E any Hilbert space, $L_2(I, E)$ is also a Hilbert space. Since V, H and V^* are Hilbert spaces, so are the function spaces $L_2(I,V)$, $L_2(I,H)$, and $L_2(I,V^*)$. We note that the dual of $L_2(I,V)$ is $L_2(I,V^*)$ and that, due to property (a1), it follows from Schwartz's inequality that $A\psi$, defined by $t \to (A\psi)(t) = A(t)\psi(t)$, belongs to $L_2(I,V^*)$. Thus, loosely speaking, for f and y we can choose the function spaces $L_2(I,V^*)$ and $L_2(I,V)$, respectively. In order that the equality in (5.1.3) be satisfied in some sense, y must be differentiable at least in the distribution sense (see Section 1.2.6) and dy/dt must belong to $L_2(I,V^*)$.

Let $\mathfrak{D}'(I,V)$ denote the space of V-valued distributions on I and $\mathfrak{D}(I)$ the space of C^∞-functions with compact support on I. For each $f \in L_2(I,V)$ we define

$$f'(\psi) \equiv \int_I f(t)\psi(t)\, dt \in V$$

mapping $\mathfrak{D}(I)$ into V. Clearly, the mapping $f \to f'$ from $L_2(I,V)$ into $\mathfrak{D}'(I,V)$ is a continuous injection. Thus we may identify f' with f and consider $L_2(I,V) \subset \mathfrak{D}'(I,V)$. Therefore, any element of $L_2(I,V)$ can be differentiated in the sense of distribution as many times as required. For $y \in L_2(I,V)$ and n an arbitrary positive integer, $(D^n y)(\psi) = (-1)^n y(D^n\psi)$, where $D^n = d^n/dt^n$. We introduce the set

$$X = \{y : y \in L_2(I,V) \text{ and } Dy \in L_2(I,V^*)\}. \tag{5.1.4}$$

The set X endowed with the norm

$$\|y\|_X \equiv \left(\|y\|^2_{L_2(I,V)} + \|Dy\|^2_{L_2(I,V^*)} \right)^{1/2}$$

is a Hilbert space. Indeed, if $\{y_n\}$ is a Cauchy sequence in X, then there exist $y_0 \in L_2(I,V)$ and $z_0 \in L_2(I,V^*)$ such that $y_n \to y_0$ in $L_2(I,V)$ and $Dy_n \to z_0$ in $L_2(I,V^*)$. Clearly, both Dy_0 and z_0 belong to $\mathfrak{D}'(I,V)$, and it follows from the equality

$$(Dy_0 - z_0)(\psi) = (-1)(y_0 - y_n)(D\psi) + (Dy_n - z_0)(\psi), \qquad \psi \in \mathfrak{D}(I,V), \tag{5.1.5}$$

that $(Dy_0 - z_0)$ vanishes on $\mathfrak{D}(I,V)$. Since $\mathfrak{D}(I,V)$ is dense in $L_2(I,V)$, $Dy_0 = z_0$, and consequently, $y_0 \in X$. We know that the set $X \subset C^0(\bar{I}, H)$ (see Section 1.2.6). Thus, for any $y \in X$, $y(0)$ is well defined and belongs to H.

Therefore, for $y_0 \in H$, $y \in X$, and $f \in L_2(I, V^*)$, the problem (5.1.3) makes sense. We prove in the following section that the evolution equation (5.1.3) indeed has a solution in X.

5.1.3. Existence and Uniqueness of Solution

In this section we wish to study the problem of existence and uniqueness of solution of the evolution equation (5.1.3). In this regard, we note that as long as the time interval $I=(0, T)$ is finite, we can take λ of (5.1.2) equal to zero without any loss of generality. Indeed, setting $y(t)=e^{\lambda t}z(t)$ and substituting in (5.1.3), we obtain

$$\frac{dz}{dt}+(\lambda I+A(t))z=\tilde{f}$$

with $\tilde{f}(t)=e^{-\lambda t}f(t)$. Thus, for finite time interval, property (a2) is entirely equivalent to the one with $\lambda=0$. Since throughout this section we shall be concerned with finite time problems, we shall assume this property.

Theorem 5.1.1. *Suppose V, H, V^*, and X are as described in Section 5.1.2 with $V \subset H \subset V^*$, V separable and dense in H. Let $\{A(t), t \in I\} \subset \mathcal{L}(V, V^*)$ be a given family of operators with the corresponding bilinear forms $\{a(t, \psi, \phi), t \in I\}$, $\psi, \phi \in V$, satisfying the properties* (a1) *and* (a2). *Then for each $y_0 \in H$ and $f \in L_2(I, V^*)$, the evolution equation* (5.1.3) *has a unique solution $y \in X$, and the mapping $\{y_0, f\} \to y$ from $H \times L_2(I, V^*)$ into $L_2(I, V)$ is linear and continuous.*

PROOF. First we prove the uniqueness. Suppose y_1, $y_2 \in X$ are any two solutions of (5.1.3). Then $y \equiv (y_1 - y_2)$ is a solution of the homogeneous problem $(dy/dt)+A(t)y=0$, $y(0)=0$. Therefore

$$\int_0^T \left\langle \frac{dy}{dt}, y(t) \right\rangle dt + \int_0^T a(t, y(t), y(t))\, dt = 0. \tag{5.1.6}$$

Since $y \in X$, the first duality bracket reduces to $\frac{1}{2}\int_0^T (d/dt)(y(t), y(t))\, dt$, and consequently, due to property (a2), we have

$$|y(T)|_H^2 + 2\alpha \int_0^T \|y(t)\|_V^2 dt \le 0. \tag{5.1.7}$$

Thus $y_1 = y_2$, proving uniqueness.

Now we prove existence. Since V is assumed to be separable, there exists a countable set that is dense in V, and consequently, V has a basis $\{v_i\}$ such that the closure of the linear span of $\{v_i\}$ is V. Since the initial state y_0 belongs to H and V is dense in H, there exists a sequence of coefficients $\{\eta_i\}$ such that $y_0^m \equiv \sum_{i=1}^m \eta_i v_i$ converges strongly to y_0 in H. We can now construct an approximate solution of (5.1.3) by defining

$$y^m(t) \equiv \sum_{i=1}^{m} \xi_i^m(t) v_i \tag{5.1.8}$$

and choosing the functions $\{\xi_i^m, 1 \le i \le m\}$ in such a way that, for $t \in I$,

$$\begin{aligned} &\left\langle \frac{d}{dt} y^m(t), v_i \right\rangle + \langle A(t) y^m(t), v_i \rangle = \langle f(t), v_i \rangle, \qquad 1 \le i \le m, \\ &y^m(0) = \sum_{1 \le i \le m} \xi_i^m(0) v_i \equiv \sum_{1 \le i \le m} \eta_i v_i. \end{aligned} \tag{5.1.9}$$

This problem is equivalent to the finite-dimensional system

$$\begin{aligned} &\beta^m \frac{d}{dt} \xi^m(t) + a^m(t) \xi^m(t) = f^m(t), \\ &\xi^m(0) = \eta^m, \end{aligned} \tag{5.1.10}$$

where the superscript m denotes the order of approximation and the elements of the matrices β^m and a^m are given by $\beta_{ij}^m = (v_i, v_j)$, $a_{ij}^m(t) = \langle A(t) v_i, v_j \rangle$, and $f^m(t) = (f_1(t), \ldots, f_m(t))'$, $\eta^m = (\eta_1, \ldots, \eta_m)'$ with $f_i(t) = \langle f(t), v_i \rangle$. This is a finite-dimensional linear system and consequently has a unique solution ξ^m that is absolutely continuous on I. We show that the sequence $\{y^m\}$ has a limit and that the limiting function is a solution of the problem (5.1.3) and that it belongs to X. Multiplying the first equation of (5.1.9) by $\xi_i^m(t)$ and summing over $1 \le i \le m$, we obtain

$$\left\langle \frac{d}{dt} y^m(t), y^m(t) \right\rangle + a(t, y^m(t), y^m(t)) = \langle f(t), y^m(t) \rangle,$$

which is equivalent to

$$\frac{d}{dt} |y^m(t)|_H^2 + 2a(t, y^m(t), y^m(t)) = 2\langle f(t), y^m(t) \rangle. \tag{5.1.11}$$

Integrating (5.1.11) over the interval $(0, \theta)$, $\theta \le T$, we have

$$\begin{aligned} &|y^m(\theta)|_H^2 + 2\int_0^\theta a(t, y^m(t), y^m(t))\, dt \\ &\quad = |y^m(0)|_H^2 + 2\int_0^\theta \langle f(t), y^m(t) \rangle\, dt. \end{aligned} \tag{5.1.12}$$

Thus, it follows from the coercivity property (a2), and the Schwartz inequality applied to the last term of (5.1.12) that

$$\begin{aligned} &|y^m(\theta)|_H^2 + 2\alpha \int_0^\theta \| y^m(t) \|_V^2 dt \\ &\quad \le |y^m(0)|_H^2 + 2\left(\int_0^\theta \| f(t) \|_{V^*}^2 dt \right)^{1/2} \left(\int_0^\theta \| y^m(t) \|_V^2 dt \right)^{1/2} \end{aligned} \tag{5.1.13}$$

for $\theta \in [0, T]$. Using the Cauchy inequality

$$ab \leq \frac{1}{2\varepsilon} a^2 + \frac{\varepsilon}{2} b^2, \qquad a, b \in R, \quad \varepsilon > 0,$$

in the last term of (5.1.13) for $\varepsilon = \alpha$, we obtain

$$|y^m(\theta)|_H^2 + \alpha \int_0^\theta \| y^m(t) \|_V^2 dt \leq |y^m(0)|_H^2 + \frac{1}{\alpha} \int_0^\theta \| f(t) \|_{V^*}^2 dt, \theta \in [0, T]. \tag{5.1.14}$$

Since $y^m(0) = y_0^m$ converges strongly to y_0 in H, it is clear that

$$|y_0^m|_H \leq |y_0|_H. \tag{5.1.15}$$

As a consequence of the assumptions that $f \in L_2(I, V^*)$ and $y_0 \in H$, it follows from (5.1.14) and (5.1.15) that the sequence $\{y^m\}$ is contained in a bounded subset of $L_2(I, V) \cap L_\infty(I, H)$. Note that $L_2(I, V)$ is a Hilbert space and consequently is a reflexive Banach space. Recalling that a reflexive Banach space is weakly complete and a bounded subset therein is conditionally weakly sequentially compact, there exists a subsequence of the sequence $\{y^m\}$, relabeled as $\{y^m\}$, and an element $\tilde{y} \in L_2(I, V)$ such that $y^m \to \tilde{y}$ weakly in $L_2(I, V)$. Clearly, due to (5.1.9),

$$\left\langle \frac{d}{dt} y^m(t), v_i \right\rangle + \langle A(t) y^m(t), v_i \rangle = \langle f(t), v_i \rangle$$

for all $m \geq i$. Taking ψ from $C^1(I)$ with $\psi(T) = 0$, defining $\psi_i(t) = \psi(t) v_i$, and multiplying the above equation by $\psi(t)$, we obtain

$$\left\langle \frac{d}{dt} y^m(t), \psi_i(t) \right\rangle + \langle A(t) y^m(t), \psi_i(t) \rangle = \langle f(t), \psi_i(t) \rangle. \tag{5.1.16}$$

Since $\psi_i \in C^1(I, V)$, we can integrate (5.1.16) by parts and obtain

$$-\int_0^T \left(y^m(t), \frac{d\psi_i}{dt} \right) dt + \int_0^T \langle y^m(t), A^*(t) \psi_i(t) \rangle \, dt$$

$$= (y_0^m, \psi_i(0)) + \int_0^T \langle f(t), \psi_i(t) \rangle \, dt \tag{5.1.17}$$

where $A^*(t)$ is the operator adjoint to the operator $A(t)$ that maps from V into V^* because V is reflexive. Note that the adjoint operator $A^*(t)$ is well defined since $A(t)$ is a bounded operator from V into V^* [property (a1)]. Clearly $(d/dt)\psi_i \in C^0(I, V) \subset L_2(I, V) \subset L_2(I, H)$, $A^* \psi_i \in L_2(I, V^*)$, and $\psi_i(0) \in V \subset H$. Since $y_0^m \to y_0$ strongly in H and $y^m \to \tilde{y}$ weakly in $L_2(I, V)$, on passing to the limit in (5.1.17), we obtain

$$-\int_0^T \left(\tilde{y}(t), \frac{d}{dt} \psi_i(t) \right) dt + \int_0^T \langle \tilde{y}(t), A^*(t) \psi_i(t) \rangle \, dt$$

$$= (y_0, \psi_i(0)) + \int_0^T \langle f(t), \psi_i(t) \rangle \, dt. \tag{5.1.18}$$

By taking ψ from $\mathfrak{D}(I)$, Equation (5.1.18) can be rewritten as

$$\int_0^T\left\langle \frac{d}{dt}\tilde{y}(t), v_i\right\rangle \psi(t)\,dt + \int_0^T \langle A(t)\tilde{y}(t), v_i\rangle \psi(t)\,dt$$
$$= \int_0^T \langle f(t), v_i\rangle \psi(t)\,dt; \tag{5.1.19}$$

since this is true for all $\psi \in \mathfrak{D}(I)$, we conclude that

$$\left\langle \frac{d}{dt}\tilde{y}(t) + A(t)\tilde{y}(t) - f(t), v_i\right\rangle \equiv l_t(v_i) = 0 \tag{5.1.20}$$

in the sense of distribution. Since i is arbitrary, this equality holds for all $v \in \operatorname{span}\{v_i\} \equiv V_0$, where V_0 is dense in V. If a linear functional on V vanishes on a dense subset V_0 of V, then it vanishes on all of V. Thus $l_t(w) = 0$ for all $w \in V$, and consequently, $l_t = 0$. This means that

$$\frac{d}{dt}\tilde{y} + A(t)\tilde{y} = f \tag{5.1.21}$$

in the sense of distribution. Further $f \in L_2(I, V^*)$ and $A\tilde{y} \in L_2(I, V^*)$, and consequently, $(d/dt)\tilde{y} \in L_2(I, V^*)$ also. Thus, $\tilde{y} \in X$ [see definition (5.1.4)]. Taking $\psi_i(t) \equiv \psi(t)v_i$ with $\psi \in C^1(I), \psi(T) = 0$, and forming the $V^* - V$ scalar product between (5.1.21) and ψ_i and integrating by parts (which is permissible since $\tilde{y} \in X$), we obtain

$$-\int_0^T\left(\tilde{y}(t), \frac{d}{dt}\psi_i(t)\right) dt + \int_0^T \langle A(t)\tilde{y}(t), \psi_i(t)\rangle\, dt$$
$$= (\tilde{y}(0), \psi_i(0)) + \int_0^T \langle f(t), \psi_i(t)\rangle\, dt. \tag{5.1.22}$$

Comparing (5.1.22) with (5.1.18), we have $(y_0, \psi_i(0)) = (\tilde{y}(0), \psi_i(0))$, and consequently, $(y_0 - \tilde{y}(0), v_i) = 0$ for all i. Since V_0 is dense in V, which, in turn, is dense in H by assumption, we conclude that $\tilde{y}(0) = y_0$. Thus, we have proved that $\tilde{y} \in X$ and is a solution of the problem (5.1.3), that is,

$$\frac{d}{dt}\tilde{y} + A(t)\tilde{y} = f,$$
$$\tilde{y}(0) = y_0.$$

Since the solution is unique, we have $\tilde{y} = y$; consequently, we can replace the statement $y^m \to \tilde{y}$ weakly in $L_2(I, V)$ by $y^m \to y$ weakly in $L_2(I, V)$.

We prove now the continuity of the mapping $\{y_0, f\} \to y$ from $H \times L_2(I, V^*)$ into $L_2(I, V)$. Since $y \in L_2(I, V)$, there exists a $\nu \in L_2(I, V^*)$ with $\|\nu\|_{L_2(I, V^*)} = 1$ such that $\|y\|_{L_2(I,V)} = \int_I \langle \nu, y\rangle\, dt$. Clearly, we can write

$$\|y\|_{L_2(I,V)} = \int_I \langle \nu, y - y^m\rangle\, dt + \int_I \langle \nu, y^m\rangle\, dt$$
$$\leq \left|\int_I \langle \nu, y - y^m\rangle\, dt\right| + \|y^m\|_{L_2(I,V)}. \tag{5.1.23}$$

By virtue of the inequalities (5.1.14) and (5.1.15), one can find a constant $\gamma > 0$ such that

$$\|y^m\|_{L_2(I,V)} \leq \gamma\big(|y_0|_H + \|f\|_{L_2(I,V^*)}\big) \tag{5.1.24}$$

for all m. Since $y^m \to y$ weakly in $L_2(I,V)$, it follows from (5.1.23) and (5.1.24) that

$$\|y\|_{L_2(I,V)} \leq \gamma\big(|y_0|_H + \|f\|_{L_2(I,V^*)}\big). \tag{5.1.25}$$

Since the mapping $\{y_0, f\} \to y$ is obviously linear and is defined on all of $H \times L_2(I,V^*)$, its continuity follows from its boundedness [see (5.1.25)]. This completes the proof of the theorem. □

In the above results, we have shown that the mapping $\{y_0, f\} \to y$ is continuous from $H \times L_2(I,V^*)$ into $L_2(I,V)$. In fact, a stronger version of the continuity is true. This is shown in the following.

Corollary 5.1.1. *The mapping $\{y_0, f\} \to y$ is a continuous linear transformation from $H \times L_2(I,V^*)$ into X.*

PROOF. Since $f \in L_2(I,V^*)$ and, for $y \in L_2(I,V)$, $Ay \in L_2(I,V^*)$, we have $(d/dt)y \in L_2(I,V^*)$, and consequently, for each $\xi \in L_2(I,V)$ we have

$$\int_I \left\langle \frac{dy}{dt}, \xi \right\rangle dt = -\int_I \langle Ay, \xi \rangle\, dt + \int_I \langle f, \xi \rangle\, dt. \tag{5.1.26}$$

Using property (a1) [inequality (5.1.1)] and the estimate (5.1.25), we deduce from (5.1.26) that

$$\begin{aligned} \left| \int_I \left\langle \frac{dy}{dt}, \xi \right\rangle dt \right| &\leq \big(c\gamma |y_0|_H + (1+c\gamma)\|f\|_{L_2(I,V^*)}\big)\|\xi\|_{L_2(I,V)} \\ &\leq (1+c\gamma)\big(|y_0|_H + \|f\|_{L_2(I,V^*)}\big)\|\xi\|_{L_2(I,V)} \end{aligned}$$

for all $\xi \in L_2(I,V)$. This inequality implies that there exits a γ' such that

$$\left\| \frac{dy}{dt} \right\|_{L_2(I,V^*)} \leq \gamma'\big(|y_0|_H + \|f\|_{L_2(I,V^*)}\big). \tag{5.1.27}$$

Therefore, it follows from (5.1.25) and (5.1.27) that there exists a γ'' such that

$$\|y\|_X \leq \gamma''\big(|y_0|_H + \|f\|_{L_2(I,V^*)}\big). \tag{5.1.28}$$

Since linearity is obvious, this completes the proof. □

There is also some interest in the following result, which is stated without proof.

Corollary 5.1.2. *The mapping of* $\{y_0, f\} \to y(T)$ *is a continuous linear transformation from* $H \times L_2(I, V^*)$ *into* H.

5.1.4. *Formulation of Some Control Problems*

State Equation and Admissible Controls. Let E be a Hilbert space and suppose the control policies $u(t)$, $t \in I$, take their values in E. We assume that the control space is given by the Hilbert space $\mathcal{U} \equiv L_2(I, E)$ with the scalar product denoted by $(\cdot,\cdot)_{\mathcal{U}}$. The class of admissible controls is given by a subset $\mathcal{U}_a$ of $\mathcal{U}$. Suppose the control system is governed by the evolution equation

$$\frac{dx}{dt} + A(t)x = f + Bu, \qquad t \in (0, T) \equiv I,$$

$$x(0) = x_0, \tag{5.1.29}$$

where A satisfies properties (a1) and (a2) (Section 5.1.2), f is a free term belonging to $L_2(I, V^*)$, $B \in \mathcal{L}(\mathcal{U}, L_2(I, V^*))$, $x_0 \in H$, and $x(t)$, $t \in I$, is the state with values in H.

Output Equation. The state is not directly observable; it is measured and observed indirectly through the output y given by

$$y = Cx.$$

The output space is assumed to be given by another Hilbert space $\mathcal{H} \equiv L_2(I, F)$, where F itself is a Hilbert space and the output $y(t)$ takes its values in F. The scalar product in $\mathcal{H}$ will be denoted by $(\cdot,\cdot)_{\mathcal{H}}$.

Performance Measure. For convenience, we shall denote by $x(u)$ and $y(u)$ the state and the output, respectively, corresponding to the control policy $u \in \mathcal{U}$. The values of $x(u)$ $[y(u)]$ may be denoted by $x(u)(t)$ $[y(u)(t)]$, $t \in I$. Let y_d be the desired output and $N \in \mathcal{L}(\mathcal{U})$ a bounded self-adjoint positive operator in $\mathcal{U}$. We wish to find a control $u_0 \in \mathcal{U}_a$ that imparts a minimum to the cost functional

$$J(u) = \int_I \big((Cx(u))(t) - y_d(t), (Cx(u))(t) - y_d(t)\big)_F dt$$

$$+ \int_I (N(t)u(t), u(t))_E dt$$

$$= (Cx(u) - y_d, Cx(u) - y_d)_{\mathcal{H}} + (Nu, u)_{\mathcal{U}}, \tag{5.1.30}$$

where $C \in \mathcal{L}(X, \mathcal{H})$ and $y_d \in \mathcal{H}$.

Similarly, we wish to consider a terminal cost functional given by

$$J(u) = (Cx(u)(T) - y_d, Cx(u)(T) - y_d)_F + (Nu, u)_{\mathcal{U}}, \tag{5.1.31}$$

where the output operator C now belongs to $\mathcal{L}(H, F)$ and $y_d \in F$. We can also consider a combination of the two, or more general, cost functionals. In this section we limit our attention to the above two only and leave the more general cases until the final section of this chapter. Further, we may also note that if one wishes to construct an optimal feedback control, it is only the linear quadratic case that can be handled satisfactorily.

A control $u_0 \in \mathcal{U}_a$ at which J attains its minimum will be called an *optimal control.*

5.1.5. Existence of Optimal Controls

In this section, we wish to prove the existence of optimal controls for these problems.

Problem (5.1.P1). Minimize (5.1.30) subject to the dynamic constraint (5.1.29).

Problem (5.1.P2). Minimize (5.1.31) subject to the dynamic constraint (5.1.29). With this end in view, we present the following basic result.

Lemma 5.1.1. *Let $\mathcal{U}$ be a reflexive Banach space with $\mathcal{U}_a$ a closed convex subset of $\mathcal{U}$. Suppose J is a weakly lower semicontinuous functional on $\mathcal{U}_a$ bounded from below: $J(u) \geq -\eta > -\infty$ for all $u \in \mathcal{U}_a$, and $J(u) \to +\infty$ as $\|u\|_{\mathcal{U}} \to +\infty$. Then J attains its minimum in $\mathcal{U}_a$, and further, if J is strictly convex, then this minimum is attained at a unique point in $\mathcal{U}_a$.*

PROOF. Since J is bounded from below, it is clear that $\inf\{J(u), u \in \mathcal{U}_a\} = \gamma$ is defined and that $\gamma \geq -\eta$. Let $\{u_n\} \subset \mathcal{U}_a$ be a minimizing sequence such that $\lim J(u_n) = \gamma$. Since $-\eta \leq \gamma < \infty$ and $J(u) \to +\infty$ as $\|u\|_{\mathcal{U}} \to +\infty$, it is clear that the sequence $\{u_n\}$ is contained in a bounded subset of $\mathcal{U}$. Owing to the fact that a bounded subset of a reflexive Banach space is conditionally weakly sequentially compact (Theorem 1.1.7), we can extract a subsequence from the sequence $\{u_n\}$, again denoted by $\{u_n\}$, and find an element $u_0 \in \mathcal{U}$ such that $u_n \to u_0$ weakly. Since $\mathcal{U}_a$ is closed and convex, it follows from the Hahn–Banach (separation) theorem (e.g., Corollary 1.1.2) that $u_0 \in \mathcal{U}_a$. Thus, $J(u_0) \geq \gamma$. Further, since J is weakly lower semicontinuous, we have

$$J(u_0) \leq \underline{\lim}\, J(u_n) \leq \lim J(u_n) = \gamma.$$

From this we conclude that $J(u_0) = \gamma$. This proves that there exists a $u_0 \in \mathcal{U}_a$ at which the infimum is attained. The uniqueness is proved by contradiction. Suppose there exists another element u_0' at which J attains the minimum; that is, $J(u_0') = \gamma$. Then, since J is strictly convex,

$$J\left(\frac{u_0}{2} + \frac{u_0'}{2}\right) < \frac{1}{2}J(u_0) + \frac{1}{2}J(u_0') = \gamma,$$

and further, because of the convexity of $\mathcal{U}_a$, $\tilde{u}_0 \equiv \frac{1}{2} u_0 + \frac{1}{2} u'_0 \in \mathcal{U}_a$, implying $J(\tilde{u}_0) \geq \gamma$. This leads to a contradiction. Thus we have completed the proof. □

For the optimal control problem (5.1.P1) we have the following result. Let $\mathcal{L}_s(K)$ denote the class of bounded linear self-adjoint operators in the Hilbert space K.

Theorem 5.1.2. *Suppose that the operator A satisfies properties* (a1) *and* (a2), *$B \in \mathcal{L}(\mathcal{U}, L_2(I, V^*))$, $f \in L_2(I, V^*)$, $x_0 \in H$ and the class of admissible controls $\mathcal{U}_a$, containing the zero element, is a closed convex subset of $\mathcal{U} \equiv L_2(I, E)$, where E is a Hilbert space. Suppose that the output operator $C \in \mathcal{L}(X, \mathcal{H})$, $y_d \in \mathcal{H}$, $N \in \mathcal{L}_s(\mathcal{U})$, and there exists a constant $\beta > 0$ such that $(Nu, u)_{\mathcal{U}} \geq \beta \|u\|^2_{\mathcal{U}}$ for all $u \in \mathcal{U}_a$. Then the optimal control problem* (5.1.P1) *has a unique solution $u_0 \in \mathcal{U}_a$.*

PROOF. Clearly, $\mathcal{U}$ is a reflexive Banach space since E is Hilbert. Thus it is only required to prove that J satisfies the conditions of Lemma 5.1.1. Since $B \in \mathcal{L}(\mathcal{U}, L_2(I, V^*))$, it follows from Theorem 5.1.1 that for each control $u \in \mathcal{U}_a$, the evolution equation (5.1.29) has a unique solution $x(u) \in X$ and, further, $u \to x(u)$ is an affine continuous mapping from $\mathcal{U}$ into X. Let $x(0)$ denote the response corresponding to the control $u = 0$. Then the mapping $u \to (x(u) - x(0))$ is a continuous linear mapping from $\mathcal{U}$ into X. We can then write

$$J(u) = Q(u, u) + l(u) + \delta, \tag{5.1.32}$$

where

$$Q(u, v) = (C(x(u) - x(0)), C(x(v) - x(0)))_{\mathcal{H}} + (Nu, v)_{\mathcal{U}},$$

$$l(u) = 2(C(x(u) - x(0)), Cx(0) - y_d)_{\mathcal{H}},$$

and

$$\delta = \|Cx(0) - y_d\|^2_{\mathcal{H}}.$$

Clearly, $Q(u, u) \geq \beta \|u\|^2_{\mathcal{U}}$. Since $C \in \mathcal{L}(X, \mathcal{H})$ and $u \to x(u) - x(0)$ is continuous and linear, l is also a continuous linear functional on $\mathcal{U}$. Thus, there exists a constant $k > 0$ such that

$$J(u) \geq \beta \|u\|^2_{\mathcal{U}} - k \|u\|_{\mathcal{U}} + \delta, \tag{5.1.33}$$

and consequently, $J(u) \to +\infty$ as $\|u\|_{\mathcal{U}} \to \infty$ and there exists a constant $\eta > 0$ such that $J(u) > -\eta$ for all $u \in \mathcal{U}$. Further, J is weakly lower semicontinuous on $\mathcal{U}$. Indeed, if $u_n \to u_0$ weakly in $\mathcal{U}$, then, from the inequality

$$J(u_0) \leq J(u_n) + 2Q(u_0, u_0 - u_n) + l(u_0 - u_n) \tag{5.1.34}$$

and the fact that a linear continuous functional is necessarily weakly continuous, it follows that $\overline{\lim}(J(u_0)-J(u_n))\leq 0$, and consequently, $J(u_0)\leq \underline{\lim} J(u_n)$. Thus, J is weakly lower semicontinuous, since it is quadratic with $N>0$, it is also strictly convex. Therefore, the cost functional J and the admissible controls $\mathfrak{U}_{\mathrm{a}}$ satisfy all the conditions of Lemma 5.1.1. Consequently, there exists a unique optimal control. This completes the proof of the theorem. □

For the terminal control problem (5.1.P2), we have the following result.

Theorem 5.1.3. *Consider Problem* (5.1.P2) *and suppose A, B, f, x_0, N, and $\mathfrak{U}_{\mathrm{a}}$ satisfy the assumptions of Theorem* 5.1.2. *Suppose that y_{d} is a given element of the Hilbert space F and $C\in\mathfrak{L}(H,F)$. Then the control problem* (5.1.P2) *has a unique solution $u_0\in\mathfrak{U}_{\mathrm{a}}$.*

PROOF. The proof is similar to that of Theorem 5.1.2. We only note that, since the mappings $u\to x(u)$, $x(u)\to x(u)(t)$, $t\in I$, are continuous and $C\in\mathfrak{L}(H,F)$, the composition mapping $u\to C(x(u)(T)-x(0)(T))$ is a continuous linear mapping from $\mathfrak{U}$ into F. Thus, J is weakly lower semicontinuous and satisfies an inequality similar to (5.1.33).

In the preceding results the operator N was assumed to be strictly positive. This hypothesis can be removed by introducing the assumption that $\mathfrak{U}_{\mathrm{a}}$ is also bounded in addition to being closed and convex.

Theorem 5.1.4. *Suppose the assumptions of Theorems* 5.1.2–5.1.3 *hold except that $(Nu,u)_{\mathfrak{U}}\geq 0$, $u\in\mathfrak{U}_{\mathrm{a}}$. If $\mathfrak{U}_{\mathrm{a}}$ is a bounded, closed, convex subset of $\mathfrak{U}$, then the set*

$$\mathfrak{U}_0\equiv\left\{u\in\mathfrak{U}_{\mathrm{a}}: J(u)=\inf_{w\in\mathfrak{U}_{\mathrm{a}}} J(w)\right\}$$

is a nonempty closed convex subset of $\mathfrak{U}_{\mathrm{a}}$.

PROOF. Since J is weakly lower semicontinuous and $\mathfrak{U}_{\mathrm{a}}$ is a weakly compact subset of the Hilbert space $\mathfrak{U}$, it follows as in Lemma 5.1.1 that $\mathfrak{U}_0$ is a nonempty set. For the closure, let $\{u_n\}$ be a sequence from $\mathfrak{U}_0$, and suppose that $u_n\to u_0$ weakly. Let $\gamma=\inf\{J(w), w\in\mathfrak{U}_{\mathrm{a}}\}$; then $J(u_n)=\gamma$ for all n, and, since J is weakly lower semicontinuous, $J(u_0)\leq \underline{lim}_n J(u_n)=\gamma$. Further, $u_0\in\mathfrak{U}_{\mathrm{a}}$, and consequently, $J(u_0)\geq\gamma$. Thus, $u_0\in\overline{\mathfrak{U}}_0$, proving the closure. For the proof of convexity, let $u_1,u_2\in\mathfrak{U}_0$; then $\alpha u_1+(1-\alpha)u_2\in\mathfrak{U}_{\mathrm{a}}$ for $0\leq\alpha\leq 1$ with $J(\alpha u_1+(1-\alpha)u_2)\geq\gamma$. Due to convexity of the functional J, we also have $J(\alpha u_1+(1-\alpha)u_2)\leq\gamma$. Thus, $\alpha u_1+(1-\alpha)u_2\in\mathfrak{U}_0$ for $0\leq\alpha\leq 1$ whenever $u_1,u_2\in\mathfrak{U}_0$. This completes the proof. □

REMARK 5.1.1. An interesting case in which the optimal control is unique even though the set $\mathfrak{U}_{\mathrm{a}}$ is unbounded and $N\equiv 0$ is provided by the following example:

$$\frac{dx}{dt}+A(t)x=f+u,$$
$$x(u)(0)=x_0\in H, \tag{5.1.35}$$
$$J(u)=\|x(u)-x_1\|^2_{L_2(I,V)}+\|\frac{d}{dt}x(u)-x_2\|^2_{L_2(I,V^*)}=\min,$$

where $x_1\in L_2(I,V)$, $x_2\in L_2(I,V^*)$, $\mathfrak{U}\equiv L_2(I,V^*)$, and $\mathfrak{U}_a$ is a closed convex subset of $\mathfrak{U}$.

In this case, we define an operator τ by setting $\tau z=[d/dt+A]z$ with domain $D(\tau)\equiv X_0=\{z: z\in L_2(I,V),\ dz/dt\in L_2(I,V^*),\ z|_{t=0}=0\}$. The set X_0 endowed with the norm

$$\|z\|_{X_0}=\left(\|z\|^2_{L_2(I,V)}+\left\|\frac{dz}{dt}\right\|^2_{L_2(I,V^*)}\right)^{1/2}$$

is a Hilbert space. By Theorem 5.1.1, the Cauchy problem

$$\tau z=u, \qquad z(0)=0$$

has a unique solution $z\in X_0$ for each given $u\in\mathfrak{U}$, and for each $z\in X_0$, $\tau z\in L_2(I,V^*)$. Thus, τ is an isomorphism of X_0 onto $L_2(I,V^*)$ and it has an inverse $L\in\mathfrak{L}(L_2(I,V^*),X_0)$. Owing to property (a1), there exists a finite positive number k such that $\|\tau z\|_{L_2(I,V^*)}\leq k\|z\|_{X_0}$ for all $z\in X_0$. Therefore, for each $u\in\mathfrak{U}$,

$$\|Lu\|_{X_0}\geq(1/k)\|u\|_{L_2(I,V^*)}. \tag{5.1.36}$$

We can write

$$J(u)=\|x(u)-x_1\|^2_{L_2(I,V)}+\|\frac{d}{dt}x(u)-x_2\|^2_{L_2(I,V^*)}$$
$$=\|x(u)-x(0)\|^2_{L_2(I,V)}+\|\frac{d}{dt}(x(u)-x(0))\|^2_{L_2(I,V^*)}+l(u)+c, \tag{5.1.37}$$

where $x(u)$ and $x(0)$ are the solutions of the evolution equation of the problem (5.1.35) corresponding to the controls u and 0, respectively, l is a bounded linear functional on $\mathfrak{U}$, and c is a constant. Clearly, $\tau(x(u)-x(0))=u$ and $(x(u)-x(0))=Lu$. Therefore, it follows from (5.1.36) and (5.1.37) that there exists a k', $0<k'<\infty$, such that

$$J(u)=\|Lu\|^2_{X_0}+l(u)+c\geq k'\|u\|^2_{L_2(I,V^*)}+l(u)+c. \tag{5.1.38}$$

Thus, J is coercive and, being quadratic, is strictly convex. Consequently, by Lemma 5.1.1. the problem (5.1.35) has a unique optimal control.

5.1.6. Necessary Conditions for Optimality

In this section we develop the necessary and sufficient conditions for optimality for several control problems including Problems (5.1.P1) and (5.1.P2).

First we consider Problem (5.1.P1). We assume that the input and output spaces are again given by the Hilbert spaces $\mathcal{U} \equiv L_2(I, E)$ and $\mathcal{H} \equiv L_2(I, F)$, respectively. Let $\mathcal{U}^*$ and $\mathcal{H}^*$ denote the corresponding adjoint spaces. Let $\Lambda_{\mathcal{U}}(\Lambda_{\mathcal{H}})$ denote the canonical isomorphism of $\mathcal{U}(\mathcal{H})$ onto $\mathcal{U}^*(\mathcal{H}^*)$ such that $\|\Lambda_{\mathcal{U}} u\|_{\mathcal{U}^*} = \|u\|_{\mathcal{U}}$ for $u \in \mathcal{U}$ and $\|\Lambda_{\mathcal{H}} x\|_{\mathcal{H}^*} = \|x\|_{\mathcal{H}}$ for $x \in \mathcal{H}$.

Theorem 5.1.5. *Consider Problem* (5.1.P1) *consisting of the evolution equation* (5.1.29) *and the cost function* (5.1.30). *Suppose that the operator A satisfies the properties* (a1) *and* (a2), $f \in L_2(I, V^*)$, $B \in \mathcal{L}(\mathcal{U}, L_2(I, V^*))$, $x_0 \in H$, $C \in \mathcal{L}(L_2(I, V), \mathcal{H})$, $y_d \in \mathcal{H}$, *and* $N \in \mathcal{L}(\mathcal{U})$ *with* $N = N^*$ *and* $(Nu, u)_{\mathcal{U}} \geq \beta \|u\|^2_{\mathcal{U}}$ *for* $\beta > 0$. *Let $\mathcal{U}_a$ be a closed convex subset of $\mathcal{U}$. Then, in order that $u_0 \in \mathcal{U}_a$ be an optimal control, it is necessary and sufficient that there exist a $\psi_0 \in X$ such that*

$$\left(\Lambda_{\mathcal{U}}^{-1} B^* \psi_0 + N u_0, w - u_0\right)_{\mathcal{U}} \geq 0 \quad \textit{for all } w \in \mathcal{U}_a. \tag{5.1.39}$$

PROOF. Since $B \in \mathcal{L}(\mathcal{U}, L_2(I, V^*))$, the mapping $u \to x(u)$ is affine continuous (Corollary 5.1.1). Thus, J is quadratic in u and convex and, further, also Gateaux differentiable at every point of $\mathcal{U}_a$. The Gateaux differential of J at $u_0 \in \mathcal{U}_a$ in the direction $(w - u_0)$, $w \in \mathcal{U}_a$, is given by

$$J'(u_0, w - u_0) = 2(C(x(w) - x(u_0)), Cx(u_0) - y_d)_{\mathcal{H}} + 2(Nu_0, w - u_0)_{\mathcal{U}}. \tag{5.1.40}$$

Under the hypotheses of the theorem, Problem (5.1.P1) has a unique solution $u_0 \in \mathcal{U}_a$ (Theorem 5.1.2). That is, u_0 is the unique control that solves Problem (5.1.P1). On the other hand, it follows from Theorem 1.1.23 that for the element $u_0 \in \mathcal{U}_a$ to be optimal, it is necessary and sufficient that $J'(u_0, w - u_0) \geq 0$ for all $w \in \mathcal{U}_a$. Therefore,

$$(C(x(w) - x(u_0)), Cx(u_0) - y_d)_{\mathcal{H}} + (Nu_0, w - u_0)_{\mathcal{U}} \geq 0 \tag{5.1.41}$$

for all $w \in \mathcal{U}_a$. Denoting $x(w) - x(u_0)$ by $\hat{x}(w - u_0)$, we observe that $\hat{x}(w - u_0)$ is the unique (weak) solution of the evolution equation

$$\frac{d}{dt}\hat{x}(w - u_0) + A(t)\hat{x}(w - u_0) = B(w - u_0),$$

$$\hat{x}(w - u_0)(0) = 0. \tag{5.1.42}$$

Let Y denote the space $L_2(I, V)$ and Y^* its adjoint, which is $L_2(I, V^*)$.

Using the canonical isomorphism $\Lambda_{\mathcal{H}}$, we can rewrite (5.1.41) as

$$(\hat{x}(w-u_0), C^*\Lambda_{\mathcal{H}}(Cx(u_0)-y_d))_{Y-Y^*}+(w-u_0, Nu_0)_{\mathcal{U}}\geq 0, \qquad w\in\mathcal{U}_a, \tag{5.1.43}$$

where C^* is the adjoint of the operator C and belongs to $\mathcal{L}(\mathcal{H}^*, L_2(I, V^*))$. The inequality (5.1.43) can be simplified further by introducing the so-called Lagrange multiplier. Toward this end, we introduce the following adjoint problem:

$$-\frac{d}{dt}\psi+A^*(t)\psi=C^*\Lambda_{\mathcal{H}}(Cx(u_0)-y_d),$$
$$\psi(T)=0. \tag{5.1.44}$$

Since $C^*\Lambda_{\mathcal{H}}(Cx(u_0)-y_d)\in Y^*$ and A^* has exactly the same properties as those of A, this problem, with the reversal of the flow of time $(t\to T-t)$, is entirely equivalent to the problem (5.1.3). Thus, by virtue of Theorem 5.1.1, the problem (5.1.44) has a unique solution in X, which we denote by ψ_0. Since $X\subset C^0(\bar{I}, H)$, we can integrate by parts the duality pairing between ψ_0 and the first equality of (5.1.42) to obtain

$$\left(\hat{x}(w-u_0), -\frac{d\psi_0}{dt}+A^*\psi_0\right)_{Y-Y^*}=(\psi_0, B(w-u_0))_{Y-Y^*}. \tag{5.1.45}$$

Using (5.1.44) and (5.1.45) in the inequality (5.1.43), we obtain the inequality

$$(\psi_0, B(w-u_0))_{Y-Y^*}+(Nu, w-u_0)_{\mathcal{U}}\geq 0, \tag{5.1.46}$$

which holds true for all $w\in\mathcal{U}_a$. Since $B^*\in\mathcal{L}(Y, \mathcal{U}^*)$, the inequality (5.1.46) can be rewritten as

$$(B^*\psi_0, w-u_0)_{\mathcal{U}^*-\mathcal{U}}+(Nu_0, w-u_0)_{\mathcal{U}}\geq 0, \qquad w\in\mathcal{U}_a. \tag{5.1.47}$$

Using now the canonical mapping $\Lambda_{\mathcal{U}}$, the above inequality reduces to

$$(\Lambda_{\mathcal{U}}^{-1}B^*\psi_0, w-u_0)_{\mathcal{U}}+(Nu_0, w-u_0)_{\mathcal{U}}$$
$$=(\Lambda_{\mathcal{U}}^{-1}B^*\psi_0+Nu_0, w-u_0)_{\mathcal{U}}\geq 0 \quad \text{for all } w\in\mathcal{U}_a. \tag{5.1.48}$$

This completes the proof of the theorem. □

REMARK 5.1.2. The inequality (5.1.48) is equivalent to the following inequality:

$$(B^*\psi_0+\Lambda_{\mathcal{U}}Nu_0, w-u_0)_{\mathcal{U}^*-\mathcal{U}}\geq 0, \qquad w\in\mathcal{U}_a.$$

REMARK 5.1.3. If the control constraint is removed by choosing for the admissible class $\mathcal{U}_a$ the whole space $\mathcal{U}$, then the necessary condition given by the inequality (5.1.39) reduces to

$$u_0 = -N^{-1}\Lambda_{\mathfrak{U}}^{-1}B^*\psi_0. \tag{5.1.49}$$

We note that, according to the inequality (5.1.39), u_0 depends on the adjoint state ψ_0 and, due to (5.1.44), ψ_0 depends on the state x, which, in turn, depends on the control u_0. This circular dependence of the variables $\{u, x, \psi\}$ requires that the inequality (5.1.39) and the state and the costate equations (5.1.29) and (5.1.44) be solved simultaneously to determine the optimal control. Later in this section we discuss an iterative technique that bypasses this difficulty.

Next we consider a pointwise necessary condition for optimality. Let E_a be a closed convex subset of the Hilbert space E and $\mathfrak{U}_a \equiv \{u \in L_2(I, E): u(t) \in E_a \text{ a.e.}\}$ the class of admissible controls. Let Λ_E denote the canonical isomorphism of E onto E^* such that $|\Lambda_E e|_{E^*} = |e|_E$. Let B_1, B_2 be arbitrary Banach spaces, and let $L_\infty(I, \mathcal{L}(B_1, B_2))$ denote the class of essentially bounded strongly measurable functions on I with values in the space of bounded linear operators from B_1 into B_2. We assume that this space is equipped with the topology

$$|||K||| = \operatorname*{ess\,sup}_{t \in I} \|K(t)\|_{\mathcal{L}(B_1, B_2)}.$$

If $B_1 = B_2 = B$, then we shall denote this space simply by $L_\infty(I, \mathcal{L}(B))$.

Corollary 5.1.3. *Suppose that the hypotheses of Theorem 5.1.5 hold and that $\mathfrak{U}_a \subset L_2(I, E)$, as defined above, is the class of admissible controls. Let $N \in L_\infty(I, \mathcal{L}(E))$ and $B \in L_\infty(I, \mathcal{L}(E, V^*))$. Then, in order that $u_0 \in \mathfrak{U}_a$ be an optimal control, it is necessary and sufficient that there exist a $\psi_0 \in X$ such that*

$$\left(\Lambda_E^{-1}B^*(t)\psi_0(t) + N(t)u_0(t), v - u_0(t)\right)_E \geq 0 \tag{5.1.50}$$

for all $v \in E_a$ and for almost all $t \in I$.

PROOF. It suffices to demonstrate the inequality (5.1.50). Let σ be any measurable subset of I containing the point t and contracting to the one-point set $\{t\}$, where t is an arbitrary point of I. Let $v \in E_a$ be arbitrary and define

$$\tilde{u}(\theta) = \begin{cases} v & \text{for } \theta \in \sigma \\ u_0(\theta) & \text{for } \theta \in I \setminus \sigma \end{cases}. \tag{5.1.51}$$

Clearly $\tilde{u} \in \mathfrak{U}_a$. Under the given hypotheses of the corollary, we can rewrite the inequality (5.1.39) in the equivalent form

$$\int_I \left(\Lambda_E^{-1}B^*(\theta)\psi_0(\theta) + N(\theta)u_0(\theta), w(\theta) - u_0(\theta)\right)_E d\theta \geq 0, \tag{5.1.52}$$

which holds for all $w \in \mathfrak{U}_a$. Substituting $\tilde{u}$ for w in (5.1.52) and dividing the resulting expression by $\mu(\sigma)$, the Lebesgue measure of the set σ, we obtain

$$\frac{1}{\mu(\sigma)}\int_{\sigma}\left(\Lambda_E^{-1}B^*(\theta)\psi_0(\theta)+N(\theta)u_0(\theta),v-u_0(\theta)\right)_E d\theta\geq 0. \tag{5.1.53}$$

Since all the elements inside the scalar product are strongly measurable, the scalar product is Lebesgue measurable, and further, due to our assumptions on the operators B and N, it is also Lebesgue integrable. Since almost all points of I are Lebesgue density points with respect to any measurable function on I, we obtain the inequality (5.1.50) by letting $\mu(\sigma)\to 0$. □

We now consider the terminal cost problem (5.1.P2) consisting of the evolution equation (5.1.29) and the cost functional (5.1.31). The output space is the Hilbert space F, y_d is a desired element of F, and C is the output operator that belongs to $\mathcal{L}(H, F)$. Note C could also be an element of $\mathcal{L}(L_2(I,V), F)$.

Theorem 5.1.6. *Consider Problem* (5.1.P2) *which consists of the evolution equation* (5.1.29) *and the cost function* (5.1.31). *Suppose that the operator A satisfies the properties* (a1) *and* (a2), $f\in L_2(I,V^*)$, $B\in\mathcal{L}(\mathcal{U}, L_2(I,V^*))$, $x_0\in H$, $C\in\mathcal{L}(H,F)$, $y_d\in F$, *and* $N\in\mathcal{L}(\mathcal{U})$ *with* $N=N^*$ *and* $(Nu,u)_{\mathcal{U}}\geq\beta\|u\|^2_{\mathcal{U}}$, $\beta>0$. *Let $\mathcal{U}_a$ be a closed convex subset of $\mathcal{U}$. Then, in order that $u_0\in\mathcal{U}_a$ be an optimal control, it is necessary and sufficient that there exist a $\psi_0\in X$ such that*

$$\left(\Lambda_{\mathcal{U}}^{-1}B^*\psi_0+Nu_0,w-u_0\right)_{\mathcal{U}}\geq 0 \quad \textit{for all } w\in\mathcal{U}_a. \tag{5.1.54}$$

PROOF. The arguments are similar to those in the proof of Theorem 5.1.5. We only give an outline of the proof. By Corollary 5.1.2, the mapping $u\to x(u)(T)$ is affine continuous since $B\in\mathcal{L}(\mathcal{U}, L_2(I,V^*))$. Thus, the mapping $u\to(x(u_0+u)-x(u_0))(T)$ is linear continuous from $\mathcal{U}$ into H. Therefore, J has Gateaux differential at each point $u_0\in\mathcal{U}_a$ and in an arbitrary direction $(w-u_0)$, $w\in\mathcal{U}_a$. This differential is given by

$$J'(u_0,w-u_0)=2(C\hat{x}(w-u_0)(T),Cx(u_0)(T)-y_d)_F+2(Nu_0,w-u_0)_{\mathcal{U}}.$$

Let Λ_F denote the canonical isomorphism of F onto F^*. Then, in order that u_0 be the optimal control (existence assured by Theorem 5.1.2), it is necessary and sufficient that

$$(\hat{x}(w-u_0)(T),C^*\Lambda_F(Cx(u_0)(T)-y_d))_H+(w-u_0,Nu_0)_{\mathcal{U}}\geq 0 \tag{5.1.55}$$

for all $w\in\mathcal{U}_a$.

At this point, we introduce the adjoint equation

$$-\frac{d\psi}{dt}+A^*(t)\psi=0,$$

$$\psi(T)=C^*\Lambda_F(Cx(u_0)(T)-y_d). \tag{5.1.56}$$

Since $C^*\Lambda_F(Cx(u_0)(T)-y_d)\in H$, by Theorem 5.1.1, this equation has a unique solution $\psi_0\in X$. Scalar multiplying the first equation of (5.1.56) by $\hat{x}(w-u_0)$, with ψ replaced by ψ_0, and integrating by parts, we obtain

$$(\hat{x}(w-u_0)(T),C^*\Lambda_F(Cx(u_0)(T)-y_d))_H$$

$$=\int_0^T\langle B(w-u_0),\psi_0\rangle_{V^*-V}dt. \tag{5.1.57}$$

The inequality (5.1.54) now follows from (5.1.55) and (5.1.57). □

REMARK 5.1.4. We note that for the terminal control problem the optimality system consists of the state equation (5.1.29), the adjoint equation (5.1.56), and the inequality (5.1.54).

REMARK 5.1.5. It is clear from Theorems 5.1.5 and 5.1.6 that a combination of the terminal and distributed cost functions can also be treated similarly. For a cost function

$$J(u)=\alpha_1(C_1x(u)-y_1,C_1x(u)-y_1)_{\mathcal{H}}+\alpha_2(C_2x(u)(T)-y_2,$$

$$C_2x(u)(T)-y_2)_F+(Nu,u)_{\mathcal{U}},\qquad \alpha_1,\alpha_2\geq 0,$$

the optimal control is characterized by the inequality (5.1.54), the state equation (5.1.29), and the adjoint equation

$$-\frac{d\psi}{dt}+A^*(t)\psi=\alpha_1C_1^*\Lambda_{\mathcal{H}}(C_1x(u_0)-y_1),$$

$$\psi(T)=\alpha_2C_2^*\Lambda_F(C_2x(u_0)(T)-y_2), \tag{5.1.58}$$

where $C_1\in\mathcal{L}(L_2(I,V),\mathcal{H})$, $C_2\in\mathcal{L}(H,F)$, $y_1\in\mathcal{H}$, and $y_2\in F$.

Another interesting problem arises when one wishes to monitor the output at a set of discrete instants of time in addition to the terminal time and adopt a control policy that optimizes the performance over all the instants of observation. This problem can be formulated as follows.

Suppose that the system is again described by the evolution equation

$$\frac{dx}{dt}+A(t)x=f+Bu,$$

$$x(0)=x_0 \tag{5.1.59}$$

and that the cost functional is given by

$$J(u)=\sum_{i=1}^{n}|C_ix(u)(t_i)-y_i|_F^2+(Nu,u)_{\mathcal{U}}, \tag{5.1.60}$$

where $t_i \in I$, $i=1,2,\ldots,n$, are the instants of observation with $0<t_1<t_2\cdots<t_{n-1}<t_n=T$; $C_i \in \mathcal{L}(H,F)$, $i=1,2,\ldots,n$, are the output operators and $y_i \in F$, $i=1,2,\ldots,n$, are the desired outputs. As usual, the problem is to find a control u_0 that minimizes the cost functional. For this problem, we have the following result.

Theorem 5.1.7. *Consider the evolution equation* (5.1.59) *with A satisfying the properties* (a1) *and* (a2), *$f \in L_2(I,V^*)$, $B \in \mathcal{L}(\mathcal{U}, L_2(I,V^*))$, and $x_0 \in H$. For the cost functional* (5.1.60), *suppose that $C_i \in \mathcal{L}(H,F)$, $i=1,2,\ldots,n$, $y_i \in F$, $i=1,2,\ldots,n$, $N=N^* \in \mathcal{L}(\mathcal{U})$ with $N \geq \beta I$, $\beta>0$, and $\mathcal{U}_a$ is a closed convex subset of $\mathcal{U}$. Then, the optimality system consists of the evolution equation* (5.1.59) *with u_0 replacing u, the inequality*

$$\left(\Lambda_{\mathcal{U}}^{-1}B^*\psi_0+Nu_0, w-u_0\right)_{\mathcal{U}} \geq 0 \quad \textit{for all } w \in \mathcal{U}_a, \tag{5.1.61}$$

and the adjoint equation

$$\begin{aligned} -\frac{d}{dt}\psi_0+A^*(t)\psi_0 &= \sum_{i=1}^{n-1} C_i^*\Lambda_F(C_i x(u_0)(t)-y_i)\delta(t-t_i), \\ \psi_0(T)=\psi_0(t_n) &= C_n^*\Lambda_F(C_n x(u_0)(t_n)-y_n), \end{aligned} \tag{5.1.62}$$

where Λ_F is the canonical isomorphism of F onto F^*.

PROOF. As in the preceding results, it is easily verified that the necessary and sufficient conditions for optimality are given by

$$\sum_{i=1}^{n} \left(\hat{x}(w-u_0)(t_i), C_i^*\Lambda_F(C_i x(u_0)(t_i)-y_i)\right)_H + (w-u_0, Nu_0)_{\mathcal{U}} \geq 0 \tag{5.1.63}$$

for all $w \in \mathcal{U}_a$, where u_0 is the optimal control and $\hat{x}(w-u_0)$ is the (unique) solution of the equation

$$\begin{aligned} \frac{d\xi}{dt}+A(t)\xi &= B(w-u_0), \\ \xi(0) &= 0. \end{aligned} \tag{5.1.64}$$

At this point, we introduce the adjoint equation

$$\begin{aligned} -\frac{d\psi}{dt}+A^*(t)\psi &= \sum_{i=1}^{n-1} C_i^*\Lambda_F(C_i x(u_0)(t)-y_i)\delta(t-t_i) \\ \psi(T) &= C_n^*\Lambda_F(C_n x(u_0)(T)-y_n). \end{aligned} \tag{5.1.65}$$

Define $\tau\xi=[(d/dt)+A]\xi$ and

$$X_0=\left\{\xi : \xi \in L_2(I,V), \frac{d\xi}{dt} \in L_2(I,V^*), \xi(0)=0\right\}.$$

The set X_0 endowed with the norm

$$\|\xi\|_{X_0}=\left(\|\xi\|^2_{L_2(I,V)}+\|\dot{\xi}\|^2_{L_2(I,V^*)}\right)^{1/2}$$

is a Hilbert space, and τ is an isomorphism of X_0 onto $L_2(I,V^*)$. We can transpose this isomorphism to solve the problem (5.1.65). Define

$$l_0(\xi)=\sum_{i=1}^{n}\left(C_i^*\Lambda_F(C_ix(u_0)(t_i)-y_i),\xi(t_i)\right)_H \tag{5.1.66}$$

and note that, as $C_i^*\Lambda_F(C_ix(u_0)(t_i)-y_i)\in H$, there exists a constant $\gamma>0$ such that $|l_0(\xi)|\leq\gamma\|\xi\|_{C^0(I,H)}$. Further, since $X_0\subset C^0(I,H)$ with continuous injection (Theorem 1.2.15), there exists a $\gamma'>0$ such that $|l_0(\xi)|\leq\gamma'\|\xi\|_{X_0}$. Thus l_0 is a continuous linear functional on X_0. In order to prove the existence of a solution of the problem (5.1.65) we first show that there exists a (unique) $\psi_0\in L_2(I,V)$ such that

$$\int_0^T\langle\psi_0,\tau\xi\rangle_{V-V^*}\,dt=l_0(\xi)\quad\text{for all }\xi\in X_0. \tag{5.1.67}$$

By Corollary 5.1.1, the mapping $h\to\xi(h)$ is a continuous linear transformation from $L_2(I,V^*)$ into X_0. In other words $\tau^{-1}\in\mathcal{L}(L_2(I,V^*),X_0)$, and consequently, the mapping $y\to(l_0\tau^{-1})(y)$ is a continuous linear functional on $L_2(I,V^*)$. Therefore, we can write (5.1.67) in the following equivalent form:

$$\int_0^T\langle\psi_0,y\rangle_{V-V^*}\,dt=(l_0\tau^{-1})(y)\quad\text{for all }y\in L_2(I,V^*). \tag{5.1.68}$$

Recalling that V is a reflexive Banach space ($V=V^{**}$), we conclude from the Riesz–Phillips representation theorem that there exists a $\nu\in(L_2(I,V^*))^*=L_2(I,V)$ such that

$$(l_0\tau^{-1})(y)=\int_0^T\langle\nu,y\rangle_{V-V^*}\,dt. \tag{5.1.69}$$

Thus, we can choose $\psi_0=\nu$, thereby proving existence. Uniqueness follows from the fact that τ is an isomorphism of X_0 onto $L_2(I,V^*)$. We now show that ψ_0 is a solution of the problem (5.1.65) in the sense of distributions. Let $\mathfrak{D}(I,V)$ denote the space of C^∞-functions on I with values in V and having compact support on I. Since $\mathfrak{D}(I,V)\subset X_0$, it follows from (5.1.67), for $\xi\in\mathfrak{D}(I,V)$, that ψ_0 satisfies the first equation of (5.1.65) with the derivatives taken in the sense of distributions; that is, $(d/dt)\psi_0\in\mathfrak{D}'(I,V)$. Then, scalar multiplying in $L_2(I,V)-L_2(I,V^*)$ the first equation of (5.1.65) by an arbitrary $\xi\in X_0$, and equating like terms with those of (5.1.67), we obtain the second equality of (5.1.65). Since the solution $\hat{x}(w-u_0)$ of Equation (5.1.64) belongs to X_0, we can substitute $\hat{x}$ for ξ in (5.1.67) and obtain

$$\int_0^T\langle\psi_0,B(w-u_0)\rangle\,dt=\sum_{i=1}^{n}\left(C_i^*\Lambda_F(C_ix(u_0)(t_i)-y_i),\hat{x}(w-u_0)(t_i)\right)_H. \tag{5.1.70}$$

Using this equality in (5.1.63) and making use of the canonical map $\Lambda_{\mathcal{U}}$, we obtain the necessary condition (5.1.61). □

It is interesting to observe that in all the problems considered above, the necessary condition for optimality is given by one and the same inequality [see (5.1.39), (5.1.50), (5.1.54), and (5.1.61)] independently of the class of observation. The difference appears in the adjoint equation, where, for distributed observation, this equation has a nonzero forcing term and zero final state, and for terminal observation, the adjoint equation has a zero forcing term and nonzero final condition. A combination of both, as in (5.1.58) and (5.1.62), appears when both distributed and terminal observations are made.

REMARK 5.1.6. If we consider the class of admissible controls $\mathcal{U}_{\mathrm{a}}$ to be given by the ball $S_\beta = \{u \in \mathcal{U}: \|u\|_{\mathcal{U}} \le \beta\}$, then the optimal control is characterized by the nonlinear functional equation

$$u_0 = -\beta \frac{\Gamma(u_0)}{\|\Gamma(u_0)\|_{\mathcal{U}}}, \tag{5.1.71}$$

where $\Gamma(u) = Nu + \Lambda_{\mathcal{U}}^{-1} B^* \psi(u)$ and $\psi(u)$ is the adjoint state corresponding to the state $x(u)$. Indeed, it follows from the necessary condition for optimality that

$$\inf_{v \in S_\beta} (\Gamma(u_0), v)_{\mathcal{U}} = (\Gamma(u_0), u_0)_{\mathcal{U}},$$

and consequently, the functional $l(v) = (\Gamma(u_0), v)_{\mathcal{U}}$, being linear, attains its minimum on the boundary of S_β as given by (5.1.71). For the pointwise necessary condition (5.1.50), a similar result holds. Indeed, let $\mathcal{U}_{\mathrm{a}} = \{u \in \mathcal{U}: u(t) \in E_{\mathrm{a}} \text{ a.e.}\}$, where $E_{\mathrm{a}} \equiv \{w \in E: |w|_E \le \beta\}$, and define $K(u_0)(t) = \Lambda_E^{-1} B^*(t) \psi(u_0)(t) + N(t) u_0(t)$. Then it follows from the necessary condition (5.1.50) that

$$u_0(t) = -\beta \frac{K(u_0)(t)}{|K(u_0)(t)|_E} \quad \text{for all } t \in (0, T) \tag{5.1.72}$$

for which $K(u_0)(t) \neq 0$ and otherwise

$$u_0(t) = -N^{-1}(t) \Lambda_E^{-1} B^*(t) \psi(u_0)(t). \tag{5.1.73}$$

5.1.7. Optimal Feedback Control

It has been known for a long time that the optimal control for linear finite-dimensional systems with quadratic cost function can be realized by a linear or affine feedback control law. The key equation that must be solved to synthesize the feedback controller is known as the matrix Riccati (differential) equation. Under appropriate technical assumptions, similar results hold for infinite-dimensional problems, and in this case one is required to solve an operator Riccati equation. This is usually a differential equation on the Banach algebra of linear operators. The infinite-dimensional problem has been extensively studied by Lions [Li.2], Curtain and Pritchard [CP.1], and others [Ah.7]. It is important to mention that a key assumption in these developments is that complete information about the current state is available to the controller. Without this assumption the problem becomes very difficult, and with this assumption the solution has only limited application. In this section, we wish to present some of the results mentioned above.

Let X, Y be a pair of separable Banach spaces with duals X^*, Y^*, and $\mathcal{L}(X,Y)$ the Banach space of bounded linear operators from X into Y. Let $L_\infty(I, \mathcal{L}(X,Y))$ denote the class of operator-valued functions on I with values in $\mathcal{L}(X,Y)$ such that for $B \in L_\infty(I, \mathcal{L}(X,Y))$ the function $t \to \langle B(t)x, y^*\rangle_{Y-Y^*}$ is measurable for each $x \in X$ and $y^* \in Y^*$ and $\operatorname{ess\,sup}_{t\in I} \|B(t)\|_{\mathcal{L}(X,Y)} < \infty$.

For convenience of discussion, we focus our attention on Problem (5.1.P1) and the corresponding necessary conditions for optimality given by Theorem 5.1.5 and Corollary 5.1.3. Let E and F denote the input and output spaces as usual, and suppose that both are separable Hilbert spaces with the canonical maps Λ_E and Λ_F, respectively. We assume throughout this section, unless mentioned otherwise, that $\mathcal{U} = L_2(I, E)$, $N \in L_\infty(I, \mathcal{L}(E))$, $B \in L_\infty(I, \mathcal{L}(E, V^*))$, $C \in L_\infty(I, \mathcal{L}(V, F))$, and $y_d \in L_2(I, F)$. By removing the constraints on the control space, that is, choosing $\mathcal{U}_a = \mathcal{U} = L_2(I, E)$, it is easy to verify that the necessary condition for optimality (5.1.39) (Theorem 5.1.6) or (5.1.50) (Corollary 5.1.3) reduces to

$$u_0(t) = -N^{-1}(t)\Lambda_E^{-1}B^*(t)\psi_0(t), \qquad t \in I = (0, T). \tag{5.1.74}$$

The adjoint equation (5.1.44) can be rewritten as

$$\begin{gathered} -\frac{d}{dt}\psi_0 + A^*(t)\psi_0 = C^*(t)\Lambda_F\big(C(t)x^0(t) - y_d(t)\big), \\ \psi_0(T) = 0, \end{gathered} \tag{5.1.75}$$

where $x^0 = x(u_0)$ is the response of the system (5.1.29) corresponding to the control policy u_0. Substituting (5.1.74) into Equation (5.1.29), we obtain

$$\begin{gathered} \frac{d}{dt}x^0 + A(t)x^0 = f(t) - \big(B(t)N^{-1}(t)\Lambda_E^{-1}B^*(t)\big)\psi_0(t), \\ x^0(0) = x_0. \end{gathered} \tag{5.1.76}$$

Defining

$$\begin{aligned} K &\equiv BN^{-1}\Lambda_E^{-1}B^*, \\ L &\equiv C^*\Lambda_F C, \\ g &\equiv -C^*\Lambda_F y_d \end{aligned} \tag{5.1.77}$$

and substituting into Equations (5.1.76) and (5.1.75), the optimality conditions are transformed into the following equivalent two-point boundary value problem:

$$\frac{d}{dt}x^0+A(t)x^0+K(t)\psi_0=f, \qquad x(0)=x_0,$$

$$-\frac{d}{dt}\psi_0+A^*(t)\psi_0-L(t)x^0=g, \qquad \psi_0(T)=0. \tag{5.1.78}$$

Note that since $B\in L_\infty(I,\mathfrak{L}(E,V^*))$, $B^*\in L_\infty(I,\mathfrak{L}(V,E^*))$, and consequently, $K\in L_\infty(I,\mathfrak{L}(V,V^*))$. Here we have used the fact that V is reflexive. Similarly we can verify that $L\in L_\infty(I,\mathfrak{L}(V,V^*))$ and $g\in L_2(I,V^*)$. Once the system of equations (5.1.78) is solved, the optimal control is obtained from the expression (5.1.74). Since we are interested in the feedback control, and the optimal control u_0 is given in terms of the adjoint state ψ_0, our aim is to find an expression for ψ_0 in terms of the state x^0. Loosely speaking, we wish to show that, in principle, we can construct an operator $P\in L_\infty(I,\mathfrak{L}(H))$ and a function $\gamma\in L_2(I,V)\cap L_\infty(I,H)$ such that

$$\psi_0(t)=P(t)x^0(t)+\gamma(t), \qquad t\in I. \tag{5.1.79}$$

Once this is established, Equations (5.1.74) and (5.1.79) give us the optimal feedback controller.

According to Bellman's principle of optimality, it is required that a part of an optimal trajectory and control be again optimal with respect to the corresponding part of the time interval. Suppose that x^0 is the optimal state trajectory, ψ_0 the optimal adjoint trajectory, and u_0 the optimal control, given by $u_0(t)=-N^{-1}(t)\Lambda_E^{-1}B^*(t)\psi_0(t)$, $t\in I$. Let $\Gamma(x^0)=\{(t,x^0(t)),\ t\in I\}\subset I\times H$ denote the graph of the state trajectory, suppose that $(s,h)\in\Gamma(x^0)$, and consider the system of evolution equations

$$\frac{d}{dt}\xi+A(t)\xi+K(t)\psi=f, \qquad \xi(s)=h,$$

$$-\frac{d}{dt}\psi+A^*(t)\psi-L(t)\xi=g, \qquad \psi(T)=0 \tag{5.1.80}$$

where $t\in(s,T)$. In order that the principle of optimality is not violated, it is required that this system of equations have a unique solution for arbitrary $(s,h)\in\Gamma(x^0)\subset I\times H$. This will guarantee that $\xi(t)=x^0(t)$, $\psi(t)=\psi_0(t)$ for $t\in(s,T)$, and consequently, $u_0(t)=-N^{-1}(t)\Lambda_E^{-1}B^*(t)\psi_0(t)$ for $t\in(s,T)$, thereby preserving the principle of optimality. Since the initial state $x_0\in H$ is arbitrary, we require uniqueness of solution of the above system of equations for arbitrary $(s,h)\in I\times H$. Further, in order that the optimal control be given by a continuous affine transformation of the state, we require that the mapping $h\to\psi(s)$, arising from (5.1.80), be also affine continuous from H into H; for if this is so, then there exists an operator $P_s\in\mathfrak{L}(H)$ and $\gamma_s\in H$ such that $\psi(s)=P_sh+\gamma_s$, and hence, if $(s,h)\in\Gamma(x^0)$, then $u^0(s)=-N^{-1}(s)\Lambda_E^{-1}B^*(s)[P_sh+\gamma_s]$. After we have resolved the above questions, we shall consider the problem of constructing P and r as suitable functions on I with values in $\mathfrak{L}(H)$ and H, respectively.

REMARK 5.1.7. We note that for finite-dimensional problems, the requirement that $h \to \psi(s)$ be affine continuous is automatically satisfied. In other words, this question simply does not arise in the case of equivalent finite-dimensional problems. For infinite-dimensional problems, this question, however; is crucial.

In the sequel we shall make use of the previous assumptions. We also recall that the set $X \equiv \{\xi : \xi \in L_2(I, V),\ \dot{\xi} \in L_2(I, V^*)\}$ equipped with the norm

$$\|\xi\|_X = \left(\|\xi\|^2_{L_2(I,V)} + \|\dot{\xi}\|^2_{L_2(I,V^*)}\right)^{1/2}$$

is a Hilbert space. For $s \in (0, T)$ we write $X(s, T)$ for the restriction of X to the interval (s, T).

Lemma 5.1.2. *For each $h \in H$, the system*

$$\begin{aligned} &\frac{d}{dt}\xi + A(t)\xi + K(t)\psi = f, \qquad \xi(s) = h \\ -&\frac{d}{dt}\psi + A^*(t)\psi - L(t)\xi = g, \qquad \psi(T) = 0, \end{aligned} \tag{5.1.80$'$}$$

where $t \in (s, T)$, has a unique solution $\xi, \psi \in L_2(s, T; V) \cap L_\infty(s, T; H)$. Further, it follows also that $\psi \in C^0(s, T; H)$.

PROOF. The optimality system (5.1.80)′ arises from the following control problem:

$$\begin{cases} \dfrac{d}{dt}\xi(v) + A(t)\xi(v) = f + Bv, \quad t \in (s, T), \\ \qquad \xi(v)(s) = h, \end{cases} \tag{5.1.81}$$

$$J(s, h, v) = \int_s^T |C\xi(v) - y_d|_F^2 dt + \int_s^T (Nv, v)_E dt = \min,$$

where J is the cost function dependent on the initial time and state and the control policy used over the interval (s, T). The minimum is to be taken over the class $\mathfrak{U}^s \equiv L_2(s, T; E)$. It follows from the existence Theorem 5.1.2 with I replaced by the interval (s, T) that this problem has a unique solution $u \in \mathfrak{U}^s$ with $\xi(u) \in L_2(s, T; V) \cap L_\infty(s, T; H)$. By Theorem 5.1.5 or Corollary 5.1.3, this control is given by

$$u(t) = -N^{-1}(t)\Lambda_E^{-1}B^*(t)\psi(t), \qquad t \in (s, T), \tag{5.1.82}$$

where ψ satisfies the system of equations

$$-\frac{d}{dt}\psi+A^*(t)\psi-L(t)\xi(u)=g, \qquad t\in(s,T),$$
$$\psi(T)=0. \tag{5.1.83}$$

Since $L\xi(u)+g\in L_2(s,T;V^*)$, system (5.1.83) has a unique solution $\psi=\psi(u)\in L_2(s,T;V)$. This follows from Theorem 5.1.1 by reversal of the flow of time. Further, from (5.1.83) it follows that $(d/dt)\psi$ exists in the distribution sense and belongs to $L_2(s,T;V^*)$, and consequently, $\psi\in C^0(s,T;H)$ (Theorem 1.2.15). Scalar multiplying (5.1.83) by ψ and integrating, it is easily verified also that $\psi\in L_\infty(s,T;H)$. This completes the proof of the lemma. □

Lemma 5.1.3. *Consider the optimality system* (5.1.80) *for* $s\in(0,T)$ *arbitrary. Then* $h\to\{\xi,\psi\}$ *is a continuous mapping from the strong topology of* H *into the weak topology of* $X(s,T)$.

PROOF. Let $\{h_n\}\in H$ and suppose $h_n\to h$ strongly. Let u be an arbitrary element of $\mathcal{U}^s\equiv L_2(s,T;E)$ and let $\xi_n(v)$ be the solution of the problem

$$\frac{d}{dt}\xi_n(v)+A\xi_n(v)=f+Bv,$$
$$\xi_n(v)(s)=h_n \tag{5.1.84}$$

over the time interval (s,T). Since $\{h_n\}$ is bounded in H, it follows from Theorem 5.1.1 that $\{\xi_n(v)\}$ is contained in a bounded subset of $L_2(s,T;V)\cap L_\infty(s,T;H)$. In fact, due to Corollary 5.1.1, $\{\xi_n(v)\}$ is contained in a bounded subset of $X(s,T)\equiv$ the restriction of X to the interval (s,T). Since $X(s,T)$ is a reflexive Banach space, there exists a subsequence, again denoted by $\{\xi_n\}$, and an element $\xi\in X(s,T)$ such that $\xi_n(v)\to\xi$ weakly. Following the same procedure as in the proof of Theorem 5.1.1, we conclude that $\xi=\xi(v)$ is the solution of the problem (5.1.84) with h_n replaced by h and that $\xi(v)\in L_\infty(s,T;H)$. In fact, for a fixed $v\in\mathcal{U}^s$, as $h_n\to h$ strongly in H, it follows from the inequality

$$|\xi_n(v)(t)-\xi(v)(t)|_H^2+2\alpha\int_s^t|\xi_n(v)(\theta)-\xi(v)(\theta)|_V^2\,d\theta\le|h_n-h|_H^2$$

that $\xi_n(v)\to\xi(v)$ strongly in $L_2(s,T;V)\cap L_\infty(s,T;H)$. Further, by assumption, C is a bounded linear operator from $L_2(I,V)$ to $L_2(I,F)$. Therefore, it follows from the definition of the function $J(\cdot,\cdot,\cdot)$, as given in (5.1.81), that whenever $h_n\to h$ strongly in H, $J(s,h_n,v)\to J(s,h,v)$. Thus, for a fixed $v\in\mathcal{U}^s$,

$$\lim_n J(s,h_n,v)=J(s,h,v). \tag{5.1.85}$$

Let $\{u_n\}\in\mathcal{U}^s$ be the sequence of controls such that

$$J(s,h_n,u_n)=\inf\{J(s,h_n,v):v\in\mathcal{U}^s\}. \tag{5.1.86}$$

In other words, $\{u_n\}$ is the sequence of optimal controls corresponding to the sequence of control problems consisting of (5.1.84) and the cost functional $J(s, h_n, v)$. By virtue of Theorem 5.1.2, such controls exist. Let u be the optimal control for the problem

$$\begin{aligned} &\frac{d}{dt}\xi(v)+A\xi(v)=f+Bv, \qquad t\in(s,T), \\ &\xi(v)(s)=h \\ &J(s,h,v)=\text{minimum with respect } v\in\mathcal{U}^s. \end{aligned} \tag{5.1.87}$$

Then clearly,

$$J(s,h_n,u_n)\le J(s,h_n,u). \tag{5.1.88}$$

Consider the system (5.1.84) with $v=u$ and let $\xi_n(u)$ denote the corresponding solution. Clearly, $\{\xi_n(u)\}$ is a bounded sequence in $X(s,T)$. This coupled with the assumption that C is a bounded linear operator implies that $\{J(s,h_n,u)\}$ is a bounded sequence of numbers. Consequently, it follows from (5.1.88) that $\sup\{J(s,h_n,u_n),\ n=1,2,\ldots\}<\infty$. On the other hand,

$$J(s,h_n,u_n)\ge\int_s^T(Nu_n,u_n)_E\,dt\ge\beta\|u_n\|^2_{\mathcal{U}^s}. \tag{5.1.89}$$

Therefore the sequence of controls $\{u_n\}$ is contained in a bounded subset of $\mathcal{U}^s$. Since $\mathcal{U}^s=L_2(s,T;E)$ is a Hilbert space, there exists a subsequence of the sequence $\{u_n\}$, again denoted by $\{u_n\}$, and an element $\tilde{u}\in\mathcal{U}^s$ such that $u_n\to\tilde{u}$ weakly. Consequently, by virtue of Corollary 5.1.1, the sequence of solutions $\{\xi_n(u_n)\}$ of Equation (5.1.84) with $v=u_n$ is contained in a bounded subset of $X(s,T)$. Therefore there exists a subsequence of the sequence $\{\xi_n(u_n)\}$, again denoted by $\{\xi_n\}$, and an element $\xi\in X(s,T)$ such that $\xi_n\to\xi$ weakly in $X(s,T)$. Further, $\xi_n\to\xi$ also in the weak* topology of $L_\infty(s,T;H)$. Note that for the problem (5.1.84) with v replaced by u_n, $\xi_n(u_n)$ is its weak solution in the sense that

$$\begin{aligned} -\int_s^T\langle\xi_n(u_n),\dot{\psi}\rangle\,dt+\int_s^T\langle\xi_n(u_n),A^*(t)\psi\rangle\,dt&=(h_n,\psi(s))_H \\ &+\int_s^T\langle f+Bu_n,\psi\rangle\,dt \end{aligned} \tag{5.1.90}$$

for all $\psi\in C^1(s,T;V)$ with $\psi(T)=0$. Using the appropriate subsequences in the equality (5.1.90) and taking the limit, we obtain

$$-\int_s^T\langle\xi,\dot{\psi}\rangle\,dt+\int_s^T\langle\xi,A^*(t)\psi\rangle\,dt=(h,\psi(s))_H+\int_s^T\langle f+B\tilde{u},\psi\rangle\,dt. \tag{5.1.91}$$

This expression implies that $\xi=\xi(\tilde{u})$ is the weak solution of the problem (5.1.84) with v replaced by $\tilde{u}$ and h_n replaced by h. Since J is quadratic in ξ

and u, it is weakly lower semicontinuous on $L_2(I,V)\times L_2(I,E)$, and consequently, as $u_n \to \tilde{u}$ weakly in $L_2(s,T;E)\equiv \mathfrak{U}^s$ and $\xi_n \to \xi(\tilde{u})$ weakly in $X(s,T)$ and hence in $L_2(s,T;V)$, we have

$$J(s,h,\tilde{u}) \leq \varliminf_n J(s,h_n,u_n). \tag{5.1.92}$$

Therefore, it follows from (5.1.85), (5.1.88), and (5.1.92) that

$$\begin{aligned} J(s,h,\tilde{u}) &\leq \varliminf_n J(s,h_n,u_n) \\ &\leq \varlimsup_n J(s,h_n,u_n) \leq \varlimsup_n J(s,h_n,u) = J(s,h,u). \end{aligned} \tag{5.1.93}$$

Since u is the optimal control for the problem (5.1.87), it follows from (5.1.93) that $\tilde{u}$ must be equal to u, and consequently, the original sequence $\{u_n\}$ itself converges weakly to u. As a result, $\xi(u)$ is the weak limit of $\xi_n(u_n)$. Using this fact for the (adjoint) problem (5.1.83) with u_n replacing u and $\xi_n(u_n)$ replacing $\xi(u)$, it is easily verified that $\psi_n(u_n) \to \psi(u)$ weakly in $X(s,T)$. This completes the proof of the lemma. □

Theorem 5.1.8. *Let $\{\xi,\psi\}$ be the solution of the optimality problem* (5.1.80) *corresponding to $h\in H$ and arbitrary $s\in(0,T)$. Then, $h\to\psi(s)$ is a continuous affine mapping of H into H and there exists $P(s)\in\mathfrak{L}(H)$ and $\gamma(s)\in H$ such that $\psi(s)=P(s)h+\gamma(s)$.*

PROOF. By Lemma 5.1.3, the mapping $h\to\{\xi,\psi\}$ is affine continuous from H into $X(s,T)\times X(s,T)$. Clearly the mapping $\{\xi,\psi\}\to\psi$ is linear and continuous from $X(s,T)\times X(s,T)$ into $X(x,T)$. We have seen that the injection map $X(s,T)\hookrightarrow C^0([s,T],H)$, is continuous (see Theorem 1.2.15). Further, the mapping $\psi\to\psi(s)$ from $C^0([s,T],H)$ into H is obviously continuous. Thus the composition map $h\to\psi(s)$ from H into H is affine continuous. This result implies that there exists an operator $P(s)\in\mathfrak{L}(H)$ and a vector $\gamma(s)\in H$ such that $\psi(s)=P(s)h+\gamma(s)$. □

REMARK 5.1.8. Since $s\in(0,T)$ is arbitrary, it follows from the above result and the existence of unique solution of the problem (5.1.78), a consequence of Lemma 5.1.2, that

$$\psi(t)=P(t)x^0(t)+\gamma(t) \quad \text{for each } t\in(0,T) \tag{5.1.94}$$

with the functions P (operator valued) and γ (vector valued) defined by

$$P(t)x^0(t)=\tilde{\psi}(t),$$

$$\gamma(t)=\tilde{\tilde{\psi}}(t), \tag{5.1.95}$$

where $\tilde{\psi}$ is the solution of the homogeneous equation

$$\frac{d}{d\theta}\tilde{\xi}+A(\theta)\tilde{\xi}+K(\theta)\tilde{\psi}=0, \qquad \tilde{\xi}(t)=x^0(t),$$

$$-\frac{d}{d\theta}\tilde{\psi}+A^*(\theta)\tilde{\psi}-L(\theta)\tilde{\xi}=0, \qquad \tilde{\psi}(T)=0 \tag{5.1.96}$$

for $\theta\in(t,T)$, and $\tilde{\tilde{\psi}}$ is the solution of the nonhomogeneous equation

$$\frac{d}{d\theta}\tilde{\tilde{\xi}}+A(\theta)\tilde{\tilde{\xi}}+K(\theta)\tilde{\tilde{\psi}}=f, \qquad \tilde{\tilde{\xi}}(t)=0,$$

$$-\frac{d}{d\theta}\tilde{\tilde{\psi}}+A^*(\theta)\tilde{\tilde{\psi}}-L(\theta)\tilde{\tilde{\xi}}=g, \qquad \tilde{\tilde{\psi}}(T)=0 \tag{5.1.97}$$

for $\theta\in(t,T)$.

Theorem 5.1.9. *The operator P satisfies the following properties*:

i. $P\in L_\infty(I,\mathfrak{L}(H))$.
ii. $(P(t)h,h)\geq 0$ *for all* $t\in I$ *and* $h\in H$.
iii. $P(t)=P^*(t)$ (*self-adjoint*).

PROOF. The optimality system (5.1.96) arises from the control problem consisting of the evolution equation

$$\frac{d}{d\theta}\xi(v)+A(\theta)\xi(v)=Bv(\theta), \qquad \theta\in(t,T),$$

$$\xi(v)(t)=x^0(t)\equiv h \tag{5.1.98}$$

and the cost function

$$J(t,h,v)=\int_t^T\langle L\xi(v)(\theta),\xi(v)(\theta)\rangle_{V^*-V}d\theta+\int_t^T(N(\theta)v,v)_E d\theta, \tag{5.1.99}$$

where $\xi(v)$ is the response of the system (5.1.98) corresponding to the control $v\in L_2(t,T;E)$.

The optimal control is given by

$$u(\theta)=-N^{-1}(\theta)\Lambda_E^{-1}B^*(\theta)\tilde{\psi}(\theta), \qquad \theta\in[t,T]. \tag{5.1.100}$$

Substituting this in the cost functional (5.1.99), we obtain

$$J(t,h,u)=\int_t^T\langle L\tilde{\xi},\tilde{\xi}\rangle_{V^*-V}d\theta+\int_t^T\langle BN^{-1}\Lambda_E^{-1}B^*\tilde{\psi},\tilde{\psi}\rangle_{V^*-V}d\theta$$

$$=\int_t^T\langle L\tilde{\xi},\tilde{\xi}\rangle_{V^*-V}d\theta+\int_t^T\langle K\tilde{\psi},\tilde{\psi}\rangle_{V^*-V}d\theta, \tag{5.1.101}$$

where the operator K, given by (5.1.77), belongs to $L_\infty(I,\mathfrak{L}(V,V^*))$. Using the second equality of (5.1.96) in the first term of (5.1.101) and integrating by parts, we obtain

$$J(t,h,u)=\big(\tilde{\psi}(t),\tilde{\xi}(t)\big)_H+\int_t^T\left\langle \tilde{\psi},\frac{d}{d\theta}\tilde{\xi}+A(\theta)\tilde{\xi}\right\rangle_{V-V^*}d\theta$$
$$+\int_t^T\langle K\tilde{\psi},\tilde{\psi}\rangle_{V^*-V}d\theta. \tag{5.1.102}$$

Thus, it follows from (5.1.95), (5.1.96), (5.1.98), and (5.1.102) that

$$J(t,h,u)=\big(\tilde{\psi}(t),\tilde{\xi}(t)\big)_H=(P(t)h,h). \tag{5.1.103}$$

Since the control u is optimal,

$$J(t,h,u)\le J(t,h,0)=\int_t^T\langle L\xi(0)(\theta),\xi(0)(\theta)\rangle_{V^*-V}d\theta, \tag{5.1.104}$$

where $\xi(0)$ is the solution of the equation (5.1.98) corresponding to $v\equiv 0$ and $\xi(0)(t)=h$. Since this equation has a unique solution for each $h\in H$ and is continuously dependent on the data, there exists an evolution operator $\Gamma(\theta,\tau)$, $0\le\tau\le\theta\le T$, such that for any fixed $\tau\ge 0$, $\theta\to\Gamma(\theta,\tau)h$ is strongly continuous on $[\tau,T]$ and $\xi(0)(\theta)=\Gamma(\theta,t)h$ for all $h\in H$ with $\Gamma(\cdot,t)h\in L_2(t,T;V)$. Thus,

$$J(t,h,0)=\int_t^T\langle L(\theta)\Gamma(\theta,t)h,\Gamma(\theta,t)h\rangle_{V^*-V}d\theta. \tag{5.1.105}$$

Since both L and Γ are bounded operators, there exists a constant $\delta>0$, independent of $t\in I=(0,T)$, and $h\in H$ such that

$$J(t,h,0)\le\delta|h|_H^2. \tag{5.1.106}$$

Clearly, it follows from (5.1.103)–(5.1.106) that

$$\operatorname*{ess\,sup}_{t\in I}(P(t)h,h)\le\delta|h|_H^2. \tag{5.1.107}$$

Thus, $P\in L_\infty(I,\mathcal{L}(H))$, which proves (i). Since $J(t,h,u)\ge 0$, it follows from (5.1.103) that $(P(t)h,h)\ge 0$ for all $t\in I$, which proves (ii). It remains to prove the self-adjointness of $P(t)$. For fixed but arbitrary $t\in(0,T)$ and initial states $h_1,h_2\in H$, let $\{u_1,u_2\}$ be the optimal controls, $\{\tilde{\xi}_1,\tilde{\xi}_2\}$ the optimal trajectories, and $\{\tilde{\psi}_1,\tilde{\psi}_2\}$ the corresponding adjoint trajectories for the problem (5.1.98)–(5.1.99). Define

$$\nu(h_1,h_2)\equiv\int_t^T\langle L\tilde{\xi}_1,\tilde{\xi}_2\rangle_{V^*-V}d\theta+\int_t^T(Nu_1,u_2)_E d\theta.$$

Using the expressions for the optimal controls $u_i=-N^{-1}\Lambda_E^{-1}B^*\tilde{\psi}_i$, $i=1,2$, in the above expression, we obtain $\nu(h_1,h_2)=(P(t)h_1,h_2)_H$ and $\nu(h_2,h_1)=(P(t)h_2,h_1)_H$. Since F is a Hilbert space, $\Lambda_F=\Lambda_F^*$, and consequently, $L=L^*$ [see Equations (5.1.77)]. By assumption, $N=N^*$; thus $\nu(h_1,h_2)=\nu(h_2,h_1)$, and consequently, $(P(t)h_1,h_2)_H=(P(t)h_2,h_1)_H$ for all $h_1,h_2\in H$. This implies $P(t)=P^*(t)$. □

In the following proposition we show that the operator-valued function P and the vector-valued function γ can be determined by solving certain abstract differential equations.

Proposition 5.1.1. The operator-valued function P and the vector-valued function γ are formally governed by the following differential equations:

$$\left\langle\left(-\frac{d}{d\theta}P+PA+A^*P+PKP-L\right)h,\eta\right\rangle=0 \quad \text{for all } h,\eta\in V,\theta\in[0,T),$$
$$P(T)=0 \tag{5.1.108}$$

and

$$\left\langle-\frac{d}{d\theta}\gamma+(A^*+PK)\gamma-Pf-g,\eta\right\rangle=0 \quad \text{for all } \eta\in V,\theta\in[0,T),$$
$$\gamma(T)=0. \tag{5.1.108$'$}$$

PROOF. Let ψ be an arbitrary element from $\mathfrak{D}(I,V)$. Multiplying the second equation of (5.1.96) by $\psi(\theta)$ to form the V^*-V duality pairing and integrating over (t,T), we obtain

$$\int_t^T\left\langle-\frac{d}{d\theta}\tilde{\psi}+A^*\tilde{\psi}-L\tilde{\xi},\psi\right\rangle_{V^*-V}d\theta=0. \tag{5.1.109}$$

In (5.1.96), we take $\tilde{\xi}(t)=h\in V$ arbitrary. Substituting $\tilde{\psi}(\theta)=P(\theta)\tilde{\xi}$ in Equation (5.1.109), we obtain

$$\int_t^T\left\langle\left(-\frac{d}{d\theta}P+PA+A^*P+PKP-L\right)\tilde{\xi}(\theta),\psi(\theta)\right\rangle d\theta=0 \tag{5.1.110}$$

for all $\psi\in\mathfrak{D}(I,V)$. Choosing $\psi(\theta)=\nu(\theta)\eta$ with $\eta\in V$ arbitrary and $\nu\in\mathfrak{D}(I)$, it follows from (5.1.110) that

$$\int_t^T\left\langle\left(-\frac{d}{d\theta}P+PA+A^*P+PKP-L\right)\tilde{\xi}(\theta),\eta\right\rangle\nu(\theta)\,d\theta=0 \tag{5.1.111}$$

for all $\nu\in\mathfrak{D}(I)$. This implies that

$$\left\langle\left(-\frac{d}{d\theta}P+PA+A^*P+PKP-L\right)\tilde{\xi}(\theta),\eta\right\rangle=0 \tag{5.1.112}$$

in the sense of distribution for all $\theta\in[t,T)$, $\eta\in V$. Since $\tilde{\xi}(t)=h$, it follows from the above equality that at $\theta=t$

$$\langle(-\dot{P}+PA+A^*P+PKP-L)h,\eta\rangle=0 \quad \text{for all } h,\eta\in V, \tag{5.1.113}$$

and consequently, this equality is true for all $t\in[0,T)$. For the differential equation for γ, consider the duality product between ϕ and the second equation of (5.1.80)′ over the interval (t, T) with $\xi(t)=h\in H$:

$$\int_t^T \langle -\dot{\psi}+A^*\psi-L\xi,\phi\rangle\, d\theta=\int_t^T \langle g,\phi\rangle\, d\theta. \tag{5.1.114}$$

Substituting $P(\theta)\xi+\gamma(\theta)$ for $\psi(\theta)$ in Equation (5.1.114) and utilizing the first equation of (5.1.80)′, we obtain

$$\int_t^T \langle(-\dot{P}+PA+A^*P+PKP-L)\xi,\phi\rangle\, d\theta$$
$$+\int_t^T \langle -\dot{\gamma}+(A^*+PK)\gamma-(Pf+g),\phi\rangle\, d\theta=0 \tag{5.1.115}$$

for all $\phi\in\mathfrak{D}(I,V)$, where $\cdot\equiv d/dt$. Due to (5.1.113), the first term of (5.1.115) vanishes, leaving

$$\int_t^T \langle -\dot{\gamma}+(A^*+PK)\gamma-(Pf+g),\eta\rangle\nu(\theta)\, d\theta=0$$

for all $\eta\in V$ and $\nu\in\mathfrak{D}(I)$. Thus, Equation (5.1.108)′ holds in the sense of distribution. This completes the proof. □

For the terminal conditions, we note that if $\tilde{\xi}(T)\equiv\tilde{\xi}(T,h)$ describes a dense subspace of H whenever h describes H, then it follows from the equality $0=\tilde{\psi}(T)=P(T)\tilde{\xi}(T,h)$ that $P(T)=0$. Hence, due to the equality $0=\psi(T)=P(T)\xi(T)+\gamma(T)$, $\gamma(T)=0$. The question of denseness mentioned above is equivalent to the question of controllability of the evolution equation

$$\frac{d}{dt}\xi+A\xi=0,\qquad \xi(0)=h, \tag{5.1.116}$$

$t\in(0,T)$, with respect to the initial state h and is connected to the backward uniqueness property of the corresponding adjoint equation

$$-\frac{d\phi}{dt}+A^*\phi=0,\qquad \phi(T)=w. \tag{5.1.117}$$

Let $(\xi(T,h),w)_H=0$ for all $h\in H$, where $\xi(T,h)$ is the state at time $t=T$ with $\xi(0,h)=h$. Then

$$0=\int_0^T\left\langle -\frac{d\phi}{dt}+A^*\phi,\xi(t,h)\right\rangle dt=(\phi(0),h)-(w,\xi(T,h)),$$

and consequently, $(w,\xi(T,h))=(\phi(0),h)$. Since $(w,\xi(T,h))=0$ for all $h\in H$, the previous equality implies that $\phi(0)=0$. If the adjoint equation satisfies the backward uniqueness property, that is, if the generally ill-posed problem

$$\left.\begin{aligned} -\frac{d\phi}{dt}+A^*\phi=0 \\ \phi(s)=0 \end{aligned}\right\} \qquad t\in(s,T), \quad 0\le s\le T,$$

satisfies the condition that $\phi(t)\equiv 0$ for $t\in(s,T)$ whenever $\phi(s)=0$, then we can conclude that $w=\phi(T)=0$. Thus, the strong closure of the set $\{\xi(T,h),\, h\in H\}$ is dense in H.

Solution of Equations (5.1.108) and (5.1.108)′. In order to conclude that the feedback control given by

$$u_1(t)=-N^{-1}(t)\Lambda_E^{-1}B^*(t)\big(P(t)x^0(t)+\gamma(t)\big) \tag{5.1.118}$$

is the optimal control, we must show that Equations (5.1.108) and (5.1.108)′ have unique solutions. First, we show that the operator Riccati equation (5.1.108) has a unique solution. The question of existence of solutions of the operator Riccati equation in infinite-dimensional spaces has received considerable attention in recent years [Li.2, Da.1, T.1, CP.1]. We present here a method developed by Temam [T.1], by whom Equation (5.1.108) is solved directly in the space of Hilbert–Schmidt operators, giving uniqueness as well.

Indeed, let V, H, V^* be as defined in Section 5.1.2. Under the assumption that the injection map $V\hookrightarrow H$ is compact, one can construct a common basis for V, H, and V^*. Let Λ be the canonical isomorphism of V onto V^* such that $\langle \Lambda x, y\rangle_{V^*-V}=(x,y)_V$ for all $x,y\in V$. Then Λ^{-1} is linear continuous from V^* into V and consequently from H into V; therefore, it is a linear self adjoint compact operator in H. As a result, there exists a sequence of orthonormal vectors $\{e_i\}$ in H forming the eigenvectors of the operator Λ^{-1} such that

$$\begin{aligned} \Lambda e_i &= \rho_i^2 e_i, \\ (e_i, e_j)_H &= \delta_{ij}, \qquad \rho_i^2>0, \end{aligned} \tag{5.1.119}$$

$i,j=1,2\ldots$. The sequence $\{e_i\}$ also forms an orthogonal basis for V and V^* such that

$$\begin{aligned} (e_i,e_j)_V &= \langle \Lambda e_i, e_j\rangle_{V^*-V}=\rho_i^2\delta_{ij}, \\ (e_i,e_j)_{V^*} &= \langle \Lambda^{-1}e_i, e_j\rangle_{V-V^*}=\frac{1}{\rho_i^2}\delta_{ij}. \end{aligned} \tag{5.1.120}$$

Let X and Y be two Hilbert spaces and let $X\otimes Y$ denote the tensor product space with the inner product

$$(x_1\otimes y_1, x_2\otimes y_2)_{X\otimes Y}=(x_1,x_2)_X\cdot(y_1,y_2)_Y \tag{5.1.121}$$

for $x_1\otimes y_1, x_2\otimes y_2\in X\otimes Y$. Let $X\hat{\otimes} Y$ denote the completion of the pre-

Hilbert space $X\otimes Y$ in the scalar product defined above. It is clear that to every element of $X\otimes Y$ one can associate a continuous linear operator from X into Y by writing $\phi\rightarrow(x\otimes y)(\phi)\equiv(\phi, x)_X\cdot y$, $\phi\in X$. The mapping $X\otimes Y\rightarrow\mathfrak{L}(X, Y)$ is linear continuous and injective with norm equal to or less than unity. It has a continuous extension onto all of $X\hat{\otimes} Y$ with norm ≤ 1. Since this map is injective, one can identify $X\hat{\otimes} Y$ as a sub-Hilbert space of the Hilbert space $\mathfrak{L}(X, Y)$ with

$$\|P\|_{X\hat{\otimes} Y}\leq\|P\|_{\mathfrak{L}(X,Y)} \quad \text{for all } P\in X\hat{\otimes} Y. \tag{5.1.122}$$

The set of all continuous linear operators from X into Y that belong to $X\hat{\otimes} Y$ is called the Hilbert–Schmidt operators. If $P\in X\hat{\otimes} Y$, then $P^*\in Y^*\hat{\otimes} X^*$,where P^* denotes the adjoint of the operator P (this should cause no confusion). It is known that if $P\in\mathfrak{L}(X, Y)$ and X has an orthonormal basis $\{\phi_i\}$, then $P\in X\hat{\otimes} Y$ if and only if $\Sigma_i|P(\phi_i)|_Y^2<\infty$. In that case

$$\|P\|_{X\hat{\otimes} Y}^2=\sum_i|P(\phi_i)|_Y^2 \tag{5.1.123}$$

and it is independent of the choice of the basis vectors $\{\phi_i\}$. By utilizing the common basis $\{e_i\}$ of V, H, and V^* and the above property, we may show that every $P\in H\hat{\otimes} H$ has the representation

$$P=\sum_{i,j}\tau_{ij}e_i\otimes e_j \tag{5.1.124}$$

with

$$\|P\|_{H\hat{\otimes} H}^2=\sum_{i,j}|\tau_{ij}|^2<\infty. \tag{5.1.125}$$

We introduce the following spaces of Hilbert–Schmidt operators associated to V, H, and V^*: $V\hat{\otimes} H$, $H\hat{\otimes} H$, $V^*\hat{\otimes} H$, $H\hat{\otimes} V$, $H\hat{\otimes} V^*$, with $V^*\hat{\otimes} H$ and $H\hat{\otimes} V^*$ the duals of $H\hat{\otimes} V$ and $V\hat{\otimes} H$, respectively. It is clear that

$$\begin{aligned} &V^*\hat{\otimes} H\subset H\hat{\otimes} H\subset V\hat{\otimes} H,\\ &H\hat{\otimes} V\subset H\hat{\otimes} H\subset H\hat{\otimes} V^*. \end{aligned} \tag{5.1.126}$$

As in (5.1.124) and (5.1.125), we have

$$\begin{aligned} &P\in V^*\hat{\otimes} H \quad \text{if and only if } \|P\|_{V^*\hat{\otimes} H}^2=\sum_{i,j}|\tau_{ij}|^2\rho_i^2<\infty,\\ &P\in V\hat{\otimes} H \quad \text{if and only if } \|P\|_{V\hat{\otimes} H}^2=\sum_{i,j}|\tau_{ij}|^2(1/\rho_i^2)<\infty,\\ &P\in H\hat{\otimes} V \quad \text{if and only if } \|P\|_{H\hat{\otimes} V}^2=\sum_{i,j}|\tau_{ij}|^2\rho_j^2<\infty,\\ &P\in H\hat{\otimes} V^* \quad \text{if and only if } \|P\|_{H\hat{\otimes} V^*}^2=\sum_{i,j}|\tau_{ij}|^2(1/\rho_j^2)<\infty. \end{aligned} \tag{5.1.127}$$

From (5.1.126), it follows that

$$V^* \hat{\otimes} H \cap H \hat{\otimes} V \subset H \hat{\otimes} H \subset V \hat{\otimes} H \cap H \hat{\otimes} V^* \tag{5.1.128}$$

and that the injection map from the first to the second is compact, since it is so for the injection map $V \hookrightarrow H$.

With this preparation, we are now ready to consider the questions of existence and uniqueness of solutions for the problems (5.1.108) and (5.1.108)′. We consider these problems in the strong form:

$$\begin{aligned} \frac{dP}{dt} &= PA + A^*P + PKP - L, \\ P(T) &= 0 \end{aligned} \tag{5.1.129}$$

and

$$\begin{aligned} \frac{d}{dt}\gamma &= (A^* + PK)\gamma - (Pf + g), \\ \gamma(T) &= 0. \end{aligned} \tag{5.1.130}$$

For the solution of the problem (5.1.129), we need the following results. Let $\mathfrak{L}^+(H)$ denote the class of bounded positive linear operators in H, $[\cdot]$ the norm in $H \hat{\otimes} H$, and $[\cdot,\cdot]$ the bilinear forms in $V^* \hat{\otimes} H$, $H \hat{\otimes} V$, and $H \hat{\otimes} H$.

Proposition 5.1.2. Let $P \in H \hat{\otimes} H$ and $R, Q \in \mathfrak{L}^+(H)$. Then $[QP, P] \geq 0$, $[PQ, P] \geq 0$, and if Q or R is self-adjoint, then $[QPR, P] \geq 0$.

PROOF. Let

$$P = \sum_{i,j}^{\infty} \tau_{ij} e_i \otimes e_j \quad \text{and} \quad P_n = \sum_{i,j}^{n} \tau_{ij} e_i \otimes e_j .$$

Then it is easily verified that

$$[QP_n, P_n] = \sum_i \left(Q\left(\sum_{j=1}^n \tau_{ij} e_j \right), \left(\sum_{j=1}^n \tau_{ij} e_j \right) \right) \geq 0.$$

Since $P_n \to P$ in the uniform operator topology, it follows from the above that $[QP, P] \geq 0$. Similarly, it follows from the inequality

$$[P_n Q, P_n] = \sum_{j=1}^n \left(\sum_{i=1}^n \tau_{ij} e_i, Q\left(\sum_{i=1}^n \tau_{ij} e_i \right) \right) \geq 0$$

that $[PQ, P] \geq 0$. Suppose that Q is self-adjoint with $\{e_i\}$ the orthonormal eigenvectors of Q and λ_i the corresponding eigenvalues. Then

$$[QP_nR, P_n] = \sum_{i,j,k,l=1}^{n} \tau_{ij}\tau_{kl}(R^*e_i, e_k)(Qe_j, e_l)$$

$$= \sum_{i,j,k=1}^{n} \lambda_j \tau_{ij}\tau_{kj}(Re_k, e_i)$$

$$= \sum_{j=1}^{n} \lambda_j \left(R\left(\sum_{k=1}^{n} \tau_{kj}e_k \right), \sum_{k=1}^{n} \tau_{kj}e_k \right).$$

Since $\lambda_i \geq 0$ and $P_n \to P$ in the uniform operator topology, it follows from the above equality that $[QPR, P] \geq 0$. □

Proposition 5.1.3. Suppose $\langle Av, v\rangle \geq \alpha |v|_V^2$ for some $\alpha > 0$ and for all $v \in V$. Then

i. $[P \cdot A, P] \geq \alpha \| P \|_{V^* \hat{\otimes} H}^2$ for all $P \in V^* \hat{\otimes} H$ and
ii. $[A^* \cdot P, P] \geq \alpha \| P \|_{H \hat{\otimes} V}^2$ for all $P \in H \hat{\otimes} V$.

PROOF. Let

$$P = \sum_{i,j=1}^{\infty} \tau_{ij} e_i \otimes e_j \quad \text{and} \quad P_n = \sum_{i,j=1}^{n} \tau_{ij} e_i \otimes e_j.$$

Then

$$[P_nA, P_n] = \left[\sum_{l,j=1}^{n} \tau_{ij} A^* e_i \otimes e_j, \sum_{k,l=1}^{n} \tau_{kl} e_k \otimes e_l \right]$$

$$= \sum_{i,j,k,l=1}^{n} \tau_{ij}\tau_{kl} \langle A^*e_i, e_k \rangle_{V^*-V} (e_j, e_l)_H$$

$$= \sum_{i,j,k=1}^{n} \tau_{ij}\tau_{kj} \langle A^*e_i, e_k \rangle_{V^*-V}$$

$$= \sum_{j=1}^{n} \left\langle A^* \left(\sum_{i=1}^{n} \tau_{ij} e_i \right), \sum_{k=1}^{n} \tau_{kj} e_k \right\rangle$$

$$\geq \alpha \sum_{j=1}^{n} \left| \sum_{i=1}^{n} \tau_{ij} e_i \right|_V^2. \tag{5.1.131}$$

Since

$$\left| \sum_{i=1}^{n} \tau_{ij} e_i \right|_V^2 = \sum_{i,k=1}^{n} \tau_{ij}\tau_{kj}(e_i, e_k)_V,$$

it follows from (5.1.120) that

$$\left|\sum_{i=1}^{n} \tau_{ij} e_i\right|_V^2 = \sum_{i=1}^{n} |\tau_{ij}|^2 \rho_i^2. \tag{5.1.132}$$

Due to (5.1.131), (5.1.132), and (5.1.127), we have $[P_n A, P_n] \geq \alpha \| P_n \|^2_{V^* \hat{\otimes} H}$. Again by the same argument as in Proposition 5.1.2 we conclude that $[P \cdot A, P] \geq \alpha \| P \|^2_{V^* \hat{\otimes} H}$ for all $P \in V^* \hat{\otimes} H$. This proves the inequality (i). The inequality (ii) follows from similar arguments. □

Theorem 5.1.10. *Consider the operator Riccati equation* (5.1.129) *with the operators A and N satisfying the basic assumption. Further, suppose that $B \in L_\infty(I, \mathcal{L}(E, H))$ and $C \in L_\infty(I, \mathcal{L}(H, F))$, so that $L \equiv C^* \Lambda_F C \in L_\infty(I, H \hat{\otimes} H)$. Then the equation* (5.1.129) *has a unique solution $P \in L_2(I, V^* \hat{\otimes} H) \cap L_2(I, H \hat{\otimes} V)$ satisfying $P(t) \geq 0$, $P(t) = P^*(t)$ for $t \in I$, and (except for a set of measure zero on I) $P \in C(\bar{I}, H \hat{\otimes} H)$. Further, $P \in L_\infty(I, H \hat{\otimes} H)$.*

PROOF. We present only a brief outline of the proof following the method given by Temam [T.1, Theorem 1, Remark 5.4, p. 96]. For uniqueness, suppose that P and Q are two solutions and define $R = P - Q$. Then it follows from (5.1.129) that

$$\frac{d}{dt} R = RA + A^* R + PKR + RKQ,$$

$$R(T) = 0.$$

Scalar multiplying the above equation by $R(t)$ in $H \hat{\otimes} H$ and integrating, we obtain

$$[R(t)]^2 + 2\int_t^T \{[RA, R] + [A^* R, R] + [PKR, R] + [RKQ, R]\}\, d\theta = 0. \tag{5.1.133}$$

Since $B \in L_\infty(I, \mathcal{L}(E, H))$, it follows from the defining relation $K \equiv BN^{-1}\Lambda_E^{-1} B^*$ that $K \in L_\infty(I, \mathcal{L}^+(H))$ and that for any self-adjoint $\Gamma \in \mathcal{L}(H)$, $(K(t)\Gamma h, h) = (\Gamma K(t) h, h)$ for all $h \in H$ and for any $t \in I$. That is, K commutes with every self-adjoint operator $\Gamma \in \mathcal{L}(H)$. Further, it is known that for any two self-adjoint commuting operators $K_1, K_2 \in \mathcal{L}(H)$, $K_1 \cdot K_2 = K_2 \cdot K_1 \geq 0$ whenever $K_1, K_2 \geq 0$. Since $P(t), Q(t)$ are self-adjoint and belong to $\mathcal{L}^+(H)$, it follows from the above that $P(t)K(t) \geq 0$, $K(t)Q(t) \geq 0$ and consequently, by Proposition 5.1.2, the last two integrals in (5.1.133) are nonnegative. Therefore, by virtue of Proposition 5.1.3, it follows from Equation (5.1.133) that

$$[R(t)]^2 + 2\alpha \int_t^T \{\| R \|^2_{V^* \hat{\otimes} H} + \| R \|^2_{H \hat{\otimes} V}\}\, d\theta \leq 0, \qquad t \in I. \tag{5.1.134}$$

Since $\alpha > 0$, this implies $R \equiv 0$, and consequently, uniqueness follows. For the proof of existence, we use the so-called fractional steps method. Partition the interval $I=[0,T]$ into m equal subintervals, each of length $\delta = \delta(m) = (T/m)$, with the nth interval given by $[(n-1)\delta, n\delta]$, $1 \leq n \leq m$. Assuming that the value P^n of $P(t)$ at time $t = n\delta$ is known, one determines P^{n-1} in three steps:

$$\begin{cases} \text{(i)} & P^{n-1/3} - P^n + \delta P^n K_n P^{n-1/3} = 0 \\ \text{(ii)} & P^{n-2/3} - P^{n-1/3} + \delta P^{n-2/3} A_n - \delta L_n = 0 \\ \text{(iii)} & P^{n-1} - P^{n-2/3} + \delta A_n^* P^{n-1} = 0, \end{cases} \tag{5.1.135}$$

where

$$K_n = \frac{1}{\delta} \int_{(n-1)\delta}^{n\delta} K(\theta)\, d\theta, \qquad L_n = \frac{1}{\delta} \int_{(n-1)\delta}^{n\delta} L(\theta)\, d\theta,$$

$$A_n = \frac{1}{\delta} \int_{(n-1)\delta}^{n\delta} A(\theta)\, d\theta. \tag{5.1.136}$$

The operators $P^{n-1/3}$, $P^{n-2/3}$, P^{n-1} are well defined. Indeed, since $K_n \geq 0$ and commutes with every self-adjoint operator $\Gamma \in \mathcal{L}(H)$, it is clear that $P^n K_n \geq 0$ whenever P^n is self-adjoint and positive. Thus, under the hypothesis that P^n is self-adjoint and positive, $(I + \delta P^n K_n)$ is invertible in $\mathcal{L}(H)$, and consequently, $P^{n-1/3}$ is uniquely determined by the expression $P^{n-1/3} = (I + \delta P^n K_n)^{-1} P^n$ as an element of $H \hat{\otimes} H$ whenever $P^n \in H \hat{\otimes} H$. Further $\|(I + \delta P^n K_n)^{-1}\|_{\mathcal{L}(H)} \leq 1$. Since $\langle A(t)v, v\rangle \geq \alpha |v|_V^2$, independently of $t \in I$, it is easy to verify that $\langle A_n v, v\rangle \geq \alpha |v|_V^2$ for all $n \in \{1, 2, \ldots m\}$; consequently,

$$\langle (I + \delta A_n)v, v\rangle \geq |v|_H^2 + \delta\alpha |v|_V^2 \geq |v|_H^2.$$

Thus, $(I + \delta A_n)$ is invertible in $\mathcal{L}(H)$, and further, $\|(I + \delta A_n)^{-1}\|_{\mathcal{L}(H)} \leq 1$. Similarly, $\|(I + \delta A_n^*)^{-1}\|_{\mathcal{L}(H)} \leq 1$. Therefore, from (5.1.135) we obtain

$$\begin{aligned} P^{n-1/3} &= (I + \delta P^n K_n)^{-1} P^n \in H \hat{\otimes} H, \\ P^{n-2/3} &= (P^{n-1/3} + \delta L_n)(I + \delta A_n)^{-1} \in V^* \hat{\otimes} H, \\ P^{n-1} &= (I + \delta A_n^*)^{-1} P^{n-2/3} \in H \hat{\otimes} V, \end{aligned} \tag{5.1.137}$$

and

$$[P^{n-1}] \leq [P^m] + \|L\|_{L_1(I, H \hat{\otimes} H)} = [P(T)] + \|L\|_{L_1(I, H \hat{\otimes} H)} \tag{5.1.138}$$

for all $n = 1, 2, \ldots m$. Using the fact that the injection map $V^* \hat{\otimes} H \cap H \hat{\otimes} V \hookrightarrow H \hat{\otimes} H$ is continuous (and compact), it follows from (5.1.137) and (5.1.138) that

$$\{P^{n-i/3}, i = 0, 1, 2, 3\} \in H \hat{\otimes} H \quad \text{for all } n = 1, 2, \ldots m. \tag{5.1.139}$$

Scalar multiplying (i) of (5.1.135) by $P^{n-1/3}$, (ii) of (5.1.135) by $P^{n-2/3}$, and (iii) of (5.1.135) by P^{n-1}, and using the results of Propositions 5.1.2 and 5.1.3, we obtain the inequalities

$$\begin{aligned}
&[P^{n-1/3}-P^n]^2+[P^{n-1/3}]^2-[P^n]^2\le 0,\\
&[P^{n-2/3}-P^{n-1/3}]^2+[P^{n-2/3}]^2-[P^{n-1/3}]^2+2\delta\alpha[P^{n-2/3}]^2_{V^*\hat{\otimes}H}\\
&\qquad\le 2\delta[L_n, P^{n-2/3}],\\
&[P^{n-1}-P^{n-2/3}]^2+[P^{n-1}]^2-[P^{n-2/3}]^2+2\delta\alpha[P^{n-1}]^2_{H\hat{\otimes}V}\le 0.
\end{aligned}\tag{5.1.140}$$

Adding the inequalities (5.1.140), summing from 1 to m, and recalling that for arbitrary $\varepsilon>0$

$$2[L_n, P^{n-2/3}]\le\frac{1}{\varepsilon}[L_n]^2_{V^*\hat{\otimes}H}+\varepsilon[P^{n-2/3}]^2_{V^*\hat{\otimes}H},\tag{5.1.141}$$

we obtain

$$\begin{aligned}
&[P^0]^2+\sum_{n=1}^{m}\left\{[P^n-P^{n-1/3}]^2+[P^{n-1/3}-P^{n-2/3}]^2+[P^{n-2/3}-P^{n-1}]^2\right\}\\
&\quad+\alpha\delta\sum_{n=1}^{m}\left\{[P^{n-2/3}]^2_{V^*\hat{\otimes}H}+[P^{n-1}]^2_{H\hat{\otimes}V}\right\}\\
&\le[P^m]^2_{H\hat{\otimes}H}+\frac{\alpha_1^2}{\alpha}\|L\|^2_{L_2(I,H\hat{\otimes}H)},
\end{aligned}\tag{5.1.142}$$

where $\alpha_1\ge 0$ is the norm of the injection map $V^*\hat{\otimes}H\hookrightarrow H\hat{\otimes}H$; that is, for $\Gamma\in V^*\hat{\otimes}H$, $\|\Gamma\|_{H\hat{\otimes}H}\le\alpha_1\|\Gamma\|_{V^*\hat{\otimes}H}$. Thus, there exists a constant $\alpha_2\ge 0$ such that

$$\begin{aligned}
&\text{(i)}\ \ \delta\sum_{n=1}^{m}[P^{n-2/3}]^2_{V^*\hat{\otimes}H}\le\alpha_2,\\
&\text{(ii)}\ \ \delta\sum_{n=1}^{m}[P^{n-1}]^2_{H\hat{\otimes}V}\le\alpha_2,\\
&\text{(iii)}\ \ \sum_{n=1}^{m}\left\{[P^n-P^{n-1/3}]^2+[P^{n-1/3}-p^{n-2/3}]^2+[P^{n-2/3}-P^{n-1}]^2\right\}\le\alpha_2.
\end{aligned}\tag{5.1.143}$$

We define the functions $P_{i,\delta}$, $i=0,1,2,3$, as

$$P_{i,\delta}(t)\equiv P^{n-i/3}\quad\text{for } t\in((n-1)\delta, n\delta],\qquad n=1,2,\ldots,m\tag{5.1.144}$$

and

$$P_\delta(t) \equiv \left(\frac{t}{\delta} - n\right)(P^n - P^{n-1}) + P^n \quad \text{for } t \in ((n-1)\delta, n\delta], \qquad n = 1, 2, \ldots, m. \tag{5.1.145}$$

From this definition, it follows that $P_{0,\delta}(t-\delta) = P_{3,\delta}(t)$ and consequently $P_{0,\delta}(t) = P^0$ for $t \in [-\delta, 0]$ and $P_{3,\delta}(t) = P^m = P(T)$ for $t \in [T, T+\delta]$. Given that $P^m \equiv P(T)$ is self-adjoint, positive, and belongs to $H \hat{\otimes} H$, it is clear from (5.1.137), (5.1.138), and (5.1.144) that $P_{i,\delta} \in L_\infty(I, H \hat{\otimes} H)$ and describes a bounded subset of $L_\infty(I, H \hat{\otimes} H)$ as $\delta \to 0$. Similarly, $P_\delta \in L_\infty(I, H \hat{\otimes} H)$ and describes a bounded subset of $L_\infty(I, H \hat{\otimes} H)$ as $\delta \to 0$. Further, it follows from (iii) of (5.1.143) that, for $i = 0, 1, 2$,

$$\sum_{n=1}^{m} \left[P^{n-i/3} - P^{n-(i+1)/3} \right]^2 \delta \le \delta \alpha_2,$$

which is equivalent to

$$\int_0^T [P_{i,\delta}(t) - P_{i+1,\delta}(t)]^2 \, dt \le \delta \alpha_2.$$

Thus,

$$\| P_{i,\delta} - P_{i+1,\delta} \|_{L_2(I, H \hat{\otimes} H)} \le \sqrt{\alpha_2 \delta}, \qquad i = 0, 1, 2. \tag{5.1.146}$$

Similarly, using (5.1.144) and (5.1.145), one can easily verify that there exists a constant $\alpha_3 \ge 0$ such that

$$\| P_{3,\delta} - P_\delta \|_{L_2(I, H \hat{\otimes} H)} \le \sqrt{\alpha_3 \delta}. \tag{5.1.147}$$

Clearly, it follows from (i) and (ii) of (5.1.143) that

$$\begin{aligned} \| P_{2,\delta} \|^2_{L_2(I, V^* \hat{\otimes} H)} &\le \alpha_2, \\ \| P_{3,\delta} \|^2_{L_2(I, H \hat{\otimes} V)} &\le \alpha_2. \end{aligned} \tag{5.1.148}$$

Since P^{n-1} is self-adjoint, $\| P^{n-1} \|_{H \hat{\otimes} V} = \| P^{n-1} \|_{V^* \hat{\otimes} H}$, and consequently, it follows from (ii) of (5.1.143) that

$$\delta \sum_{n=1}^{m} \| P^{n-1} \|^2_{H \hat{\otimes} V} = \delta \sum_{n=1}^{m} \| P^{n-1} \|^2_{V^* \hat{\otimes} H} \le \alpha_2. \tag{5.1.149}$$

From the definition of P_δ, we have

$$\int_{(n-1)\delta}^{n\delta} \| P_\delta(t) \|^2_{V^* \hat{\otimes} H} dt \le 4\delta \left(\| P^n \|^2_{V^* \hat{\otimes} H} + \| P^{n-1} \|^2_{V^* \hat{\otimes} H} \right),$$

which, after summation from $n = 1$ to $n = m - 1$, gives

$$\int_0^{T-\delta} \| P_\delta(t) \|^2_{V^* \hat{\otimes} H} dt \le 8\alpha_2 \equiv \alpha_4 \quad \text{for all } \delta > 0. \tag{5.1.150}$$

Similarly, we have

$$\int_0^{T-\delta} \| P_\delta(t)\|^2_{H\hat{\otimes} V} dt \le \alpha_4 \quad \text{for all } \delta > 0. \tag{5.1.151}$$

By addition of the equalities (i)–(iii) of (5.1.135) and using the definitions (5.1.144) and (5.1.145), we obtain

$$\begin{aligned} &-\frac{d}{dt}P_\delta + P_{2,\delta}A_\delta + A_\delta^* P_{3,\delta} + P_{0,\delta}K_\delta P_{1,\delta} - L_\delta = 0,\\ &P_\delta(T) = P^m \equiv P(T), \end{aligned} \tag{5.1.152}$$

where

$$\left.\begin{matrix} A_\delta(t) \\ K_\delta(t) \\ L_\delta(t) \end{matrix}\right\} = \left\{\begin{matrix} A_n \\ K_n \\ L_n \end{matrix}\right. \quad \text{for } t \in ((n-1)\delta, n\delta] \tag{5.1.153}$$

and A_n, K_n, L_n are as defined in (5.1.136). It follows from boundedness of the sequence $\{P_{i,\delta}, P_\delta\}_{\delta\downarrow 0}$ as elements of $L_\infty(I, H\hat{\otimes} H)$ [see remarks following Equation (5.1.145)], the estimates (5.1.148), and the boundedness of the sequence $\{d/dt)P_\delta\}_{\delta\downarrow 0}$ as elements of $L_2(I, V\hat{\otimes} H) + L_2(I, H\hat{\otimes} V^*)$ [see Equation (5.1.152)] that we can extract a subsequence of the sequence $\{\delta\}$, again denoted by $\{\delta\}$, such that

$$\begin{aligned} &P_{i,\delta} \overset{w^*}{\to} P_i && \text{in } L_\infty(I, H\hat{\otimes} H),\\ &P_\delta \overset{w^*}{\to} P && \text{in } L_\infty(I, H\hat{\otimes} H),\\ &P_{2,\delta} \overset{w}{\to} P_2 && \text{in } L_2(I, V^*\hat{\otimes} H),\\ &P_{3,\delta} \overset{w}{\to} P_3 && \text{in } L_2(I, H\hat{\otimes} V),\\ &\frac{d}{dt}P_\delta \overset{w}{\to} \frac{d}{dt}P && \text{in } L_2(I, V\hat{\otimes} H) + L_2(I, H\hat{\otimes} V^*). \end{aligned} \tag{5.1.154}$$

Due to (5.1.150) and (5.1.151), we can extract a new subsequence of the sequence $\{\delta\}$, again denoted by $\{\delta\}$, such that

$$P_\delta \overset{w}{\to} P \quad \text{in } L_2(0, T-\delta; V^*\hat{\otimes} H) \cap L_2(0, T-\delta; H\hat{\otimes} V). \tag{5.1.155}$$

But, since the injection map $V^*\hat{\otimes} H \cap H\hat{\otimes} V \hookrightarrow H\hat{\otimes} H$ is continuous and compact,

$$P_\delta \overset{s}{\to} P \quad \text{(strongly) in } L_2(0, T-\delta; H\hat{\otimes} H). \tag{5.1.156}$$

Thus, it follows from the results (5.1.146), (5.1.147), (5.1.154), and (5.1.156) that $P_0 = P_1 = P_2 = P_3 = P$ and, for $i = 0, 1, 2, 3$, $P_{i,\delta} \overset{s}{\to} P$ (strongly) in $L_2(0, T-\delta; H\hat{\otimes} H)$ as $\delta \to 0$. Further, by construction, $A_\delta(K_\delta, L_\delta)$ converges strongly to $A(K, L)$ in $L_2(I, \mathcal{L}(V, V^*))(L_2(I, \mathcal{L}(H)), L_2(I, H\hat{\otimes}H))$. For arbitrary $h, \eta \in V$, we can write (5.1.152) as

$$-\langle \dot{P}_\delta(t)h, \eta\rangle + \langle A_\delta(t)h, P^*_{2,\delta}(t)\eta\rangle + \langle P_{3,\delta}(t)h, A_\delta(t)\eta\rangle$$
$$+\langle K_\delta(t)P_{1,\delta}(t)h, P_{0,\delta}(t)\eta\rangle - \langle L_\delta(t)h, \eta\rangle = 0.$$

Multiplying this by $\psi \in C_0^\infty(0, T-\delta)$ and then integrating, we obtain

$$\int_0^{T-\delta}\{ -\langle \dot{P}_\delta h, \eta\rangle + \langle A_\delta h, P^*_{2,\delta}\eta\rangle + \langle P_{3,\delta}h, A_\delta\eta\rangle$$
$$+\langle K_\delta P_{1,\delta}h, P_{0,\delta}\eta\rangle - \langle L_\delta h, \eta\rangle\}\psi(t)\,dt = 0. \tag{5.1.157}$$

Letting $\delta \to 0$ in the above equation and using the previous results, it follows from the Lebesgue dominated convergence theorem that

$$\int_0^{T}\{ -\langle \dot{P}h, \eta\rangle + \langle Ah, P^*\eta\rangle + \langle Ph, A\eta\rangle$$
$$+\langle KPh, P\eta\rangle - \langle Lh, \eta\rangle\}\psi(t)\,dt = 0 \tag{5.1.158}$$

for arbitrary $\psi \in C_0^\infty(0, T-\delta)$ and arbitrary $\delta > 0$.

That P satisfies the differential equation (5.1.129), follows from (5.1.158). From (5.1.154), we have $P \in L_2(I, V^*\hat{\otimes}H) \cap L_2(I, H\hat{\otimes}V) \cap L_\infty(I, H\hat{\otimes}H)$. In view of the fact that the injection map $V^*\hat{\otimes}H \cap H\hat{\otimes}V \hookrightarrow H\hat{\otimes}H$ is continuous and compact, it can be shown that $P \in C(\bar{I}, H\hat{\otimes}H)$ (neglecting a set of measure zero on I). Since $P^m \equiv P(T)$ is self-adjoint and positive, by assumption, it is easy to verify from the relations (5.1.137) and (5.1.77) that, for all $n = 1, 2, \ldots, m$, P^n is self-adjoint and positive. Using this fact and the definition of P_δ [see Equation (5.1.145)], we have

$$(P_\delta(t)h, h) = \left(1 + \frac{t}{\delta} - n\right)(P^n h, h) + \left(n - \frac{t}{\delta}\right)(P^{n-1}h, h) \geq 0 \tag{5.1.159}$$

for all $t \in [(n-1)\delta, n\delta]$, $n = 1, 2, \ldots, m$. This is true for all $\delta > 0$, and consequently, the limit $(P(t)h, h) \geq 0$, $h \in H$. This completes the proof of the theorem. □

Theorem 5.1.11. *Suppose the assumptions of Theorem* 5.1.10 *hold and* $f \in L_2(I, H)$. *Then the evolution equation*

$$-\frac{d}{dt}\gamma + (A^* + PK)\gamma = Pf + g,$$
$$\gamma(T) = 0$$

has a unique solution $\gamma \in X$.

PROOF. Since $P, K \geq 0$ and P self-adjoint and commutes with K, the product $PK \geq 0$, and consequently, $\langle (A^*(t) + P(t)K(t))v, v\rangle \geq \alpha |v|_V^2$ uniformly in $t \in I$. On the other hand, since P is an element of $L_\infty(I, H \hat{\otimes} H) \subset L_\infty(I, \mathfrak{L}(H))$ and $f \in L_2(I, H)$, it is clear that $Pf \in L_2(I, H)$. Owing to the assumption of Theorem 5.1.10, $g \in L_2(I, H)$. Thus, after reversal of the flow of time, the present theorem is a corollary of Theorem 5.1.1. □

We now verify in the following theorem that the control u given by the feedback law (5.1.118) is optimal.

Theorem 5.1.12. *Under the assumptions of Theorem* 5.1.11, *the control* $u \equiv N^{-1}\Lambda_E^{-1}B^*(P\xi' + \gamma)$, *where* ξ' *is the state (trajectory) corresponding to the control* u, *is optimal.*

PROOF. As in the preceding Theorems 5.1.10 and 5.1.11, let P and γ denote the solutions of Equations (5.1.129) and (5.1.130). Substituting the expression for u into the equation $(d/dt)\xi' + A\xi' = Bu + f$ and recalling the definition of K, we obtain the equation $(d/dt)\xi' + (A + KP)\xi' + K\gamma = f$. Let $s \in (0, T)$ be arbitrary and consider the problem

$$\text{S}' \quad \begin{cases} \dfrac{d\xi'}{dt} + (A + KP)\xi' + K\gamma = f, & t \in (s, T), \\ \xi'(s) = h \in H. \end{cases}$$

Since $K \in L_\infty(I, \mathfrak{L}(H))$ and $\gamma \in L_2(I, V)$, we have $K\gamma \in L_2(I, H)$; similarly, $KP \in L_\infty(I, \mathfrak{L}(H))$. Thus, by Theorem 5.1.1, the problem S′ has a unique solution $\xi' \in L_2(s, T; V) \cap L_\infty(s, T; H)$. Define $p' = (P\xi' + \gamma)$. Then, using the equations (5.1.129) and (5.1.130) in the equality

$$\frac{dp'}{dt} = \frac{dP}{dt}\xi' + P\frac{d\xi'}{dt} + \frac{d\gamma}{dt},$$

one obtains

$$\text{S}'' \quad \begin{cases} \dfrac{dp'}{dt} = A^*p' - L\xi' - g, \\ p'(T) = 0. \end{cases}$$

These two evolution equations are identical to those of (5.1.80)′ of Lemma 5.1.2. Since, by Lemma 5.1.2, these problems have unique solutions, we conclude that $\{\xi', p'\} = \{\xi, \psi\}$. Therefore, it follows from Lemma 5.1.2 and equation (5.1.82) that

$$u = -N^{-1}\Lambda_E^{-1}B^*\psi = -N^{-1}\Lambda_E^{-1}B^*p' = -N^{-1}\Lambda_E^{-1}B^*(P\xi' + \gamma)$$

is the optimal control. Since $s \in (0, T)$ and $h \in H$ are arbitrary, this proves the theorem. □

An Example. In this section, we consider a special case and conclude with an example from thermal diffusion process.

For $y_d \equiv 0$, and $f \equiv 0$ we have $\gamma \equiv 0$. The feedback control is given by

$$u(t) = -(N^{-1}\Lambda_E^{-1}B^*P)y \equiv -G(t)y,$$

where y and P are the solutions of the equations

$$\frac{dy}{dt} + A(t)y = B(t)u,$$

$$-\frac{dP}{dt} + PA + A^*P + PKP = L, \qquad P(T) = 0;$$

here $K \equiv BN^{-1}\Lambda_E^{-1}B^*$ and $L \equiv C^*\Lambda_F C$. In case A, B, C and N are all time invariant, the operator Riccati equation can be simplified into a pair of operator equations directly involving the feedback operator G. Indeed, differentiating the equation in P once, we obtain

$$-(P_t)_t + P_t(A + KP) + (A^* + PK)P_t = 0,$$

$$P_t(T) = -L.$$

Writing for the solution of the above equation in the form

$$P_t \equiv -S^*(t)\Lambda_F S(t)$$

and substituting into the equation, we find that

$$(S_t - SA - SKP)^*\Lambda_F S + S^*\Lambda_F(S_t - SA - SKP) = 0.$$

Clearly this is satisfied if S is chosen so that

$$S_t = S(A + KP) = S(A + BG),$$

$$S(T) = C.$$

Taking into account the definition for the operator G, we arrive at the following pair of operator equations:

$$\frac{dG}{dt} = -(N^{-1}\Lambda_E^{-1}B^*)S^*(t)\Lambda_F S(t), \qquad G(T) = 0,$$

$$\frac{dS}{dt} = S(t)(A + BG(t)), \qquad S(T) = C, \tag{5.1.160}$$

where it is easy to verify that $G(t) \in \mathfrak{L}(V^*, E)$ and $S(t) \in \mathfrak{L}(V^*, F)$ for each $t \in (0, T)$.

In case the control and the output (measurement) spaces E and F are of finite dimension ($E = R^m$, $F = R^n$), the decomposition (5.1.160) considerably reduces the complexity of the problem. In this case, $\Lambda_E = I_m$, $\Lambda_F = I_n$, $N \in \mathfrak{L}(R^m, R^m)$, $B \in \mathfrak{L}(R^m, V^*)$, $C \in \mathfrak{L}(V, R^n)$, and consequently $G(t) \in \mathfrak{L}(V^*, R^m)$ and $S(t) \in \mathfrak{L}(V^*, R^n)$. Then Equation (5.1.160) has the kernel representation

$$\frac{\partial}{\partial t}g(t,x) = -\left(\int_\Omega N^{-1}b(\xi)\gamma(t,\xi)\,d\xi\right)\gamma'(t,x), \qquad g(T,x) = 0,$$

$$\frac{\partial}{\partial t}\gamma(t,x) = A^*\gamma(t,x) + g'(t,x)\int_\Omega b(\xi)\gamma(t,\xi)\,d\xi, \qquad \gamma(T,x) = c(x), \tag{5.1.160$'$}$$

for $x \in \Omega$ with the boundary conditions on $\partial\Omega$ being the same as those of the state equation. The functions b and c are defined from the operators B and C by

$$(Bu(t))(x) = b'(x)u(t) \quad \text{and} \quad Cy(t) = \int_\Omega c(\xi)y(t,\xi)\,d\xi. \tag{5.1.161}$$

The optimal control is given by

$$u(t) = -\int_\Omega g(t,\xi)y(t,\xi)\,d\xi.$$

For a specific example, consider a rod of length l with heat supplied at the midpoint and ends held to a constant temperature (say 0°). The temperature y is governed by the heat equation

$$\frac{\partial y(t,x)}{\partial t} = k\frac{\partial^2 y(t,x)}{\partial x^2} + \delta\left(x - \frac{l}{2}\right)u(t), \qquad x \in \Omega \equiv (0,l),$$

where k is the diffusion constant. In this case,

$$H = \{y : y \in L_2(\Omega), y(0) = y(l) = 0\}$$

and

$$V = \{y : y \in H, y_x \in L_2(\Omega), y(0) = y(l) = 0\} = H_0^1.$$

The control and the output spaces are $E = F = R$. The output z is given by the average temperature

$$z(t) = \int_0^l y(t,x)c(x)\,dx.$$

The problem is to choose a control u that minimizes the cost function

$$J(u) = \int_0^T \{z^2(t) + u^2(t)\}\,dt.$$

In this case, the Riccati equation (5.1.160)′ is given by

$$\frac{\partial}{\partial t}g(t,x) = -\gamma\left(t,\frac{l}{2}\right)\gamma(t,x), \qquad g(T,x) = 0, \quad x \in (0,l),$$

$$\frac{\partial}{\partial t}\gamma(t,x) = -k\frac{\partial^2}{\partial x^2}\gamma(t,x) + g(t,x)\gamma\left(t,\frac{l}{2}\right), \qquad \gamma(T,x) = c(x),$$

$$x \in (0,l),$$

$$g(t,0) = g(t,l) = 0, \qquad \gamma(t,0) = \gamma(t,l) = 0, \tag{5.1.160$''$}$$

and the optimal control is given by

$$u^*(t) = -\int_0^l g(t,\xi)y(t,\xi)\,d\xi. \tag{5.1.162}$$

5.2. Linear Evolution Equations Using Semigroup Approach

5.2.1. *Introduction*

In the previous section, we considered that the operator $-A(t)$, arising in the evolution equation $(d/dt)x=A(t)x+f$, be coercive. Even though this class covers a large variety of systems problems involving partial differential equations and delay differential equations, yet it is not general enough to cover general diffusion problems. For example, evolution equations with the operator $A(t)$ given by

$$A(t)\psi \equiv \sum_{i,j=1}^{n} a_{ij}(t,\cdot)\psi_{x_i,x_j} - \sum_{i=1}^{n} a_i(t,\cdot)\psi_{x_i} - a(t,\cdot)\psi \tag{5.2.1}$$

is not covered unless the coefficients a_{ij} are sufficiently smooth to allow the operator $A(t)$ to be written in the divergence form (Section 3.2). However, in case the operator $A(t)=A$ is time invariant, this case is well covered under the general setup of the semigroup technique. In this section, we wish to study briefly some well-established results of control theory based on semigroup approach. In passing, we note that the semigroup theory also plays a major role in constructing solutions for nonautonomous (time varying) evolution equations.

5.2.2. *System Description*

Suppose that the state of the system is governed by the evolution equation

$$\frac{d}{dt}x(t)=Ax(t)+B(t)u(t), \qquad t\in I=(0,\tau), \quad \tau<\infty,$$

$$x(0)=x_0, \tag{5.2.2}$$

where x_0 is a given element of H, which is the state space for this problem, and the state $x(t)$ naturally takes values in H. We assume that A is the infinitesimal generator of a strongly continuous semigroup $T(t)$, $t\geq 0$, in H and $B(t)$, $t\geq 0$, is a family of bounded linear operators with values $B(t)\in\mathcal{L}(E,H)$, where both E and H are a suitable pair of self-adjoint Hilbert spaces. We assume that $B\in L_\infty(I,\mathcal{L}(E,H))$, that is, $\operatorname{ess\,sup}_{t\in I}\|B(t)\|_{\mathcal{L}(E,H)}<\infty$, and $u\in L_2(I,E)$. We choose for the class of admissible controls a closed convex subset $\mathcal{U}_{\mathrm{a}}$ of $L_2(I,E)$. For discussions on Semigroups see Section 1.3.

5.2.3. *Existence and Uniqueness of Solutions*

For the existence and uniqueness of solutions of the evolution equation (5.2.2), we have the following result.

Theorem 5.2.1. *Consider the state equation* (5.2.2) *and suppose that A is the infinitesimal generator of a strongly continuous semigroup $T(t)$, $t \geq 0$, in H, $B \in L_\infty(I, \mathfrak{L}(E, H))$, and $x_0 \in H$. Then, for every $u \in \mathfrak{U}_a \subset L_2(I, E)$, the problem* (5.2.2) *has a unique mild (weak) solution $x \in C(I, H) \cap L_2(I, H)$ and is given by*

$$x(t) = T(t)x_0 + \int_0^t T(t-\theta)B(\theta)u(\theta)\, d\theta, \qquad t \in I. \tag{5.2.3}$$

PROOF. The proof is an immediate consequence of Theorem 2.5.2. □

5.2.4. Formulation of Some Control Problems

Let I be an interval from the real line, X a self-adjoint Hilbert space, and $\mathfrak{L}^+(X)$ the class of positive symmetric bounded linear operators in X. Let $R \in L_\infty(I, \mathfrak{L}^+(H))$, $N \in L_\infty(I, \mathfrak{L}^+(E))$, where E and H are the two self-adjoint Hilbert spaces introduced earlier. Let y be a fixed element of $L_2(I, H)$ and define the functional

$$J(u) \equiv \int_I (R(t)(x(t)-y(t)), x(t)-y(t))_H\, dt + \int_I (N(t)u(t), u(t))_E\, dt, \tag{5.2.4}$$

where x is the weak solution of (5.2.2) corresponding to the control $u \in \mathfrak{U}_a \subset L_2(I, E)$.

Problem (5.2.P1). The problem is to find a control $u^0 \in \mathfrak{U}_a$ such that the cost functional $J(u)$ attains its minimum on $\mathfrak{U}_a$ at $u = u^0$.

Problem (5.2.P2). Another problem we wish to consider is the time optimal problem. Let $x(u) \equiv \{x(t, u), t \geq 0\}$ denote the state trajectory corresponding to the control $u \in \mathfrak{U}_a$. Suppose that x_0 and x_1 are a given pair of states from H. Define $\mathfrak{U}_0 \equiv \{u \in \mathfrak{U}_a : x(0, u) = x_0$ and $x(t, u) = x_1$ for some $t \in I \equiv [0, \tau]\}$. We take τ to be any finite positive number. For nonempty $\mathfrak{U}_0$ we can define the transition time to be the first time $\tilde{t}(u) = \tilde{t}(u, x_0, x_1)$, so that $x(\tilde{t}, u) = x_1$. The problem is to find a control $u^0 \in \mathfrak{U}_0$ such that $\tilde{t}(u^0) \leq \tilde{t}(u)$ for all $u \in \mathfrak{U}_0$. A control satisfying this criterion is called the *time optimal control*. For a given set of the parameters $\tau \in [0, \infty)$, x_0, $x_1 \in H$, and $\mathfrak{U}_a \subset L_2(0, \tau; H)$, a natural question is whether the set $\mathfrak{U}_0$ is nonempty or not. This is called the *problem of controllability*. Once this question is settled in the affirmative, one becomes interested in the question of existence of a time optimal control. Again, an affirmative answer to this question leads to the problem of characterizing the optimal control so that it can be computed.

5.2.5. Existence of Optimal Controls

For Problem (5.2.P1) we have the following existence theorem.

Theorem 5.2.2. *Consider the system* (5.2.2) *with the cost function given by* (5.2.4). *Suppose that the assumptions of Theorem* 5.2.1 *hold, the admissible controls consist of a closed convex subset* $\mathcal{U}_a$ *of* $\mathcal{U} \equiv L_2(I, E)$, y *is a given element of* $L_2(I, H)$, $R \in L_\infty(I, \mathcal{L}^+(H))$, *and* $N \in L_\infty(I, \mathcal{L}^+(E))$ *with* $(N(t)e, e)_E \geq \gamma |e|_E^2$, $\gamma > 0$. *Then there exists a unique control policy* $u^0 \in \mathcal{U}_a$ *such that* $J(u^0) \leq J(u)$ *for all* $u \in \mathcal{U}_a$.

PROOF. Define $\mathcal{H} \equiv L_2(I, H)$, $\mathcal{U} \equiv L_2(I, E)$ with the obvious scalar products denoted by $(\cdot, \cdot)_{\mathcal{H}}$ and $(\cdot, \cdot)_{\mathcal{U}}$, respectively. Let L denote the transformation $u \to \int_0^t T(t-\theta)B(\theta)u(\theta)\,d\theta$ from $\mathcal{U}$ into $\mathcal{H}$. Clearly, $L \in \mathcal{L}(\mathcal{U}, \mathcal{H})$. Define z by $z(t) = T(t)x_0$; then $x = z + Lu$ and J can be written as

$$J(u) = (Mu, u)_{\mathcal{U}} + 2(g, u)_{\mathcal{U}} + c, \tag{5.2.5}$$

where $M \equiv L^*RL + N$, $g = L^*R(z-y)$, and $c \equiv (R(z-y), z-y)_{\mathcal{H}}$. L^* is the adjoint of the bounded linear operator L and maps from $\mathcal{H}$ into $\mathcal{U}$ since they are self-adjoint. Because $L^*RL \geq 0$ and $N \geq \gamma > 0$, it is clear that $M \geq \gamma$; the linear functional $u \to (g, u)_{\mathcal{U}}$ is bounded and c is finite. Thus $J(u) \to \infty$ as $\|u\|_{\mathcal{U}} \to +\infty$. These facts coupled with the hypothesis that $\mathcal{U}_a$ is a closed convex subset of $\mathcal{U}$ imply the existence and uniqueness of the optimal control (see Theorem 5.1.2). □

Theorem 5.2.3. *Consider the system* (5.2.2) *with the terminal cost function* $J(u) \equiv (R(x(T)-y), x(T)-y)_H + \int_I (N(t)u, u)_E\,dt$. *Suppose that* $R \in \mathcal{L}^+(H)$, $y \in H$, *and the rest of the assumptions of Theorem* 5.2.2 *hold. Then there is a unique control that minimizes the terminal cost.*

PROOF. The proof is similar to that of Theorem 5.1.3. □

We now consider Problem (5.2.P2). Here we shall be concerned only with the existence problem. First we note that if $\mathcal{U}_a = L_2(0, \tau; E)$ and $\mathcal{U}_0 (\subset \mathcal{U}_a)$ is nonempty, the problem does not have a solution, since in that case $\inf\{\tilde{t}(u), u \in \mathcal{U}_0\} = 0$ and there is no control $u^0 \in \mathcal{U}_0$ for which $\tilde{t}(u^0) = 0$. This fact will be discussed in Section 2.2.8. Thus, let us consider $\mathcal{U}_a$ to be a closed bounded convex subset of $L_2(0, \tau; E)$, or more generally, $L_p(0, \tau; E)$, $1 < p \leq \infty$. In this case, we can show that a time optimal control exists.

Theorem 5.2.4 (Existence of time optimal controls). *Consider the system* (5.2.2) *with the operators* A *and* B *satisfying the assumptions of Theorem* 5.2.1. *Let* $x_0, x_1 \in H$ *be given,* $0 \leq \tau < \infty$; $\mathcal{U}_a$ *is a closed convex bounded subset of* $L_p(0, \tau; E)$ *for* $1 < p \leq \infty$, *and suppose that* $\mathcal{U}_0$ $(\subset \mathcal{U}_a)$, *as defined* *in Problem* (5.2.P2), *is nonempty. Then there exists an optimal control* u^0 *that transfers the state from* x_0 *to* x_1 *in minimum time.*

PROOF. Since $x \in C(0, \tau; H)$, the transition time $\tilde{t}(u)$ is well defined for each $u \in \mathcal{U}_0$, assuming that $\mathcal{U}_0$ is nonempty. Define $K \equiv \{\tilde{t}(u), u \in \mathcal{U}_0\}$ and $t^* = \inf K$. Let $\{t_n\}$ be a sequence from the set K such that $\{t_n\}$ is nonincreasing and $\lim_n t_n = t^*$. We show that $t^* \in K$. Since $t_n \in K$, there exists a $u_n \in \mathcal{U}_0$ such that $t_n = \tilde{t}(u_n)$. Define, for $0 \leq t \leq t_n \leq \tau$,

$$x_n(t) = T(t)x_0 + \int_0^t T(t-\theta)B(\theta)u_n(\theta)\,d\theta;$$

then

$$x_1 = x_n(t_n) = T(t_n)x_0 + \int_0^{t_n} T(t_n - \theta)B(\theta)u_n(\theta)\,d\theta.$$

since $\{u_n\} \subset \mathcal{U}_0 \subset \mathcal{U}_a$, and $\mathcal{U}_a$, by hypothesis, is a closed convex bounded set, there exists a subsequence $\{u_{n_k}\}$ of the sequence $\{u_n\}$ and a $u^0 \in \mathcal{U}_a$ such that

$$\begin{aligned} u_{n_k} &\overset{\mathrm{w(w^*)}}{\to} u^0 \quad \text{in } L_p(L_\infty), \\ t_{n_k} &\to t^* \quad \text{in } [0, \tau], \end{aligned} \tag{5.2.6}$$

where we have used $\overset{\mathrm{w}}{\to}$ to denote weak convergence in $L_p(0, \tau; E)$, $1 < p < \infty$, and $\overset{\mathrm{w^*}}{\to}$ to denote w*-convergence in $L_\infty(0, \tau; E)$. Clearly,

$$x_1 = T(t_{n_k})x_0 + \int_0^{t_{n_k}} T(t_{n_k} - \theta)B(\theta)u_{n_k}(\theta)\,d\theta. \tag{5.2.7}$$

Let $h \in H$ and form the scalar product in H:

$$(x_1, h) = \big(T(t_{n_k})x_0, h\big) + \int_0^{t_{n_k}} \big(T(t_{n_k} - \theta)B(\theta)u_{n_k}(\theta), h\big)\,d\theta. \tag{5.2.8}$$

Clearly, (5.2.8) is well defined and, since $t_{n_k} \geq t^*$, we can rewrite (5.2.8) as

$$\begin{aligned} (x_1, h) = \big(T(t_{n_k})x_0, h\big) &+ \int_0^{t^*} \big(T(t^* - \theta)B(\theta)u_{n_k}(\theta), T^*(t_{n_k} - t^*)h\big)\,d\theta \\ &+ \int_{t^*}^{t_{n_k}} \big(T(t_{n_k} - \theta)B(\theta)u_{n_k}(\theta), h\big)\,d\theta. \end{aligned} \tag{5.2.9}$$

By virtue of the facts that $T(t)$ is a strongly continuous semigroup in H and H is Hilbert, hence reflexive, $T^*(t)$ is also a strongly continuous semigroup in H with the infinitesimal generator A^* (see Theorem 1.3.11 and the remark following the theorem).

Thus as $k \to \infty$

$$\begin{cases} \text{(i)} & T(t_{n_k})x_0 \overset{s}{\to} T(t^*)x_0 (\text{s} = \text{strongly}) \text{ in } H \\ \text{(ii)} & T^*(t_{n_k} - t^*)h \overset{s}{\to} T^*(0)h = h \text{ strongly in } H(H = H^*) \\ \text{(iii)} & T(t^* - \cdot)B(\cdot)u_{n_k}(\cdot) \overset{w(w^*)}{\to} T(t^* - \cdot)B(\cdot)u^0(\cdot) \text{ in } L_p(L_\infty). \end{cases} \tag{5.2.10}$$

Since $B \in L_\infty(0, \tau; \mathfrak{L}(E, H))$, we have

$$\operatorname*{ess\,sup}_{t \in (0, \tau)} \| B(t) \|_{\mathfrak{L}(E, H)} \le b < \infty;$$

and since T is a c_0-semigroup

$$\sup_{0 \le t \le \tau} \| T(t) \|_{\mathfrak{L}(H)} \le M < \infty$$

for $\tau < \infty$ (see Theorem 1.3.1). Therefore, for $1 < p < \infty$,

$$\left| \int_{t^*}^{t_{n_k}} \left(T(t_{n_k} - \theta) B(\theta) u_{n_k}(\theta), h \right) d\theta \right| \le Mb |h|_H \int_{t^*}^{t_{n_k}} |u_{n_k}(\theta)|_E \, d\theta$$

$$\le Mb |h|_H \left(\int_{t^*}^{t_{n_k}} |u_{n_k}(\theta)|_E^p \, d\theta \right)^{1/p} \left(t_{n_k} - t^* \right)^{1/q}, \tag{5.2.11}$$

and for $p = \infty$,

$$\left| \int_{t^*}^{t_{n_k}} \left(T(t_{n_k} - \theta) B(\theta) u_{n_k}(\theta), h \right) d\theta \right|$$

$$\le Mb |h|_H \cdot \operatorname*{ess\,sup}_{0 \le \theta \le \tau} |u_{n_k}(\theta)|_E \cdot \left(t_{n_k} - t^* \right). \tag{5.2.12}$$

Since $\mathfrak{U}_a$ is a bounded subset of $L_p(0, \tau; E)$, $1 < p \le \infty$, it follows from (5.2.11) and (5.2.12) that as $t_{n_k} \to t^*$,

$$\int_{t^*}^{t_{n_k}} T(t_{n_k} - \theta) B(\theta) u_{n_k}(\theta), h \big)_H \, d\theta \to 0. \tag{5.2.13}$$

Using (5.2.10) and (5.2.13) in (5.2.9), we obtain, as $k \to \infty$,

$$(x_1, h) = (T(t^*)x_0, h) + \int_0^{t^*} \left(T(t^* - \theta) B(\theta) u^0(\theta), h \right)_H d\theta. \tag{5.2.14}$$

Since h is arbitrary, we have

$$x_1 = T(t^*)x_0 + \int_0^{t^*} T(t^* - \theta) B(\theta) u^0(\theta) \, d\theta. \tag{5.2.15}$$

The equality (5.2.15) implies that $u^0 \in \mathfrak{U}_0$. Since, by definition, $\tilde{t}(u^0) = \min\{t \in [0, \tau] : x(t, u^0) = x_1\}$ and, due to (5.2.15), $t^* \in \{t \in [0, \tau] : x(t, u^0) = x_1\}$, it is clear that $\tilde{t}(u^0) \le t^*$. On the other hand, $t^* \equiv \inf K$, and consequently, $t^* \le \tilde{t}(u^0)$. Thus, $\tilde{t}(u^0) = t^*$. This proves the theorem. □

For problems in which the target is a set, varying with time, we can formulate a similar time optimal control problem. Let K be a closed convex bounded set in H with $K^{\varepsilon} \equiv \{z \in H : \inf_{y \in K} |z-y|_H < \varepsilon\}$ its ε-neighbourhood. We need the following.

Definition 5.2.1. A set-valued map $t \to \sigma(t)$ with values from the class of closed convex bounded sets in H is said to be *upper semicontinuous with respect to inclusion* on $[0, \tau] = I$ if for each $t^* \in I$ and any sequence t_n converging to t^* and arbitrary $\varepsilon > 0$ there exists a number $n_0 = n_0(t^*, \varepsilon)$ such that $\sigma(t_n) \subset \sigma^{\varepsilon}(t^*)$ for all $n > n_0$.

We state the following result, whose proof differs trivially from that of the previous theorem.

Theorem 5.2.5. *Consider the system* (5.2.2) *with the operators A and B satisfying the assumptions of Theorem* 5.2.1. *Let $x_0 \in H \backslash \sigma(0)$, where $t \to \sigma(t)$ is a set-valued map with values in the space of closed convex bounded sets in H and upper semicontinuous on* $[0, \tau]$ *with respect to inclusion. Suppose the set $\mathcal{U}_0 \equiv \{u \in \mathcal{U}_{\mathrm{a}} : x(0, u) = x_0,\ x(t, u) \in \sigma(t)$ for some $t \in [0, \tau]\}$ is non-empty. Then there exists an optimal control u^0 that transfers the system from the state x_0 to the target set $\sigma(t^*)$ in minimum time $t^* = \tilde{t}(u^0) \leq \tilde{t}(u)$, $u \in \mathcal{U}_0$.*

Controllability. In the previous two theorems, we assumed the existence of controls that transfer the system from a given state to a desired target. This leads to the important concept known as controllability. We shall consider this question only briefly and refer the reader to the current literature for deeper study. Let X and E be a pair of separable Banach spaces and A a closed linear operator in X with domain $D(A)$ dense in X. Let $T(t)$, $t \geq 0$, be a strongly continuous semigroup in X with A as its generator. Let $B \in \mathcal{L}(E, X)$ and $\mathcal{U}_{\mathrm{a}} \subset L_p^{\mathrm{loc}}(E)$, $1 \leq p \leq \infty$, where $L_p^{\mathrm{loc}}(E)$ denotes the equivalence classes of strongly measurable and locally pth-power summable E-valued functions on $[0, \infty)$. Consider the system

$$S \quad \begin{cases} \dfrac{d}{dt} x(t) = & Ax(t) + Bu(t), \qquad x(0) = x_0 \in H, \qquad t \geq 0, \\ u \in & \mathcal{U}_{\mathrm{a}} \subset L_p^{\mathrm{loc}}(E) \end{cases}$$

with the state trajectory given by

$$x(t, u) = T(t)x_0 + \int_0^t T(t-\theta)Bu(\theta)\, d\theta, \qquad u \in \mathcal{U}_{\mathrm{a}}. \tag{5.2.16}$$

Define

$$K(t) \equiv \left\{ y \in X : y = T(t)x_0 + \int_0^t T(t-\theta)Bu(\theta)\, d\theta,\ u \in \mathcal{U}_{\mathrm{a}} \right\}.$$

Definition 5.2.2. The system S is said to be *approximately* [*exactly*] *controllable over the time interval* $[0, t]$ if $K(t)$ is dense in X $[K(t)=X]$. The system S is said to be *approximately* [*exactly*] *controllable in finite time if* $\bigcup_{t>0} K(t)$ *is dense in* X $[\bigcup_{t>0} K(t)=X]$.

Depending on the choice of the class of admissible controls $\mathcal{U}_a$, the solution x, given by (5.2.16), is either a strong or a mild solution. If, for each $u \in \mathcal{U}_a$, the trajectory x is a strong solution, then $x(t, u) \in D(A)$ for all $t>0$ and, since for unbounded operators A the domain $D(A)$ is only dense in X, it is clear that exact controllability is out of question. Even for mild (weak) solutions this conclusion is true if the operator $T(t)B$ is compact for $t \geq 0$. We present here only an elementary but fundamental criterion for controllability. Since for a given $x_0 \in X$ and fixed t, $T(t)x_0$ is fixed, it suffices for our purpose to consider the set

$$K_0(t)=\left\{y \in X: \int_0^t T(t-\theta)Bu(\theta)\,d\theta,\ u \in L_p(0, t; E)\right\}.$$

Define

$$Lu \equiv \int_0^t T(t-\theta)Bu(\theta)\,d\theta, \qquad u \in L_p(0, t; E).$$

Clearly, L is a bounded linear operator from $L_p(0, t; E)$ into X and its adjoint $L^* \in \mathcal{L}(X^*, L_q(0, t; E^*))$, where $(1/p)+(1/q)\equiv 1$ and $1 \leq p < \infty$.

Theorem 5.2.6. *The necessary and sufficient condition for the system S to be approximately controllable over the time interval* $[0, t]$, $0<t<\infty$, *is that* $\{x^* \in X^*: L^*(x^*)=0\} \equiv \operatorname{Ker} L^* = \{0\}$, *or equivalently*, $x^*(T(\theta)BE)=0$ *for* $\theta \in [0, t]$ *and* $x^* \in X^*$ *implies* $x^*=0$

PROOF. Necessary condition: If the system is approximately controllable over the interval $[0, t]$, then, by definition, the set $K_0(t)$ and hence $K(t)$ is dense in X, and consequently, $x^*(y)=0$ for all $y \in K_0(t)$ implies $x^*=0$. The statement that $x^*(y)=0$ for all $y \in K_0(t)$ is equivalent to $(x^*, Lu)_{X^*-X}=0$ for all $u \in L_p(0, t; E)$, and consequently, $L^*x^*=0$. Thus $x^* \in \operatorname{Ker} L^*$ and, due to denseness of $K_0(t)$, $x^*=0$. Therefore $\operatorname{Ker} L^*=\{0\}$. Equivalently,

$$x^*\left(\int_0^t T(t-\theta)Bu(\theta)\,d\theta\right)=0 \quad \text{for all} \quad u \in L_p(0, t; E) \Rightarrow x^*=0,$$

or

$$x^*\left(\int_0^t (T(\theta)Bv(\theta)\,d\theta\right)=0 \quad \text{for all} \quad v \in L_p(0, t; E) \Rightarrow x^*=0. \tag{5.2.17}$$

Let $\tau \in (0, t)$, $J(\subset [0, t])$ contain the point τ in its interior, and the Lebesgue measure $\mu(J)\to 0$. Define for $e\in E$

$$v(\theta)=\begin{cases} e, & \theta\in J \\ 0 & \text{elsewhere}\end{cases}$$

Then (5.2.17) is equivalent to

$$x^*\left(\frac{1}{\mu(J)}\int_J (T(\theta)Be)\,d\theta\right)=0 \quad \text{for all} \quad e\in E. \tag{5.2.18}$$

Since the semigroup $T(t)$ is strongly continuous, it follows from (5.2.18), on letting $\mu(J)\to 0$, that $x^*(T(\tau)Be)=0$ for all $e\in E$, or equivalently, $x^*(T(\tau)BE)=0$. Since $\tau\in[0,t]$ is arbitrary and $T(t)$ strongly continuous, $x^*(T(\theta)BE)=0$ for all $\theta\in[0,t]$. This implies the equivalence of the necessary conditions.

Sufficient condition: Let $\operatorname{Ker} L^*=\{0\}$ and suppose that $K_0(t)$ is not dense in X. Then there exists at least one $x\in X$ at a positive distance from the linear subspace $K_0(t)$. Therefore, as a consequence of the Hahn–Banach theorem (see Sections 1.1.5, 1.1.6), there exists a nonzero functional $x^*\in X^*$ such that $x^*(K_0(t))=0$ and $|x^*|_{X^*}>0$. But $x^*(K_0(t))=0$ implies $(x^*, Lu)=0$ for all $u\in L_p(0,t;E)$, or equivalently, $L^*x^*=0$ as an element of $(L_p(0,t;E))^*$. Thus, $x^*\in\operatorname{Ker} L^*$, and consequently, $x^*=0$, which is a contradiction. Therefore, $K_0(t)$ is dense in X. This completes the proof. □

In the Hilbert space situation, we have the following result.

Corollary 5.2.1. *If both X and E are self-adjoint separable Hilbert spaces and* $\mathfrak{U}_a=L_2^{\text{loc}}(E)$, *the following conditions are equivalent*:

i. *The system* S *is approximately controllable in the interval* $[0,t]$.
ii. $\operatorname{Ker} L^*=\{0\}$.
iii. $LL^*\in\mathfrak{L}^+(H)$ (*space of self-adjoint positive operators in* H).

PROOF. The equivalence of (i) and (ii) follows from Theorem 5.2.6. We show (ii)⇒(iii)⇒(i). For $h\in H$, $u\in L_2(0,t;E)$,

$$(Lu,h)_H=(u,L^*h)_{L_2(0,t;E)}\equiv\int_0^t (u(\theta), B^*T^*(t-\theta)h)_E\,d\theta. \tag{5.2.19}$$

It is clear that for every $h\in H$, $(L^*h)(\cdot)\equiv B^*T^*(t-\cdot)h$ is an element of $L_2(0,t;E)$; thus, for $u=L^*h$, we have

$$(LL^*h,h)_H=\|L^*h\|^2_{L_2(0,t;E)}. \tag{5.2.20}$$

Since $\operatorname{Ker} L^*=\{0\}$, it follows from (5.2.20) that $LL^*\in\mathfrak{L}^+(H)$. Assuming that $LL^*\in\mathfrak{L}^+(H)$, we show that $K_0(t)$ is dense in H, where $K_0(t)=\{Lu, u\in\mathfrak{U}\}$. For $h\in H$, $h(Lu)\equiv(Lu,h)_H=(u,L^*h)_{L_2(0,t;E)}=0$ for all $u\in$

$L_2(0, t; E)$ implies $L^*h=0$, and consequently, $(LL^*h, h)_H=0$. But since $LL^* \in \mathfrak{L}^+(H)$, this equality implies that $h=0$. Thus, range L is dense in H, or equivalently, $K_0(t)$ is dense in H, which is (i) by definition. □

REMARK 5.2.1. We note that, according to Theorem 5.2.6, a necessary and sufficient condition for approximate controllability is that $\bigcup_{\theta \in [0, t]}(T(\theta)B)$ be dense in X. It is a well-known fact for finite-dimensional problems that this condition is equivalent to range$\{A^k B, k=0,1,\ldots, n-1\}=R^n$, or rank$\{B, AB,\ldots, A^{n-1}B\}=n$. In general, for infinite-dimensional problems there is no such simple and elegant result. However, analogous to the rank condition for finite-dimensional problems, we have the following result for infinite-dimensional problems.

Theorem 5.2.7. *Consider the system* S *with initial condition* $x_0=0$. *Suppose that* A *is the generator of a strongly continuous semigroup* (c_0-*semigroup*) $T(t)$, $t>0$, *and* $B \in \mathfrak{L}(E, X)$. *Then, for approximate controllability of the system* S, *the condition*

$$\operatorname{cl\,span}\{A^n BE, n=0,1,2,\ldots.\}=X \tag{5.2.21}$$

(*where* cl span *is the closure of the linear span*) *is* (i) *sufficient if* $BE \subset D_\infty(A) \equiv \bigcap_{n=1}^{\infty} D(A^n)$ *and* (ii) *both necessary and sufficient if* $T(t)$, $t>0$, *is a holomorphic semigroup and* $BE \subset X_0 \equiv \bigcup_{t>0} T(t)X$.

PROOF. We give the proof by establishing contradiction. For part (i) suppose that cl span$\{A^n BE, n=0,1,2\ldots.\}=X$ and the system is not controllable. Then by virtue of Theorem 5.2.6 there exists a nontrivial $x^* \in X^*$ such that $x^*(T(t)Be)=0$ for all $t>0$ and $e \in E$. Since $T(t)$, $t>0$, is a c_0-semigroup, it follows from this that $x^*(Be)=0$. By assumption, $BE \subset D_\infty(A)$, and consequently, for any positive integer n, $(d^n/dt^n)x^*(T(t)Be) = x^*(T(t)A^n Be)$ is well defined and equals zero for all $t>0$. Since $T(t)$, $t>0$, is a c_0-semigroup, $\lim_{t \downarrow 0} x^*(T(t)A^n Be)=x^*(A^n Be)$. Thus, $x^*(A^n Be) =0$ for all nonnegative integers n and for all $e \in E$. In other words, $x^*(\text{span}\{A^n Be, n=0,1,2,\ldots.\})=0$. Since span$\{A^n BE, n=0,1,2,\ldots\}$ is dense in X, this implies that $x^*=0$. This leads to a contradiction proving that the condition cl span$\{A^n BE, n=0,1,2\ldots\}=X$ implies (approximate) controllability.

For part (ii), first we note that a holomorphic semigroup is characterized by the property that $T(t)X \subset D(A)$ for all $t>0$. This, in fact, implies (Theorem 1.3.4) that $T(t)X \subset D(A^n)$ for all $n \geq 1$, and consequently, $T(t)X \subset D_\infty(A) \equiv \bigcap_{n=1}^{\infty} D(A^n)$ for $t>0$. In other words, $X_0 \equiv \bigcup_{t>0} T(t)X \subset D_\infty(A)$. Therefore, sufficiency of the condition follows from part (i). For the proof of the necessary condition, suppose the system is approximately controllable over finite time. For any $e \in E$, under the given assumption, $Be \in X_0$, and consequently, there exists a $t_e>0$ and $x_e \in X$ such that $Be=T(t_e)x_e$. Suppose that span$\{A^n BE, n=0,1,2,\ldots\}$ is not dense in X.

Then, as a consequence of Hahn–Banach theorem, there exists a nonzero $x^* \in X^*$ such that $x^*(A^n Be)=0$ for all nonnegative integers n and all $e \in E$. Since $T(t)$, $t>0$, is a holomorphic semigroup, it follows from Theorem 1.3.4 that $(d^n/dt^n)T(t)=A^n T(t)$ for all integers $n \geq 0$; and for each $t>0$, $A^n T(t)$ is a bounded linear operator in X and $t \to A^n T(t)$ is continuous in the uniform operator topology. Thus,

$$\frac{d^n}{dt^n} x^*(T(t)x_e) = x^*(T^{(n)}(t)x_e) = x^*(A^n T(t)x_e)$$

and $t \to x^*(A^n T(t)x_e)$ is continuous on $(0, \infty)$ for all $n \geq 0$. But $x^*(A^n T(t_e)x_e) = x^*(A^n Be) = 0$. In summary, the function $t \to x^*(T(t)x_e)$ is C^∞ on $(0, \infty)$ and along with its derivatives (of all orders) vanishes at $t=t_e$. Therefore, $x^*(T(t)x_e) \equiv 0$ for $t>0$. For $t \geq t_e$,

$$x^*(T(t)x_e) = x^*(T(t-t_e)T(t_e)x_e) = x^*(T(t-t_e)Be) = 0,$$

or equivalently, $x^*(T(t)Be)=0$ for $t>0$. Since the system is approximately controllable and e is an arbitrary element of E, it follows from the above equality that $x^*=0$. The contradiction proves the theorem. □

REMARK 5.2.2. We have briefly touched upon the question of (approximate) controllability of a system described by first-order evolution equations. Naturally, similar questions arise for higher-order evolution equations. Recently Triggiani [Trig.2] has discovered several interesting connections between the first- and second-order systems regarding approximate controllability. In fact, he has proved the existence of a set $X_0 \subset X$, dense in X, such that if $BE \subset X_0$, then the second-order system $(d^2/dt^2)x(t) = Ax + Bu$ is approximately controllable if and only if the first-order system $(d/dt)x(t) = Ax + Bu$ is. We refer the reader to the original literature for further details.

The relation between controllability and normality for finite-dimensional systems is well known. For infinite-dimensional systems, this question has been discussed by Knowles [Kn.3] specially for the class of equations of the form $(d/dt)x = Ax + gu$, where g is a fixed element of X and the control u is a scalar function. Such a system is said to be *normal in X* if, for all $\tau > 0$,

$$x^*(T(\tau-\theta)g)=0 \quad \text{for} \quad \theta \in J \subset (0,\tau), \qquad \mu(J)>0,$$

implies $x^*=0$.

In the sequel, we shall have occasion to study the bang-bang principle and its relation to the concept of normality. At this point, we simply note that if for the given g, $\theta \to T(\theta)g$ is analytic and the system $(d/dt)x = Ax + gu$ is approximately controllable, then it is also normal in X. Indeed, suppose that for $\tau > 0$, $x^* \in X^*$, and $J \subset (0, \tau)$ with $\mu(J) > 0$, we have $x^*(T(\tau-\theta)g) = 0$ for $\theta \in J$. Then since $\theta \to T(\theta)g$ is analytic, it is clear that $\theta \to x^*(T(\tau-\theta)g)$ is an analytic scalar function and, since it vanishes on J

(having positive measure), it vanishes for all $\theta \in (0, \tau)$, and consequently $x^*(T(t)g) = 0$ for all $t \in (0, \tau)$. But the approximate controllability of the system then implies that $x^* = 0$. Thus, according to the definition, the system is normal. Later in this section we shall see that under certain situations normality assures the existence of bang-bang controls.

5.2.6. Necessary Conditions for Optimality

In the preceding section, we have discussed the questions of controllability and existence of optimal controls. In the present section, we wish to develop the necessary and sufficient conditions for optimality that can be used to construct the optimal control if one exists. We recall that $(Lf)(t) \equiv \int_0^t T(t-\theta)B(\theta)f(\theta)\,d\theta$, $z(t) = T(t)x_0$, and y is a given element of $L_2(I, H)$.

Theorem 5.2.8. *Consider the system* (5.2.2) *with the cost function given by* (5.2.4). *Suppose that the assumptions of Theorem* 5.2.2 *hold. Then a necessary and sufficient condition for* $u_0 \in \mathfrak{U}_a$ *to be an optimal control is that*

$$((L^*RL+N)u_0 + L^*R(z-y), u-u_0)_{\mathfrak{U}} \geq 0 \tag{5.2.22}$$

for all $u \in \mathfrak{U}_a$.

In the unconstrained case (*i.e.*, $\mathfrak{U}_a = \mathfrak{U}$), *the optimal control is given by*

$$u_0 = -(N+L^*RL)^{-1}L^*R(z-y). \tag{5.2.23}$$

PROOF. Since J is quadratic in u, it is strictly convex, and thus, a necessary and sufficient condition for $u_0 \in \mathfrak{U}_a$ to be optimal (Theorem 1.1.23) is that $J'_{u_0}(u-u_0) \geq 0$ for all $u \in \mathfrak{U}_a$, where $J'_{u_0}(v)$ is the Gateaux differential of J at u_0 in the direction v. Performing the Gateaux differential of J, using (5.2.5), we obtain the result immediately. In the unconstrained case, the inequality of (5.2.22) reduces to equality, and consequently, (5.2.23) follows.

In the unconstrained case, it is readily shown that u_0 given by (5.2.23) minimizes $J(u)$. For, let $M = (L^*RL+N)$, $g = L^*R(z-y)$, and $c = (R(z-y), z-y)$; then $J(u) = (Mu, u)_{\mathfrak{U}} + 2(g, u)_{\mathfrak{U}} + c$ and

$$J(u) - J(u_0) = (M(u-u_0), u-u_0)_{\mathfrak{U}}. \tag{5.2.24}$$

Since $L^*RL \geq 0$ and $N > 0$, it is clear that $M > 0$, and consequently, $J(u) \geq J(u_0)$ for all $u \in \mathfrak{U}$. The minimum is given by

$$J(u_0) = (R(z-y), x^0 - y)_{\mathfrak{X}},$$

where $x^0 = Lu_0 + z$ is the solution of the evolution equation (5.2.2) corresponding to the control u_0.

It may be tempting to conclude that the necessary condition given by (5.2.22) is explicit and does not require the adjoint (costate) equation. This conclusion is not true, however, since it is only in some trivial cases that the semigroup can be explicitly determined. Thus, for any practical application it is almost essential to introduce the adjoint equation as in the previous section (Section 5.1.6) and write the necessary conditions of optimality in terms of the adjoint state. We can do this easily by using the necessary condition (5.2.22). □

Theorem 5.2.9. *Under the assumptions of Theorem* 5.2.8, *the necessary conditions of optimality are given by*

$$\begin{cases} \text{(i)} & \begin{cases} \dfrac{dx^0}{dt} - Ax^0 = Bu^0, \quad 0 < t \leq \tau, \quad x^0(0) = x_0 \end{cases} \\ \text{(ii)} & \begin{cases} \dfrac{dp}{dt} + A^*p = -R(x^0 - y), \quad 0 \leq t < \tau, \quad p(\tau) = 0 \end{cases} \\ \text{(iii)} & (B^*p + Nu^0, \quad u - u^0)_{\mathcal{U}} \geq 0 \quad \text{for all} \quad u \in \mathcal{U}_a. \end{cases} \tag{5.2.25}$$

PROOF. Let u^0 denote the optimal control. Define

$$p(t) = \int_t^\tau T^*(s-t)R(s)(x^0(s) - y(s))\, ds, \qquad t \in (0, \tau) \tag{5.2.26}$$

where x^0 is the weak solution (Definition 2.5.2) of the problem (i) corresponding to the control u^0. Then, for any $h \in D(A)$,

$$(p(t), h)_H \equiv \int_t^\tau (R(s)(x^0(s) - y(s)), T(s-t)h)_H\, ds$$

is well defined for $t \in (0, \tau)$. Since $T(t)$, $t > 0$ is a c_0-semigroup and $R(x^0 - y) \in L_2(0, \tau; H)$, $t \to (p(t), h)$, is absolutely continuous. Thus, for each t for which $R(x^0 - y)$ is defined, we have

$$\frac{d}{dt}(p(t), h)_H = -(R(t)(x^0(t) - y(t)), h)_H$$
$$-\left(\int_t^\tau T^*(s-t)R(s)(x^0(s) - y(s))\, ds, Ah\right)_H$$

Consequently, for all $h \in D(A)$,

$$\frac{d}{dt}(p(t), h) + (p(t), Ah) = -(R(t)(x^0(t) - y(t)), h) \tag{5.2.27}$$

for almost all $t \in [0, \tau]$ and $p(\tau) = 0$. Thus, p, as defined by (5.2.26), solves the problem (ii) in the weak sense. From the definition of L it follows that its adjoint L^* is given by

$$(L^*z)(t) = \int_t^\tau B^*T^*(s-t)z(s)\, ds \tag{5.2.28}$$

for $z \in \mathcal{H}$ with $z(\theta) \in D(A^*)$, $\theta \in (0, \tau)$. Therefore,

$$B^*p = L^*R(x^0 - y). \tag{5.2.29}$$

For u^0 to be optimal it is necessary that the inequality (5.2.22) be satisfied; that is,

$$\left(L^*R(Lu^0+z-y)+Nu^0, u-u^0\right)_{\mathcal{U}} \geq 0 \quad \text{for all} \quad u \in \mathcal{U}_a. \tag{5.2.30}$$

Since $Lu^0+z=x^0$, we have, using (5.2.29) and (5.2.30),

$$\left(B^*p+Nu^0, u-u^0\right)_{\mathcal{U}} \geq 0 \quad \text{for all} \quad u \in \mathcal{U}_a. \tag{5.2.31}$$

This completes the proof. □

5.2.7. Optimal Feedback Control

In terms of the adjoint state p, (5.2.23) becomes

$$u^0(t) = -N^{-1}(t)B^*p(t). \tag{5.2.23$'$}$$

Thus, in the case of the unconstrainted problem, we can substitute (5.2.23)′ into (i), the evolution equation (5.2.25) and obtain the coupled system of equations

$$\left.\begin{aligned} &\frac{dx^0}{dt} - Ax^0(t) + BN^{-1}(t)B^*p(t) = 0, \qquad && x^0(0) = x_0 \\ &\frac{d}{dt}p + A^*p(t) + R(t)\left(x^0(t) - y(t)\right) = 0, \qquad && p(\tau) = 0 \end{aligned}\right\} \quad t \in (0, \tau). \tag{5.2.32}$$

This is entirely similar to the two-point boundary value problem encountered in Section 5.1.7. As shown in that section, we can generate the optimal control as an affine transformation of the state x^0. That is,

$$u^0(t) = -N^{-1}(t)B^*\left(P(t)x^0(t) + \gamma(t)\right), \tag{5.2.33}$$

where $P(t)$, $t \in (0, \tau)$, is a family of positive self-adjoint bounded linear operators in H. Substituting $p(t) = P(t)x^0 + \gamma(t)$ in the state equation, we obtain

$$\begin{aligned} &\frac{dx^0}{dt} = (A - BN^{-1}B^*P)x^0 - BN^{-1}B^*\gamma, \\ &x^0(0) = x_0. \end{aligned} \tag{5.2.34}$$

By virtue of Theorem 2.4.4, we know that the generator of a transition semigroup (evolution operator) is stable under bounded additive perturbation. Therefore, $(A - BN^{-1}B^*P)$ is the generator of a strongly continuous evolution (or transition) operator ϕ satisfying the properties

$$\begin{aligned} &\phi(t,t) = I, \\ &\phi(t,s)\phi(s,\theta) = \phi(t,\theta) \quad \text{for} \quad 0 \leq \theta < s < t \leq \tau, \\ &\frac{\partial}{\partial s}\phi(t,s)\eta = -\phi(t,s)\left(A - BN^{-1}B^*P(s)\right)\eta \end{aligned} \tag{5.2.35}$$

for $0 < s < t < \tau$ and $\eta \in D(A)$.

Thus, the solution of the problem (5.2.34) is given by

$$x^0(t)=\phi(t,0)x_0-\int_0^t\phi(t,\theta)(BN^{-1}(\theta)B^*)\gamma(\theta)\,d\theta,\qquad t\in(0,\tau). \tag{5.2.36}$$

From (5.2.29) and (5.2.23)′, we have

$$u^0=-N^{-1}L^*R(x^0-y),$$

or equivalently,

$$u^0(t)=-N^{-1}(t)B^*\int_t^\tau T^*(s-t)R(s)(x^0(s)-y(s))\,ds. \tag{5.2.37}$$

Substituting (5.2.36) into (5.2.37), we obtain

$$u^0(t)=-N^{-1}(t)B^*\left(\int_t^\tau ds\,T^*(s-t)R(s)\phi(s,t)\right)x^0(t)$$

$$+N^{-1}(t)B^*\int_t^\tau ds T^*(s-t)R(s)\left\{\int_t^s d\theta\,\phi(s,\theta)BN^{-1}(\theta)B^*\gamma(\theta)+y(s)\right\}. \tag{5.2.37}$$

Comparing (5.2.37)′ with (5.2.33), we obtain

$$P(t)h=\int_t^\tau ds\,T^*(s-t)R(s)\phi(s,t)h\quad\text{for}\quad h\in D(A) \tag{5.2.38}$$

and

$$\gamma(t)+\int_t^\tau d\theta\,T^*(\theta-t)P(\theta)BN^{-1}(\theta)B^*\gamma(\theta)+\int_t^\tau d\theta\,T^*(\theta-t)R(\theta)y(\theta)=0 \tag{5.2.39}$$

for $t\in(0,\tau)$.

These are functional (or integral) equations in the unknowns P and γ. It is convenient to transform these equations into the usual differential forms. Setting $y\equiv0$ in (5.2.32) and considering the resulting Cauchy problem over $t<s<\tau$ with $x^0(t)=h$, it can be shown, as in the previous section, that the mapping $h\to p(t)$ is a bounded linear transformation in H with $p(t)=P(t)h$. Thus, $(A-BN^{-1}(t)B^*P(t))$ generates a strongly continuous evolution operator ϕ as given in (5.2.35).

Proposition 5.2.1. Under the assumptions of Theorem 5.2.8, P and γ satisfy the following differential equations:

$$\frac{d}{dt}(P(t)h,\eta)+(P(t)h,A\eta)+(Ah,P(t)\eta)$$

$$-(P(t)BN^{-1}(t)B^*P(t)h,\eta)+(R(t)h,\eta)=0,$$

$$P(\tau)=0\quad\text{for almost all (a.a.) } t\in(0,\tau),\quad h,\eta\in D(A), \tag{5.2.40a}$$

$$\frac{d}{dt}(\gamma(t),h)+(\gamma(t),Ah)-(P(t)BN^{-1}(t)B^*\gamma(t),h)$$
$$-(R(t)y(t),h)=0$$
$$\gamma(\tau)=0,\quad \text{for a.a. } t\in(0,\tau),\quad h\in D(A), \tag{5.2.40b}$$

PROOF. The proof is formal. Scalar multiplying (5.2.38) by $\eta \in D(A)$, and (5.2.39) by $h\in D(A)$ and differentiating the resulting expressions, we use (5.2.35) to obtain (5.2.40a) and (5.2.40b). □

Solution of Equations (5.2.40). We have discussed the questions of existence and uniqueness of solutions of similar equations in Section 5.1.7. There we used the so-called fractional step method [compare Equation (5.1.135)], which gives a proof of the existence of a solution and also a method for constructing it. In this section we wish to briefly indicate another constructive approach, which is based on the principle of linearization and successive approximation. Equation (5.2.40a) can be written as

$$\frac{d}{dt}(P(t)h,\eta)+(A-KP)h,P\eta)+(Ph,(A-KP)\eta)+((PKP+R)h,\eta)=0,$$
$$P(\tau)=0,\qquad t\in(0,\tau),\quad h,\eta\in D(A), \tag{5.2.40a)$'$}$$

where $K\equiv BN^{-1}B^*$.

For linearization assume P_{n-1} $(\in L_\infty(I,\mathcal{L}(H)))$ is already known and then for P_n solve the linearized version of (5.2.40a)′ given by

$$\frac{d}{dt}(P_n(t)h,\eta)+((A-KP_{n-1})h,P_n\eta)+(P_nh,(A-KP_{n-1})\eta)$$
$$+((P_{n-1}KP_{n-1}+R)h,\eta)=0,\qquad t\in I=(0,\tau),$$
$$P_n(\tau)=0,\qquad h,\eta\in D(A),\quad n=1,2,3,\ldots. \tag{5.2.41}$$

Since $B\in L_\infty(I,\mathcal{L}(E,H))$, $N^{-1}\in L_\infty(I,\mathcal{L}(E))$, and E, H are self-adjoint, it is clear that $K\in L_\infty(I,\mathcal{L}(H))$, and therefore, $KP_{n-1}\in L_\infty(I,\mathcal{L}(H))$. Thus, by the perturbation theory (Theorem 2.4.4), $(A-KP_{n-1})$ is the generator of an evolution operator $\phi_n(t,s)$, $0<s\le t<\tau$, satisfying the properties (5.2.35). Using this operator we can solve (5.2.41):

$$(P_n(t)h,\eta)=\int_t^\tau d\theta(\phi_n^*(\theta,t)[R(\theta)+P_{n-1}(\theta)K(\theta)P_{n-1}(\theta)]\phi_n(\theta,t)h,\eta). \tag{5.2.42}$$

This way we obtain a sequence $\{P_n\}\in L_\infty(I,\mathcal{L}(H))$ and our task reduces to showing that P_n has a limit that solves (5.2.40a)′. At this point we note that if P_n has a limit P and ϕ_n has a limit ϕ that is the transition operator corresponding to the generator $(A-KP)$, then P satisfies the functional equation

$$(P(t)h,\eta)=\int_t^\tau d\theta(\phi^*(\theta,t)[R+PKP]\phi(\theta,t)h,\eta). \tag{5.2.43}$$

Differentiating (5.2.43) with respect to t and using the properties (5.2.35), one can easily check that P solves (5.2.40a)′.

For the proof of existence of a solution of the problem (5.2.40) we need the following definition. We denote by $\mathcal{L}'(H)$ the space of linear (not necessarily bounded) operators in H.

Definition 5.2.3. A transition operator $\phi(t,s)$, $s,t\in\Delta=\{0<s\le t<\tau\}$, mapping H into H is said to be a *mild evolution operator* if (i) $\phi(r,s)\phi(s,t)=\phi(r,t)$ for $0<t\le s\le r<\tau$ and (ii) $\phi(r,t)$ is weakly continuous in t on $(0,r)$ and in r on (t,τ). If, in addition, there exists an operator $A\in L_\infty(I,\mathcal{L}'(H))$ with closed values $A(t)$, $t\in I$, such that for $0<t\le r<\tau$, $(f,\phi(r,t)h-h)=\int_t^r d\theta(f,\phi(r,\theta)A(\theta)h)$ for $h\in\cap D(A(\theta))$ and $f\in H$, then ϕ is called a *quasi-evolution* operator.

Theorem 5.2.10. *Suppose that $I=(0,\tau)$ is a finite interval, A the generator of a c_0-semigroup, $B\in L_\infty(I,\mathcal{L}(E,H))$, $N\in L_\infty(I,\mathcal{L}(E))$ with $(N(t)u,u)_E\ge \nu|u|_E^2$ for $t\in I$ and $\nu>0$, and $R\in L_\infty(I,\mathcal{L}(H))$ with $R\ge 0$. Then there exists a unique $P\in L_\infty(I,\mathcal{L}(H))$ that satisfies the integral equation* (5.2.43) [*and in turn* (5.2.40a)] *and $P(t)$ is self-adjoint and positive.*

PROOF. For $y=0$ [Equation (5.2.32)] the operator P is entirely determined by the coupled system of equations

$$\left.\begin{aligned}&\frac{d\xi^0}{d\theta}-A\xi^0+(BN^{-1}B^*)p^0=0, && \xi^0(t)=h\\ &\frac{dp^0}{d\theta}+A^*p^0+R\xi^0=0, && p^0(\tau)=0\end{aligned}\right\}\quad \text{for}\quad \theta\in(t,\tau), \tag{5.2.44}$$

where P is such that $p^0(t)=P(t)h$. This system arises from the optimization problem

$$\frac{dx(\theta)}{d\theta}=Ax(\theta)+Bu(\theta),\qquad x(t)=h,\quad \theta\in(t,\tau),$$

$$\min J(t,h,u)=\min\int_t^\tau\{(R(\theta)x(\theta),x(\theta))_H+(N(\theta)u(\theta),u(\theta))_E\}\,d\theta,$$

$$u\in\mathcal{U}_a=L_2(t,\tau;E). \tag{5.2.45}$$

Therefore it suffices to construct a sequence of feedback controls $\{u_n\}$ in terms of certain positive self-adjoint operators $\{P_n\}\in L_\infty(I,\mathcal{L}(H))$ such that while u_n converges to the optimal control for the problem (5.2.45), P_n converges to a self-adjoint positive operator $P\in L_\infty(I,\mathcal{L}(H))$ that satisfies the functional equation (5.2.43).

Define for positive integers n

$$\begin{cases} \text{(i)} & F_n = -N^{-1}B^*P_{n-1},\ P_0 = 0 \\ \text{(ii)} & u_n = F_n x \ (x \text{ is the state corresponding to the control } u_n) \\ \text{(iii)} & L_n = BF_n = -BN^{-1}B^*P_{n-1} = -KP_{n-1} \\ \text{(iv)} & M_n = R + P_{n-1}KP_{n-1} \\ \text{(v)} & (P_n(t)h,\eta) = \int_t^\tau d\theta(\phi_n^*(\theta,t) \\ & \times[R(\theta)+P_{n-1}(\theta)K(\theta)P_{n-1}(\theta)]\phi_n(\theta,t)h,\eta), \end{cases} \tag{5.2.46}$$

where ϕ_n is the evolution operator corresponding to the generator $A+BF_n = A - KP_{n-1} \equiv A_n$. Since $KP_{n-1} \in L_\infty(I, \mathcal{L}(H))$, it follows from the previous discussion that for each n, ϕ_n is a mild evolution operator in H. From (5.2.45) and (5.2.46) it is easy to verify that

$$(P_n(t)h, h) = \int_t^\tau \{(Rx_n, x_n) + (Nu_n, u_n)_E\}\, d\theta > 0, \tag{5.2.47}$$

where x_n is the solution of the evolution equation of (5.2.45) corresponding to $u = u_n$. Note that (v) of Equation (5.2.46) is identical to Equation (5.2.42), in which P_n is the solution of the problem (5.2.41).

Define

$$\begin{aligned} P_n - P_{n-1} &= Q_n, \\ A - KP_{n-1} &= A_n. \end{aligned} \tag{5.2.48}$$

Then we obtain from (5.2.41) and (5.2.48)

$$\frac{d}{dt}(Q_n h, \eta) + (A_n h, Q_n \eta) + (Q_n h, A_n \eta) - (KQ_{n-1}h, Q_{n-1}\eta) = 0$$

$$Q_n(\tau) = 0. \tag{5.2.49}$$

The solution for this problem is given by

$$(Q_n(t)h,\eta) = -\int_t^\tau (Q_{n-1}(\theta)K(\theta)Q_{n-1}(\theta)\phi_n(\theta,t)h, \phi_n(\theta,t)\eta)\, d\theta.$$

Since K is a nonnegative operator from $L_\infty(I, \mathcal{L}(H))$ and $\{Q_n\}$ is symmetric, we have $(Q_n(t)h, h) \leq 0$ for $h \neq 0$, and consequently,

$$(P_n(t)h, h) \leq (P_{n-1}(t)h, h) \quad \text{for all} \quad t \in I, \quad h \in H. \tag{5.2.50}$$

Thus, $\{P_n\}$ is a monotone nonincreasing sequence of elements from $L_\infty(I, \mathcal{L}^+(H))$, [$\mathcal{L}^+(H) \equiv$ positive self adjoint operators in H]. Since $R \in L_\infty(I, \mathcal{L}^+(H))$, $T(t)$, $t \geq 0$, is a c_0-semigroup, and τ is finite, there exists a positive number β such that

$$\begin{aligned} 0 \leq (P_1(t)h, h) &= \int_t^\tau d\theta (R(\theta)T(\theta - t)h, T(\theta - t)h) \\ &\leq \beta |h|_H^2 \quad \text{for all} \quad t \in (0, \tau). \end{aligned} \tag{5.2.51}$$

It follows from the inequalities (5.2.50) and (5.2.51) that

$$\sup_{t\in I}(P_n(t)h,h)\le\beta|h|_H^2 \quad \text{for all} \quad n\ge 1. \tag{5.2.52}$$

Thus for each $t\in I$, $P_n(t)$ converges in the strong operator topology to a positive self-adjoint operator $P(t)$ in H and $P\in L_\infty(I,\mathcal{L}^+(H))$. In fact, the convergence is uniform in t on $[0,\tau]$. We must show that this P solves the problem (5.2.43).

Since $P_n(t)\to P(t)$ in the strong operator topology uniformly on I, it follows from (i) of (5.2.46) that there exists an $F\in L_\infty(I,\mathcal{L}(H,E))$ such that $F_n(t)\to F(t)$ in the strong operator topology a.e. on I. Thus $L_n(t)\to L(t)\equiv -K(t)P(t)$ strongly a.e. and $M_n(t)\to M(t)\equiv R(t)+P(t)K(t)P(t)$ in the weak operator topology a.e. on I and $M\in L_\infty(I,\mathcal{L}^+(H))$. We note that the (mild) evolution operator $\phi_n(\cdot,\cdot)$ corresponding to the generator $A_n=A-KP_{n-1}$ is the solution of the integral equation

$$\phi_n(r,t)h=T(r-t)h+\int_t^r T(r-\theta)(BF_n)(\theta)\phi_n(\theta,t)h\,d\theta. \tag{5.2.53}$$

Since $\{BF_n=-KP_{n-1}\}$ is confined in a bounded subset of $L_\infty(I,\mathcal{L}(H))$, due to (5.2.52) and the hypotheses on B and N, it follows from Gronwall's lemma that there exists a finite number $\delta>0$ such that

$$\sup_{0<t\le r<\tau}|\phi_n(r,t)h|_H<\delta|h|_H \quad \text{independently of} \quad n. \tag{5.2.54}$$

Let ϕ be the solution of the integral equation

$$\phi(r,t)h=T(r-t)h+\int_t^r T(r-\theta)(BF)(\theta)\phi(\theta,t)h\,d\theta \tag{5.2.55}$$

corresponding to the generator $(A+BF)=(A-KP)$. Then ϕ also satisfies the estimate (5.2.54), and it follows from the Lebesgue dominated convergence theorem and the equality

$$(\phi_n h-\phi h)=\int_t^r TB(F_n-F)\phi h\,d\theta+\int_t^r TBF_n(\phi_n h-\phi h)\,d\theta \tag{5.2.56}$$

that ϕ_n converges to ϕ strongly (strong operator topology) uniformly on $0<t\le r<\tau$. Since $P_n\to P$ strongly uniformly on I, $\phi_n\to\phi$ strongly uniformly on $0<t\le r\le\tau$, and $M_n\to M\equiv R+PKP$ weakly a.e. on I, again by virtue of the Lebesgue dominated convergence theorem we obtain

$$\int_t^\tau d\theta(M_n(\theta)\phi_n(\theta,t)h,\phi_n(\theta,t)\eta)\to\int_t^\tau d\theta(M(\theta)\phi(\theta,t)h,\phi(\theta,t)\eta). \tag{5.2.57}$$

Therefore it follows from (v) of (5.2.46) and (5.2.57) that

$$(P(t)h,\eta)=\int_t^\tau d\theta(\phi^*(\theta,t)[R(\theta)+P(\theta)K(\theta)P(\theta)]\phi(\theta,t)h,\eta) \tag{5.2.58}$$

for all $h, \eta \in D(A)$. Thus P satisfies the integral equation (5.2.43), $P(t) \geq 0$, and is self-adjoint. We show that P satisfies the differential equation (5.2.40). Since $BF_n \in L_\infty(I, \mathcal{L}(H))$ and A is the generator of a c_0-semigroup, $A_n \equiv A + BF_n$ is the generator of a quasi-evolution operator ϕ_n satisfying

$$(f, \phi_n(r,t)h - h) = \int_t^r (f, \phi_n(r,\theta)A_n(\theta)h)\, d\theta \quad \text{for} \quad h \in D(A) \text{ and } f \in H. \tag{5.2.59}$$

Since $BF_n \to BF = -KP$ in the strong operator topology almost everywhere on I and $\phi_n \to \phi$ in the strong operator topology uniformly on Δ, it follows from (5.2.59) that

$$(f, \phi(r,t)h - h) = \int_t^r (f, \phi(r,\theta)\tilde{A}(\theta)h)\, d\theta \quad \text{for} \quad h \in D(A), \tag{5.2.60}$$

where $\tilde{A}(t) = (A + (BF)(t)) = (A - (KP)(t))$. Thus, ϕ is also a quasi-evolution operator with $\tilde{A}$ as its generator. Therefore,

$$\frac{\partial}{\partial t}(f, \phi(r,t)h) = -(f, \phi(r,t)\tilde{A}(t)h) \tag{5.2.60$'$}$$

and $P(t)$ is absolutely continuous in the weak operator topology in the sense that for $h, \eta \in D(A)$, $t \to (P(t)h, \eta)$ is absolutely continuous. As a consequence, for $h, \eta \in D(A)$, it follows from (5.2.60)′ that

$$\begin{aligned} \frac{d}{dt}(P(t)h, \eta) &= -(M(t)h, \eta) + \int_t^\tau \frac{\partial}{\partial t}(M(\theta)\phi(\theta,t)h, \phi(\theta,t)\eta)\, d\theta \\ &= -(M(t)h, \eta) - (P(t)h, \tilde{A}(t)\eta) - (\tilde{A}(t)h, P(t)\eta) \quad \text{a.e.,} \end{aligned} \tag{5.2.61}$$

where $M(t) \equiv R(t) + (PKP)(t)$. Thus P satisfies the differential equation (5.2.40a)′ and hence (5.2.40a). This proves existence. For uniqueness, suppose that $Q \in L_\infty(I, \mathcal{L}(H))$ is another solution of (5.2.40a) and define $\tilde{P} = P - Q$. Then one can easily verify that $\tilde{P}$ satisfies the differential equation

$$\begin{aligned} &\frac{d}{dt}(\tilde{P}(t)h, \eta) + (\tilde{P}(t)h, \tilde{A}(t)\eta) + (\tilde{A}(t)h, \tilde{P}(t)\eta) \\ &\quad + (\tilde{P}(t)K(t)\tilde{P}(t)h, \eta) = 0, \\ &\tilde{P}(\tau) = 0, \qquad h, \eta \in D(A), \quad t \in (0, \tau), \end{aligned} \tag{5.2.62}$$

and

$$(\tilde{P}(t)h, \eta) = \int_t^\tau (\phi^*(\theta,t)(\tilde{P}(\theta)K(\theta)\tilde{P}(\theta))\phi(\theta,t)h, \eta)\, d\theta, \qquad t \in (0, \tau), \tag{5.2.63}$$

for all $h, \eta \in D(A)$. Since

$$\sup_{0<t\leq\theta<\tau} \|\phi(\theta,t)\|_{\mathcal{L}(H)} < \delta,$$

we have

$$|(\tilde{P}(t)h, h)| \leq |h|_H^2 \int_t^\tau \delta^2 \|K(\theta)\|_{\mathcal{L}(H)} \|\tilde{P}(\theta)\|_{\mathcal{L}(H)}^2 \, d\theta \qquad (5.2.64)$$

for all $h \in D(A)$ and $t \in [0, \tau]$. Since $D(A)$ is dense in H, it follows from (5.2.64) that

$$\|\tilde{P}(t)\|_{\mathcal{L}(H)} \leq \delta^2 \int_t^\tau \|K(\theta)\|_{\mathcal{L}(H)} \|\tilde{P}(\theta)\|_{\mathcal{L}(H)}^2 \, d\theta \quad \text{for all} \quad t \in [0, \tau]. \qquad (5.2.65)$$

Thus $\tilde{P}(t) \equiv 0$, as implied by the following lemma. That $P(t)$ is positive and self-adjoint is obvious from (5.2.43). This completes the proof of the theorem. □

Lemma 5.2.1. *Let* $0 \leq x(t) \leq \alpha + \int_0^t h(\theta)\psi(x(\theta))\, d\theta$, $t \in (0, \tau)$, *where* $\alpha \geq 0$, $0 \leq h \in L_1$, *and* ψ *is a nonnegative, nondecreasing continuous function on* $[0, \infty]$. *Then if* $\sup_{\theta \in [0,\tau]} x(\theta) = b \geq \alpha$, *we have*

$$\int_\alpha^b \frac{d\xi}{\psi(\xi)} \leq \int_0^\tau h(\theta)\, d\theta < \infty. \qquad (5.2.66)$$

Further if $\alpha = 0$ *and* $\int_0^\varepsilon [d\xi/\psi(\xi)] = \infty$ *for every* $\varepsilon > 0$, *then* $x(t) \equiv 0$ *for* $t \in [0, \tau]$.

PROOF. See [C.1, Lemma 4.7, Corollary 4.8, p. 127]. □

We are now left with Equation (5.2.40b). For the existence and uniqueness of a solution of this problem we have the following result.

Theorem 5.2.11. *Under the assumptions of Theorem* 5.2.10 *the adjoint evolution equation* (5.2.40b) *has a unique mild solution for every* $y \in L_2(I, H)$.

PROOF. Due to the previous theorem, there exists a unique $P \in L_\infty(I, \mathcal{L}(H))$, with $P(t) = P^*(t)$ and $P(t) \geq 0$ that solves Equation (5.2.40a) and $(PK)^* = KP$. Therefore, Equation (5.2.40b) is equivalent to

$$\frac{d}{dt}(r(t), h) + (r(t), \tilde{A}(t)h) - (R(t)y(t), h) = 0,$$
$$r(\tau) = 0, \qquad t \in (0, \tau), \qquad (5.2.67)$$

where $\tilde{A}(t) = A - (KP)(t)$. Since $\tilde{A}$ is the generator of the quasi-evolution operator ϕ, Equation (5.2.67) has the unique mild solution given by

$$(r(t),h) = -\int_t^\tau (R(\theta)y(\theta),\phi(\theta,t)h)\,d\theta \quad \text{for} \quad t\in(0,\tau) \tag{5.2.68}$$

for all $h\in D(A)$. □

5.2.8. *Time Optimal Control*

Let E be an arbitrary Banach space and suppose that A is a closed linear operator in E with domain $D(A)$ dense in E. Suppose that A admits a strongly continuous semigroup (compare Section 2.4) $T(t)$, $t\geq 0$. Consider the system

$$\begin{aligned} &\frac{d}{dt}x = Ax+u(t), \qquad t\in I\equiv(0,\tau), \quad \tau<\infty, \\ &x(0)=x_0\in E, \end{aligned} \tag{5.2.69}$$

and $u\in L_p(I,E)$, $1\leq p\leq\infty$. We know (compare Section 2.4) that under these general conditions Equation (5.2.69) has only a mild solution given by

$$x(t) = T(t)x_0 + \int_0^t T(t-\theta)u(\theta)\,d\theta. \tag{5.2.70}$$

Let $\mathcal{U}_a$ denote the class of admissible controls and let $x_0, x_1\in E$. A trajectory x is said to be admissible if $x(0)=x_0$ and $x(t)=x_1$ for some finite $t>0$. Since the trajectories $\{x\}$ are continuous E-valued functions, there exists a smallest t for which $x(t)=x_1$, provided x is admissible. Let $\mathcal{U}_0\subset\mathcal{U}_a$ denote the class of controls that generate the admissible trajectories as described above.

Let $x(u)$, with values $x(u)(t)\in E$, $t\geq 0$, denote the trajectory corresponding to the control $u\in\mathcal{U}_a$ and initial state $x(u)(0)=x_0\in E$. For $u\in\mathcal{U}_0$, let $\sigma(u)\equiv\inf\{t\geq 0,\ x(u)(t)=x_1\}$ denote the transition time corresponding to the control u, and define $\sigma^*=\inf\{\sigma(u),\ u\in\mathcal{U}_0\}$. The problem is to find an $u^*\in\mathcal{U}_0$ such that $\sigma(u^*)=\sigma^*$. This problem raises three fundamental questions: (i) For given $x_0, x_1\in E$ and arbitrary but finite τ, is $\mathcal{U}_0\neq\varnothing$ (nonempty)? (ii) Given that $\mathcal{U}_0\neq\varnothing$, does there exist a control $u^*\in\mathcal{U}_0$ such that $\sigma^*=\sigma(u^*)$? (iii) Given that such a control exists, can it be characterized by necessary (and sufficient) conditions through which it may be determined? Question (i) is equivalent to the question of controllability, which we have briefly discussed in Section 5.2.5. In the same section we considered the question of existence of time optimal controls (Theorem 5.2.4) for E a Hilbert space. In fact, this result holds also for reflexive Banach spaces, as discussed below. Our main concern in this section is with question (iii), which deals with the necessary (and sometimes sufficient) conditions of optimality. We shall also discuss the bang-bang principle. For the existence of time optimal control we have the following result.

Theorem 5.2.12 (Existence of time optimal control). *Consider the system (5.2.69) with given initial and desired final states $x_0, x_1 \in E$, where E is a reflexive Banach space. Let the admissible controls be given by $\mathfrak{U}_a = \{u \in L_\infty(I, E) : u(t) \in B_1$ a.e.$\}$, where B_1 is the unit ball in E. Suppose that $\mathfrak{U}_0$, denoting the class of controls from $\mathfrak{U}_a$ that generate the admissible trajectories, is nonempty. Then there exists an optimal control.*

PROOF. Since E is a reflexive Banach space, $(L_1(I, E^*))^* = L_\infty(I, E)$ and $T^*(t)$, $t \geq 0$, is also a strongly continuous semigroup on E^* with generator A^*. These facts imply that the proof given for Theorem 5.2.4 also holds for this case, provided we interpret the scalar products appearing in Equations (5.2.8)–(5.2.14) as the $E-E^*$ duality products with $h \in E^*$. □

REMARK 5.2.3. We note that the above theorem remains true also for the admissible controls given by $\mathfrak{U}_a \equiv \{u : u \in L_p(I, E),\ \int_I |u(t)|_E^p \, dt \leq r < \infty\}$ for $p > 1$.

In the rest of this section we consider question (iii) concerning the necessary conditions of optimality. We develop certain maximum principles similar to those for finite-dimensional systems [PBGM.1, HL.1].

It is well known [HL.1, p. 65] that under the conditions of "normality," the bang-bang principle holds for finite-dimensional systems. Fattorini has developed [Fa.4, Fa.6] certain maximum principles for infinite-dimensional systems and has shown [Fa.4] that under certain conditions also the bang-bang principle holds. We discuss some of these results below. The material of this section is based on the recent results of Fattorini [Fa.6].

Let E be a separable Banach space and define for each $\tau > 0$ and $p \in [1, \infty]$ the set $K_p^\tau = \{y : y = \int_0^\tau T(\tau - \theta) u(\theta) \, d\theta,\ u \in L_p(0, \tau; E)\}$. It is clear that for finite τ, $K_p^\tau \subset E$. In general we have the following result.

Lemma 5.2.2. *Let E be a separable Banach space and A the generator of a strongly continuous semigroup $T(t)$, $t \geq 0$, on E with domain $D(A)$ dense in E. The family K_p^τ, $\tau \geq 0$, $p \in [1, \infty]$, has the following properties:*

i. *for any $\tau, \tau' \in (0, \infty)$, $p \in [1, \infty]$, $K_p^\tau = K_p^{\tau'} = K_p$;*
ii. *for $1 \leq p \leq p'$, $K_{p'} \subset K_p$;*
iii. *if for some $t > 0$, $T(t)E = E$, then $K_p = E$; and*
iv. *$D(A) \subset K_\infty$.*

PROOF. (i) Clearly, for $0<\tau'<\tau$, $K_p^{\tau'}\subset K_p^{\tau}$. For any $u\in L_p(0,\tau;E)$,

$$\int_0^\tau T(\tau-\theta)u(\theta)d\theta=\int_0^{\tau-\tau'}T(\tau-\theta)u(\theta)d\theta+\int_{\tau-\tau'}^{\tau}T(\tau-\theta)u(\theta)d\theta$$

$$=T(\tau')\left(\int_0^{\tau-\tau'}T(\tau-\tau'-\theta)u(\theta)d\theta\right)+\int_{\tau-\tau'}^{\tau}T(\tau-\theta)u(\theta)d\theta$$

$$=\int_0^{\tau'}\frac{T(\tau'-t)T(t)}{\tau'}\left(\int_0^{\tau-\tau'}T(\tau-\tau'-\theta)u(\theta)d\theta\right)dt$$

$$+\int_{\tau-\tau'}^{\tau}T(\tau-\theta)u(\theta)d\theta$$

$$=\int_0^{\tau'}T(\tau'-t)\left\{\frac{T(t)}{\tau'}\int_0^{\tau-\tau'}T(\tau-\tau'-\theta)u(\theta)d\theta+u(t+\tau-\tau')\right\}dt$$

$$=\int_0^{\tau'}T(\tau'-t)v(t)dt$$

where

$$v(t)\equiv\frac{T(t)}{\tau'}\int_0^{\tau-\tau'}T(\tau-\tau'-\theta)u(\theta)d\theta+u(t+\tau-\tau'),\qquad t\in(0,\tau').$$

Since $T(t)$, $t\geq 0$, is a c_0-semigroup and $\tau<\infty$, there exists a number $M>0$ such that $\sup_{0\leq t\leq\tau}\|T(t)\|_{\mathcal{L}(E)}\leq M$. Thus one can easily verify that v, as defined above, belongs to $L_p(0,\tau';E)$, and we conclude that for every pair τ,τ' ($\tau'<\tau$, say) and $u\in L_p(0,\tau;E)$, there exists a $v\in L_p(0,\tau';E)$ such that

$$\int_0^\tau T(\tau-\theta)u(\theta)d\theta=\int_0^{\tau'}T(\tau'-\theta)v(\theta)d\theta.$$

Consequently, $K_p^\tau\subset K_p^{\tau'}$. Combining the two results, we have $K_p^\tau=K_p^{\tau'}$ for any $\tau,\tau'\in(0,\infty)$, and hence, $K_p^\tau=K_p$, a constant for $\tau>0$. (ii) This follows from the fact that $L_{p'}(0,\tau;E)\subset L_p(0,\tau;E)$ whenever $0<\tau<\infty$ and $p'>p$. (iii) Obviously $K_p\subset E$; we show that $E\subset K_p$. Let $y\in E$; then due to the fact that $T(t)E=E$, there exists an $x\in E$ such that $T(t)x=y$. By the semigroup property we have

$$y=T(t)x=\int_0^t T(t-\theta)\left(\frac{T(\theta)x}{t}\right)d\theta,\tag{5.2.71}$$

and since, for finite t, $\sup\{\|T(\theta)\|_{\mathcal{L}(E)},\ 0\leq\theta\leq t\}\leq M_t<\infty$, the function v, given by $v(\theta)=T(\theta)x/t$, belongs to $L_p(0,t;E)$. Consequently, $y\in K_p^t=K_p$ and $E\subset K_p$. (iv) Let $u\in D(A)$ and define for finite τ

$$z\equiv\int_0^\tau T(\tau-\theta)\left\{\frac{1}{\tau}(u-\theta Au)\right\}d\theta\quad\text{and}\quad v(\theta)\equiv\frac{1}{\tau}(u-\theta Au).\tag{5.2.72}$$

Clearly $v\in L_\infty(0,\tau;E)$, and consequently, $z\in K_\infty$. Evaluating the integral in (5.2.72) and recalling the fact that, for $u \in D(A)$, $(d/dt)(T(t)u) = T(t)Au = AT(t)u$ (Theorem 1.3.2), we obtain

$$z = \frac{1}{\tau}\int_0^\tau T(\theta)u\,d\theta - \int_0^\tau \frac{d}{d\theta}(T(\theta)u)d\theta + \frac{1}{\tau}\int_0^\tau \theta\frac{d}{d\theta}(T(\theta)u)d\theta$$
$$= u.$$

Thus $z = u \in K_\infty$, and consequently, $D(A) \subset K_\infty$. □

REMARK 5.2.4. In general, $K_\infty \not\subset D(A)$, and hence, $K_\infty \neq D(A)$ except when A is a bounded operator.

For convenience of notation we use $(\cdot,\cdot)$ to denote the duality product in $E^* - E$.

Theorem 5.2.13. *Let B_1 denote the closed unit ball in E, define $\mathcal{U}_a \equiv \{u \in L_p^{\mathrm{loc}}(E): u(t) \in B_1$ a.e.$\}$, $p > 1$, and consider the system (5.2.69) with the initial and final states $x_0, x_1 \in E$ and admissible controls $\mathcal{U}_a$. Suppose there exists a $t > 0$ such that $T(t)E = E$ and a control $u \in \mathcal{U}_a$ that transfers the system from the state x_0 to the state x_1 in finite time. Then, if u_0 is the optimal control and τ the corresponding transition time, there exists a nonzero $e^* \in E^*$ such that*

$$(T^*(\tau - t)e^*, u_0(t)) = \sup_{v \in B_1}(T^*(\tau - t)e^*, v) = |T^*(\tau - t)e^*|_{E^*} \quad \text{a.e.} \tag{5.2.73}$$

on $[0, \tau]$. In other words, u_0 satisfies the maximum principle.

PROOF. Define

$$\Omega(\tau) = \left\{y : y = \int_0^\tau T(\tau - \theta)u(\theta)d\theta,\ u \in \mathcal{U}_a\right\}. \tag{5.2.74}$$

It is clear that $\Omega(\tau)$ is a closed, bounded, and convex subset of E. We show that $\Omega(\tau)$ has nonempty interior. Since $T(t)$ is surjective, it follows from the closed graph theorem (Theorem 1.1.2) that there exists a constant $c > 0$ such that for all $y, x \in E$ for which $y = T(t)x$, $|x|_E \le c|y|_E$. The constant c may depend on t. For $t > \tau$ we can write

$$y = T(t)x = \int_0^\tau T(\tau - \theta)\left(\frac{1}{\tau}T(t + \theta - \tau)x\right)d\theta. \tag{5.2.75}$$

Since $T(\theta)$, $\theta \ge 0$ is a c_0-semigroup, there exists a finite $M > 0$ such that $\sup\{\|T(\theta)\|_{\mathcal{L}(E)} : 0 \le \theta \le \tau\} \le M$. Defining

$$u(\theta) = \frac{T(t + \theta - \tau)x}{\tau}, \tag{5.2.76}$$

we have

$$|u(\theta)|_E \le \frac{M}{\tau}|x|_E \le \frac{cM}{\tau}|y|_E \quad \text{for all } \theta \in [0, \tau]. \tag{5.2.77}$$

Thus for $|y|_E<(\tau/cM)$ we have $|u(\theta)|<1$ for all $\theta\in[0,\tau]$. Therefore, $y\in\operatorname{int}\Omega(\tau)$. Similarly, for $t<\tau$

$$y=T(t)x=\int_0^t T(t-\theta)\left(\frac{T(\theta)x}{t}\right)d\theta=\int_{\tau-t}^{\tau} T(\tau-\xi)\left\{\frac{T(\xi+t-\tau)x}{t}\right\}d\xi$$
$$\equiv\int_0^{\tau} T(\tau-\xi)\tilde{u}(\xi)d\xi, \tag{5.2.78}$$

where

$$\tilde{u}(\xi)=\begin{cases}0 & \text{for } 0\le\xi<\tau-t\\ \dfrac{T(\xi-(\tau-t))x}{t} & \text{for } \tau-t\le\xi\le\tau.\end{cases}$$

Again for $|y|<(t/cM)$ we have $\operatorname{ess\,sup}_{0\le\xi\le\tau}|\tilde{u}(\xi)|_E<1$.

Therefore, $y\in\operatorname{int}\Omega(\tau)$, and we conclude that $\operatorname{int}\Omega(\tau)\neq\varnothing$. In fact, every $y\in E$ with sufficiently small norm belongs to $\Omega(\tau)$. Since u_0 is the optimal control and τ the optimal transition time we have,

$$x_1=T(\tau)x_0+\int_0^{\tau} T(\tau-\theta)u_0(\theta)d\theta \tag{5.2.79}$$

and $\omega_0\equiv(x_1-T(\tau)x_0)\in\Omega(\tau)$. Moreover, $\omega_0\in\partial\Omega(\tau)$, the boundary of $\Omega(\tau)$. Suppose not. Then, since the semigroup $T(t)$, $t\ge 0$, is strongly continuous, there exists an $\eta>0$ and a $\beta>1$ such that for $\tau-\eta<t<\tau$, $\beta(x_1-T(t)x_0)\in\operatorname{int}\Omega(\tau)$. Therefore, for $\tau-\eta<t<\tau$,

$$\beta(x_1-T(t)x_0)=\int_0^{\tau} T(\tau-\theta)u(\theta)d\theta \quad \text{for some } u\in\mathcal{U}_a. \tag{5.2.80}$$

Defining

$$v(\xi)=\frac{T(\xi)}{t}\int_0^{\tau-t} T(\tau-t-\theta)u(\theta)d\theta+u(\xi+\tau-t)$$

for $\xi\in[0,t]$, we can write

$$\int_0^{\tau} T(\tau-\theta)u(\theta)d\theta=\int_0^t T(t-\xi)v(\xi)d\xi. \tag{5.2.81}$$

Thus,

$$\beta(x_1-T(t)x_0)=\int_0^t T(t-\xi)v(\xi)d\xi. \tag{5.2.82}$$

Since $u\in\mathcal{U}_a$ and $\|T(\theta)\|_{\mathcal{L}(E)}\le M$ for all $\theta\in[0,t]$ we have

$$\operatorname{ess\,sup}\{|v(\xi)|_E:\xi\in[0,t]\}\le(1+\eta M^2/t).$$

Choosing η sufficiently small, so that $0<\eta<(\beta-1)t/M^2$, and defining $\tilde{u}(\xi)=(1/\beta)v(\xi)$, we find that $\tilde{u}\in\mathcal{U}_a$. Thus, for sufficiently small η it follows from

$$(x_1-T(t)x_0)=\int_0^t T(t-\xi)\tilde{u}(\xi)d\xi \tag{5.2.83}$$

that $x_1-T(t)x_0\in\Omega(t)$.

This contradicts the fact that τ is the optimal transition time and u_0 the optimal control. Therefore, $\omega_0 \in \partial\Omega(\tau)$. Since $\Omega(\tau)$ is a closed bounded convex set with nonempty interior and $\omega_0 \in \partial\Omega(\tau)$, it follows from the separation theorem (Corollary 1.1.4) that ω_0 is a support point of the set $\Omega(\tau)$. Thus there exists a support hyperplane, or equivalently, an e^* $(\neq 0) \in E^*$ such that

$$(e^*, \omega) \leq (e^*, \omega_0) \quad \text{for all } \omega \in \Omega(\tau). \tag{5.2.84}$$

That is,

$$\left(e^*, \int_0^\tau T(\tau-\theta)u(\theta)d\theta\right) \leq \left(e^*, \int_0^\tau T(\tau-\theta)u_0(\theta)d\theta\right) \tag{5.2.85}$$

for all $u \in \mathcal{U}_a$. Since $\|T(t)\|_{\mathcal{L}(E)}$ is bounded on bounded intervals, it is clear that for $\nu \in \mathcal{U}_a$, $T(\tau - \cdot)\nu(\cdot) \in L_\infty(0, \tau; E) \subset L_1(0, \tau; E)$, and for $e^* \in E^*$, $e^*(T(\tau - \cdot)\nu(\cdot)) \equiv (e^*, T(\tau - \cdot)\nu(\cdot)) \in L_1(0, \tau; R)$. Consequently, (5.2.85) is equivalent to

$$\int_0^\tau (e^*, T(\tau-\theta)u(\theta))d\theta \leq \int_0^\tau (e^*, T(\tau-\theta)u_0(\theta))d\theta \tag{5.2.86}$$

for all $u \in \mathcal{U}_a$. Using the standard Lebesgue density arguments, we can deduce from (5.2.86) that

$$(e^*, T(\tau-t)v) \leq (e^*, T(\tau-t)u_0(t)), \quad \text{a.e. in } [0, \tau], \tag{5.2.87}$$

for all $v \in B_1$. Therefore, for almost all $t \in [0, \tau]$,

$$(T^*(\tau-t)e^*, v) \leq (T^*(\tau-t)e^*, u_0(t)) \quad \text{for all } v \in B_1,$$

and consequently,

$$(T^*(\tau-t)e^*, u_0(t)) = \sup_{v \in B_1} (T^*(\tau-t)e^*, v) \equiv |T^*(\tau-t)e^*|_{E^*} \tag{5.2.88}$$

a.e. in $[0, \tau]$. This completes the proof. □

REMARK 5.2.5. In case the set of admissible controls is given by $\mathcal{U}_a \equiv \{u \in L_p(0, \tau; E) : \|u\|_{L_p(0, \tau; E)} \leq 1\}$, we obtain an integral version of the maximum principle from (5.2.86), and it is given by

$$\sup_{\|u\|_{L_p(0,\tau;E)} \leq 1} \int_0^\tau (T^*(\tau-\theta)e^*, u(\theta))d\theta = \int_0^\tau (T^*(\tau-\theta)e^*, u_0(\theta))d\theta. \tag{5.2.86$'$}$$

Corollary 5.2.2 (Bang-bang control)

(a) *Suppose that E is a reflexive Banach space and that there is a continuous map ν: $E^* \backslash \{0\} \to E$ such that for each $x^* \in E^*$ there is a* *unique* $x=\nu(x^*)\in E$ *with* $|\nu(x^*)|_E=1$ *and* $|x^*|_{E^*}=(x^*,x)=(x^*,\nu(x^*))$. *Then the optimal control is bang-bang and is given by*

$$u_0(t)=\nu(T^*(\tau-t)e^*), \qquad t\in[0,\tau]. \tag{5.2.89}$$

(b) *If* E *is also strictly convex* ($x_1,x_2\in E$, $|x_1|_E=|x_2|_E=1$, $|\frac{1}{2}(x_1+x_2)|_E=1$ *imply* $x_1=x_2$), *then the optimal control is bang-bang and also unique and is given by* (5.2.89).

(c) *If* E *is a self-adjoint Hilbert space* H, *then the optimal control is also bang-bang and unique and is given by*

$$u_0(t)=\frac{T^*(\tau-t)e^*}{|T^*(\tau-t)e^*|_H}. \tag{5.2.90}$$

PROOF

(a) If E is reflexive, then for each $x^*\in E^*$ there exists an $x_0\in\partial B_1=\{e\in E:|e|_E=1)$ such that $(x^*,x_0)=|x^*|_{E^*}=\sup_{x\in B_1}|(x^*,x)|$. Thus under the given assumption $x_0=\nu(x^*)$, and it follows from the maximum principle (5.2.73) (Theorem 5.2.13) that $u_0(t)=\nu(T^*(\tau-t)e^*)$ a.e. in $[0,\tau]$, and consequently, the control is bang-bang that is, $|u_0(t)|_E=1$ a.e.

(b) If u_1 and u_2 are two optimal controls (of course with the same transition time τ), then $\frac{1}{2}(u_1+u_2)$ is also an optimal control. Since strict convexity implies reflexivity, all these controls are bang-bang, and consequently, $|u_1(t)|_E=|u_2(t)|_E=|\frac{1}{2}(u_1(t)+u_2(t))|$ a.e. Therefore, due to strict convexity we have $u_1(t)=u_2(t)$ a.e. on $[0,\tau]$.

(c) If $E=H$ (self-adjoint) then $\nu(x)=x/|x|_H$, and hence,

$$u_0(t)=\nu(T^*(\tau-t)e^*)=\frac{T^*(\tau-t)e^*}{|T^*(\tau-t)e^*|_H}.$$

Clearly, the optimal control is bang-bang. □

REMARK 5.2.6

(a) In case E is a reflexive Banach space, $T^*(t)$, $t\geq 0$, is also a c_0-semigroup with A^* as its generator (Theorem 1.3.11). In that case u_0 is actually continuous whenever ν is a continuous map from $E^*\backslash\{0\}$ into E. For example, in the case of Hilbert space, it follows from (5.2.90) that u_0 is continuous since, for $e^*\neq 0$, the denominator never vanishes.

(b) Define $\psi(t)=T^*(\tau-t)e^*$; then for $e\in D(A)$ we have $(\psi(t),e)=(e^*,T(\tau-t)e)$ and

$$\frac{d}{dt}(\psi(t),e)=-(e^*,T(\tau-t)Ae)=-(\psi(t),Ae).$$

Thus ψ satisfies the differential equation

$$\begin{aligned}&\frac{d\psi}{dt}+A^*\psi=0, \qquad t\in(0,\tau),\\ &\psi(\tau)=e^*\end{aligned} \tag{5.2.91}$$

in the weak sense, that is, for all $e \in D(A)$,

$$\frac{d}{dt}(\psi(t), e) + (\psi(t), Ae) = 0 \quad \text{a.e. in } (0, \tau),$$

$$\psi(\tau) = e^*. \tag{5.2.91$'$}$$

The optimal control $u_0(t) = \nu(\psi(t))$, and the optimality system is given by

$$\frac{d}{dt}x(t) = Ax(t) + \nu(\psi(t)), \qquad x(0) = x_0, \quad x(\tau) = x_1,$$

$$\frac{d}{dt}\psi(t) = -A^*\psi(t), \qquad \psi(\tau) = e^*. \tag{5.2.92}$$

Since we do not know whether e^* belongs to $D(A^*)$, in general we cannot expect ψ to be a strong solution of (5.2.91).

(c) If $E = H$ (a Hilbert space) and $A^* = iA$ (skew adjoint), then $T(t)$ and $T^*(t)$ are unitary operators in H for all t and $T^*(t) = T^{-1}(t) = T(-t)$ and $|T(t)e|_H = |e|_H$ for all $e \in H$. It follows from (5.2.90) that, in this case, the optimal control has the form

$$u_0(t) = T(t-\tau)e^*$$

and the trajectory x^0 corresponding to the control u_0 is given by

$$\begin{aligned} x^0(t) &= T(t)x_0 + \int_0^t T(t-\theta)u_0(\theta)\,d\theta \\ &= T(t)x_0 + \int_0^t T(t-\theta)T(\theta-\tau)e^*\,d\theta \\ &= T(t)x_0 + tT(t-\tau)e^*; \end{aligned}$$

consequently, $x_1 = T(\tau)x_0 + \tau e^*$, so that $e^* = (1/\tau)(x_1 - T(\tau)x_0)$. For $x_0 = 0$, $e^* = (1/\tau)x_1$ and $u_0(t) = (1/\tau)(T(t-\tau)x_1)$. Since $|e^*|_H = 1$, $\tau = |x_1|_H$, and thus the optimal transition time τ equals the radius of the ball on which x_1 lies and the optimal trajectory $x^0(t) = (t/\tau)T(t-\tau)x_1$, $t \in [0, \tau]$. Similarly, for $x_1 = 0$, $x_0 \neq 0$, we have $\tau = |x_0|_H$, $u^0(t) = -(1/\tau)T(t)x_0$, and $x^0(t) = [1-(t/\tau)]T(t)x_0$.

(d) The crucial point in the proof of Theorem 5.2.13 is that the set $\Omega(\tau)$ has nonempty interior. Note also that $T(t)$ is not a compact operator, since it is assumed to be surjective and consequently $\Omega(\tau)$ is not a compact set.

(e) Since all Hilbert spaces and L_p-spaces for $1 < p < \infty$ are strictly convex, the optimal control is unique.

We saw that the proof of the maximum principle (Theorem 5.2.13) was critically dependent on whether or not the set $\Omega(\tau)$ could be separated from a point on its boundary by a hyperplane. According to the separation theorem, it is required that the set have nonvoid interior. The assumption $T(t)E = E$ for some $t > 0$ guarantees this condition. In case E is a Hilbert space, the converse is also true, that is, if $\operatorname{int}\Omega(\tau) \neq \varnothing$, then $T(t)E = E$ for all $t > 0$. This is formally presented in the following lemma.

Lemma 5.2.3. *Suppose E is a Hilbert space and for some $\tau>0$, $int\,\Omega(\tau)\neq\varnothing$, or more generally, $K_2=E$. Then $T(t)E=E$ for all $t\geq 0$.*

PROOF. First we note that if $\text{int}\,\Omega(\tau)\neq\varnothing$, then $K_2=E$. By Lemma 5.2.2, $K_\infty\subset K_2\subset E$; therefore, it suffices to show that $K_\infty=E$. If $E\not\subset K_\infty$, there exists a $k\in E\backslash K_\infty$ and, due to the fact that $\Omega(\tau)\subset K_\infty$ and $\text{int}\,\Omega(\tau)\neq\varnothing$, there exists an $h\in E$ with $|h|_E=1$ such that $(h,w)=0$ for all $w\in\Omega(\tau)$ and $(h,k)=|k|_E$. That is,

$$\left(h,\int_0^\tau T(\tau-\theta)v(\theta)\,d\theta\right)=0 \quad \text{for all } v\in L_\infty(0,\tau;E) \text{ with } |v(t)|_E\leq 1. \tag{5.2.93}$$

Define $u(t)=(T^*(\tau-t)h/M)$, $t\in[0,\tau]$, where $M=\sup\{\|T(t)\|_{\mathcal{L}(E)},\ t\in[0,\tau]\}$. Clearly, u, is strongly measurable since, for reflexive Banach spaces, T^* is a c_0-semigroup, and consequently, $u(t)$ is even strongly continuous. Thus $u\in L_\infty(0,\tau;E)$ and $|u(t)|_E\leq 1$, $t\in[0,\tau]$. Using this u in (5.2.93), we obtain

$$\int_0^\tau |T^*(\tau-\theta)h|_E^2\,d\theta=0, \tag{5.2.94}$$

which implies that $T^*(\tau-t)h=0$ a.e. in $(0,\tau]$. Since T^* is strongly continuous, $T^*(\tau-t)h=0$ for all $t\in[0,\tau]$, and therefore, $h=0$. This is a contradiction. Therefore, $K_\infty=E$, and consequently, $K_2=E$. We now show that if $K_2=E$, then $T(t)E=E$ for all $t\geq 0$. First we show this for $t\in[0,\tau]$, $\tau<1$. By definition, for τ such that $0<\tau<1$,

$$K_2=\left\{y\in E: y=\int_0^\tau T(\tau-\theta)v(\theta)\,d\theta,\ v\in L_2(0,\tau;E)\right\}.$$

For each $y\in K_2$ define

$$\|y\|_{K_2}\equiv\inf\left\{\|v\|=\left(\int_0^\tau |v(\theta)|_E^2\,d\theta\right)^{1/2}: y=\int_0^\tau T(\tau-\theta)v(\theta)\,d\theta,\ v\in L_2\right\}. \tag{5.2.95}$$

It is easy to verify that this is a norm. Equipped with this norm, $K_2=(K_2,\|\cdot\|_{K_2})$ is a Banach space. Indeed, if $\{y_n\}$ is a Cauchy sequence in K_2, there exists a sequence $\{u_n\}\in L_2(0,\tau;E)$ such that $y_n=\int_0^\tau T(\tau-\theta)u_n(\theta)\,d\theta$ and $\|y_n\|_{K_2}=\|u_n\|_{L_2(0,\tau;E)}$ for all $n\geq 1$. Thus $\{u_n\}$ is also a Cauchy sequence in $L_2(0,\tau;E)$, and consequently, there exists a u_0 such that $u_n\to u_0$ strongly in $L_2(0,\tau;E)$. By Schwarz's inequality, $|y|_E\leq M\|y\|_{K_2}$ for all $y\in K_2$; therefore, $\{y_n\}$ is a Cauchy sequence in E whenever it is a Cauchy sequence in K_2. Since E is a Banach space, there exists a $y_0\in E$ such that $y_n\to y_0$ in E. For arbitrary $h\in E$ we have $(h,y_0)=\lim_n(h,y_n)=\lim_n(h,\int_0^\tau T(\tau-\theta)u_n(\theta)\,d\theta)$, and since $u_n\to u_0$ strongly in $L_2(0,\tau;E)$, we have, $(h, y_0)=(h, \int_0^\tau T(\tau-\theta)u_0(\theta)d\theta)$. Since h is arbitrary and $u_0 \in L_2(0, \tau; E)$, $y_0=\int_0^\tau T(\tau-\theta)u_0(\theta)d\theta \in K_2$. Thus K_2 is a Banach space in either of the norms $\|\cdot\|_{K_2}$ and $|\cdot|_E$. Therefore, the two norms are equivalent, that is, there exists a second constant $M'>0$ such that $\|y\|_{K_2} \le M'|y|_E$ for all $y \in K_2=E$. Let $e \in E$ be arbitrary and define

$$\hat{e}=\int_0^\tau T(\tau-\theta)T^*(\tau-\theta)e\,d\theta=\int_0^\tau T(\theta)T^*(\theta)e\,d\theta. \tag{5.2.96}$$

Clearly, $\hat{e} \in K_2$ and there exists a $v \in L_2(0, \tau; E)$ such that $\hat{e}=\int_0^\tau T(\theta)v(\theta)d\theta$. Then for every $\nu \in E$, $\int_0^\tau (T(\theta)(T^*(\theta)e-v(\theta)), \nu)d\theta=0$, and consequently, for $\nu=e$ we have

$$\int_0^\tau (T^*(\theta)e-v(\theta), T^*(\theta)e)d\theta=0,$$

which is equivalent to

$$\int_0^\tau |T^*(\theta)e|_E^2\,d\theta=\int_0^\tau (T^*(\theta)e, v(\theta))d\theta. \tag{5.2.97}$$

This leads to the inequality

$$\int_0^\tau |T^*(\theta)e|_E^2\,d\theta \le \int_0^\tau |v(\theta)|_E^2\,d\theta, \tag{5.2.98}$$

and therefore, it follows from the definition of the norm $\|\cdot\|_{K_2}$ that

$$\|\hat{e}\|_{K_2}^2=\int_0^\tau |T^*(\theta)e|_E^2\,d\theta. \tag{5.2.99}$$

Using (5.2.96), (5.2.99), and the fact that the two norms $\|\cdot\|_{K_2}$ and $|\cdot|_E$ are equivalent, we obtain

$$\left|\int_0^\tau T(\theta)T^*(\theta)e\,d\theta\right|_E^2 \ge \beta\int_0^\tau |T^*(\theta)e|_E^2\,d\theta \tag{5.2.100}$$

for some $\beta(=(1/M')^2)>0$.

Defining

$$Le=\int_0^\tau T(\theta)T^*(\theta)e\,d\theta \tag{5.2.101}$$

and noting that $L=L^*$, we have from the inequality (5.2.100) that $(L^2e, e) \ge \beta(Le, e)$ for all $e \in E$. Thus $L \ge \beta I$, or equivalently,

$$\int_0^\tau (T(\theta)T^*(\theta)e, e)d\theta \ge \beta|e|_E^2, \qquad e \in E. \tag{5.2.102}$$

Note that in the above inequality we can choose β to be an arbitrarily small positive number. Using this inequality with sufficiently small β and defining

$$F(e)=\{t \in (0, \tau) : (T(t)T^*(t)e, e)<\beta|e|_E^2\},$$

we obtain

$$\beta|e|_E^2 \le \int_{F(e)} (T(\theta)T^*(\theta)e,e)d\theta + \int_{(0,\tau)\backslash F(e)\equiv F'(e)} (T(\theta)T^*(\theta)e,e)d\theta$$

$$<\{(\beta-M^2)\mu(F(e))+\tau M^2\}|e|_E^2, \tag{5.2.103}$$

where $M=\sup\{\|T(t)\| : t\in[0,\tau]\}$. Since we can choose β such that $0<\beta<(M^2 \wedge \tau M^2)$, we have $\mu(F(e))<(\tau M^2-\beta)/(M^2-\beta)\equiv\eta<\tau$. Thus $F'(e) \neq\emptyset$ with the Lebesgue measure $\mu(F'(e))>0$, and consequently, there exists a $t_0\in(0,\tau)\cap F'(e)$ such that

$$(T(t_0)T^*(t_0)e,e)\ge\beta|e|^2 \quad \text{for all } e\in E. \tag{5.2.104}$$

Since E is a Hilbert space, this implies that $T(t_0)E=E$. If not, there exists a $\xi\in E\backslash T(t_0)E$ such that $(T(t_0)e,\xi)=0$ for all $e\in E$, and consequently, $T^*(t_0)\xi=0$. Therefore, $(T^*(t_0)\xi, T^*(t_0)\xi)=(T(t_0)T^*(t_0)\xi,\xi)=0$, which contradicts (5.2.104) unless $\xi=0$. We now show that $T(t)E=E$ for all $t\in(0,t_0)$. Since $T(t_0)=T(t_0-t)T(t)=T(t)T(t_0-t)$, we have $(T^*(t_0)e, T^*(t_0)e)\le M^2(T^*(t)e, T^*(t)e)$, and as a consequence $(T(t)T^*(t)e,e)\ge(\beta/M^2)|e|_E^2$. Thus, by the same argument as above, we have $T(t)E=E$ for all $t\in(0,t_0]$. For $t\in(t_0,\tau]$ there exists a point $r\in(0,t_0]$ and a positive integer m such that $mr=t$. Since $T(t)=T(mr)=T^m(r)$,

$$(T(t)T^*(t)e,e)=(T(mr)T^*(mr)e,e)\ge(\beta')^m|e|_E^2, \tag{5.2.105}$$

where $\beta'\equiv(\beta/M^2)>0$. Thus, again by the same argument, we have $T(t)E=E$ for all $t\in[t_0,\tau]$. For $t>\tau$ we use the same argument as 5.2.105. This completes the proof. □

REMARK 5.2.7.

(a) If E is reflexive and $1<p<\infty$, then $L_p(0,\tau;E)$ is a reflexive Banach space and $L_p(0,\tau;E)=(L_q(0,\tau;E^*))^*$ for $p^{-1}+q^{-1}=1$. In this case, there exists a mapping ν: $L_q(0,\tau;E^*)\to L_p(0,\tau;E)$ such that for each $y^*\in L_q(0,\tau;E^*)$, $\|\nu(y^*)\|_{L_p(0,\tau;E)}=1$ and $(y^*,\nu(y^*))_{L_q(0,\tau;E^*)-L_p(0,\tau;E)}=\|y^*\|_{L_q(0,\tau;E^*)}$. Then according to the maximum principle (5.2.86)′ (Remark 5.2.5., Theorem 5.2.13) the optimal control is $u_0=\nu(y^*)$, where $y^*(t)=T^*(\tau-t)e^*$ and

$$\int_0^\tau (T^*(\tau-\theta)e^*, u_0(\theta))d\theta=(y^*,\nu(y^*))_{L_q(0,\tau;E^*)-L_p(0,\tau;E)}$$

$$=\left(\int_0^\tau |T^*(\tau-t)e^*|_{E^*}^q dt\right)^{1/q}. \tag{5.2.106}$$

(b) In case E is a self-adjoint Hilbert space $(E=H)$, the "pointwise" maximum principle (5.2.73) (Theorem 5.2.13) holds and the optimal control is given by

$$u_0(t)=y^*(t)|y^*(t)|_E^{q-2}/\|y^*\|_{L_q(0,\tau;E)}^{q-1}. \tag{5.2.106$'$}$$

In Theorem 5.2.13 it was shown that the maximum principle holds if the semigroup $T(t)$, $t \geq 0$, satisfies the property $T(t)E = E$ for some $t > 0$. In the following result we show that this assumption can be dispensed with if the initial and the desired final states lie in the domain of the generator A.

Theorem 5.2.14. *Suppose $x_0, x_1 \in D(A) \subset E$. Let $\mathfrak{U}_a$ denote the class of admissible controls (as given in Theorem 5.2.13) and u_0 the optimal control (time optimal) with transition time τ. Then for any $\lambda \in \rho(A)$ ($\equiv$resolvent set of A), there exists an $e^* \in E^*$ such that $T^*(t)e^* \in D(A^*)$ for all $t > 0$ and*

$$((\lambda I - A^*)T^*(\tau - t)e^*, u_0(t)) = \sup_{|e|_E \leq 1} ((\lambda I - A^*)T^*(\tau - t)e^*, e)$$

$$= |(\lambda I - A^*)T^*(\tau - t)e^*|_{E^*} \qquad (5.2.107)$$

for almost all $t \in (0, \tau)$.

PROOF. For the optimal transition time τ define

$$\Omega(\tau) = \left\{ y \in E : y = \int_0^\tau T(\tau - \theta)u(\theta)d\theta,\ u \in \mathfrak{U}_a \right\}$$

and $\tilde{\Omega}(\tau) = \Omega(\tau) \cap D(A)$. Clearly, $\tilde{\Omega}(\tau)$ is a convex set, and, since $D(A)$ is a Banach space with the graph norm

$$|y|_{D(A)} \equiv |y|_E + |Ay|_E,$$

the set $\tilde{\Omega}(\tau)$ has nonempty interior. Indeed, from the relation

$$y = \int_0^\tau T(\tau - \theta)\left(\frac{y - \theta Ay}{\tau}\right)d\theta$$

one can easily check that for any $y \in D(A)$ with $|y|_{D(A)} < (\tau/1 \vee \tau)$ ($1 \vee \tau \equiv \max\{1, \tau\}$), we have $y \in \operatorname{int} \tilde{\Omega}(\tau)$. Thus, $\tilde{\Omega}(\tau)$ is a convex subset of the Banach space $D(A)$ with nonempty interior. Since $x_0, x_1 \in D(A)$ and $T(t)x \in D(A)$ whenever $x \in D(A)$, it follows that

$$x_1 - T(\tau)x_0 = \int_0^\tau T(\tau - \theta)u_0(\theta)d\theta \in D(A) \cap \Omega(\tau) = \tilde{\Omega}(\tau).$$

In fact, by an argument similar to that given in Theorem 5.2.13, we can deduce that $x_1 - T(\tau)x_0 \in \partial\tilde{\Omega}(\tau)$.

Thus again, by virtue of the Hahn–Banach separation theorem (see Corollary 1.1.4), there exists a support hyperplane in $D(A)$ through $x_1 - T(\tau)x_0$. In other words, there exists a continuous linear functional ν on $D(A)$ such that

$$\nu(\omega) \leq \nu(x_1 - T(\tau)x_0) \quad \text{for all } \omega \in \tilde{\Omega}(\tau). \qquad (5.2.108)$$

Let $R(\lambda, A)$ denote the resolvent of the operator A corresponding to $\lambda \in \rho(A)$. Clearly, $v \rightarrow \nu(R(\lambda, A)v)$ is a continuous linear functional on E, and consequently, there exists an $e^* = e^*_\nu \in E^*$ such that

$$(e^*, v) = \nu(R(\lambda, A)v) \quad \text{for all } v \in E.$$

Since the range of $R(\lambda, A)\subset D(A)$, writing $\omega=R(\lambda, A)v$ gives

$$(e^*,(\lambda I-A)\omega)=\nu(\omega) \quad \text{for all } \omega\in D(A). \tag{5.2.109}$$

Using (5.2.108) and (5.2.109) and the fact that

$$x_1-T(\tau)x_0=\int_0^\tau T(\tau-\theta)u_0(\theta)d\theta,$$

we have

$$\left(e^*,(\lambda I-A)\int_0^\tau T(\tau-\theta)u(\theta)d\theta\right)\leq\left(e^*,(\lambda I-A)\int_0^\tau T(\tau-\theta)u_0(\theta)d\theta\right)<\infty \tag{5.2.110}$$

for all $u\in\mathfrak{U}_a$ such that $\int_0^\tau T(\tau-\theta)u(\theta)d\theta\in D(A)$. At this point it is not clear whether or not $e^*=e_\nu^*\in D(A^*)$. To overcome this difficulty we introduce for $\mu\in\rho(A)$ the operator $B_\mu\equiv\mu R(\mu, A)$. Clearly, $B_\mu E\subset D(A)$ and, for μ large enough $[\mu>\mu_0\equiv\lim_{t\uparrow\infty}(\log\|T(t)\|/t)]$, there exists a number $\alpha>0$ such that $\|B_\mu\|\leq\alpha$ independently of $\mu>\mu_0$. Thus

$$\int_0^\tau T(\tau-\theta)\alpha^{-1}B_\mu u(\theta)d\theta\in\tilde{\Omega}(\tau)$$

for all $\mu>\mu_0$ and all $u\in L_\infty(0,\tau;E)$ satisfying $|u(t)|_E\leq 1$ a.e. on $[0,\tau]$. Further, for each $x\in E$, $B_\mu x \underset{s}{\rightarrow} x$ as $\mu\to\infty$ (Theorem 1.3.6). Therefore, it follows from the inequality (5.2.110) that

$$j_\mu\equiv\left(e^*,(\lambda I-A)\int_0^\tau T(\tau-\theta)\alpha^{-1}B_\mu u(\theta)d\theta\right) \tag{5.2.111}$$

is uniformly bounded independently of $\mu\in\rho(A)$, $u\in L_\infty(0,\tau;E)$, and $|u(t)|_E\leq 1$ a.e. on $[0,\tau]$.

Clearly, for $u\in\mathfrak{U}_a$ such that $\int_0^\tau T(\tau-\theta)u(\theta)d\theta\in D(A)$, we have

$$\begin{aligned} j_\mu&=\left(e^*,(\lambda I-A)\int_0^\tau T(\tau-\theta)\alpha^{-1}B_\mu u(\theta)d\theta\right)\\ &=\alpha^{-1}\left(e^*, B_\mu(\lambda I-A)\int_0^\tau T(\tau-\theta)u(\theta)d\theta\right). \end{aligned} \tag{5.2.111$'$}$$

Since $D(B_\mu^*)=E^*$,

$$j_\mu=\alpha^{-1}\left(B_\mu^*e^*,(\lambda I-A)\int_0^\tau T(\tau-\theta)u(\theta)d\theta\right)$$

for all u such that $\int_0^\tau T(\tau-\theta)u(\theta)d\theta\in D(A)$. Again, because range $B_\mu^*\equiv R(B_\mu^*)\subset D(A^*)$ and $D(T^*(t))=E^*$ for all $t\in[0,\tau]$, we can write

$$j_\mu=\alpha^{-1}\int_0^\tau(T^*(\tau-\theta)(\lambda I-A)^*B_\mu^*e^*, u(\theta))d\theta.$$

We note that the expression on the right-hand side of this equality is defined for all $u\in\mathfrak{U}_a$. Since on $D(A^*)$, $T^*(t)$ commutes with A^*, we can write

$$j_\mu = \alpha^{-1}\int_0^\tau ((\lambda I - A)^* T^*(\tau-\theta) B_\mu^* e^*, u(\theta))d\theta$$
$$= \alpha^{-1}\int_0^\tau ((\lambda I - A)^* B_\mu^* T^*(\tau-\theta) e^*, u(\theta))d\theta. \qquad (5.2.111)''$$

Assuming for the moment that $T^*(t)e^* \in D(A^*)$ for $t \in (0, \tau)$, we have

$$j_\mu = \alpha^{-1}\int_0^\tau (B_\mu^*(\lambda I - A)^* T^*(\tau-\theta) e^*, u(\theta))d\theta$$
$$= \alpha^{-1}\int_0^\tau ((\lambda I - A)^* T^*(\tau-\theta) e^*, B_\mu u(\theta))d\theta, \qquad (5.2.112)$$

which is defined for all $u \in \mathcal{U}_a$. Consequently, due to (5.2.111) and (5.2.112) and strong convergence of B_μ to I, we have

$$\left(e^*, (\lambda I - A)\int_0^\tau T(\tau-\theta)u(\theta)d\theta\right) = \int_0^\tau ((\lambda I - A)^* T^*(\tau-\theta)e^*, u(\theta))d\theta. \qquad (5.2.113)$$

Thus the expressions (5.2.110) can be written as

$$\int_0^\tau ((\lambda I - A^*)T^*(\tau-\theta)e^*, u(\theta))d\theta \le \int_0^\tau ((\lambda I - A^*)T^*(\tau-\theta)e^*, u_0(\theta))d\theta, \qquad (5.2.114)$$

which holds for all $u \in \mathcal{U}_a$.

From (5.2.114) we can obtain the maximum principle (5.2.107) as stated in the theorem. We now return to prove that $v^* \equiv T^*(\tau - t)e^* \in D(A^*)$ for $0 \le t < \tau$. From the uniform boundedness of j_μ for $\mu > \mu_0$ and $u \in \mathcal{U}_a$ it follows from (5.2.111)″ that for all $\mu > \mu_0$

$$\int_0^\tau |(\lambda I - A^*)B_\mu^* T^*(\tau-\theta)e^*|_{E^*} d\theta \quad \text{is finite.}$$

Therefore, by Fatou's lemma the function

$$t \to \underline{\lim}_\mu |(\lambda I - A^*)B_\mu^* T^*(\tau - t)e^*|_{E^*}$$

is integrable on $[0, \tau]$ and is finite a.e. As a result there exists a set $J \subset [0, \tau]$, of Lebesgue measure τ, on which $|A^* B_\mu^* T^*(\tau - t)e^*|_{E^*}$ is bounded. In other words,

$$|\mu A^* R(\mu, A^*)v^*|_{E^*} < \infty \qquad (5.2.115)$$

uniformly with respect to $\mu > \mu_0$, where $v^* = T^*(\tau - t)e^*$, $t \in J$. Since

$$\mu(\mu R(\mu, A^*)v^* - v^*) = \mu A^* R(\mu, A^*)v^*, \qquad (5.2.116)$$

it follows from (5.2.115) that $\mu R(\mu, A^*)v^* - v^* \to 0$ as $\mu \to \infty$. Thus, from the initial value theorem of Laplace transform theory and Equation

(5.2.117),

$$\mu R(\mu, A^*)v^* - v^* = \mu \int_0^{\infty} e^{-\mu t}(T^*(t) - I)v^* \, dt, \qquad (5.2.117)$$

we conclude that $(T^*(t)-I)v^* \to 0$ strongly as $t \to 0^+$. Consequently, $v^* \in E_0^*$, where E_0^* is the closure in E^* of $D(A^*)$. Let $\{T_0^*(t), t \geq 0\}$ denote the restriction of $\{T^*(t), t \geq 0\}$ to E_0^*. Since E_0^* is invariant with respect to $R(\mu, A^*)$ we have

$$\begin{aligned} |T_0^*(t)v^* - v^*|_{E^*} &= \lim_{\mu \to \infty} |(T^*(t) - I)\mu R(\mu, A^*)v^*|_{E^*} \\ &= \lim_{\mu \to \infty} \left| \int_0^t \mu T^*(\theta) A^* R(\mu, A^*) v^* \, d\theta \right|_{E^*}. \end{aligned} \qquad (5.2.118)$$

By virtue of (5.2.115) and the bounds of $\|T^*(\theta)\|_{\mathcal{L}(E^*)}$ over finite intervals, it follows from (5.2.118) that

$$|T_0^*(t)v^* - v^*|_{E^*} = O(t). \qquad (5.2.119)$$

Therefore, as a consequence of the saturation theorem of Bernes and Butzer [Theorem 1.3.12(ii)], we have $v^* \in D(A^*)$. This completes the proof of the theorem. □

REMARK 5.2.8.

(a) The maximum principle (Theorem 5.2.14) also implies the bang-bang principle if $T^*(\tau - t)e^* \neq 0$ a.e. on $[0, \tau]$ whenever $e^* \neq 0$. This later condition holds whenever the semigroup $t \to T^*(t)$ is analytic.

(b) If $e^* \in D(A^*)$, then $T^*(\tau - t)e^* \in D(A^*)$, and consequently, $(\lambda I - A^*)$ commutes with $T^*(\tau - t)$ on $D(A^*)$ and we have $((\lambda I - A^*)T^*(\tau - t)e^*, e) = (T^*(\tau - t)(\lambda I - A^*)e^*, e)$. This implies that the maximum principle of Theorem 5.2.14 can be restated as

$$(T^*(\tau - t)v^*, u_0(t)) = \sup_{|e|_E \leq 1} (T^*(\tau - t)v^*, e) = |T^*(\tau - t)v^*|_{E^*}$$

for a nonzero $v^* \in E^*$. Thus, under the given condition, the maximum principle given by the Theorem 5.2.14 reduces to that of Theorem 5.2.13.

We have seen that $D(A) \subset K_\infty$; it is not clear, however, whether the equality holds or not. Under certain assumptions we can show that if the maximum principle (Theorem 5.2.13) holds, then necessarily x_0, x_1 must be in $D(A)$. Thus, if $D(A)$ is a proper subset of K_∞ and $x_0, x_1 \in K_\infty \backslash D(A)$, then the above maximum principle may not hold. This is illustrated by the following result presented without proof [Fa.6, Lemma 4.2, p. 175].

Corollary 5.2.3. *Suppose that E is a Hilbert space and the operator A is self-adjoint. Let u_0 be the time optimal control transferring the state from $x_0 = 0$ to $x_1 \in K_\infty$. Then u_0 satisfies the maximum principle (Theorem 5.2.13) only if $x_1 \in D(A)$.*

From this result it is clear that if $D(A) \subsetneq K_\infty$ and $x_1 \in K_\infty \backslash D(A)$, then the maximum principle (Theorem 5.2.13) does not hold; for if it did, then x_1 must be in $D(A)$, leading to a contradiction. Indeed, under certain conditions, $D(A)$ is a proper subset of K_∞ if A is unbounded. This follows from a result again due to Fattorini [Fa.6, Lemma 4.3, p. 176].

Corollary 5.2.4. *Let E be a reflexive Banach space such that $(L_1(0, \tau; E^*))^* = L_\infty(0, \tau; E)$ for some $\tau > 0$. Then if $D(A) = K_\infty$, A must be a bounded operator in E.*

An Example. In the existence theorems for time optimal controls (Theorems 5.2.4, 5.2.12), the state space was assumed to be a reflexive Banach space. But for the problems of heat transfer, the natural state space is $E = C(\Omega)$ where Ω is a suitable open bounded domain in R^3 or, in general, R^n, and for diffusion problems the state space is $E = L_1(\Omega)$. None of these spaces are reflexive, and therefore, the theorems mentioned above do not apply. Fattorini [Fa.6] has discussed these problems in connection with the equation

$$\frac{\partial \phi}{\partial t} = A\phi + u,$$

where $A = k\Delta$, $k > 0$, and Δ is the Laplacian operator given by $\Delta\phi = \Sigma_{i=1}^n (\partial^2\phi / \partial x_i^2)$. He has shown that at least for this problem we can have both the existence theorem and the necessary conditions of optimality similar to those already given.

Let us take E to be equal to $C_0(\bar{\Omega}) \equiv \{\psi \in C(\bar{\Omega}) : \psi|_{\partial\Omega} = 0\}$ and define $D(A) = \{\psi \in C_0(\bar{\Omega})\} : A\psi \in C_0(\bar{\Omega})\}$. Then A is the infinitesimal generator of a c_0-semigroup $T(t)$, $t \geq 0$, on E defined by

$$(T(t)\psi)(x) \equiv \int_\Omega G(t; x, \xi)\psi(\xi)\,d\xi, \qquad \psi \in E, \quad x \in \Omega, \tag{5.2.120}$$

where G is the Green's function for the cylinder $(0, \infty) \times \Omega$ given by

$$G(t; x, \xi) = \sum_{n=1}^{\infty} e^{k\lambda_n t} \phi_n(x)\phi_n(\xi), \tag{5.2.121}$$

and $\{\phi_n\}$ are the eigenfunctions and λ_n the corresponding eigenvalues of the Laplacian operator Δ. That is, $\Delta\phi_n = \lambda_n\phi_n$, $\phi_n \in C_0(\Omega)$, $n = 1, 2, \ldots$. Note that $\lambda_n < 0$. It is known [Fr.1, Chapter 3] that G satisfies the following properties:

$$\begin{cases} \text{(i)} & G(t; x, \xi) > 0 \text{ for } t > 0 \\ \text{(ii)} & G(t; x, \xi) = 0 \text{ for } x \in \partial\Omega,\ t > 0 \\ \text{(iii)} & \int_\Omega G(t; x, \xi)\,d\xi \leq 1 \text{ for } x \in \Omega,\ t > 0. \end{cases} \tag{5.2.122}$$

Thus $T(t)$, $t \geq 0$, is a strongly continuous contraction semigroup on E. Since the dual E^* of the Banach space E is the space $\mathfrak{M}_0(\bar{\Omega})$ of all regular countably additive measures μ on $\bar{\Omega}$ such that $|\mu|(\partial\Omega)=0$ ($|\mu|$ is the total variation of μ), the dual semigroup $\{T^*(t), t \geq 0\}$ has the representation

$$(T^*(t)\mu)(x) \equiv \int_\Omega G(t; x, \xi)\mu(d\xi), \qquad \mu \in E^*, \quad x \in \Omega. \quad (5.2.123)$$

We note that both the semigroups $T(t)$, $T^*(t)$, $t \geq 0$, are analytic in t for $t>0$ but that $T^*(t)$, $t \geq 0$, is not necessarily strongly continuous at $t=0$. Further, $T^*(t)E^* \subseteq F \equiv L_1(\Omega)$. For $t>0$ we define

$$(S(t)\psi)(x) \equiv \int_\Omega G(t; x, \xi)\psi(\xi)d\xi, \qquad \psi \in F, \quad x \in \Omega. \quad (5.2.124)$$

Due to the symmetry of the Green's function and (iii) of property (5.2.122), we have

$$|S(t)\psi|_F \leq |\psi|_F \quad \text{for all } t>0. \quad (5.2.125)$$

Since for $\psi \in E$, $S(t)\psi \to \psi$ as $t \downarrow 0$ and $S(t+\theta)\psi = S(t)S(\theta)\psi = S(\theta)S(t)\psi$ and E is dense in F, $S(t)$ is a strongly continuous (c_0-s.g.) contraction semigroup on F. The corresponding dual semigroup $S^*(t)$, $t \geq 0$, is defined on $F^* = L_\infty(\Omega)$ and has the representation

$$(S^*(t)\psi)(x) = \int_\Omega G(t; x, \xi)\psi(\xi)d\xi, \psi \in F^*, \quad x \in \Omega. \quad (5.2.126)$$

Both the semigroups $S(t)$, $S^*(t)$, $t \geq 0$, are analytic for $t>0$; but $S^*(t)$ is not necessarily strongly continuous at the origin ($t=0$). We collect the following properties of the above semigroups:

$$\begin{cases} \text{(i)} \quad T(t)E \subset D(A) \subseteq E,\ S(t)F \subseteq F \\ \text{(ii)} \quad T^*(t)E^* \subseteq F,\ S^*(t)F^* \subseteq E,\ t>0 \\ \text{(iii)} \quad T(t) = S^*(t)|_E,\ S(t) = T^*(t)|_F, \end{cases} \quad (5.2.127)$$

where $S^*(t)|_E \equiv$ restriction of $S^*(t)$ to E and $T^*(t)|_F \equiv$ restriction of $T^*(t)$ to F.

Let $L_\infty^\omega(F^*)$ denote the class of w^*-measurable F^*-valued functions on $(0, \infty)$, and define the admissible controls as

$$\mathfrak{U}_a^\omega \equiv \{u \in L_\infty^\omega(F^*) : |u(t)|_{F^*} \leq 1\}.$$

Then for $u \in \mathfrak{U}_a^\omega$, the solution of the nonhomogeneous problem

$$\frac{\partial\phi}{\partial t} = A\phi + u, \qquad t>0,$$

$$\phi(0) = \phi_0 \in E \quad (5.2.128)$$

is given by

$$\phi(t) = T(t)\phi_0 + \int_0^t S^*(t-\theta)u(\theta)d\theta, \qquad t \geq 0, \quad (5.2.129)$$

where the integral is defined in the weak sense (Pettis integral, Section 1.1.8) as the unique element of F^* such that for each $h \in F$

$$\left(\int_0^t S^*(t-\theta)u(\theta)d\theta, h \right)_{F^*-F} = \int (u(\theta), S(t-\theta)h)_{F^*-F} d\theta. \tag{5.2.130}$$

Using the properties (5.2.127) and the fact that an analytic semigroup is uniformly continuous on any compact interval $[a, b]$ for $a>0$, we can prove that the trajectory ϕ, given by (5.2.129), is a continuous E-valued function [Fa.6, Lemma 6.1]. As in Theorems 5.2.4 and 5.2.12, we use this fact to prove the following result.

Theorem 5.2.15 (Existence of time optimal control). *Let $\phi_0, \phi_1 \in E$ and suppose there exists a control in $\mathfrak{U}_a^\omega$ that transfers the state of the system* (5.2.128) *from ϕ_0 to ϕ_1 in finite time. Then there exists an optimal control.*

PROOF. The proof follows from exactly the same line of reasoning as in Theorems 5.2.4 and 5.2.12. We note the important differences. Let $\{t_n\}\downarrow$ be a minimizing sequence of transition times with $\{v_n\} \in \mathfrak{U}_a^\omega$ the corresponding sequence of controls. By hypothesis there exists a finite number $\tau>0$ such that $\{t_n\} \leq \tau$. The corresponding sequence of trajectories is given by

$$\phi_n(t) = T(t)\phi_0 + \int_0^t S^*(t-\theta)v_n(\theta)d\theta,$$

and then

$$\begin{aligned} \phi_1 = \phi_n(t_n) &= T(t_n)\phi_0 + \int_0^{t_n} S^*(t_n-\theta)v_n(\theta)d\theta \\ &= T(t_n)\phi_0 + \int_0^{\tau} S^*(t_n-\theta)u_n(\theta)d\theta, \end{aligned} \tag{5.2.131}$$

where

$$u_n(\theta) \equiv \begin{cases} v_n(\theta) & \text{for } 0 \leq \theta \leq t_n \\ 0 & \text{for } t_n < \theta \leq \tau. \end{cases}$$

Clearly, $u_n \in \mathfrak{U}_a^\omega$. Due to a result of Dieudonné used by Fattorini [Fa.6], it is known that for a separable Banach space F, $(L_1(F))^* = L_\infty^\omega(F^*)$. Thus, for $h \in F$

$$(\phi_1, h) = (T(t_n)\phi_0, h) + \int_0^{\tau} (u_n(\theta), S(t_n-\theta)h)_{F^*-F} d\theta. \tag{5.2.132}$$

Since $\mathfrak{U}_a^\omega$ is a closed bounded subset of $L_\infty^\omega(F^*)$ and $\{u_n\} \in \mathfrak{U}_a^\omega$, there exists a subsequence of the sequence $\{u_n\}$, again denoted by $\{u_n\}$, and a $u_0 \in \mathfrak{U}_a^\omega$ such that $u_n \xrightarrow{w^*} u_0$ (weakly*) in $\mathfrak{U}_a^\omega$. Taking the limit, while recalling the fact that both the semigroups $T(t)$ and $S(t)$, $t \geq 0$, are strongly continuous, we obtain

$$(\phi_1, h) = (T(t^*)\phi_0, h) + \int_0^{t^*} (u_0(\theta), S(t^*-\theta)h)_{F^*-F}\, d\theta \quad \text{for all } h \in F,$$

where t^* $(=\lim_n t_n)$ is the optimal transition time and u_0 the optimal control. □

For the time optimal control of the system (5.2.128) we have the following maximum principle.

Theorem 5.2.16 (Maximum principle). *Consider the system* (5.2.128) *with* $\mathcal{U}_a^\omega$ *as the set of admissible controls. Suppose that the initial and final states* ϕ_0 *and* ϕ_1 *belong to* $D(A)$ *and that there exists an optimal control* $u_0 \in \mathcal{U}_a^\omega$ *that transfers* ϕ_0 *to* ϕ_1 *in (optimal) time* τ. *Then there exists a* $\mu^* \in E^*$ *such that* $T^*(t)\mu^* \in D(A^*)$ *and for any* $\lambda \in \rho(A)$

$$\begin{aligned} &((\lambda I - A^*)T^*(\tau - t)\mu^*, u_0(t))_{F-F^*} \\ &\quad = \sup_{|v|_{F^*} \le 1} ((\lambda I - A^*)T^*(\tau - t)\mu^*, v)_{F-F^*} \\ &\quad = |(\lambda I - A^*)T^*(\tau - t)\mu^*|_F \quad \text{a.e. } in\ [0, \tau]. \end{aligned} \tag{5.2.133}$$

PROOF. The proof follows from exactly similar arguments as given for Theorem 5.2.14 and is thus omitted. □

In the context of the heat equation discussed above the maximum principle (5.2.133), Theorem 5.2.16, has a familiar interpretation. Define

$$\Psi_{\mu^*}(t, x) = (\lambda I - A^*) \int_\Omega G(\tau - t; x, \xi)\mu^*(d\xi). \tag{5.2.134}$$

Since the controls in the class $\mathcal{U}_a^\omega$ can be identified with the essentially bounded measurable functions $\{u\}$ on $(0, \tau) \times \Omega$, with $\operatorname{ess\,sup}|u| \le 1$, the maximum principle can be interpreted as

$$\int_\Omega \Psi_{\mu^*}(t, x)u(t, x)\,dx \le \int_\Omega \Psi_{\mu^*}(t, x)u_0(t, x)\,dx = \int_\Omega |\Psi_{\mu^*}(t, x)|\,dx \tag{5.2.135}$$

for all $u \in \mathcal{U}_a^\omega$ and a.e. in $(0, \tau)$. The optimal control u_0 then satisfies the bang-bang principle; that is,

$$u_0(t, x) = \operatorname{sign} \Psi_{\mu^*}(t, x), \qquad (t, x) \in (0, \tau) \times \Omega, \tag{5.2.136}$$

where Ψ_{μ^*} satisfies the adjoint equation

$$\begin{aligned} \frac{\partial}{\partial t}\Psi_\mu^* &= -A^*\Psi_{\mu^*}, && (t, x) \in (0, \tau) \times \Omega, \\ \Psi_\mu^* &= 0, && (t, x) \in (0, \tau) \times \partial\Omega, \\ \Psi_\mu^*(\tau, \cdot) &= (\lambda I - A^*)\mu^*, && x \in \Omega, \end{aligned} \tag{5.2.137}$$

since for $\mu^* \neq 0$ the zeros of the function Ψ_{μ^*} on $(0, \tau) \times \Omega$ has (Lebesgue) measure zero, the optimal control u_0 given by (5.2.136) is well defined almost everywhere on $(0, \tau) \times \Omega$, and further, it is unique.

5.3. Linear Evolution Equations Using Penalty Function Approach

5.3.1. Introduction

In the previous two sections of this chapter we have considered the solutions of optimization problems using the adjoint method. The necessary conditions given in Sections 5.1.6 and 5.2.6 involve a coupled system of evolution equations [Section 5.1.6: Theorem 5.1.5, Equations (5.1.29), (5.1.44), and (5.1.39); Section 5.2.6, Theorem 5.2.9, Equations (5.2.25)(i)–(5.2.25)(iii)] with mixed initial and terminal conditions. These equations must be solved to determine the optimal control. This is a difficult task requiring sophisticated numerical techniques and a considerable amount of computer time.

In many practical situations it may be satisfactory to have a suboptimal control that closely approximates the optimal strategy. In that situation it is possible to bypass the difficult task of solving two-point boundary value problems or even solving differential equations. We discuss in this section one such method, known as the penalty function method. The method consists in replacing all the dynamic and any other side constraints by suitably augmenting the cost functional through additional terms designed to penalize deviations from the given constraints. Let (X, ρ) be a metric space with X_0 a subset, f a real-valued functional defined on X with $f(x) > -\infty$ for $x \in X$, and F a mapping from X into X. The problem is to find an $x^* \in X_0 \cap \{x \in X : F(x) = 0\}$ such that $f(x^*) \leq f(x)$ for all $x \in X_0 \cap \{x \in X : F(x) = 0\}$. Under suitable conditions this constrained extremal problem can be reduced to a family of unconstrained problems. For $\varepsilon > 0$, define

$$f_\varepsilon(x) \equiv f(x) + (1/\varepsilon)\{\rho(F(x), 0) + \rho(x, X_0)\}$$

and find $x_\varepsilon \in X$ such that $f_\varepsilon(x_\varepsilon) \leq f_\varepsilon(x)$ for all $x \in X$, where $\rho(x, K) \equiv \inf_{y \in K} \rho(x, y)$, for any subset $K \subset X$. If such an x_ε exists and, as $\varepsilon \downarrow 0$, $x_\varepsilon \to x_0$, $f_\varepsilon(x_\varepsilon) \to f(x_0)$, and $f(x_0) = \inf\{f(x) : x \in X_0 \cap \{x : F(x) = 0\}\}$, then we observe that for sufficiently small $\varepsilon > 0$ the solution of the unconstrained ε-problem is a good approximation to the true extremal x_0. If the above scheme of things works, there are several advantages to be gained: (i) the difficult task of solving two-point boundary value problems involving a coupled system of evolution equations is completely bypassed; (ii) all types of constraints (dynamic, algebraic, etc.) can be handled in the same way; and (iii) the method suggests an algorithm for actual compution of the extremal.

5.3.2. System Description

Let H_1 and H_2 be real self-adjoint Hilbert spaces with their norms denoted by $|\cdot|_1$ and $|\cdot|_2$, respectively and $I\equiv[0,\tau]$ a fixed time interval. Define $Y\equiv L_2(I, H_1)$ and $\mathcal{U}\equiv L_2(I, H_2)$ with their respective norms denoted by $\|\cdot\|_1$ and $\|\cdot\|_2$. We wish to consider two types of problems: (i) linear problems with distributed control, and (ii) linear problems with boundary control.

For the first type of problems, we consider that the system is described by the linear evolution equation

$$\frac{dx}{dt}=Ax+Bu, \qquad x(0)=x_0, \tag{5.3.1}$$

where A is an unbounded linear operator with domain and range in H_1 and $D(A)$ is dense in H_1. B is a bounded linear operator from H_2 into H_1, that is, $B\in\mathcal{L}(H_2, H_1)$. The controls belong to $\mathcal{U}$. For the second type of problems we consider that the system is governed by the evolution equation

$$\begin{aligned}&\frac{dx}{dt}=Ax, \qquad x(0)=x_0,\\ &Cx=u,\end{aligned} \tag{5.3.2}$$

where A is a linear operator as described before, and C is a linear operator (not necessarily bounded) with domain $D(C)\subset Y$ and range in $\mathcal{U}$ and $u\in\mathcal{U}$. For simplicity we assume that $x_0\in D(A)\subset H_1$; then the system equations can be reduced to

$$\begin{aligned}&\frac{dy}{dt}=Ay+Bu+f,\\ &y(0)=0\end{aligned} \tag{5.3.1$'$}$$

and

$$\begin{aligned}&\frac{dy}{dt}=Ay+f, \qquad y(0)=0,\\ &Cy=u,\end{aligned} \tag{5.3.2$'$}$$

where for $f=Ax_0$ we obtain the original systems.

5.3.3. Formulation of Some Control Problems

Define $Ly\equiv[(d/dt)-A]y$ by $(Ly)(t)=[(d/dt)y(t)-Ay(t)]$, $t\in I$, and $D(L)=\{y: y\in Y, y(0)=0, Ly\in Y\}$. We introduce the following assumptions.

Assumption (5.3.A1). $f\in Y$, $B\in\mathcal{L}(H_2, H_1)$.

Assumption (5.3.A2). $D(L)$ is dense in Y.

Assumption (5.3.A3). The operator A is such that L is closable with $\bar{L}$ the smallest closed extension of L and $D(\bar{L})$ the domain in Y.

Since $D(L)$ is dense in Y, L^* exists and is unique. Moreover, since L is assumed to be closable and H_1 is a Hilbert space, L^* is closed, $D(L^*)$ is dense in $Y^*=Y$, and we have $L^{**}=\bar{L}$ [K.2, Theorem 5.29, p. 168]. We consider the solutions of (5.3.1)′ or (5.3.2)′ in the weak sense, that is, $y\in Y$ is a weak solution if $(y, L^*z)=(g, z)$ for all $z\in D(L^*)$, where $g=Bu+f$ or f. This is equivalent to

$$\bar{L}y=g, \ (g=Bu+f \text{ or } f). \tag{5.3.3}$$

Optimization Problems. Suppose that $\mathcal{U}_a$ is a closed convex subset of $\mathcal{U}$ and J: $\mathcal{U}_a\times Y\to R$ satisfies the following properties:

J1. $J(u, y)\geq 0$ for $(u, y)\in\mathcal{U}_a\times Y$.
J2. J is weakly lower semicontinuous on $\mathcal{U}_a\times Y$.
J3. J is radially unbounded in the sense that $J(u_n, x_n)\to+\infty$ whenever $\|u_n\|_2+\|x_n\|_1\to+\infty$.

We consider the following control problems.

Problem (5.3.P1). Given the system (5.3.1)′, the cost functional J, and the class of admissible controls $\mathcal{U}_a$, find a control $u^0\in\mathcal{U}_a$ and $y^0\in D(\bar{L})\subset Y$ such that

$$J(u^0, y^0)=j_0,$$

where

$$j_0\equiv\inf\{J(u, y):(u, y)\in\mathcal{U}_a\times Y, \bar{L}y=Bu+f\}.$$

Problem (5.3.P2). Given the system (5.3.2)′, the cost function J, and the admissible controls $\mathcal{U}_a$, find a control $u^0\in\mathcal{U}_a$ and $y^0\in D(\bar{L})$ such that

$$J(u^0, y^0)=j_0$$

where

$$j_0\equiv\inf\{J(u, y):(u, y)\in\mathcal{U}_a\times Y, \bar{L}y=f, Cy=u\}.$$

5.3.4. *Existence of Optimal Controls*

First we consider Problem (5.3.P1). As discussed in the introduction, we convert this constrained optimization problem into a family of unconstrained optimization problems. For $\varepsilon>0$ we define

$$J_\varepsilon(u, y)\equiv J(u, y)+(1/\varepsilon)\{\|\bar{L}y-Bu-f\|_1^2\} \tag{5.3.4}$$

and

$$j_\varepsilon \equiv \inf\{J_\varepsilon(u, y) : (u, y) \in \mathcal{U}_a \times Y\}. \tag{5.3.5}$$

Then we find $(u_\varepsilon, y_\varepsilon) \in \mathcal{U}_a \times Y$ such that

$$J_\varepsilon(u_\varepsilon, y_\varepsilon) = j_\varepsilon. \tag{5.3.6}$$

We call this Problem $(5.3.\text{P1})_\varepsilon$ and show that under suitable conditions the problem has a solution $(u_\varepsilon, y_\varepsilon)$. Then, in order that we can consider this solution to be an approximate solution of the original problem (5.3.P1), we prove that $(u_\varepsilon, y_\varepsilon) \to (u^0, y^0) \in \mathcal{U}_a \times Y$, where (u^0, y^0) solves the original problem. That is,

$$J(u^0, y^0) = \inf\{J(u, y) : (u, y) \in \mathcal{U}_a \times Y, \bar{L}y = Bu + f\} \equiv j_0. \tag{5.3.7}$$

Theorem 5.3.1. *Consider Problem* $(5.3.\text{P1})_\varepsilon$ *with* $\varepsilon > 0$ *and suppose that Assumptions* (5.3.A1)–(5.3.A3) *hold,* $\mathcal{U}_a$ *is a closed convex subset of* $\mathcal{U}$, *and the cost functional J satisfies properties* (J1)–(J3). *Then there exists* $(u_\varepsilon, y_\varepsilon) \in \mathcal{U}_a \times D(\bar{L})$ *that solves Problem* $(5.3.\text{P1})_\varepsilon$.

PROOF. Let $\{(u_n, y_n)\}$ be a minimizing sequence of J_ε, that is, $J_\varepsilon(u_n, y_n) \downarrow j_\varepsilon$ as $n \uparrow \infty$. Being a decreasing sequence, $J_\varepsilon(u_n, y_n)$ is bounded and always nonnegative due to (J1) and the definition of J_ε (c.f. (5.3.4)). Therefore, $J(u_n, y_n)$ is also bounded, and hence, due to the property (J3), there exists a constant $0 < \alpha < \infty$ such that $\|u_n\|_2 \leq \alpha$ and $\|y_n\|_1 \leq \alpha$. Further, since $J_\varepsilon(u_n, y_n)$ is a nonincreasing sequence converging to $j_\varepsilon \geq 0$, there exists a constant $0 < \beta < \infty$ such that $(1/\varepsilon)\{\|\bar{L}y_n - Bu_n - f\|_1^2\} \leq \beta$ for all $n \geq 1$. Since $\|u_n\|$ is bounded and B is a bounded linear operator, $\|Bu_n\|$ is also bounded, and consequently, there exists a constant $0 \leq \gamma < \infty$ such that $\|\bar{L}y_n\|_1 \leq \gamma$ for all $n \geq 1$. Thus $\{u_n\} \subset \mathcal{U}$, $\{y_n\} \subset Y$, and $\{\bar{L}y_n\} \subset Y$ are bounded subsets. Since both $\mathcal{U}$ and Y are Hilbert spaces and bounded subsets of Hilbert spaces are conditionally weakly (sequentially) compact, there exist subsequences, relabled as $\{u_n\}$, $\{y_n\}$, and $\{\bar{L}y_n\}$, $u_\varepsilon \in \mathcal{U}$, $y_\varepsilon \in Y$, and $z_\varepsilon \in Y$ such that

$$\begin{aligned} &u_n \xrightarrow{w} u_\varepsilon && \text{(weakly) in } \mathcal{U}, \\ &y_n \xrightarrow{w} y_\varepsilon && \text{in } Y, \\ &\bar{L}y_n \xrightarrow{w} z_\varepsilon && \text{in } Y. \end{aligned} \tag{5.3.8}$$

Since $\mathcal{U}_a$ is a closed convex set, it is also weakly closed, and consequently, $u_\varepsilon \in \mathcal{U}_a$. Since the operator $\bar{L}$ is closed, its graph $\Gamma(\bar{L}) \equiv \{(x, \bar{L}x) : x \in D(\bar{L})\}$ is a closed linear manifold of $Y \times Y$. In fact, $\Gamma(\bar{L}) = \overline{\Gamma(L)}$. We may assume that $Y \times Y$ is endowed with the topology $\|(y, z)\| \equiv (\|y\|_Y^2 + \|z\|_Y^2)^{1/2}$, which makes it into a Banach space. Since a closed linear manifold in a Banach space is also weakly closed, we conclude that

$$\mathrm{w}\lim_{n\to\infty}\left(y_n, \bar{L}y_n\right) = (y_\varepsilon, z_\varepsilon) \in \Gamma(\bar{L}). \tag{5.3.9}$$

Therefore we have $y_\varepsilon \in D(\bar{L})$ and $z_\varepsilon = \bar{L}y_\varepsilon$.

Since the norm $\|\bar{L}y - Bu - f\|_1$ is weakly lower semicontinuous and, by assumption (J2), J is weakly lower semicontinuous, we have

$$\begin{aligned} j_\varepsilon = \lim_n J_\varepsilon(u_n, y_n) &= \lim_n J(u_n, y_n) + \lim_n \frac{1}{\varepsilon}\{\|\bar{L}y_n - Bu_n - f\|_1^2\} \\ &\geq \underline{\lim}_n J(u_n, y_n) + \underline{\lim}_n \frac{1}{\varepsilon}\{\|\bar{L}y_n - Bu_n - f\|_1^2\} \\ &\geq J(u_\varepsilon, y_\varepsilon) + \frac{1}{\varepsilon}\{\|\bar{L}y_\varepsilon - Bu_\varepsilon - f\|_1^2\} \equiv J_\varepsilon(u_\varepsilon, y_\varepsilon). \end{aligned} \tag{5.3.10}$$

By the definition (5.3.5), j_ε is the infimum, and consequently, $j_\varepsilon \leq J_\varepsilon(u_\varepsilon, y_\varepsilon)$. Therefore the equality must hold. This proves the existence of a pair $(u_\varepsilon, y_\varepsilon) \in \mathcal{U}_a \times D(\bar{L})$ that solves Problem $(5.3.\mathrm{P1})_\varepsilon$. □

We now prove that the original optimization problem (5.3.P1) has a solution to which the solution of the ε-problem $(5.3.\mathrm{P1})_\varepsilon$ converges as $\varepsilon \to 0^+$.

Theorem 5.3.2. *Let $\{\varepsilon_n\}$ be a decreasing sequence of positive real numbers converging to zero and $\{(u_{\varepsilon_n}, y_{\varepsilon_n})\}$ the optimal pair for Problem* $(5.3.\mathrm{P1})_{\varepsilon_n}$. *Then there exists a subsequence of the sequence $\{\varepsilon_n\}$, again denoted by $\{\varepsilon_n\}$, such that*

$$u_{\varepsilon_n} \xrightarrow{\mathrm{w}} u^0 \qquad (\textit{weakly}) \textit{ in } \mathcal{U},$$

$$y_{\varepsilon_n} \xrightarrow{\mathrm{w}} y^0 \qquad \textit{in } Y, \tag{5.3.11}$$

$$\bar{L}y^0 = Bu^0 + f,$$

$$J(u^0, y^0) = j_0 \quad [\textit{as defined for Problem } (5.3.\mathrm{P1})], \tag{5.3.12}$$

and

$$\lim_n J_{\varepsilon_n}(u_{\varepsilon_n}, y_{\varepsilon_n}) = J(u^0, y^0). \tag{5.3.13}$$

PROOF. Define $K \equiv \{(u, y) \in \mathcal{U}_a \times Y : y \in D(\bar{L}),\ \bar{L}y = Bu + f\}$. Then it is clear that

$$\begin{aligned} \inf\{J_\varepsilon(u, y) : (u, y) \in \mathcal{U}_a \times Y\} &\leq \inf\{J_\varepsilon(u, y) : (u, y) \in K\} \\ &= \inf\{J(u, y) : (u, y) \in K\} \equiv j_0. \end{aligned} \tag{5.3.14}$$

Therefore, by Theorem 5.3.1, there exists a pair $(u_\varepsilon, y_\varepsilon)$ such that

$$j_0 \geq \inf\{J_\varepsilon(u, y) : (u, y) \in \mathcal{U}_a \times Y\} = J_\varepsilon(u_\varepsilon, y_\varepsilon)$$

for every $\varepsilon > 0$. Let $\{\varepsilon_n\}$ be a decreasing sequence of positive real numbers converging to zero and consider the sequence $\{J_{\varepsilon_n}(u_{\varepsilon_n}, y_{\varepsilon_n})\}$, which is clearly nonnegative and bounded above by j_0. Since j_0 is a finite positive number, the sequence $\{J_{\varepsilon_n}(u_{\varepsilon_n}, y_{\varepsilon_n})$ is bounded, and consequently, the sequence $\{J(u_{\varepsilon_n}, y_{\varepsilon_n})\}$ is also bounded and, due to (J1), positive. Consequently, due to the property (J3) there exists a finite positive number α such that

$$\| u_{\varepsilon_n} \|_2 \leq \alpha, \qquad \| y_{\varepsilon_n} \|_1 \leq \alpha,$$

and, since B is a bounded linear operator and $f \in Y$, we have $\| \bar{L} y_{\varepsilon_n} \|_1 \leq \alpha$. Thus there exists a subsequence of the sequence $\{\varepsilon_n\}$, relabeled as $\{\varepsilon_n\}$, and $u^0 \in \mathcal{U}$, $y^0 \in Y$, and $z^0 \in Y$ such that

$$
\begin{aligned}
u_{\varepsilon_n} &\xrightarrow{w} u^0 \qquad \text{(weakly) in } \mathcal{U}, \\
y_{\varepsilon_n} &\xrightarrow{w} y^0 \qquad \text{in } Y, \\
\bar{L} y_{\varepsilon_n} &\xrightarrow{w} z^0 \qquad \text{in } Y.
\end{aligned}
\tag{5.3.15}
$$

Therefore, by the same arguments, as given in the proof of Theorem 5.3.1, we can deduce that $u^0 \in \mathcal{U}_a$ and $(y^0, z^0) \in \Gamma(\bar{L})$, that is,

$$y^0 \in D(\bar{L}) \quad \text{and} \quad z^0 = \bar{L} y^0. \tag{5.3.16}$$

Since

$$\infty > j_0 \geq J_{\varepsilon_n}(u_{\varepsilon_n}, y_{\varepsilon_n}) \quad \text{for all } n \geq 1, \tag{5.3.17}$$

there exists a number $\beta > 0$ such that

$$\| \bar{L} y_{\varepsilon_n} - B u_{\varepsilon_n} - f \|_1 \leq \beta \sqrt{\varepsilon_n} \quad \text{for all } n \geq 1, \tag{5.3.18}$$

and consequently,

$$\lim_{n \to \infty} \| \bar{L} y_{\varepsilon_n} - B u_{\varepsilon_n} - f \|_1 = 0. \tag{5.3.19}$$

Since, for all $n \geq 1$,

$$2\left(\bar{L} y_{\varepsilon_n} - B u_{\varepsilon_n} - f, \bar{L} y^0 - B u^0 - f \right)_Y \leq \| \bar{L} y_{\varepsilon_n} - B u_{\varepsilon_n} - f \|_1^2 + \| \bar{L} y^0 - B u^0 - f \|_1^2,$$

it follows from (5.3.19) that

$$2 \lim_n \left(\bar{L} y_{\varepsilon_n} - B u_{\varepsilon_n} - f, \bar{L} y^0 - B u^0 - f \right)_Y \leq \| \bar{L} y^0 - B u^0 - f \|_1^2. \tag{5.3.20}$$

Due to boundedness of the operator B it follows from (5.3.15), (5.3.16), and (5.3.20) that $2 \| \bar{L} y^0 - B u^0 - f \|_1^2 \leq \| \bar{L} y^0 - B u^0 - f \|_1^2$, and consequently, $\| \bar{L} y^0 - B u^0 - f \|_1 = 0$, or equivalently,

$$(u^0, y^0) \in K, \text{ and in particular } \bar{L} y^0 = B u^0 + f. \tag{5.3.21}$$

Thus we have proved (5.3.11). For the proof of (5.3.12) and (5.3.13) we note that, since $(u^0, y^0) \in K$ and $j_0 \equiv \inf\{J(u, y) : (u, y) \in K\}$,

$$J(u^0, y^0) \geq j_0. \tag{5.3.22}$$

On the other hand, due to (5.3.17) and the weak lower semicontinuity of J, we have

$$\begin{aligned} j_0 &\geq \overline{\lim_n}\, J_{\varepsilon_n}(u_{\varepsilon_n}, y_{\varepsilon_n}) \geq \lim_n J_{\varepsilon_n}(u_{\varepsilon_n}, y_{\varepsilon_n}) \geq \underline{\lim_n}\, J_{\varepsilon_n}(u_{\varepsilon_n}, y_{\varepsilon_n}) \\ &\geq \underline{\lim_n}\, J(u_{\varepsilon_n}, y_{\varepsilon_n}) \geq J(u^0, y^0). \end{aligned} \tag{5.3.23}$$

Thus, $J(u^0, y^0) = j_0$ and $\lim_n J_{\varepsilon_n}(u_{\varepsilon_n}, y_{\varepsilon_n})$ exists and equals $J(u^0, y^0)$. This completes the proof of the theorem. □

Now we consider Problem (5.3.P2). Again we construct an ε-problem and show that it has a solution. Finally, we obtain the optimal control by letting the ε-problem converge to the original problem. Define $M \equiv \{(u, y) \in \mathfrak{U}_a \times Y : u = Cy\}$, $G \equiv (\mathfrak{U}_a \times D(\bar{L})) \cap M$, where, as before, $\bar{L}$ is the smallest closed extension of the operator L. Assuming G to be nonempty we define, for $\varepsilon > 0$,

$$J_\varepsilon(u, y) \equiv J(u, y) + \frac{1}{\varepsilon} \| \bar{L}y - f \|_1^2 \tag{5.3.24}$$

and find a pair $(u_\varepsilon, y_\varepsilon) \in G$ such that

$$J_\varepsilon(u_\varepsilon, y_\varepsilon) \leq J_\varepsilon(u, y) \quad \text{for all } (u, y) \in G. \tag{5.3.25}$$

We call this Problem $(5.3.\text{P2})_\varepsilon$.

Theorem 5.3.3. *Consider Problem* $(5.3.\text{P2})_\varepsilon$, *for* $\varepsilon > 0$, *with the cost function* J *satisfying these properties*:

J1′. $J(u, y) \geq 0$, $(u, y) \in \mathfrak{U} \times Y$.

J2′. *J is weakly lower semicontinuous on* $\mathfrak{U} \times Y$.

J3′. $J(u, y) \to +\infty$ *as* $\|y\|_1 \to +\infty$ *for* $(u, y) \in \mathfrak{U} \times Y$.

Further, suppose that the operator L is closable, with the minimal closed extension denoted by $\bar{L}$, and that the operator C is closed with its domain $D(C)$ containing $D(\bar{L})$ and its range $R(C)$ contained in $\mathfrak{U}$. Suppose C is also relatively bounded with respect to the operator $\bar{L}$ in the sense that there exist finite numbers $a, b \geq 0$ such that

$$\|Cy\|_2 \leq a\|y\|_1 + b\|\bar{L}y\|_1 \quad \textit{for all } y \in D(\bar{L}). \tag{5.3.26}$$

Then Problem $(5.3.\text{P2})_\varepsilon$ *has a solution.*

PROOF. For $\varepsilon > 0$ define

$$j_\varepsilon \equiv \inf\{J_\varepsilon(u, y) : (u, y) \in G\}. \tag{5.3.27}$$

Then our problem is to find a pair $(u_\varepsilon, y_\varepsilon) \in G$ such that $J_\varepsilon(u_\varepsilon, y_\varepsilon) = j_\varepsilon$. Let $\{(u_n, y_n)\} \in G$ be a minimizing sequence such that

$$\lim_{n\to\infty} J_\varepsilon(u_n, y_n) = j_\varepsilon \geq 0. \tag{5.3.28}$$

Clearly, there exists a finite number $\alpha > 0$ such that

$$J_\varepsilon(u_n, y_n) \leq \alpha, \qquad J(u_n, y_n) \leq \alpha \quad \text{for all } n \geq 1. \tag{5.3.29}$$

Therefore, due to properties (J1)′, (J3)′, and the assumption that $f \in Y$, it follows from (5.3.29) and the definition of J_ε that there exists a finite number $\beta > 0$ such that

$$\| y_n \|_1 \leq \beta, \qquad \| \bar{L} y_n \|_1 \leq \beta. \tag{5.3.30}$$

Since Y is a Hilbert space, there exists a subsequence of the sequence $\{y_n\}$, again denoted by $\{y_n\}$, and elements $y_\varepsilon, z_\varepsilon \in Y$ such that

$$\begin{aligned} y_n &\xrightarrow{w} y_\varepsilon && \text{(weakly) in } Y, \\ \bar{L} y_n &\xrightarrow{w} z_\varepsilon && \text{(weakly) in } Y. \end{aligned} \tag{5.3.31}$$

Since the graph $\Gamma(\bar{L})$ is a closed linear manifold of $Y \times Y$, it is weakly closed, and consequently, the weak limit $(y_\varepsilon, z_\varepsilon)$ belongs to $\Gamma(\bar{L})$, or equivalently,

$$y_\varepsilon \in D(\bar{L}), \qquad z_\varepsilon = \bar{L} y_\varepsilon. \tag{5.3.32}$$

Recalling that by definition the set G contains the minimizing sequence $\{(u_n, y_n)\}$, we note that

$$u_n = C y_n, \qquad u_n \in \mathcal{U}_a. \tag{5.3.33}$$

Therefore, since C is relatively bounded with respect to the operator $\bar{L}$, it follows from (5.3.26), (5.3.30), and (5.3.33) that

$$\| u_n \|_2 \leq \beta(a+b) < \infty \quad \text{for all } n \geq 1. \tag{5.3.34}$$

Therefore, there exists a subsequence of the sequence $\{u_n\}$, relabeled as $\{u_n\}$, and a $u_\varepsilon \in \mathcal{U}$ such that

$$u_n \xrightarrow{w} u_\varepsilon \quad \text{(weakly) in } \mathcal{U}. \tag{5.3.35}$$

Since $\mathcal{U}_a$ is a closed convex subset of $\mathcal{U}$, it is also weakly closed, and consequently, $u_\varepsilon \in \mathcal{U}_a$. On the other hand, since C is a closed linear operator from $D(C)$ into $\mathcal{U}$, its graph $\Gamma(C)$ is a closed linear manifold of $Y \times \mathcal{U}$ and hence weakly closed. Thus, the weak limit of the sequence $\{(y_n, u_n)\}$ from $\Gamma(C)$ belongs to $\Gamma(C)$. That is,

$$(y_\varepsilon, u_\varepsilon) \in \Gamma(C), \tag{5.3.36}$$

or equivalently, $y_\varepsilon \in D(C)$ and $u_\varepsilon = C y_\varepsilon$. Since $u_\varepsilon \in \mathcal{U}_a$, it follows that $(u_\varepsilon, y_\varepsilon) \in M$, and consequently, by (5.3.32), we obtain

$$(u_\varepsilon, y_\varepsilon) \in G. \tag{5.3.37}$$

It remains to show that $J_\varepsilon(u_\varepsilon, y_\varepsilon)=j_\varepsilon$. Since the norm is weakly lower semicontinuous and, by hypothesis (J2)′, J is weakly lower semicontinuous, it follows from (5.3.31), (5.3.32), and (5.3.35) that

$$\begin{aligned} j_\varepsilon &= \lim_n J_\varepsilon(u_n, y_n) = \lim_n J(u_n, y_n) + \frac{1}{\varepsilon} \lim_n \|\bar{L}y_n - f\|_1^2 \\ &\geq \underline{\lim}_n\, J(u_n, y_n) + \frac{1}{\varepsilon}\, \underline{\lim}_n\, \|\bar{L}y_n - f\|_1^2 \\ &\geq J(u_\varepsilon, y_\varepsilon) + \frac{1}{\varepsilon} \|\bar{L}y_\varepsilon - f\|_1^2 = J_\varepsilon(u_\varepsilon, y_\varepsilon). \end{aligned} \tag{5.3.38}$$

Since j_ε is the infimum of J_ε on G and, by (5.3.37), $(u_\varepsilon, y_\varepsilon) \in G$, it is clear that

$$J_\varepsilon(u_\varepsilon, y_\varepsilon) \geq j_\varepsilon. \tag{5.3.39}$$

Therefore it follows from (5.3.38) and (5.3.39) that $(u_\varepsilon, y_\varepsilon)$ solves Problem $(5.3.\mathrm{P}2)_\varepsilon$. This completes the proof. □

It is now required to show that the solution of the ε-problem $(5.3.\mathrm{P}2)_\varepsilon$ converges to the solution of the original problem (5.3.P2). Since the proof is similar to that of Theorem 5.3.2, we state the result without proof.

Theorem 5.3.4. *Let* $\{\varepsilon_n\}$ *be a decreasing sequence of positive real numbers converging to zero and* $\{(u_{\varepsilon_n}, y_{\varepsilon_n})\}$ *the optimal pair for Problem* $(5.3.\mathrm{P}2)_{\varepsilon_n}$. *Then there exists a subsequence of the sequence* $\{\varepsilon_n\}$, *relabeled as* $\{\varepsilon_n\}$, *such that*

$$\begin{aligned} & u_{\varepsilon_n} \xrightarrow{w} u^0, && \text{(weakly) in } \mathcal{U}, \\ & y_{\varepsilon_n} \xrightarrow{w} y^0, && \text{in } Y, \\ & \bar{L}y^0 = f, && \qquad (5.3.40) \\ & Cy^0 = u^0 && \\ & J(u^0, y^0) = j_0 && [\textit{as defined for Problem } (5.3.\mathrm{P}2)] \quad (5.3.41) \end{aligned}$$

and

$$\lim_n J_{\varepsilon_n}(u_{\varepsilon_n}, y_{\varepsilon_n}) = J(u^0, y^0). \tag{5.3.42}$$

REMARK 5.3.1. Note that for Problem (5.3.P2) we did not remove all the constraints. However, under the given assumptions, we can obtain a completely unconstrained problem by replacing the cost functional (5.3.24) by

$$J_\varepsilon(u, y) \equiv J(u, y) + \frac{1}{\varepsilon}\left\{\|\bar{L}y - f\|_1^2 + \|Cy - u\|_2^2\right\}. \tag{5.3.24$'$}$$

This modified problem is entirely equivalent to the previous one and the conclusions of Theorems 5.3.3 and 5.3.4 remain unchanged.

An Example. We consider an example illustrating (5.3.P2). Let Ω be an open bounded set in R^n with smooth boundary $\Gamma\equiv\partial\Omega$. Define $Q\equiv I\times\Omega$, $\Sigma\equiv I\times\Gamma$, and take $H_1\equiv L_2(\Omega)$, $H_2=L_2(\Gamma)$, $Y\equiv L_2(I, H_1)=L_2(Q)$, and $\mathcal{U}\equiv L_2(I, H_2)=L_2(\Sigma)$. Let m be a nonnegative integer and let $H^m(\Omega)$ denote the Sobolev space of order m on Ω defined by

$$H^m(\Omega)\equiv\{\psi: D^\alpha\psi\in L_2(\Omega) \text{ for all } \alpha, |\alpha|\le m\}$$

where $\alpha=(\alpha_1,\alpha_2\ldots\alpha_n)$, α_i nonnegative integers, $|\alpha|=\Sigma\alpha_i$, and

$$D^\alpha\equiv\frac{\partial^{\alpha_1+\alpha_2+\cdots+\alpha_n}}{\partial x_1^{\alpha_1}\cdots\partial x_n^{\alpha_n}}.$$

Similarly, we can define Sobolev spaces $H^{r,s}(Q)$ as

$$H^{r,s}(Q)=\{\psi: D_x^\alpha\psi, D_t^k\psi\in L_2(Q) \text{ for } |\alpha|\le r, 0\le k\le s\}.$$

Endowed with the norm

$$\|\psi\|_{H^m}\equiv\left(\sum_{|\alpha|\le m}|D^\alpha\psi|^2_{L_2(\Omega)}\right)^{1/2},$$

H^m is a Hilbert space. The space $H^{r,s}(Q)$ can be considered as

$$H^{r,s}(Q)=L_2(I, H^r(\Omega))\cap H^s(I, L_2(\Omega)),$$

which is a Hilbert space with the norm

$$\|\psi\|_{H^{r,s}}\equiv\left(\int_I\|\psi(t)\|^2_{H^r(\Omega)}\,dt+\int_I\sum_{0\le j\le s}\|D_t^j\psi\|^2_{L_2(\Omega)}\,dt\right)^{1/2}.$$

In fact, these spaces are well defined for all positive real numbers [LM.1, LM.2] either by the method of interpolation between two Sobolev spaces of integral order or by the use of the Fourier transform technique. The latter technique also allows to define these spaces for negative real numbers. However, we do not need this here. Let A be a second-order elliptic operator and define $L=[(\partial/\partial t)-A]$ with domain

$$D(L)=\{y: y\in Y, \partial y/\partial t\in Y, Ay\in Y\}. \tag{5.3.43}$$

Define

$$\|y\|_{D(L)}=\|y\|_Y+\|Ly\|_Y, \tag{5.3.44}$$

and the operator C by

$$Cy=\left.\frac{\partial}{\partial\nu_A}y\right|_\Sigma, \qquad D(C)=\{y\in L_2(Q): Cy\in\mathcal{U}\}, \tag{5.3.45}$$

where

$$\frac{\partial}{\partial \nu_A}\psi \equiv \sum_{i,j} a_{ij}(\cdot)\frac{\partial \psi}{\partial x_j}\cos(\nu, x_i) \quad \text{on } \Gamma;$$

$\cos(\nu, x_i)$ is ith direction cosine of ν, ν is the unit normal to Γ directed exterior to Ω, and a_{ij} are the coefficients of the principal part of the operator A.

For $y \in D(L)$, it follows from the trace theorem [LM.1, Theorem 2.1, p. 9] that $Cy \in H^{1/2,1/4}(\Sigma)$ and that the map $y \rightarrow Cy$ is continuous and linear. Consequently, there exists a $\beta > 0$ such that

$$\|Cy\|_{H^{1/2,1/4}(\Sigma)} \leq \beta \|y\|_{D(L)}. \tag{5.3.46}$$

Since $H^{1/2,1/4}(\Sigma) \subset L_2(\Sigma) \equiv \mathcal{U}$, we have $Cy \in L_2(\Sigma)$ and

$$\|Cy\|_{L_2(\Sigma)} \leq \beta' \|y\|_{D(L)} \quad \text{for a constant } \beta' > 0.$$

Clearly, the operators L and C satisfy the requirements of Theorem 5.3.3, and consequently, Problem $(5.3.\text{P2})_\varepsilon$ with the cost functional

$$J_\varepsilon(u,\psi) = J(u,\psi) + \frac{1}{\varepsilon}\left\{ \left\| \left(\frac{\partial}{\partial t} - A\right)\psi - f \right\|^2_{L_2(Q)} + \left\| \frac{\partial}{\partial \nu_A}\psi - u \right\|^2_{L_2(\Sigma)} \right\}$$

is well defined and has a solution. In other words, under the given assumptions, we can always solve the optimal control problem

$$\frac{\partial \psi}{\partial t} = A\psi + f, \qquad (t,x) \in Q,$$

$$\psi(0) = \psi_0 \in H_1, \qquad x \in \Omega,$$

$$\frac{\partial \psi}{\partial \nu_A} = u \in \mathcal{U}_a \subset L_2(\Sigma),$$

$$J(u,\psi) = \min$$

by the approximation technique given above.

5.4. A Class of Nonlinear Evolution Equations

5.4.1. *Introduction*

In many diffusion and wave propagation problems the system is governed by essentially nonlinear partial differential equations. For example, the well-known Navier–Stokes equation arising in hydrodynamical problems is a system of nonlinear diffusion equations. Similarly, in the problems of chemical (nuclear) kinetics, involving simultaneously chemical (nuclear) reactions and heat transfer, one has to deal with a system of nonlinear diffusion equations. The well known Korteweg–de Vries equation describing the dynamics of water waves is a system of nonlinear equations. There is no clearcut classification for nonlinear equations as in the linear case (elliptic, parabolic, hyperbolic). However, a large body of nonlinear diffusion and nonlinear wave propagation problems can be described by the so-called quasi-linear parabolic [LSU.1] and quasi-linear hyperbolic equations [Je.1], respectively.

The nonlinearities usually appear in the lower-order terms. A fairly large class of such problems can be covered by quasi-linear evolution equations on suitable Banach spaces, in the same way as that, we have seen, many linear diffusions and waves are covered under the semigroup formulation.

The subject of control theory for nonlinear problems is very sketchy and is wide open [Ce.4, Ce.5, Be.2, Li.2, Ah.2, AT.7]. In this section we wish to study control problems for a large class of nonlinear evolution equations. We shall be concerned mainly with the question of existence of optimal controls and briefly touch on the subject of the necessary conditions of optimality.

5.4.2. System Description

Let H be a Hilbert space with inner product (x, y) for $x, y \in H$ and E a dense linear subspace of H carrying the structure of a reflexive Banach space with continuous injection into H. Let E^* be the topological dual of E with duality bracket denoted either by $\langle y, x\rangle$ or $\langle x, y\rangle$ for $y \in E^*$ and $x \in E$. Let $|\cdot|_E$, $|\cdot|_H$, and $|\cdot|_{E^*}$ denote the norms in E, H, and E^*, respectively. Let $I \equiv [0, T]$ be any closed bounded interval of $R_+ = [0, \infty)$ and $L_p(I, E)$ and $L_q(I, E^*)$ the usual Banach spaces with $p = (p-1)q$, $1 < p < \infty$. Since E is reflexive, E^* is also reflexive, and thus, for $1 < p < \infty$ the Banach spaces $L_p(I, E)$ and $L_q(I, E^*)$ are also reflexive and $(L_p(I, E))^* = L_q(I, E^*)$. Let B be another reflexive Banach space and denote by $\mathrm{cbc}(B)$ the class of closed bounded convex subsets of B and suppose $t \to \Gamma(t)$ is a measurable function on I with values in $\mathrm{cbc}(B)$, that is, $\Gamma(t) \in \mathrm{cbc}(B)$ for each $t \in I$. Let $\mathcal{U}$ denote the class of all strongly measurable functions $\{u\}$ on I with values $u(t) \in \Gamma(t)$. We consider $\mathcal{U}$, unless stated otherwise, to be the class of admissible controls.

The control system we wish to consider is governed by the nonlinear evolution equation

$$
\begin{aligned}
&\frac{dx}{dt} = A(t)x + f(t, x, u(t)), \qquad t \in I \\
&x(0) = x_0, \\
&\quad u \in \mathcal{U},
\end{aligned}
\tag{5.4.1}
$$

where $\{A(t),\ t \in I\}$ is a family of densely defined linear operators, not necessarily bounded, with domain $D(A(t)) \subset E$ and range $R(A(t)) \subset E^*$ and $f: I \times E \times B \to E^*$. More specific hypotheses on the operators A and f will be introduced shortly.

We mention that the system (5.4.1) covers a wide variety of nonlinear distributed parameter systems including the class of strongly nonlinear parabolic systems [Ah.2, Ah.4] and the linear parabolic systems [Li.2].

Another example covered by the abstract evolution equation (5.4.1) is given by the parabolic equation

$$\frac{\partial \phi}{\partial t} = \sum_{|\alpha|,|\beta| \leq m+1} (-1)^{|\alpha|} D^{\alpha}\left(q_{\alpha,\beta}(t,\xi)\, D^{\beta}\phi\right) + \sum_{|\alpha| \leq m} D^{\alpha} F_{\alpha}(t,\xi,\phi,D^{1}\phi,\ldots,D^{m}\phi,u(t,\xi)) \tag{5.4.2}$$

for $(t,\xi)\in Q = I\times\Omega$, where Ω is an open connected bounded subset of R^n, and

$$D^{r}\psi = \partial^{r_1+\cdots+r_n}\psi/\partial x_1^{r_1}\cdots x_n^{r_n}, \qquad r_i \text{ nonnegative integers},$$

$$|r| = \sum_{i=1}^{n} r_i, \qquad r=\alpha \text{ or } \beta. \tag{5.4.3}$$

For any positive integer m, introduce the Sobolev space $H^{m,p}(\Omega) = \{f\colon D^{\alpha}f \in L_p(\Omega),\ |\alpha|\leq m\}$. Endowed with the norm

$$|f|_{H^{m,p}} \equiv \left(\sum_{|\alpha|\leq m} \| D^{\alpha} f \|^{p}_{L_p(\Omega)} \right)^{1/p}, \tag{5.4.4}$$

$H^{m,p}$ is a Banach space, and for $1<p<\infty$ it is also a reflexive Banach space. For the state space E we can choose any linear subspace of the Sobolev space $H^{m,p}$, so that it is a reflexive Banach space and $C_0^{\infty}(\Omega)\equiv \mathfrak{D}(\Omega)$, the space of C^{∞}-functions with compact support in Ω, is contained in E. For example, we can choose $E = H_0^{m,p}$, which is the completion of $\mathfrak{D}(\Omega)$ with respect to the norm topology of $H^{m,p}(\Omega)$. In that case E^* is given by $H^{-m,q}(\Omega)$, where $(1/p)+(1/q)=1$, and it is a subspace of the space of distributions $\mathfrak{D}'(\Omega)$. For the Hilbert space H we can choose $L_2(\Omega)$. In that case we may choose $E = H_0^{m,p}$ with $p\geq 2$.

5.4.3. Existence and Uniqueness of Solutions

For the study of the problems of existence of optimal controls it is required to ensure that the evolution equation describing the control system (5.4.1) have a unique solution for each admissible control. We can resolve this problem easily by use of the results of Section 2.5.

Definition 5.4.1. Consider the system (5.4.1) and suppose u is a fixed element of $\mathfrak{U}$. Then an element $x\in Y\equiv L_p(I,E)$ is said to be a *strong solution of the evolution equation*

$$\frac{dx}{dt} = A(t)x + f(t,x,u(t)), \qquad t\in I,$$

with initial condition $x(0)=x_0\in D(A(t))$, $t\in I$, if

$$(L_s x)(t)=f(t,x(t),u(t)) \quad \text{a.e. on } I,$$

where L_s is the strong extension of the operator L, given by $(L\phi)(t)\equiv[(d/dt)-A(t))\phi(t)]$ from Y to Y^* (see Section 2.5).

Theorem 5.4.1. *Consider the system* (5.4.1) *with* $u\in\mathfrak{U}$ *and* $x_0\in D(A(t))$ *for* $t\in I$. *Suppose that for each* $u\in\mathfrak{U}$ *the function* g_u, *given by* $g_u(t,x)\equiv f(t,x,u(t))$, *maps* $I\times E$ *into* E^* *and satisfies Assumptions* (2.5.A3)–(2.5.A6) *and the operator* A *satisfies Assumptions* (2.5.A1)–(2.5.A2) *and that for* $Y=L_p(I,E)$ *the strong and weak extensions of* L *from* Y *into* Y^* *coincide in the sense of Definition* 2.5.3. *Then the control system* (5.4.1) *has a unique strong solution* $x_u\in Y$ *for each* $u\in\mathfrak{U}$.

PROOF. The proof is an immediate consequence of Theorem 2.5.3. □

5.4.4. *Formulation of Some Control Problems*

With respect to the system (5.4.1) we wish to consider the following control problems.

Problem (5.4.P1) (Terminal control). Let ϕ_0 be a function on E with values in $R_+=[0,\infty)$ and S a nonempty subset of E. The problem is to find a control $u\in\mathfrak{U}$ such that $J(u)\equiv\phi_0(x(T))$ is minimum, subject to the constraint that x is the response of the system (5.4.1) corresponding to the control u and that $x(T)\in S$.

Problem (5.4.P2) (Time optimal control). Let S be a set-valued function on I with values $S(t)$ a nonempty closed bounded convex subset of E. That is, $S(t)\in\text{cbc}(E)$ for each $t\in I$. Suppose that there exist controls in $\mathfrak{U}$ that can transfer the system (5.4.1) from the state x_0 to the target set $S(t)$ for some $t\in I$. It is required to find a control $u\in\mathfrak{U}$ that does it in minimum time.

Problem (5.4.P3) (Special Bolza problem). Let $\phi_0: E\to R_+$ and $f_0: I\times E\to R_+$. The problem is to find a control $u\in\mathfrak{U}$ that minimizes the functional

$$J(u)=\phi_0(x(T))+\int_0^T f_0(t,x(t))\,dt,$$

where x is the response of the system (5.4.1) corresponding to the control $u\in\mathfrak{U}$.

Problem (5.4.P4) (Lagrange problem). Let $f^0: I\times E\times B\to R$ be a function. The problem is to find a control $u\in\mathfrak{U}$ that minimizes the functional

$$J(u)=\int_0^T f^0(t,x(t),u(t))\,dt$$

subject to the constraint that x is the response of the system (5.4.1) corresponding to the control policy $u \in \mathcal{U}$.

Clearly, these problems cannot be solved without further assumptions. We shall introduce these assumptions as we present the results.

5.4.5. *Existence of Optimal Controls*

Generally the assumptions required for the proof of existence of optimal controls are stronger than those required for the proof of existence of solutions of the evolution equation. Thus, in order to solve the control problems (5.4.P1)–(5.4.P4), we shall introduce a revised set of hypotheses. First, we recall these definitions:

$$D(L)=\{x\in L_p(I,E): x(t)\in D(A(t))\cap D(A^*(t)), \text{ for } t\in I$$
$$\text{and } t\to A(t)x(t), A^*(t)x(t), \dot{x}(t)\in C^0(I,E^*)\},$$
$$D(A)=\{g\in Y\equiv L_p(I,E):(Ag)(t)=A(t)g(t) \text{ is defined a.e. on } I \text{ and}$$
$$Ag\in Y^*\equiv L_q(I,E^*)\},$$

and similarly $D(A^*)$. Since $D(A)$ is dense in Y, A^* is well defined and $D(A^*)\subset L_p(I,E)$.

Assumption (5.4.A1). $\{A(t),\ t\in I\}$ is a family of densely defined closed linear operators (usually unbounded) in H with domain $D(A(t))\subset E$ and range $R(A(t))\subset E^*$. E_0 is a reflexive Banach space, the injection map $E_0\equiv\cup_{t\in I}D(A(t))\hookrightarrow E$ is continuous and compact, and for each $\xi\in C^0(I,E)\cap D(A)$, $A(t_n)\xi(t_n)\overset{w}{\to} A(t_0)\xi(t_0)$ in E^* whenever $t_n\to t_0$.

Assumption (5.4.A2). $\langle A(t)\eta,\eta\rangle\le 0$ for all $\eta\in D(A(t))$, $t\in I$, and the strong and weak extensions (see Definitions 2.5.1–2.5.2) of L from Y to Y^* coincide.

Assumption (5.4.A3). The function f: $I\times E\times B\to E^*$ is demicontinuous in the sense that whenever $t_n\to t$ in I, $x_n\overset{s}{\to} x$ in E, and $u_n\overset{w}{\to} u$ in B, $\langle f(t_n,x_n,u_n),e\rangle\to\langle f(t,x,u),e\rangle$ for each $e\in E$.

Assumption (5.4.A4). $\langle f(t,x,v)-f(t,y,v),x-y\rangle\le 0$ for all $x,y\in E$ and $v\in\Gamma(t)$, $t\in I$.

Assumption (5.4.A5). $|f(t,x,v)|_{E^*}\le h(t)+\alpha|x|_E^{p/q}$ for $h\in L_q(I,R_+)$, $\alpha>0$, and $x,v\in E\times\Gamma(t)$, $t\in I$.

Assumption (5.4.A6). $\langle f(t,x,v),x\rangle\le h_1(t)-\beta|x|_E^p$ for $h_1\in L_1^+(I,R)$, $\beta>0$, and $x,v\in E\times\Gamma(t)$, $t\in I$.

For the proof of closure of the set of trajectories we shall impose an additional condition on f.

Assumption (5.4.A7). For each $(t,x)\in I\times E$, the set $f(t,x,\Gamma(t))$ defined by

$$f(t,x,\Gamma(t))=\{e^*\in E^*: e^*(e)\equiv\langle e^*,e\rangle=\langle f(t,x,v),e\rangle \text{ for some } v\in\Gamma(t) \text{ and all } e\in D(A^*(t))\}$$

is a convex subset of E^*.

Properties of the Trajectories X. Denote by X the set of trajectories $\{x_u: u\in\mathfrak{U}\}$ generated by the evolution equation of (5.4.1), where each $x_u\in Y\equiv L_p(I,E)$ and satisfies

i. $x_u(0)=x_0$,
ii. $\dot{x}_u(t)=A(t)x_u(t)+f(t,x_u(t),u(t))$ a.e. on I.

Lemma 5.4.1. *Under* Assumptions (5.4.A1)–(5.4.A6), *the set X is a bounded subset of $L_p(I,E)\cap L_\infty(I,H)$.*

PROOF. Let x_u be a strong solution of the evolution equation (5.4.1) corresponding to the control $u\in\mathfrak{U}$ (see Theorem 5.4.1). Then

$$\langle\dot{x}_u(t),x_u(t)\rangle=\langle A(t)x_u(t),x_u(t)\rangle+\langle f(t,x_u(t),u(t)),x_u(t)\rangle$$

is defined a.e. as a duality product in E and E^*. Since $x_u\in L_p(I,E)$ and $\dot{x}_u\in L_q(I,E^*)$, it follows from Theorem 1.2.15 that $x_u\in C^0(I,H)$ and, further, that

$$\langle\dot{x}_u(t),x_u(t)\rangle=(\dot{x}_u(t),x_u(t))_H=\frac{1}{2}\frac{d}{dt}(|x_u(t)|_H^2). \tag{5.4.5}$$

Using Assumptions (5.4.A2) and (5.4.A6), we have

$$\frac{d}{dt}(|x_u(t)|_H^2)\le 2h_1(t)-2\beta|x_u(t)|_E^p \quad \text{a.e.} \tag{5.4.6}$$

Integrating (5.4.6), we have, for each $t\in I$,

$$|x_u(t)|_H^2+2\beta\int_0^t|x_u(\theta)|_E^p\,d\theta\le|x_0|_H^2+2\int_0^t h_1(\theta)\,d\theta. \tag{5.4.7}$$

Since the right-hand side of the above inequality is independent of $u\in\mathfrak{U}$, the assertion of the lemma follows from it. □

Lemma 5.4.2. *Suppose that $D(A^*)$, contained in $L_p(I,E)$, is a set of category* II [*Baire category theorem* 6.1.1, Lar, 1, *p.* 146] *and A and f satisfy the assumptions of Lemma* 5.4.1. *Then the set $Z\equiv\{\dot{x}_u: u\in\mathfrak{U}\}$ is a bounded subset of $L_q(I,E^*)$.*

PROOF. Let x_u be a solution of the evolution equation (5.4.1) corresponding to the control $u \in \mathcal{U}$, and let $\nu \in D(A^*)$. Then

$$\int_I \langle \dot{x}_u(t), \nu(t) \rangle \, dt$$

is defined and equals m_u, where

$$m_u = \int_I \{ \langle A^*(t)\nu(t), x_u(t) \rangle + \langle f(t, x_u(t), u(t)), \nu(t) \rangle \} \, dt. \quad (5.4.8)$$

Using Assumption (5.4.A5) and Hölder's inequality, we obtain

$$|m_u| \le \| A^*\nu \|_{L_q(I, E^*)} \| x_u \|_{L_p(I, E)} + \left\{ \| h \|_{L_q(I, R)} + \alpha \left(\| x_u \|_{L_p(I, E)} \right)^{p/q} \right\} \times \| \nu \|_{L_p(I, E)}.$$

By Lemma 5.4.1, $b \equiv \sup\{ \| x \|_{L_p(I, E)} : x \in X \}$ is finite. Thus, there exists a constant b_ν, dependent on ν, such that

$$|m_u| \le b \| A^*\nu \|_{L_q(I, E^*)} + \left\{ \| h \|_{L_q(I, R)} + \alpha b^{p/q} \right\} \| \nu \|_{L_p(I, E)} \equiv b_\nu \quad (5.4.9)$$

for all $u \in \mathcal{U}$. Consequently,

$$\sup_{u \in \mathcal{U}} \left| \int_I \langle \dot{x}_u(t), \nu(t) \rangle \, dt \right| \le b_\nu \quad \text{for each } \nu \in D(A^*).$$

This implies that the set $\{\dot{x}_u : u \in \mathcal{U}\}$ or the set $\{\dot{x} : x \in X\}$ is a family of bounded linear functionals on $D(A^*)$. By hypothesis, $D(A^*)$ is a set of category II; thus, it follows from the uniform boundedness principle (see Theorem 1.1.3 or [Lar.1, Theorem 6.2, p. 149]) that there exists a constant $c > 0$ such that

$$\sup_{u \in \mathcal{U}} \| \dot{x}_u \|_{L_q(I, E^*)} \le c. \quad (5.4.10)$$

This proves the lemma. □

Let E_w denote the space E equipped with the weak topology (i.e., E^*-topology of E) and let $C(I, E_w)$ denote the topological vector space of functions defined on I with values in the set E and continuous in the weak topology. This is a locally convex linear topological space, as can be verified easily.

Lemma 5.4.3. *Suppose that E is dense in H and H dense in E^*. Then, under Assumptions* (5.4.A1)–(5.4.A6), *the set X is an equicontinuous subset of $C(I, E_w)$ and*

$$\max\left\{ \sup_{t \in I} |x(t)|_E : x \in X \right\} < \infty.$$

PROOF. Let M denote the family of all functions $y \in L_q(I, E^*)$, $1<q<\infty$, such that

$$\int_J y(\theta)\, d\theta \in E$$

for any Borel set $J \subset I$. Since $E \subset E^*$, M is nonempty, we show that M is a closed linear subspace of $L_q(I, E^*)$. Linearity is obvious; we prove the closure.

Since in a reflexive Banach space strong closure follows from weak closure, we prove that M is weakly closed. Let $\{y_n\} \in M$, and suppose that $y_n \xrightarrow{w} y_0$ in $L_q(I, E^*)$. We must show that

$$\int_J y_0\, d\theta \in E$$

for every Borel set $J \subset I$. Note that in a reflexive Banach space the weak and weak* topologies are equivalent and that, for $1<p<\infty$, $L_p(I, E)$ and $L_q(I, E^*)$ are reflexive Banach spaces whenever E is reflexive. Let $\psi \in (L_q(I, E^*))^* = L_p(I, E)$; then

$$\int_I \langle y_n(\theta), \psi(\theta)\rangle\, d\theta \to \int_I \langle y_0(\theta), \psi(\theta)\rangle\, d\theta \quad \text{as } n \to \infty. \tag{5.4.11}$$

Define $\psi(\theta) = C_J(\theta)\nu$, where C_J is the characteristic function of the set J, a Borel set, and $\nu \in E$. Substituting this ψ in (5.4.11), we have

$$\int_I \langle y_n(\theta), C_J(\theta)\nu\rangle\, d\theta \to \int_I \langle y_0(\theta), C_J(\theta)\nu\rangle\, d\theta;$$

consequently,

$$\left\langle \int_J y_n(\theta)\, d\theta, \nu \right\rangle \to \left\langle \int_J y_0(\theta)\, d\theta, \nu \right\rangle.$$

By definition,

$$z_n \equiv \int_J y_n(\theta)\, d\theta \in E$$

and, from above, $\langle z_n, \nu\rangle \to \langle z_0, \nu\rangle$ for each $\nu \in E \subset H \subset E^*$, where

$$z_0 \equiv \int_J y_0(\theta)\, d\theta.$$

Since, by hypothesis, E is dense in H and H dense in E^*, $z_n \xrightarrow{w} z_0$. Further, being a reflexive Banach space, E is weakly complete, and thus the weak limit $z_0 \in E$. Hence, M is a closed linear subspace of $L_q(I, E^*)$, and consequently a Banach space. For the proof of equicontinuity, we note that the operator T_J, defined for each Borel set $J \subset I$ by

$$T_J y = \int_I C_J(\theta) y(\theta) d\theta,$$

is a weakly continuous linear operator from the Banach space M into the Banach space E, and hence strongly continuous (Theorem 1.1.8). Therefore

$$\|T_J\| \equiv \sup\{|T_J y|_E / \|y\|_{L_q(I, E^*)} : y \in M\} < \infty.$$

Since T_J is a bounded linear operator from M into E for each $e^* \in E^*$, $e^*(T_J y)$ is a set of scalars as y ranges through M, and $(e^* T_J)(y)$ is a continuous linear form on $M \subset L_q(I, E^*)$. Thus there exists an $f_{e^*} \in L_p(I, E)$ such that

$$e^*(T_J y) = \int_I \langle f_{e^*}(\theta), y(\theta) \rangle C_J(\theta)\, d\theta$$

and

$$|e^*(T_J y)| \leq \left(\int_J |f_{e^*}(\theta)|_E^p \, d\theta \right)^{1/p} \|y\|_{L_q(I, E^*)} \tag{5.4.12}$$

for $y \in M$. Consider $x \in X$; then $\dot{x} \in L_q(I, E^*)$ by Lemma 5.4.2; and, since $q > 1$, $L_q(I, E^*) \subset L_1(I, E^*)$ for $\mu(I) < \infty$; consequently, by a property of Bochner integrals [Yo.1, Corollary 2, p. 134], we have

$$x(t+h) - x(t) = \int_t^{t+h} \dot{x}(\theta)\, d\theta \quad \text{for } h > 0 \text{ and } t, t+h \in I.$$

Since $x_0 \in E$, it follows from the above equality that the set $Z \equiv \{\dot{x} : x \in X\} \subset M$ and $x(t+h) - x(t) = (T_J \dot{x}) \in E$, where $J = [t, t+h]$. Therefore, it follows from (5.4.12) and Lemma 5.4.2, [Equation (5.4.10)] that

$$|e^*(x(t+h) - x(t))| = |\langle x(t+h) - x(t), e^* \rangle| \leq c \left(\int_J |f_{e^*}(\theta)|_E^p \, d\theta \right)^{1/p} \tag{5.4.13}$$

independently of $x \in X$. Since $\int_J |f_{e^*}(\theta)|_E^p \, d\theta \to 0$ as $\mu(J) \to 0$ or $h \to 0$ for any arbitrary but fixed element e^* of E^*, we conclude that X is an equicontinuous subset of $C(I, E_w)$. Boundedness of X follows from that of the set Z as a subset of $L_q(I, E^*)$ and the fact that T_J, J Borel, is a bounded linear operator from M into E and that $|x_0|_E < \infty$. This concludes the proof. □

It is well known from Ascoli's theorem that the necessary and sufficient conditions for conditional compactness of a subset of the space of continuous functions on compact intervals with values in a finite-dimensional space are that the set be norm bounded and equicontinuous. Similar results also hold for more general spaces, for example, the space of continuous functions on a Hausdorff space to a Hausdorff uniform space [Wi.1, Theorem 43.15, p. 287]. We state here in the following lemma a result that is of interest to us.

Lemma 5.4.4. *A subset $D \subset C(I, E_w)$ is conditionally sequentially compact if and only if it is bounded in the norm topology for each $t \in I$ and equicontinuous on I in the E^* topology of E.*

Theorem 5.4.2. *Under the given hypotheses on A and f, the set X is a conditionally sequentially compact subset of $C(I, E_w)$.*

PROOF. The proof is an immediate consequence of Lemmas 5.4.3 and 5.4.4. □

We wish to prove that the set X is actually sequentially compact. This fact is later used for the proof of existence of optimal controls. For this purpose we introduce the concept of the so-called velocity field (also called Orientor field).

Definition 5.4.2. For each

$$(t, x) \in \Delta \equiv \{(t, x) \in I \times E : t \in I \text{ and } x \in D(A(t))\}$$

the set

$$R(t, x) \equiv \left\{ \begin{array}{l} e^* \in E^* : e^*(\phi) \equiv \langle e^*, \phi \rangle = \langle A(t)x + f(t, x, u), \phi \rangle \\ \text{for some } u \in \Gamma(t) \text{ and for all } \phi \in D(A^*(t)) \end{array} \right\},$$

where $\Gamma(t) \in \text{cbc}(B)$ for each $t \in I$, determines the field of possible directions of motion of the system (5.4.1) from the point (t, x) and is called the *velocity field*.

We intend to show that, under reasonable hypotheses, the set $R(t, x)$ for each $(t, x) \in \Delta$ is a weakly compact subset of E^*.

Lemma 5.4.5. *Suppose, for each $t \in I$, that $\Gamma(t)$ is a nonempty weakly closed bounded, not necessarily convex, subset of a reflexive Banach space B, and suppose that, for each $(t, x) \in I \times E$, $f(t, x, \cdot): \Gamma(t) \to E^*$ is weakly continuous and f satisfies Assumption* (5.4.A5). *Then the set $F(t, x) \equiv f(t, x, \Gamma(t))$, where*

$$f(t, x, \Gamma(t)) \equiv \left\{ \begin{array}{l} e^* \in E^* : e^*(\phi) = \langle f(t, x, u), \phi \rangle \\ \text{for some } u \in \Gamma(t) \text{ and all } \phi \in E \end{array} \right\},$$

is a weakly compact subset of E^* for each $(t, x) \in I \times E$.

PROOF. Since $\Gamma(t)$ is a weakly closed bounded subset of a reflexive Banach space, it is weakly sequentially compact, and hence weakly compact, by the Eberlein–Smulian theorem (Theorem 1.1.6). This, coupled with the assumption that $f(t, x, \cdot)$ is weakly continuous from $\Gamma(t)$ to E^* for each $(t, x) \in I \times E$, implies that $f(t, x, \Gamma(t))$ is a weakly closed subset of E^*. By Assumptions (5.4.A3) and (5.4.A5), $f(t, x, \Gamma(t))$ is a bounded subset of E^*, where E^* is itself reflexive, being the dual of a reflexive Banach space E. Thus, it follows again from the Eberlein–Smulian theorem that $f(t, x, \Gamma(t))$ is a weakly compact subset of E^*. □

Lemma 5.4.6. *Let f and the set-valued map Γ satisfy the hypotheses of Lemma 5.4.5, and let the family of operators $\{A(t),\ t \in I\}$ satisfy Assumption (5.4.A1). Then for each $(t, x) \in \Delta$, the set $R(t, x)$ as well as $\operatorname{cl\,co} R(t, x)$ is a weakly compact subset of E^*.*

PROOF. Under the given hypotheses, it follows from Definition 5.4.2 and Lemma 5.4.5 that $R(t, x)$ is bounded and weakly closed, and consequently a weakly compact subset of E^*. Thus, by the Krein–Smulian theorem (Theorem 1.1.14), the closed convex hull of $R(t, x)$, denoted by $\operatorname{cl\,co} R(t, x)$, is also weakly compact. □

REMARK 5.4.1. Note that in the above lemma, we use only the first part of Assumption (5.4.A1).

As indicated in Section 1.4, selection theorems play a central role in the study of existence of optimal controls. We have quoted a few results in Section 1.4. Here we present another selection theorem particularly suitable for set-valued maps with values nonempty weakly compact subsets of a reflexive Banach space. The result presented here is an extension due to Ahmed and Teo [AT.6] of a result originally due to Cole [Co.1].

Theorem 5.4.3. *Let B be a separable reflexive Banach space, $\mathrm{WC}(B)$ the class of nonempty weakly compact subsets of B, K a compact metric space with finite Lebesgue measure, and G: $K_\rho \to \mathrm{WC}(B)$ a map such that $\{G(t),\ t \in K\}$ is a bounded subset of B and*

$$\bigcap_{n=1}^{\infty} \operatorname{cl} \bigcup_{i=n}^{\infty} G(t_i) \subseteq G(t^*) \tag{5.4.14}$$

if $t_n \to t^* \in K$. Then there exists a strongly measurable selection u: $K \to B$ such that $u(t) \in G(t)$ a.e. on I.

PROOF. Since B is reflexive, separability of B implies that of B^* [DS.1, Lemma II.3.16, p. 65]. Thus there exists a countable set $\{f_n\}$ that is dense in B^*. Define $G_0(t) \equiv G(t)$, $t \in K$. We construct a decreasing sequence of set-valued maps G_n with values $G_n(t)$ according to the following procedure. Let $\max\{f_n(u): u \in G_{n-1}(t)\} \equiv c_{n-1}$, and define for all positive integers n

$$G_n(t) \equiv \{u \in G_{n-1}(t): f_n(u) = c_{n-1}\}. \tag{5.4.15}$$

For $n=1$, we have $\max\{f_1(u): u \in G_0(t)\} = c_0$ and $G_1(t) \equiv \{u \in G_0(t): f_1(u) = c_0\}$. Since $G_0(t)$ is weakly compact and $f_1 \in B^*$, it is clear that $G_1(t)$ is a nonempty bounded set. We prove by induction that the sequence of sets $\{G_n(t)\}$ are weakly compact for all positive integers n. Suppose $G_{n-1}(t)$ is weakly compact; we show that this implies that $G_n(t)$ is weakly compact. Since $G_n(t)$ is a bounded subset of a reflexive Banach space, it is conditionally weakly sequentially compact. We prove its weak closure. Let $\{z_k\} \in G_n(t)$ and $z_k \overset{w}{\to} z_0$. Then, since $z_k \in G_n(t)$,

$$f_n(z_0) = f_n(z_0 - z_k) + f_n(z_k) = f_n(z_0 - z_k) + c_{n-1}. \tag{5.4.16}$$

Letting $k \to \infty$, we obtain $f_n(z_0) = c_{n-1}$, and consequently, by (5.4.15), $z_0 \in G_n(t)$. Therefore, $G_n(t)$ is weakly closed and consequently, weakly sequentially compact. Thus, it follows from the Eberlein–Smulian theorem (Theorem 1.1.6) that $G_n(t)$ is weakly compact. Since, by hypothesis, $G_0(t)$ is weakly compact, by induction we conclude that, for each $t \in K$, $\{G_n(t), n = 0, 1, 2 \ldots\}$ is a sequence of nonempty weakly compact subsets of B and $G_0(t) \supset G_1(t) \supset \cdots G_n(t) \supset G_{n+1}(t) \cdots$ for $t \in K$. Define

$$u(t) = \bigcap_{i=0}^{\infty} G_i(t). \tag{5.4.17}$$

We show that $u(t)$ is unique. For each nonnegative integer i, we choose an x_i from $G_i(t)$ giving the sequence $\{x_i\}$. Clearly $\{x_i, i \geq n\} \in \bigcap_{i=0}^{n} G_i(t)$. Since $G_n(t)$ is weakly sequentially compact, there exists a subsequence of the sequence $\{x_i\}$, again denoted by $\{x_i\}$, and an element x^* such that $x_i \overset{w}{\to} x^*$. Since $\{x_i, i \geq n\} \subset \bigcap_{i=0}^{n} G_i(t)$ for all n, it is clear that $x^* \in \bigcap_{i=0}^{\infty} G_i(t)$. Thus $\bigcap_{i=0}^{\infty} G_i(t) \neq \emptyset$. We claim that x^* is the only element in $u(t)$. Suppose there is another element $y^* \in u(t)$; then $f_1(y^*) = \max\{f_1(y): y \in G_0(t)\} = c_0$, and similarly, $f_n(y^*) = c_{n-1}$ for all integers $n \geq 1$. This is true of x^* also, that is, $f_n(x^*) = c_{n-1}$ for all integers $n \geq 1$. Consequently, $f_i(x^* - y^*) = 0$ for all integers $i \geq 1$ and, since $\{f_i\}$ is dense in B^*, $x^* = y^*$. Thus, $u(t)$ is uniquely defined. It remains to show that u, determined by the values $u(t)$, $t \in K$, is strongly measurable. For this we use the upper semicontinuity property (5.4.14). Consider the scalar-valued function $t \to (f_1 u)(t) = f_1(u(t))$, and suppose that $\{t_i\}$ is a sequence having the limit $t^* \in K$.

Since $\bigcup_{t \in K} G_0(t)$ is bounded and contains the sequence $\{u(t_i)\}$, there exists a subsequence of the sequence $\{t_i\}$, relabeled as $\{t_i\}$, and an element ξ^* such that $u(t_i) \overset{w}{\to} \xi^*$ while $u(t_i) \in \bigcup_{j=n}^{\infty} G_0(t_j)$ for all $i \geq n$. Therefore $\xi^* \in \mathrm{cl} \bigcup_{i \geq n} G_0(t_i)$ for all integers $n \geq 1$, and consequently, $\xi^* \in \bigcap_{n \geq 1} \mathrm{cl}$

$\bigcup_{i \geq n} G_0(t_i) \subseteq G_0(t^*)$, where the last inclusion follows from our assumption (5.4.14). Thus,

$$(f_1 u)(t_i) = f_1(u(t_i)) \rightarrow f_1(\xi^*) \tag{5.4.18}$$

and

$$f_1(\xi^*) \leq \max\{f_1(v) : v \in G_0(t^*)\} = f_1(u(t^*)) \equiv (f_1 u)(t^*). \tag{5.4.19}$$

It follows from (5.4.18) and (5.4.19) that

$$\overline{\lim_i}\ (f_1 u)(t_i) \leq (f_1 u)(t^*);$$

hence, the function $t \rightarrow (f_1 u)(t)$ is upper semicontinuous, and consequently, $(f_1 u)(\cdot)$ is a measurable function on K. Thus by Lusin's theorem, for every $\varepsilon > 0$, there exists a compact set $K_1 \subset K$ such that the Lebesgue measure $\mu(K \backslash K_1) < \varepsilon/2$ and $(f_1 u)(\cdot)$ is continuous on K_1. Similarly, let $\{\tau_n\} \in K_1$ such that $\tau_n \rightarrow \tau^* \in K_1$ and $u(\tau_n) \xrightarrow{w} \eta^*$ for some $\eta^* \in B$. By continuity of $(f_1 u)(\cdot)$ on K_1, we have

$$(f_1 u)(\tau_n) \rightarrow (f_1 u)(\tau^*) \tag{5.4.20}$$

and due to weak convergence of $u(\tau_n)$ to η^* we have

$$\begin{cases} \text{(i)} & f_1(u(\tau_n)) \rightarrow f_1(\eta^*) \\ \text{(ii)} & f_2(u(\tau_n)) \rightarrow f_2(\eta^*). \end{cases} \tag{5.4.21}$$

It follows from (5.4.20) and (i) of (5.4.21) that

$$(f_1 u)(\tau^*) \equiv f_1(u(\tau^*)) = f_1(\eta^*). \tag{5.4.22}$$

Since $G_i(\tau^*)$ is the set of extremals for f_i, and $u(\tau^*) \in \bigcap_{i \geq 0} G_i(\tau^*)$, it follows from (5.4.22) that $\eta^* \in G_1(\tau^*)$. Thus,

$$f_2(\eta^*) \leq \max\{f_2(\eta), \eta \in G_1(\tau^*)\} = f_2(u(\tau^*)), \tag{5.4.23}$$

and hence, by virtue of (ii) of (5.4.21), we obtain

$$\overline{\lim_n}\ f_2(u(\tau_n)) \leq f_2(u(\tau^*)). \tag{5.4.24}$$

Thus, $(f_2 u)(\cdot)$ is upper semicontinuous on K_1 and hence measurable and again by Lusin's theorem, for every $\varepsilon > 0$, there exists a compact set $K_2 \subset K_1$ such that $\mu(K_1 \backslash K_2) < \varepsilon/2^2$ and $(f_2 u)(\cdot)$ restricted to K_2 is continuous. Carrying out this procedure we construct a sequence of compact sets $\{K_i\}$ with $K \supset K_1 \supset K_2 \cdots \supset K_n \cdots$ such that $\mu(K_{n-1} \backslash K_n) < \varepsilon/2^n$ and $(f_n u)(\cdot)$ is measurable on $K_\varepsilon \equiv \bigcap_{i \geq 1} K_i$ for all integers $n \geq 1$. Clearly,

$$\mu(K \backslash K_\varepsilon) = \sum_{n=1}^{\infty} \mu(K_{n-1} \backslash K_n) \leq \varepsilon, \tag{5.4.25}$$

where $K_0 \equiv K$. Since $\varepsilon > 0$ is arbitrary, $(f_n u)(\cdot)$ is measurable on K for all integers $n \geq 1$. Further, $\{f_n\}$ being dense in B^*, it follows that $(fu)(\cdot)$ is measurable for each $f \in B^*$, and hence, $u(\cdot)$ is weakly measurable. For separable Banach spaces weak and strong measurability are equivalent notions (Corollary 1.1.5). Since B is assumed to be separable, we have u strongly measurable. This completes the proof. □

We use this selection theorem to prove Theorem 5.4.4.

For the proof of existence of optimal controls, we follow the Filippov–Cesari approach. In this approach the control system is replaced by a multivalued differential system $\dot{\xi}(t)\in R(t,\xi(t))$, $t\in I$, where, for each $(t,x)\in I\times D(A(t))$, $R(t,x)$ with values in $\mathcal{P}(E^*)$—the power set of E^*— is the velocity field determined by the mapping $u\to A(t)x+f(t,x,u)$ and the admissible values for u at t. This formulation is more general than the differential equation formulation since, in this case, for an arbitrary set-valued map $R(t,x)$, the multivalued differential system $\dot{\xi}(t)\in R(t,\xi(t))$, $t\in I$, need not have any relation to a control system. However, if R arises from a mapping as mentioned above, then, under certain conditions, the two formulations are equivalent. Another advantage of the multivalued formulation is that the existence problem can be solved relatively easily by exploiting many powerful results on convex sets in locally convex topological vector spaces.

Denote by $\mathrm{AC}(I,E_w)$ the space of weakly absolutely continuous functions on I with values in E, and let $\mathcal{X}\equiv\mathrm{AC}(I,E_w)\cap D(L)$, where $D(L)$ is as defined in the introduction of this section. Suppose that R is given by Definition 5.4.2. Define the system

$$\begin{aligned} &\dot{\xi}(t)\in R(t,\xi(t)), \qquad t\in I,\\ &\xi(0)=x_0. \end{aligned} \tag{5.4.26}$$

An element $x\in\mathcal{X}$ is said to be a *solution of the system* (5.4.26) if $x(0)=x_0$ and $\dot{x}(t)\in R(t,x(t))$ a.e. on I.

Throughout the rest of this section we assume, unless stated otherwise, that the set valued function Γ satisfies the following property:

Assumption (5.4.A8). The set-valued function Γ from I into $\mathcal{P}(B)$ is either constant or a measurable function with values in the class of closed bounded convex subsets of B denoted by $\mathrm{cbc}(B)$. Further, for any $b^*\in B^*$, $t\to b^*(\Gamma(t))\equiv\sup\{b^*(v): v\in\Gamma(t)\}$ is continuous. That is, if $t_n\to t^*$, then $b^*(\Gamma(t_n))\to b^*(\Gamma(t^*))$ as $n\to\infty$.

Theorem 5.4.4. *Suppose that A and f satisfy Assumptions* (5.4.A1)–(5.4.A6). *Then, the system* (5.4.1) *is equivalent to the system* (5.4.26) *in the sense that every solution of the former is a solution of the latter, and conversely.*

PROOF. A solution of the system (5.4.1) is clearly a solution of the system (5.4.26). For the converse, let $x\in\mathcal{X}$ be a solution of (5.4.26); we must show that x is also a solution of (5.4.1). Let $t\in I$ at which $\dot{x}(t)$ is defined as an element of E^*, and construct the set-valued map G with values $G(t)$ given by

$$G(t)\equiv\left\{\begin{array}{l} u\in\Gamma(t):\langle\dot{x}(t),\phi\rangle=\langle A(t)x(t)+f(t,x(t),u),\phi\rangle \\ \text{for all } \phi\in D(A^*(t)) \end{array}\right\}. \tag{5.4.27}$$

Since $\dot{x}(t)\in R(t,x(t))$ a.e. and $\{\Gamma(t),\ t\in I\}$ is bounded, it follows from the general assumptions about A and f that $G(t)$ is a nonempty bounded subset of B. Since B is a reflexive Banach space, $G(t)$, for each $t\in I$, is conditionally weakly sequentially compact. For weak compactness, it is sufficient to show that $G(t)$ is weakly closed. Let $\{u_n\}\in G(t)$, and suppose that $u_n \xrightarrow{w} u_0$. We must show that $u_0\in G(t)$. Suppose the contrary; then there exists an $\varepsilon>0$ such that

$$|\dot{x}(t)-A(t)x(t)-f(t,x(t),u_0)|_{E^*}\geq\varepsilon.$$

Since E is reflexive, there exists a $\nu\in E$ with $|\nu|_E=1$ such that

$$\begin{aligned}\langle\dot{x}(t)-A(t)x(t)-f(t,x(t),u_0),\nu\rangle \\ =|\dot{x}(t)-A(t)x(t)-f(t,x(t),u_0)|_{E^*}\geq\varepsilon.\end{aligned} \tag{5.4.28}$$

Since $u_n\in G(t)$,

$$\langle\dot{x}(t)-A(t)x(t)-f(t,x(t),u_n),\nu\rangle=0. \tag{5.4.29}$$

Thus it follows from (5.4.28) and (5.4.29) that

$$\begin{aligned}&\langle f(t,x(t),u_n)-f(t,x(t),u_0),\nu\rangle \\ &\quad=\langle\dot{x}(t)-A(t)x(t)-f(t,x(t),u_n)+f(t,x(t),u_n)-f(t,x(t),u_0),\nu\rangle \\ &\quad=\langle\dot{x}(t)-A(t)x(t)-f(t,x(t),u_0),\nu\rangle\geq\varepsilon>0,\end{aligned} \tag{5.4.30}$$

which contradicts the assumption that $u\to f(t,x,u)$ is weakly continuous. Thus, $G(t)$ is weakly closed and hence weakly compact. Now we show that $G(t)$ satisfies the upper semicontinuity property (5.4.14). First we show that the graph of G,

$$\gamma(G)\equiv\{(t,u)\in I\times B: t\in I,\ u\in G(t)\},$$

is a closed subset of $I\times B$. Suppose that $(t_n,u_n)\in\gamma(G)$ and $t_n\to t_0$ and $u_n\xrightarrow{w} u_0$. Then we show that $(t_0,u_0)\in\gamma(G)$. Clearly, $t_0\in I$. Since $u_n\in G(t_n)$,

$$\langle\dot{x}(t_n),\phi\rangle=\langle A(t_n)x(t_n),\phi\rangle+\langle f(t_n,x(t_n),u_n),\phi\rangle \tag{5.4.31}$$

for all $\phi\in D(A^*(t_n))$. Since $x\in D(L)$, Ax, $\dot{x}\in C^0(I,E^*)$, and consequently, $\langle\dot{x}(t_n),\phi\rangle\to\langle\dot{x}(t_0),\phi\rangle$ and $\langle A(t_n)x(t_n),\phi\rangle\to\langle A(t_0)x(t_0),\phi\rangle$ for all $\phi\in E$. By virtue of Assumption (5.4.A1), $x(t_n)\xrightarrow{s} x(t_0)$ in E, and, since $u_n\xrightarrow{w} u_0$ and $t_n\to t_0$, it follows from the demicontinuity of f [Assumption (5.4.A3)] that $\langle f(t_n, x(t_n), u_n), \phi\rangle \to \langle f(t_0, x(t_0), u_0), \phi\rangle$ for all $\phi \in E$. Letting $n\to\infty$ in (5.4.31), we obtain

$$\langle \dot{x}(t_0), \phi\rangle = \langle A(t_0)x(t_0), \phi\rangle + \langle f(t_0, x(t_0), u_0), \phi\rangle \qquad (5.4.32)$$

for all $\phi \in D(A^*(t_0))$. Define

$$F(t) \equiv \left\{ \begin{array}{l} u\in B : \langle \dot{x}(t), \phi\rangle = \langle A(t)x(t) + f(t, x(t), u), \phi\rangle \\ \text{for all}\, \phi \in D(A^*(t)) \end{array} \right\}.$$

Clearly $G(t) = F(t) \cap \Gamma(t)$, and it follows from (5.4.32) that $u_0 \in F(t_0)$. We claim that $u_0 \in \Gamma(t_0)$. Suppose the contrary; then, since $\Gamma(t_0) \in \text{cbc}(B)$, there exist $b^* \in B^*$, constant c, and $\varepsilon > 0$ such that

$$b^*(u_0) \geq c > c - \varepsilon \geq b^*(\Gamma(t_0)). \qquad (5.4.33)$$

Since $u_n \in \Gamma(t_n)$, we also have

$$b^*(u_0) = b^*(u_0 - u_n) + b(u_n) \leq b^*(u_0 - u_n) + b^*(\Gamma(t_n)) \qquad (5.4.34)$$

for all integers $n \geq 1$. Therefore, there exists an integer $n_\varepsilon \geq 1$ such that

$$b^*(u_0) \leq \varepsilon/2 + \varlimsup_{n \geq n_\varepsilon} b^*(\Gamma(t_n)). \qquad (5.4.35)$$

Since by Assumption (5.4.A8) $\lim_n b^*(\Gamma(t_n)) = b^*(\Gamma(t_0))$, it follows from (5.4.35) and (5.4.33) that

$$b^*(u_0) \leq c - \varepsilon/2 < c. \qquad (5.4.36)$$

The contradiction verifies the claim that $u_0 \in \Gamma(t_0)$, and hence, $u_0 \in G(t_0)$ and the graph $\gamma(G)$ is weakly closed. The closure of $\gamma(G)$ implies that $t \to G(t)$ is upper semicontinuous (see Theorem 1.4.3) in the sense that whenever $t_i \to t^*$,

$$\bigcap_{n\geq 1} \text{cl} \bigcup_{i \geq n} G(t_i) \subseteq G(t^*). \qquad (5.4.37)$$

Thus, $G(t)$, as defined by (5.4.27), satisfies all the hypotheses of Theorem 5.4.3; consequently, there exists a $u_x \in \mathcal{U}$ such that $u_x(t) \in G(t)$ for every $t \in I$, that is, $\dot{x}(t) = A(t)x(t) + f(t, x(t), u_x(t))$ a.e. on I. Thus, x is a solution of (5.4.1). □

We need the following result for the proof of the closure of the set X.

Lemma 5.4.7. *Let I be a closed bounded interval, μ the Lebesgue measure, $\mathcal{Y}$ a Banach space, K a closed bounded convex subset of $\mathcal{Y}$, and $y \in L_1(I, \mathcal{Y})$. Let $y(t) \in K$ for almost all $t \in J$, where J is any measurable subset of I with $\mu(J) > 0$. Then*

$$[1/\mu(J)] \int_J y(t)\,dt \in K. \qquad (5.4.38)$$

PROOF. Since $y \in L_1(I, \mathcal{Y})$ and $y(t) \in K$ a.e. on J, there exists a sequence of simple functions $y_n \in L_1(I, \mathcal{Y})$ such that (i) $y_n(t) \in K$ for each $t \in J$ and for all integers $n \geq 1$ and (ii) $y_n \to y$ strongly in $L_1(I, \mathcal{Y})$. Clearly,

$$\int_J |y_n(t) - y(t)|_{\mathcal{Y}} \, dt \to 0 \quad \text{as } n \to \infty.$$

Since, for each n, y_n is simple, there exists a sequence of disjoint intervals $\{J_i^{(n)},\ i = 1, 2, \ldots m(n)\} \subset J$ and a sequence of elements $\{\alpha_i^{(n)},\ i = 1, 2, \ldots m(n)\} \in K \subset \mathcal{Y}$ such that

$$J = \bigcup_{i=1}^{m(n)} J_i^{(n)}$$

and

$$y_n(t) = \alpha_i^{(n)} \quad \text{for } t \in J_i^{(n)}.$$

Then

$$(1/\mu(J)) \int_J y_n(\theta) \, d\theta = (1/\mu(J)) \sum_{i=1}^{m(n)} \alpha_i^{(n)} \mu(J_i^{(n)})$$

$$= \sum_{i=1}^{m(n)} \left(\mu(J_i^{(n)})/\mu(J)\right) \alpha_i^{(n)}. \tag{5.4.39}$$

Because $\{\alpha_i^{(n)},\ i = 1, 2, \ldots m(n)\} \in K$, $(\mu(J_i^{(n)})/\mu(J)) \geq 0$ and $\sum_{i=1}^{m(n)} (\mu(J_i^{(n)})/\mu(J)) = 1$, and K is convex, it is clear that $z_n \equiv (1/\mu(J)) \int_J y_n(\theta) \, d\theta \in K$ for all integers $n \geq 1$. Clearly,

$$z_n \xrightarrow{s} z_0 \equiv (1/\mu(J)) \int_J y(\theta) \, d\theta \quad \text{(strongly) in } \mathcal{Y}.$$

We must show that $z_0 \in K$. Suppose the contrary; that is, $z_0 \notin K$. Since K is a closed convex subset of a Banach space, it follows from the Hahn–Banach separation theorem that there exists an $f^* \in \mathcal{Y}^*$, a constant c, and $\varepsilon > 0$ such that

$$f^*(z_0) \geq c > c - \varepsilon \geq f^*(z) \quad \text{for all } z \in K. \tag{5.4.40}$$

Then, since $z_n \in K$ and $f^*(z_0) = f^*(z_0 - z_n) + f^*(z_n)$, we have

$$f^*(z_0) \leq f^*(z_0 - z_n) + (c - \varepsilon) \quad \text{for all integers } n \geq 1. \tag{5.4.41}$$

Letting $n \to \infty$ in (5.4.41), we obtain $f^*(z_0) \leq (c - \varepsilon) < c$. This is a contradiction; thus $z_0 \in K$. □

We are now ready to prove the following closure theorem. We need this result for the proof of existence of optimal controls as stated in Section 5.4.4.

Theorem 5.4.5. *Suppose that A and f satisfy the assumptions of Lemma 5.4.1 and Lemma 5.4.2 and, further, f satisfies the convexity Assumption (5.4.A7). B is a separable reflexive Banach space, and $t \to \Gamma(t)$ is a measurable multifunction with values in* cbc(B) *satisfying Assumption (5.4.A8). Let $\mathcal{U}$ denote the class of all strongly measurable functions $\{u\}$ on I with values $u(t) \in \Gamma(t)$, $t \in I$. Then the set X is a closed and sequentially compact subset of $C(I, E_w)$, where $E_w = (E, \tau_{E^*})$ is the topological space E equipped with the weak topology.*

PROOF. By Theorem 5.4.2, the set X is a conditionally sequentially compact subset of $C(I, E_w)$; thus, if closed, it is sequentially compact. For closure, let $x^* \in \mathrm{cl}\ X$. Since X is a conditionally sequentially compact subset of $C(I, E_w)$, there exists a sequence $\{x_n\} \in X$ such that $x_n(t) \overset{w}{\to} x^*(t)$ uniformly on I. Clearly, $x^* \in C(I, E_w)$, that is, x^* is weakly continuous. We must show that $x^* \in X$, or equivalently, $\dot{x}^*(t) \in R(t, x^*(t))$ a.e. on I (Theorem 5.4.4). First, we observe that x^* is weakly differentiable. For any $h \in (0, T)$ and $t \in I$, define f_h as

$$\begin{aligned} f_h(t) &= [x^*(t+h) - x^*(t)]/h = \mathrm{w}\lim_n [x_n(t+h) - x_n(t)]/h \\ &= \mathrm{w}\lim_n \left[\int_0^1 \dot{x}_n(t+\theta h)\, d\theta\right], \end{aligned} \tag{5.4.42}$$

where w lim ξ_n denotes the weak limit of ξ_n. Let $\nu \in L_p(I, E)$, and consider

$$\tilde{f}_h(\nu) \equiv \int_0^{T-h} \langle f_h(t), \nu(t) \rangle\, dt \tag{5.4.43}$$

where $\langle \cdot, \cdot \rangle$ is the $E^* - E$ duality product. Clearly,

$$\tilde{f}_h(\nu) = \int_0^{T-h} \lim_n \left\langle \int_0^1 \dot{x}_n(t+\theta h)\, d\theta, \nu(t) \right\rangle dt,$$

and

$$\begin{aligned} |\tilde{f}_h(\nu)| &\leq \int_0^{T-h} \lim_n \left| \int_0^1 \dot{x}_n(t+\theta h)\, d\theta \right|_{E^*} |\nu(t)|_E\, dt \\ &\leq \overline{\lim_n} \left(\int_0^{T-h} \int_0^1 |\dot{x}_n(t+\theta h)|_{E^*}^q\, d\theta\, dt \right)^{1/q} \|\nu\|_{L_p(I,E)}. \end{aligned} \tag{5.4.44}$$

By Fubini's theorem, the right-hand expression equals

$$\overline{\lim_n} \left(\int_0^1 \int_0^{T-h} |\dot{x}_n(t+\theta h)|_{E^*}^q\, dt\, d\theta \right)^{1/q} \|\nu\|_{L_p(I,E)}.$$

Thus,

$$|\tilde{f}_h(\nu)| \le \overline{\lim_n} \left(\int_0^1 \left(\int_{\theta h}^{T-h+\theta h} |\dot{x}_n(\xi)|_{E^*}^q \, d\xi \right) d\theta \right)^{1/q} \|\nu\|_{L_p(I,E)}$$

$$\le \overline{\lim_n} \left(\int_0^T |\dot{x}_n(\xi)|_{E^*}^q \, d\xi \right)^{1/q} \|\nu\|_{L_p(I,E)}$$

independently of $h \in (0, T)$. By Lemma 5.4.2, $\{\dot{x}_n\} \in Z$, which is a bounded subset of $L_q(I, E^*)$, and

$$\sup_n \|\dot{x}_n\|_{L_q(I,E^*)} \le c < \infty .$$

Therefore, $|\tilde{f}_h(\nu)| \le c \|\nu\|_{L_p(I,E)}$ for all $h \in (0, T)$.

Thus by the uniform boundedness principle, $\lim_{h \to 0} \tilde{f}_h$ exists as an element of $(L_p(I, E))^* = L_q(I, E^*)$. In other words, $\dot{x}^* \in L_q(I, E^*)$; consequently, $\dot{x}^*(t)$ is defined a.e. and $|\dot{x}^*(t)|_{E^*}$ is finite a.e. on I. Since $q > 1$, $L_q(I, E^*) \subset L_1(I, E^*)$; consequently $\dot{x}^* \in L_1(I, E^*)$.

Further, x^* is weakly continuous on I, with $x^*(0) = x_0$; therefore, $x^*(t) = x_0 + \int_0^t \dot{x}^*(\theta)\, d\theta$. Now we show that $x^*(t) \in D(A(t))$ a.e. Since $\{x_n\} \in X$ and $\{\dot{x}_n\} \in Z$ and both X and Z are bounded subsets of $L_p(I, E)$ and $L_q(I, E^*)$, respectively, and f satisfies Assumption (5.4.A5), it is clear that $\{Ax_n\}$ is a bounded sequence from $L_q(I, E^*)$. Thus, there exists a subsequence of the sequence $\{Ax_n\}$, again denoted by $\{Ax_n\}$, and an element $z \in L_q(I, E^*)$ such that $Ax_n \to z$ weakly in $L_q(I, E^*)$. Thus, z is defined a.e. on I and $z(t) \in E^*$ a.e. Let t_0 be an element of I at which both $|z(t_0)|_{E^*}$ and $|\dot{x}^*(t_0)|_{E^*}$ are finite. By Mazur's theorem (Corollary 1.1.1) we can construct a sequence $\{z_j\}$, given by

$$z_j \equiv \sum_{i=1}^{m(j)} \alpha_{ij} A x_{n_j+i}, \qquad \alpha_{ij} \ge 0,$$

$$\sum_i^{m(j)} \alpha_{ij} = 1, \qquad m(j) < \infty, \tag{5.4.45}$$

such that $z_j \xrightarrow{s} z$ (strongly) in $L_q(I, E^*)$. Then we can select a subsequence of the sequence $\{z_j\}$, again denoted by $\{z_j\}$, such that $z_j \to z$ a.e. on I strongly in E^*. Clearly, $z_j(t_0) \to z(t_0)$ weakly in E^* also. Corresponding to the last subsequence we define

$$y_j \equiv \sum_{i=1}^{m(j)} \alpha_{ij} x_{n_j+i}. \tag{5.4.46}$$

Since $x_n(t) \in D(A(t))$, $t \in I$, and $z_j = Ay_j$, we have, for each integer $j \ge 1$, $y_j(t) \in D(A(t))$ for $t \in I$ and $y_j(t_0) \to x^*(t_0)$ weakly in E. Let $\gamma(A(t))$ denote the graph of the operator $A(t)$. Collecting the above information, we have

$$\begin{cases} \text{(i)} & \big(y_j(t_0), z_j(t_0)\big)\in\gamma(A(t_0)) \\ \text{(ii)} & y_j(t_0)\overset{w}{\rightarrow}x^*(t_0) \text{ in } E \text{ (and also strongly)} \\ \text{(iii)} & z_j(t_0)\overset{w}{\rightarrow}z(t_0) \text{ in } E^* \text{ (and also strongly).} \end{cases} \tag{5.4.47}$$

Since $A(t_0)$ is a closed linear operator [Assumption (5.4.A1)], the graph $\gamma(A(t_0))$ is a closed and hence weakly closed subspace of the product space $E\times E^*$. Thus the weak limit $(x^*(t_0), z(t_0))$ of the sequence $\{(y_j(t_0), z_j(t_0))\}$ belongs to $\gamma(A(t_0))$. Consequently, $z(t_0)=A(t_0)x^*(t_0)$. Since the set $\{t\in I: z(t), \dot{x}^*(t)\in E^*\}$ has full measure and, for t_0 in this set, $x^*(t_0)\in D(A(t_0))$, we conclude that $x^*(t)\in D(A(t))$ a.e. on I. To show that $\dot{x}^*(t)\in R(t, x^*(t))$ a.e., let $t_0\in I$ at which both $\dot{x}^*(t_0)$ and $z(t_0)=A(t_0)x^*(t_0)$ are defined as elements of E^*. Then

$$\big(x^*(t_0+h)-x^*(t_0)\big)/h = \mathrm{w}\lim_n \int_0^1 \dot{x}_n(t_0+\theta h)\,d\theta. \tag{5.4.48}$$

For any subset $\sigma\subset E^*$, denote by σ^ε the closed ε-neighborhood of the set σ. Since, by hypotheses, the injection map $D(A(t_0))\hookrightarrow E$ is continuous and compact and $x_n(t_0), x^*(t_0)\in D(A(t_0))$, it follows that $x_n(t_0)\overset{s}{\rightarrow} x^*(t_0)$ in E. By Assumption (5.4.A3), f is demicontinuous in x on E, and by Assumption (5.4.A8), $t\rightarrow\Gamma(t)$ is smooth; therefore, for every $\varepsilon>0$ there exists an integer $n_0=n_0(\varepsilon)\geq 1$ such that $F(t_0, x_n(t_0))\subset F^\varepsilon(t_0, x^*(t_0))$ for all $n\geq n_0$, and, since $x_n\in C^0(I, E)\cap D(A)$, there exists a number $h_0=h_0(\varepsilon)>0$ such that for $|h|<h_0$, $F(t_0+\theta h, x_n(t_0+\theta h))\subset F^\varepsilon(t_0, x_n(t_0))$ for all $0\leq\theta\leq 1$. Since $x_n\in X$, it follows from Theorem 5.4.4 that $\dot{x}_n(t)\in R(t, x_n(t))$ a.e. on I and, in particular, $\dot{x}_n(t_0+\theta h)\in R(t_0+\theta h, x_n(t_0+\theta h))$, or equivalently,

$$\dot{x}_n(t_0+\theta h)-A(t_0+\theta h)x_n(t_0+\theta h)\in F\big(t_0+\theta h, x_n(t_0+\theta h)\big). \tag{5.4.49}$$

Thus for $|h|<h_0$ and $n>n_0$, it follows from the above observations that

$$\dot{x}_n(t_0+\theta h)-A(t_0+\theta h)x_n(t_0+\theta h)\in F^{2\varepsilon}\big(t_0, x^*(t_0)\big) \tag{5.4.50}$$

for all $0\leq\theta\leq 1$. Therefore, by virtue of the fact that $x_n\in D(L)$ we can choose h sufficiently small such that for $|h|<h_1(\varepsilon)>0$,

$$\dot{x}_n(t_0+\theta h)\in A(t_0)x_n(t_0)+F^{3\varepsilon}\big(t_0, x^*(t_0)\big) \tag{5.4.51}$$

for $\theta\in[0,1]$ and $n>n_0$.

From (5.4.51) and the definitions (5.4.45) and (5.4.46) for z_j and y_j, respectively, we obtain, for sufficiently large j [e.g., $n_j>n_0(\varepsilon)$],

$$\dot{y}_j(t_0+\theta h)\in z_j(t_0)+F^{3\varepsilon}\big(t_0, x^*(t_0)\big) \tag{5.4.52}$$

for all $\theta\in[0,1]$ and $|h|<h_1(\varepsilon)$. Since $z_j(t_0)\overset{s}{\rightarrow} z(t_0)$ in E^* [see Equation

(5.4.47)], we can choose a positive integer $j_0 = j_0(\varepsilon)$ such that, for $j > j_0$, $|h| < \min(h_0, h_1)$, and $\theta \in [0,1]$:

$$\dot{y}_j(t_0 + \theta h) \in z(t_0) + F^{4\varepsilon}(t_0, x^*(t_0)) = R^{4\varepsilon}(t_0, x^*(t_0)). \quad (5.4.53)$$

In the last equality we have used Definition 5.4.2 (for the set R) and the fact that $z(t_0) = A(t_0)x^*(t_0)$. Clearly,

$$\left[y_j(t_0 + h) - y_j(t_0)\right]/h = \int_0^1 \dot{y}_j(t_0 + \theta h)\, d\theta$$

and

$$\begin{aligned}\left[x^*(t_0 + h) - x^*(t_0)\right]/h &= \underset{j\to\infty}{\text{w lim}} \left[y_j(t_0 + h) - y_j(t_0)\right]/h \\ &= \underset{j\to\infty}{\text{w lim}} \int_0^1 \dot{y}_j(t_0 + \theta h)\, d\theta. \end{aligned} \quad (5.4.54)$$

Thus, by virtue of Lemma 5.4.7, it follows from (5.4.53) and (5.4.54) that

$$\left[x^*(t_0 + h) - x^*(t_0)\right]/h \in R^{4\varepsilon}(t_0, x^*(t_0)) \quad (5.4.55)$$

for all h such that $|h| < \min(h_0, h_1)$. This implies that

$$\dot{x}^*(t_0) \in R^{4\varepsilon}(t_0, x^*(t_0)) \quad \text{for every } \varepsilon > 0. \quad (5.4.56)$$

Since $R(t_0, x^*(t_0))$ is weakly closed (Lemma 5.4.6) and convex by Assumption (5.4.A7), we conclude that

$$\dot{x}^*(t_0) \in R(t_0, x^*(t_0)) \quad (5.4.57)$$

for every $t_0 \in I$ for which $\dot{x}^*(t_0)$ is defined. Thus $\dot{x}^*(t) \in R(t, x^*(t))$ a.e. on I; consequently, $x^* \in X$. This completes the proof of the closure of X. □

Definition 5.4.3. For each $t \in I$, the set $K(t) \subset E$ given by $K(t) \equiv \{y \in E : y = x(t), x \in X\}$ is called the *attainable set of the system* (5.4.1).

As a corollary to Theorem 5.4.5, we have the following result.

Corollary 5.4.1. *For each $t \in I$, the attainable set $K(t)$ is a weakly compact subset of E.*

Note: One may easily verify that the *above* results are also true under the relaxed hypothesis: $\overline{\lim}_n b^*(\Gamma(t_n)) \leq b^*(\Gamma(t_0))$ whenever $b^* \in B^*$ and $t_n \to t_0$.

Existence Theorems. Using the results of Theorem 5.4.5 and Corollary 5.4.1, we can obtain the following existence results.

Theorem 5.4.6 [Terminal control (5.4.P1)]. *Suppose that all the hypotheses of Theorem 5.4.5 are satisfied, S is a weakly closed subset of the state space E, $K(T) \cap S \neq \varnothing$ (nonempty), and $\phi : E \to R$ is a nonnegative, weakly lower semicontinuous function. Then there exists a control $u^* \in \mathcal{U}$ that transfers* *the system* (5.4.1) *from the initial state* x_0 *to the target set* S *and imparts a minimum to the function* ϕ *on the set* $K(T)\cap S$.

PROOF. The proof follows from the facts that $K(T)\cap S$ is weakly compact and that ϕ is a weakly lower semicontinuous function [see Theorem 1.1.20]. □

Theorem 5.4.7 [Time optimal control (5.4.P2)]. *Suppose that all the hypotheses of Theorem* 5.4.5 *are satisfied and the target set* S *is a weakly compact subset of* E *with* $I(K)\equiv\{t\in I: K(t)\cap S\neq\varnothing\}$ *nonempty. Then there exists a control* $u^*\in\mathcal{U}$ *that transfers the system* (5.4.1) *from the initial state* x_0 *to the target set* S *in minimum time.*

PROOF. Define $\tau\equiv\inf I(K)$. It suffices to prove that $\tau\in I(K)$. Let $\{\tau_n\}$ be a sequence from the set $I(K)$ such that $\tau_n\to\tau$. Since $\tau_n\in I(K)$, there exists a sequence $\{x_n\}\subset X$ such that $x_n(\tau_n)\in S\cap K(\tau_n)$. Since X is sequentially compact and closed (Theorem 5.4.5), there exists a subsequence, again denoted by $\{x_n\}$, and an element $x^*\in X$ such that $x_n(t)\overset{w}{\to} x^*(t)$ uniformly on I. Closure of I implies that $\tau\in I$ and, by Corollary 5.4.1, $x^*(\tau)\in K(\tau)$. Since S is a weakly compact subset of E and $z_n\equiv x_n(\tau_n)\in S$, there exists a subsequence, again denoted by $\{z_n\}$, and an element $z^*\in S$ such that $z_n\overset{w}{\to} z^*$. It suffices to show that $z^*=x^*(\tau)$. For each $\nu\in E^*$, we can write

$$\langle x^*(\tau)-z^*,\nu\rangle=\langle x^*(\tau)-x_n(\tau),\nu\rangle+\langle x_n(\tau)-x_n(\tau_n),\nu\rangle+\langle z_n-z^*,\nu\rangle. \tag{5.4.58}$$

Since $x_n(t)\overset{w}{\to} x^*(t)$ uniformly on I, $\tau\in I$, and $z_n\overset{w}{\to} z^*$, the first and the last terms of (5.4.58) can be made arbitrarily small by choosing n sufficiently large. Since X is an equicontinuous subset of $C(I, E_w)$ (Lemma 5.4.3) and $\tau_n\to\tau$, the second term also can be made arbitrarily small by choosing n large enough. Since this is true for every $\nu\in E^*$, we have $x^*(\tau)=z^*$. This implies that $x^*(\tau)\in S\cap K(\tau)$; consequently, $\tau\in I(K)$ and there exists a control $u^*\in\mathcal{U}$ that satisfies the conclusion of the theorem. □

REMARK 5.4.2. For variable upper semicontinuous S, since $\tau_n\to\tau$, the sequence $S(\tau_n)$ can be confined in a closed ε-neighborhood of $S(\tau)$ by choosing n sufficiently large. Thus $z^*\in S^{\varepsilon}(\tau)$ for every $\varepsilon>0$, and the result follows.

Theorem 5.4.8 [Special Bolza problem (5.4.P3)]. *Consider the system* (5.4.1) *with* A *and* f *satisfying the hypotheses of Theorem* 5.4.5. *Let* $f_0: I\times E\to R$ *be a nonnegative function such that, for each* $t\in I$, $f_0(t,\cdot)$ *is weakly lower* *semicontinuous in E and, for each $y \in E$, $f_0(\cdot, y)$ is Lebesgue measurable on I. Let ϕ_0 satisfy the hypothesis of Theorem 5.4.6 for ϕ. Then there exists a control $u^* \in \mathcal{U}$ that minimizes the functional*

$$J(u)=\phi_0(x(T))+\int_I f_0(t, x(t))\, dt, \tag{5.4.59}$$

where x is the response of the system (5.4.1) *corresponding to the control $u \in \mathcal{U}$.*

PROOF. It is sufficient to show that the functional

$$\xi(x) \equiv \phi_0(x(T))+\int_I f_0(t, x(t))\, dt \tag{5.4.60}$$

is weakly lower semicontinuous on $C(I, E_w)$. This follows from Fatou's Lemma and the hypotheses on ϕ_0 and f_0. $\square$

For Problem (5.4.P4), consider the linear operator $L \equiv [(d/dt)-A]$ with domain $D(L) \subset L_p(I, E)$ and range $R(L) \subseteq L_q(I, E^*)$. Let L_s (L_w) denote the strong (weak) extension of L from $L_p(I, E)$ into $L_q(I, E^*)$ and assume that $L_s = L_w$ (see Definition 5.4.1).
We introduce the following assumptions for the class of admissible controls and the operators A, f and f^0.

Assumption (5.4.A9). Let B be a polish space with an equivalent metric ρ, and let $t \to \Gamma(t)$ be a measurable multifunction defined on I with nonempty closed subsets of B as its values. Let $\mathcal{U}_0$ denote the class of measurable functions $\{u\}$ on I with values $u(t) \in \Gamma(t)$ a.e.

Assumption (5.4.A1)′. L is a densely defined linear operator from $D(L) \subset L_p(I, E)$ into $L_q(I, E^*)$ with the injection map $D(L_s) \hookrightarrow L_p(I, E)$ compact.

Assumption (5.4.A2)′. Same as Assumption (5.4.A2).

Assumption (5.4.A3)′. f: $I \times E \times B \to E^*$ is demicontinuous in the sense that whenever $t_n \to t$ in I, $x_n \xrightarrow{s} x$ in E, and $v_n \xrightarrow{\rho} v$, $\langle f(t_n, x_n, v_n), e\rangle \to \langle f(t, x, v), e\rangle$ for each $e \in E$.

Assumptions (5.4.A4)′, (5.4.A5)′, and (5.4.A6)′ are the same as Assumptions (5.4.A4), (5.4.A5), and (5.4.A6), respectively.
The cost integrand f^0 maps $I \times E \times B$ into R and satisfies the following properties:

$F_0$1. $t \to f^0(t, \xi, v)$ is continuous for $(\xi, v) \in E \times B$.

$F_0$2. $\xi \to f^0(t, \xi, v)$ is continuous on E for $(t, v) \in I \times B$.

$F_0$3. $v \to f^0(t, \xi, v)$ is lower semicontinuous from B into R for $(t, \xi) \in I \times E$.

$F_0$4. There exists a $g \in L_1(I, R)$ such that $f^0(t, \xi, v) \geq g(t)$ a.e. for all $(\xi, v) \in E \times B$.

For $(t, \xi) \in \Delta \equiv \{(t, \xi) \in I \times E : t \in I, \xi \in D(A(t))\}$, define the set-valued map Q with values

$$Q(t, \xi) \equiv \left\{ \begin{array}{l} (\lambda, e^*) \in R \times E^* : \lambda \geq f^0(t, \xi, v);\ \langle e^*, \phi \rangle = \langle f(t, \xi, v), \phi \rangle \\ \text{for some } v \in \Gamma(t) \text{ and all } \phi \in D(A^*(t)) \end{array} \right\} \subset R \times E^*.$$

Let $N_\varepsilon(\xi)$ denote the closed ε-neighborhood of the point $\xi \in E$ and define

$$Q(t, N_\varepsilon(\xi)) = \bigcup_{\eta \in N_\varepsilon(\xi)} Q(t, \eta).$$

Definition 5.4.4. The set-valued map Q is said to *satisfy the weak Cesari property* if for every $(t, \xi) \in \Delta$, $Q(t, \xi)$ is nonempty and

$$\bigcap_{\varepsilon > 0} \operatorname{cl} \operatorname{co} Q(t, N_\varepsilon(\xi)) \subseteq Q(t, \xi). \tag{5.4.61}$$

Here cl means w*-closure, which is equivalent to w-closure since E is reflexive.

Theorem 5.4.9 [Lagrange problem (5.4.P4)]. *Consider the system* (5.4.1) *with the admissible controls* $\mathcal{U}_0$ *as defined above. Suppose that* A, f, *and* f^0 *satisfy Assumptions* (5.4.A1)′–(5.4.A2)′, (5.4.A3)′–(5.4.A6)′, *and properties* ($F_0$1)–($F_0$4), *respectively. Suppose that* Q *satisfies the weak Cesari property. Then there exists an optimal control for Problem* (5.4.P4).

PROOF. Let X denote the set of admissible trajectories that correspond to the admissible controls $\mathcal{U}_0$. Define $\mathcal{Q}_0 \equiv \{(x, u) \in X \times \mathcal{U}_0 : x = x_u$ for some $u \in \mathcal{U}_0\}$, where x_u is the response of the system (5.4.1) corresponding to the control u. Problem (5.4.P4) can then be rephrased as follows: Find a pair $(x^*, u^*) \in \mathcal{Q}_0$ such that $J(x^*, u^*) \leq J(x, u)$ for all $(x, u) \in \mathcal{Q}_0$. Let $\{(x_n, u_n)\} \in \mathcal{Q}_0$ be a minimizing sequence, that is,

$$\lim_n J(x_n, u_n) = \inf\{J(x, u) : (x, u) \in \mathcal{Q}_0\} \equiv m_0. \tag{5.4.62}$$

By property ($F_0$4),

$$J(x_n, u_n) = \int_I f^0(t, x_n(t), u_n(t))\, dt \geq \int_I g(t)\, dt > -\infty. \tag{5.4.63}$$

If $m_0 = +\infty$, there is nothing to prove; so we assume that $m_0 < \infty$. Since X is bounded subset of $L_p(I, E)$ (Lemma 5.4.1) and Z is a bounded subset of $L_q(I, E^*)$ (Lemma 5.4.2), it follows from Assumption (5.4.A6)′ that $\{Ax : x$

$\in X\}$ is a bounded subset of $L_q(I, E^*)$, and consequently, $\{Lx: x \in X\}$ is a bounded subset of $L_q(I. E^*)$. It is clear that, for $\{(x_n, u_n)\} \in \mathscr{Q}_0$, both the sequences $\{x_n\}$ and $\{Lx_n\}$ are bounded and belong to $L_p(I, E)$ and $L_q(I, E^*)$, respectively. Therefore, there exists a common subsequence of the sequence $\{n\}$, again denoted by $\{n\}$, and $x^* \in L_p(I, E)$ and $z^* \in L_q(I, E)$ such that

$$\begin{aligned} x_n &\overset{\text{w}}{\to} x^* \quad \text{in } L_p(I, E), \\ (Lx_n) &\overset{\text{w}}{\to} z^* \quad \text{in } L_q(I, E^*). \end{aligned} \tag{5.4.64}$$

Since $L_s = L_w$ and L_s is the strong closure of L, which is a densely defined linear operator from $D(L) \subset L_p(I, E)$ into $L_q(I, E^*)$, it is clear that the graph of L_s, denoted by $\gamma(L_s) \equiv \{(x, y) \in L_p(I, E) \times L_q(I, E^*): y = L_s x\}$, is a closed and hence a weakly closed subspace of $L_p(I, E) \times L_q(I, E^*)$. Therefore it follows from (5.4.64) that

$$(x^*, z^*) \in \gamma(L_s), \tag{5.4.65}$$

and consequently, $z^* = L_s x^*$. In other words, $x_n \overset{\text{w}}{\to} x^*$ in $L_p(I, E)$ and $Lx_n \overset{\text{w}}{\to} L_s x^*$ in $L_q(I, E^*)$. Since, by (5.4.A1)$'$, the injection map $D(L_s) \hookrightarrow L_p(I, E)$ is compact,

$$x_n \overset{\text{s}}{\to} x^* \quad \text{in } L_p(I, E). \tag{5.4.66}$$

Again, by Mazur's theorem (Corollary 1.1.1) there exists a sequence of finite convex combinations of $\{x_n\}$ and $\{Lx_n\}$ that converge strongly to x^* and $L_s x^*$, respectively. More specifically, for each integer j there exists an integer n_j, a set of integers $i = 1, 2, \ldots m(j)$, and a set of nonnegative numbers $\{\alpha_{ji}, i = 1 \ldots m(j)\}$ with $\Sigma_{i=1}^{m(j)} \alpha_{ji} = 1$ for all j such that

$$\psi_j \equiv \sum_{i=1}^{m(j)} \alpha_{ji} \left(Lx_{n_j + i} \right) \overset{\text{s}}{\to} L_s x^* \quad \text{in } L_q(I, E^*). \tag{5.4.67}$$

Clearly for each integer $i \in \{1, 2, \ldots m(j)\}$,

$$x_{n_j + i} \overset{\text{s}}{\to} x^* \quad \text{in } L_p(I, E) \text{ as } j \to \infty. \tag{5.4.68}$$

Since ψ_j converges strongly to $L_s x^*$ in $L_q(I, E^*)$ and $x_{n_j + i}$ converges strongly to x^* in $L_p(I, E)$, there exists a subsequence of the sequence $\{j\}$, again denoted by $\{j\}$, such that

$$\begin{aligned} \psi_j(t) &\overset{\text{s}}{\to} (L_s x^*)(t) \quad \text{in } E^* \text{ for almost all } t \in I, \\ x_{n_j + i}(t) &\overset{\text{s}}{\to} x^*(t) \qquad \text{in } E \text{ for almost all } t \in I. \end{aligned} \tag{5.4.69}$$

Corresponding to the sequence (5.4.69), define

$$\lambda_j(t) = \sum_{i=1}^{m(j)} \alpha_{ji} f^0\left(t, x_{n_j + i}(t), u_{n_j + i}(t)\right) \tag{5.4.70}$$

where x_{n_j+i} is the (strong) solution of the evolution equation (5.4.1) corresponding to the control u_{n_j+i}. Define

$$\lambda(t) = \underline{\lim}\, \lambda_j(t). \tag{5.4.71}$$

Since $\lambda_j(t) \geq g(t)$ a.e. (for all integers j) and $g \in L_1$, it is clear that $\liminf \lambda_j(t)$ is defined a.e. For convenience of notation let us set $f^0(t, x_{n_j+i}(t), u_{n_j+i}(t)) \equiv f^0_{n_j+i}(t)$. By Fatou's lemma,

$$\int_I \lambda(t)\, dt = \int_I \underline{\lim}\, \lambda_j(t)\, dt \leq \underline{\lim} \int_I \lambda_j(t)\, dt \tag{5.4.72}$$

and, by definition,

$$\int_I \lambda_j(t)\, dt = \sum_{i=1}^{m(j)} \alpha_{ji} J(x_{n_j+i}, u_{n_j+i}). \tag{5.4.73}$$

Due to (5.4.62), $\lim_{j\to\infty} J(x_{n_j+i}, u_{n_j+i}) = m_0$, and hence it also follows that

$$\lim_{j\to\infty} \sum_{i=1}^{m(j)} \alpha_{ji} J(x_{n_j+i}, u_{n_j+i}) = m_0. \tag{5.4.74}$$

Thus it follows from (5.4.72)–(5.4.74) that

$$\int_I \lambda(t)\, dt \leq m_0, \tag{5.4.75}$$

and, since $\lambda(t) \geq g(t)$ a.e. with $g \in L_1$, it is clear that $\lambda \in L_1$. Now we show that

$$(\lambda(t), (L_s x^*)(t)) \in Q(t, x^*(t)) \quad \text{a.e. on } I. \tag{5.4.76}$$

Let

$$I_1 \equiv \{t \in I : |\lambda(t)| < \infty\} \cap \left\{t \in I : \lim_j |\psi_j(t) - (L_s x^*)(t)|_{E^*} = 0\right\}.$$

Define $N_s \equiv \{t \in I : u_s(t) \notin \Gamma(t)\}$, $N_0 \equiv \bigcup_s N_s$, and set $I_2 \equiv (I \backslash N_0)$ and $I_3 = I_1 \cap I_2$. Due to (5.4.69) and the fact that $\lambda \in L_1$, it is clear that the Lebesgue measure $l(I \backslash I_1) = 0$; and by definition of the set of admissible controls $\mathcal{U}_0$, the Lebesgue measure $l(I \backslash I_2) = 0$. Thus I_3 has full Lebesgue measure. Then for $t \in I_3 \backslash \{0, T\}$, there exists a subsequence, possibly dependent on t, of the sequence $\{\lambda_j\}$, again denoted by $\{\lambda_j\}$, such that

$$\lambda_j(t) \to \lambda(t). \tag{5.4.77}$$

Choosing the corresponding subsequences for the sequences (5.4.69) and retaining the original labeling, we have

$$\begin{aligned} \psi_j(t) &\overset{s}{\to} (L_s x^*)(t) \quad \text{in } E^*, \\ x_{n_j+i}(t) &\overset{s}{\to} x^*(t) \qquad \text{in } E. \end{aligned} \tag{5.4.78}$$

Clearly, for every $\varepsilon > 0$, there exists an integer $j_0 = j_0(t, \varepsilon)$ such that for

$j>j_0$, $x_{n_j+i}(t)\in N_\varepsilon(x^*(t))$. Since $\xi\to f(t,\xi,v)$ is demicontinuous [Assumption (5.4.A3)′] and $\xi\to f^0(t,\xi,v)$ is continuous $((F_0 2))$, for $j>j_0$,

$$Q\big(t, x_{n_j+i}(t)\big)\subset Q\big(t, N_\varepsilon(x^*(t))\big) \quad \text{for } \varepsilon>0. \tag{5.4.79}$$

It follows from the definition of Q that

$$\big(f^0_{n_j+i}(t), (Lx_{n_j+i})(t)\big)\in Q\big(t, x_{n_j+i}(t)\big) \quad \text{for almost all } t\in I. \tag{5.4.80}$$

Thus, for sufficiently large j,

$$\big(f^0_{n_j+i}(t), (Lx_{n_j+i})(t)\big)\in Q\big(t, N_\varepsilon(x^*(t))\big) \quad \text{for all } \varepsilon>0, \tag{5.4.81}$$

and consequently,

$$\big(\lambda_j(t), \psi_j(t)\big)\in \operatorname{co} Q\big(t, N_\varepsilon(x^*(t))\big) \quad \text{for } \varepsilon>0. \tag{5.4.82}$$

Since, for $t\in I_3\backslash\{0,T\}$, $\lambda_j(t)\to\lambda(t)$ and $\psi_j(t)\overset{s}{\to}(L_s x^*)(t)$ in E^*, it follows from (5.4.82) that

$$\big(\lambda(t), (L_s x^*)(t)\big)\in \operatorname{cl}\operatorname{co} Q\big(t, N_\varepsilon(x^*(t))\big) \quad \text{for all } \varepsilon>0, \tag{5.4.83}$$

and consequently,

$$\big(\lambda(t), (L_s x^*)(t)\big)\in \bigcap_{\varepsilon>0} \operatorname{cl}\operatorname{co} Q\big(t, N_\varepsilon(x^*(t))\big). \tag{5.4.84}$$

Since Q satisfies the weak Cesari property, it follows from (5.4.84) that

$$\big(\lambda(t), (L_s x^*)(t)\big)\in Q\big(t, x^*(t)\big) \quad \text{for } t\in I_3\backslash\{0,T\}, \tag{5.4.85}$$

and hence almost everywhere in I.

It follows from (5.4.85), the definition of Q, and the assumption that $\Gamma(t)$ is nonempty for each $t\in I$, that there exists a function $\tilde{u}$ with $\tilde{u}(t)\in\Gamma(t)$ such that

$$\begin{gathered}\lambda(t)\geq f^0(t, x^*(t), \tilde{u}(t)),\\ \langle (L_s x^*)(t), e\rangle = \langle f(t, x^*(t), \tilde{u}(t)), e\rangle \quad \text{for all } e\in D(A^*(t))\end{gathered} \tag{5.4.86}$$

for almost all $t\in I$.

It is not yet clear if a measurable substitute for $\tilde{u}$ exists. At this point we use selection theorems.

For $t\in I_3\backslash\{0,T\}\equiv\tilde{I}$, define the set-valued map G^* with values

$$G^*(t)=\left\{\begin{aligned} &u\in\Gamma(t): \lambda(t)\geq f^0(t, x^*(t), u); \langle (L_s x^*)(t), e\rangle\\ &=\langle f(t, x^*(t), u), e\rangle \quad \text{for all } e\in D(A^*(t)).\end{aligned}\right\}. \tag{5.4.87}$$

We show that G^* has a measurable selection u^*, that is, there exists a measurable function u^* with values $u^*(t)\in G^*(t)$, $t\in\tilde{I}$. First we show that G^* has closed values. For $t\in\tilde{I}$, let $\{v_n\}\in G^*(t)$ and suppose that $v_n\overset{\rho}{\to} v_0$;

we show that $v_0 \in G^*(t)$. Since f is demicontinuous in v and f^0 is lower semicontinuous in v, we have

$$\langle f(t, x^*(t), v_n), e\rangle \to \langle f(t, x^*(t), v_0), e\rangle \quad \text{for } e \in E,$$
$$f^0(t, x^*(t), v_0) \le \underline{\lim}_n f^0(t, x^*(t), v_n) \le \lambda(t). \tag{5.4.88}$$

The closure of $G^*(t)$ now follows from (5.4.88) and the closure of $\Gamma(t)$. For the proof of measurability of the set-valued map G^*, we utilize the equivalence of (i) and (v) of Theorem 1.4.3. For this we need to verify that, for every $\varepsilon > 0$, there exists a closed set $I_\varepsilon \subset \tilde{I}$ such that $G_\varepsilon^* = G^*|_{I_\varepsilon}$ (restriction of G^* to I_ε) has a closed graph, that is, the graph $\gamma(G_\varepsilon^*) \equiv \{(t, v) : t \in I_\varepsilon$ and $v \in G_\varepsilon^*(t)\}$ is a closed subset of $I_\varepsilon \times B$. Since $\lambda \in L_1$, f^0 satisfies the properties (F_01)–(F_03), $t \to (L_s x^*)(t)$ is a strongly measurable E^*-valued function, and f satisfies Assumption (5.4.A3)′, it follows that, for every $\varepsilon > 0$, there exists a closed set $I_\varepsilon \subset \tilde{I}$ with Lebesgue measure $l(I_\varepsilon) \ge l(I) - \varepsilon$ such that the restrictions of these maps to the interval I_ε have these properties:

$$\begin{cases} \text{(a)} & t \to \lambda_\varepsilon(t) \text{ is continuous on } I_\varepsilon \\ \text{(b)} & (t, v) \to f_\varepsilon^0(t, x^*(t), v) \text{ is lower semicontinuous} \\ & \text{in } v \in \Gamma(t) \text{ and continuous in } t \in I_\varepsilon \\ \text{(c)} & t \to (L_s x^*)_\varepsilon(t) \text{ is strongly continuous } E^*\text{-valued function} \\ \text{(d)} & (t, v) \to \langle f_\varepsilon(t, x^*(t), v), e\rangle \text{ is continuous for each } e \in E \\ \text{(e)} & t \to \Gamma_\varepsilon(t) \text{ is continuous on } I_\varepsilon \text{ with respect to the uniformity} \\ & \text{determined by } \rho \text{ on } 2^B. \end{cases} \tag{5.4.89}$$

In the above, the subscript ε, attached to any function, indicates the restriction of the function to the interval I_ε.

Let $(t_n, v_n) \in \gamma(G_\varepsilon^*)$ and suppose that $t_n \to t_0$ and $v_n \xrightarrow{\rho} v_0$. Since I_ε is closed, $t_0 \in I_\varepsilon$. Due to (a) and (b) of (5.4.89), we have

$$\lambda(t_0) = \underline{\lim}_n \lambda(t_n) \ge \underline{\lim}_n f^0(t_n, x^*(t_n), v_n) \ge f^0(t_0, x^*(t_0), v_0), \tag{5.4.90}$$

and due to (c) and (d) of (5.4.89), we have, for each $e \in E$,

$$\langle (L_s x^*)(t_0), e\rangle = \lim_n \langle (L_s x^*)(t_n), e\rangle = \lim_n \langle f(t_n, x^*(t_n), v_n), e\rangle$$
$$= \langle f(t_0, x^*(t_0), v_0), e\rangle. \tag{5.4.91}$$

Due to (e) of (5.4.89), we have

$$v_0 \in \Gamma(t_0). \tag{5.4.92}$$

Thus it follows from (5.4.90)–(5.4.92) that $(t_0, v_0) \in \gamma(G_\varepsilon^*)$, that is, the graph of G_ε^* is closed. This, in turn, proves the measurability of the set-valued function G^*. Since B is a polish space and $G^*(t)$ is closed, it is also complete. Thus it follows from the selection theorem (Theorem 1.4.4) that there exists a measurable function u^* on I with values $u^*(t) \in G^*(t)$ a.e. Thus, in (5.4.86) we can replace $\tilde{u}$ by u^*. This completes the proof of the theorem. □

Relaxed Systems. Let B be a polish space (or complete separable metric space) and Γ a closed subset of B. Let $\mathfrak{M}(\Gamma)$ denote the space of all probability measures on the Borel σ-field of Γ. Let $C_b(\Gamma)$ denote the space of real-valued bounded continuous functions on Γ. We can topologize the space $\mathfrak{M}(\Gamma)$ by defining a base of open neighborhoods $\{N(\nu, F, \varepsilon), \varepsilon > 0, \nu \in \mathfrak{M}(\Gamma)$, and F finite subsets of $C_b(\Gamma)\}$, where

$$N(\nu, F, \varepsilon) \equiv \left\{ \mu \in \mathfrak{M}(\Gamma) : \left| \int_\Gamma f \, d\mu - \int_\Gamma f \, d\nu \right| < \varepsilon, f \in F \right\}.$$

This base of neighborhoods defines a topology on $\mathfrak{M}(\Gamma)$, and with this topology $\mathfrak{M}(\Gamma)$ becomes a topological space. This is called the *weak topology* by probabilists. A net $\{\mu_\alpha\} \in \mathfrak{M}(\Gamma)$ converges weakly to an element $\mu \in \mathfrak{M}(\Gamma)$ if and only if, for every $f \in C_b(\Gamma)$,

$$\int_\Gamma f \, d\mu_\alpha \to \int f \, d\mu. \tag{5.4.93}$$

From the point of view of functional analysts, $\mathfrak{M}(\Gamma)$ is considered as a subset of the dual space $C_b^*(\Gamma)$ of the vector space $C_b(\Gamma)$. Consequently, the topology mentioned above is called the w*-topology, and the associated convergence as the w*-convergence. This topology is metrizable and $\mathfrak{M}(\Gamma)$ with this topology becomes a separable metric space [P.1, Theorem 6.2, p. 43]. Further, if Γ is compact, then $\mathfrak{M}(\Gamma)$ is a compact metric space [P.1, Theorem 6.4, p. 45]. Since Γ is a closed subset of the polish space B, $\mathfrak{M}(\Gamma)$ is topologically complete, that is, $\mathfrak{M}(\Gamma)$ is homeomorphic to a complete metric space.

For $(t, x) \in \Delta$ (see Definition 5.4.2), define

$$R(t,x) \equiv \{ e^* \in E^* : \langle e^*, \phi \rangle = \langle A(t)x, \phi \rangle + \langle f(t,x,\sigma), \phi \rangle \text{ for some } \sigma \in \Gamma \text{ and all } \phi \in D(A^*(t)) \},$$

$$R_0(t,x) \equiv \operatorname{cl} \operatorname{co} R(t,x),$$

$$R_r(t,x) \equiv \left\{ e^* \in E^* : \langle e^*, \phi \rangle = \langle A(t)x, \phi \rangle + \int_\Gamma \langle f(t,x,\sigma), \phi \rangle \, d\mu(\sigma) \text{ for some } \mu \in \mathfrak{M}(\Gamma) \text{ and all } \phi \in D(A^*(t)) \right\}. \tag{5.4.94}$$

We consider the relaxed systems S_0 and S_r defined as

$$S_0 \qquad \begin{cases} \dot{x}(t) \in R_0(t, x(t)) & \text{a.e.} \\ x(0) = x_0, \end{cases} \tag{5.4.95}$$

$$S_r \qquad \begin{cases} \dot{x}(t) \in R_r(t, x(t)) & \text{a.e.} \\ x(0) = x_0. \end{cases} \tag{5.4.96}$$

Let X_0 and X_r denote (respectively) the corresponding set of trajectories; that is,

$$X_0 \equiv \{x \in \mathcal{X}: x(0) = x_0 \quad \text{and} \quad \dot{x}(t) \in R_0(t, x(t)) \text{ a.e.}\}$$

and

$$X_r \equiv \{x \in \mathcal{X}: x(0) = x_0 \quad \text{and} \quad \dot{x}(t) \in R_r(t, x(t)) \text{ a.e.}\}.$$

In general, we shall consider the system S_r and establish conditions for the closure of the set X_r and its relationship with the set X_0.

Theorem 5.4.10. *Let B be a polish space, Γ a closed subset of B and $\mathfrak{M}(\Gamma)$ the space of probability measures on Γ. Consider the relaxed system S_r and suppose that A satisfies the Assumptions* (5.4.A1)–(5.4.A2) *and f satisfies* (5.4.A3)″ *and* (5.4.A4)–(5.4.A6), *where* (5.4.A3) *is replaced by the following*:

Assumption (5.4.A3)″

(a) $f: I \times E \times B \to E^*$ is demicontinuous in the sense that whenever $t_n \to t$ in I, $x_n \xrightarrow{s} x$ in E, and $u_n \xrightarrow{\rho} u$, $\lim_n \langle f(t_n, x_n, u_n), e\rangle = \langle f(t, x, u), e\rangle$ for every $e \in E$.

(b) Whenever $t_n \to t$ in I and $x_n \xrightarrow{s} x$ in E, $\lim_n \langle f(t_n, x_n, u), e\rangle = \langle f(t, x, u), e\rangle$ uniformly in u on any compact subset $\Gamma_0 \subset \Gamma$ for each $e \in E$.

Then the set X_r is a closed and hence a sequentially compact subset of $C(I, E_w)$ if either one of the following conditions is satisfied:

i. for each $(t, x) \in \Delta$, $R_r(t, x)$ is a closed subset of E^*;
ii. Γ is a compact subset of B.

In case (ii) holds, $X_r = X_0$.

PROOF. Under the present hypotheses, the conclusions of Lemmas 5.4.1–5.4.4 remain valid. Thus, according to Theorem 5.4.2, X_r is a conditionally sequentially compact subset of $C(I, E_w)$. For sequential compactness, it suffices to prove its closure. As in Theorem 5.4.5, we can show that if $\{x_n\} \in X_r$ and $x_n \to x^*$ in $C(I, E_w)$, then (1) $\dot{x}^*(t)$ exists a.e. on I with $\dot{x}^* \in L_q(I, E^*)$, (2) $x^*(t) \in D(A(t))$ a.e., and (3) $\dot{x}^*(t) \in R_r(t, x^*(t))$ a.e. if either one of the conditions (i) and (ii) is satisfied. As a consequence of (3), there exists a function $t \to \mu_t$ from I into $\mathfrak{M}(\Gamma)$ such that

$$\langle \dot{x}^*(t), e \rangle = \langle A(t)x^*(t), e \rangle + \int_\Gamma \langle f(t, x^*(t), \sigma), e \rangle \, d\mu_t(\sigma) \tag{5.4.97}$$

for $t \in I$ and $e \in D(A^*(t))$.

The question that remains to be resolved is whether or not we can select a measurable substitute μ^* for μ, that is, a measurable function $t \to \mu_t^*$ that satisfies the above equality with μ^* substituted for μ. For any $t \in I$ for which $\dot{x}^*(t)$ and $A(t)x^*(t)$ exist, define the set-valued map G^* with values

$$G^*(t) \equiv \Big\{ \nu \in \mathfrak{M}(\Gamma) : \langle \dot{x}^*(t), e \rangle = \langle A(t)x^*(t), e \rangle + \int_\Gamma \langle f(t, x^*(t), \sigma), e \rangle \, d\nu(\sigma) \text{for all } e \in D(A^*(t)) \Big\}. \tag{5.4.98}$$

We show that G^* has a measurable selection. For this we note that, since Γ is a closed subspace of the polish space B, $\mathfrak{M}(\Gamma)$ is a separable metric space with a metric compatible with its topology. Thus, if we can show that $t \to G^*(t)$ is a measurable multifunction (set-valued function) with closed values, then by virtue of the selection theorem (Theorem 1.4.4) we shall be able to conclude that there exists a measurable function μ^* from I into $\mathfrak{M}(\Gamma)$ such that $\mu_t^* \in G^*(t)$ for $t \in I$. This would then imply that X_r is a (sequentially) closed subset of $C(I, E_w)$, as desired. First we show that G^* has closed values. Let $\{\mu^n\} \subset G^*(t)$ and suppose $\mu^n \xrightarrow{w} \mu^0$ in $\mathfrak{M}(\Gamma)$. We must show that $\mu^0 \in G^*(t)$. But this follows readily from the fact that for the fixed $t \in I$ and arbitrary $e \in E_w$, the function $\sigma \to \langle f(t, x^*(t), \sigma), e \rangle$ is continuous on Γ and that $\mu^n \xrightarrow{w} \mu^0$, so that

$$\int_\Gamma \langle f(t, x^*(t), \sigma), e \rangle \, d\mu^n(\sigma) \to \int_\Gamma \langle f(t, x^*(t), \sigma), e \rangle \, d\mu^0(\sigma).$$

It remains to show that G^* is a measurable multifunction. By virtue of the equivalence of conditions (i) and (v) of Theorem 1.4.3, it suffices to show that, for every $\varepsilon > 0$, there exists a closed set $I_\varepsilon \subset I$ such that the graph $\gamma(G_\varepsilon^*)$ of the set-valued map G_ε^*, which is the restriction of G^* to I_ε, is a closed subset of $I_\varepsilon \times \mathfrak{M}(\Gamma)$. Since $\dot{x}^*$, Ax^*, and $\{f(\cdot, x^*(\cdot), \sigma), \sigma \in \Gamma\} \subset L_q(I, E^*)$, for every $\varepsilon > 0$, there exists a closed set $I_\varepsilon \subset I$ with Lebesgue measure $l(I \backslash I_\varepsilon) < \varepsilon$ such that $\dot{x}^*$, Ax^*, and $f(\cdot, x^*(\cdot), \sigma)$ are continuous E^*-valued functions on I_ε. Let $(t^n, \mu^n) \subset \gamma(G_\varepsilon^*)$ such that $t^n \to t^0$ and $\mu^n \xrightarrow{w} \mu^0$; we show that $(t^0, \mu^0) \in \gamma(G_\varepsilon^*)$, or equivalently, $\mu^0 \in G_\varepsilon^*(t^0)$. Since I_ε is closed, $t^0 \in I_\varepsilon$, and further, due to continuity of the restrictions

the property that $T(t)E=E$ for some $t>0$. This condition is critical and ensures that the interior of the attainable set is nonempty, so that the separation theorem holds. In Corollary 5.2.2, we have discussed conditions under which bang-bang principle holds. In the case of strictly convex Banach spaces the bang-bang control is also unique. In Lemma 5.2.2, it is shown that in case E (the state space) is a Hilbert space, the property that $T(t)E=E$ is also a necessary condition for the interior $\Omega(\tau)$ to be nonempty. In Theorem 5.2.14, this restriction $[T(t)E=E]$ is removed by introducing the assumption that both the initial and final states lie in the domain $D(A)$ of the generator A of the semigroup $T(t)$, $t\geq 0$. As discussed in Remark 5.2.7, if the semigroup $T(t)$, $t\geq 0$, is analytic, then the maximum principle (Theorem 5.2.14) also implies the bang-bang principle. In case the state space E is a Hilbert space and A is a self-adjoint operator, the maximum principle (Theorem 5.2.13) holds only if the terminal state $x_1\in D(A)$. In case E is a reflexive Banach space and $D(A)=K_\infty$ [generally, $D(A)\subset K_\infty$], then A is necessarily a bounded operator. Thus if A is unbounded, $D(A)$ is a proper subset of K_∞. Recently, Fattorini [Fa.9] has also discussed time optimal control problems for second-order evolution equations in Banach spaces using the well known sine–cosine operators. In this case, the solution of the equation

$$\frac{d^2x}{dt^2}=Ax+u, \qquad x(0)=x_0, \qquad \dot{x}(0)=x_1$$

is given by

$$x(t)=C(t)x_0+S(t)x_1+\int_0^t S(t-\theta)u(\theta)\,d\theta,$$

where $C(t)=\cos\sqrt{-A}\ t$ and $S(t)=(-A)^{-1/2}\sin\sqrt{-A}\ t$ are sine–cosine operators. For details, the reader is referred to Fattorini [Fa.9]. The question of time optimal boundary controls for systems governed by parabolic equations has been discussed by Fattorini [Fa.8], Knowles [Kn.2] and, most recently, Washburn [Was.1]. Washburn has discussed this problem for the heat equation $\partial\phi/\partial t=\Delta\phi$ in Ω with $\phi(0)=0$ and $\phi|_{\Gamma=\partial\Omega}=u$, where Ω is an open bounded subset of R^n. Using Green's operator G arising from the Dirichlet problem $\Delta f=0$, $f|_{\partial\Omega}=u$, so that $f=Gu$, one can formulate the boundary control problem under the semigroup setting:

$$\frac{d}{dt}x=Ax-\frac{d}{dt}Gu, \qquad x(0)=0,$$

where $x=\phi-Gu$, $x\in D(A)$, $D(A)\equiv H_0^1(\Omega)\cap D(\Delta)$, $D(\Delta)\equiv\{f: f,\ \Delta f\in L_2(\Omega)\}$, and the operator A is the closed restriction of Δ on $H_0^1(\Omega)\cap D(\Delta)$. Since, under the above conditions, A is the generator of a c_0-semigroup $T(t)$, $t\geq 0$, the solution x is given by

$$x(t)=-\int_0^t T(t-\xi)\frac{d}{d\xi}(Gu)(\xi)\,d\xi,$$

and hence, $\phi(t) = -\int_0^t AT(t-\xi)(Gu)(\xi)\,d\xi$ is a weak solution of the original problem in the sense that $\phi \in L_2(0, \tau; L_2(\Omega))$ whenever $u \in L_2(0, \tau; L_2(\Gamma))$ for $0 \leq \tau < \infty$. For time optimal controls, it is required that $\phi(t)$ be defined for each $t \in [0, \tau]$. This, in turn, requires that the class of admissible controls be smaller than $L_2(0, \tau; L_2(\Gamma))$. Using the theory of intermediate spaces [Li.2, BB.1] and properties of analytic semigroups, Washburn gives a bound of the output ϕ in terms of bounds of the kernel $K(t, \xi) \equiv AT(t-\xi)G$. Letting $V = L_2(\Gamma)$ and $E = L_2(\Omega)$, one can show that if $G: V \to [D(A), E]_{1-\theta}$ [the intermediate space between $D(A)$ and E] for some $0 < \theta < 1$, then $\| AT(t)G \| = O(t^{\theta - 1})$, and consequently, the map $u \to \phi$ is a continuous linear transformation from $L_p(I, V)$ into $L_\infty(I, E)$ for $p > 1/\theta$ and $\mu(I) < \infty$. In particular, for $\theta = \frac{1}{4}$ and $p > 4$, $\phi(t) \in E$ for all $t \in I$. Thus, the time optimal control has meaning. The existence of time optimal boundary controls follows from arguments similar to those given for Theorems 5.2.4, 5.2.12. However, the necessary conditions of optimality (or equivalently, the characterization of optimal controls) require special consideration if the Dirichlet operator G is compact. In this case, the input map $u \to \phi(\tau)$ is compact, and hence, the attainable set $\mathcal{Q}(\tau)$ is a compact subset of E. This implies that the interior of $\mathcal{Q}(\tau)$ is empty, and hence, every point of $\mathcal{Q}(\tau)$ is its boundary point. In a linear topological space, every boundary point of a convex set with nonempty interior is a support point. Since $\mathcal{Q}(\tau)$ is convex without interior points, its boundary points are not necessarily its support points. However, since support points of a closed convex set in a Banach space are dense in its boundary points, for every point $x^0 \in \mathcal{Q}(\tau) = \partial \mathcal{Q}(\tau)$, there exists a sequence of support points $\{x_n\}$ of $\mathcal{Q}(\tau)$ such that $x_n \to x^0$. Thus, if x^0 is the target point reachable in the optimal time τ, then there exists a sequence $\{x_n\}$ of support points of $\mathcal{Q}(\tau)$ converging to x^0. In general, the bang-bang principle does not hold in infinite-dimensional problems. However, in this case, under the assumption that the semigroup $T(t)$, $t \geq 0$, is analytic, one can show that the optimal control u^0 is the w*-limit of a sequence of bang-bang controls. For further details the reader is referred to the paper of Washburn [Was.1].

Recently, White [Wh.1] has considered the optimal control problem for a class of pseudoparabolic initial value problems of the form $My_t + Ly = u$ in $Q = I \times \Omega$, where both M and L are second-order symmetric uniformly strongly elliptic operators in Ω. For C^∞-coefficients, M is an isomorphism of $H_0^1(\Omega) \cap H^2(\Omega)$ onto $L_2(\Omega)$ and $M^{-1}L$ is a bounded linear operator in $H_0^1(\Omega) \cap H^2(\Omega)$. Thus, for $u \in L_2(Q)$, the system can be equivalently described as $y_t + M^{-1}Ly = M^{-1}u$ with $y(0) = y_0 \in L_2(\Omega)$. The solution of this problem can be explicitly written as

$$y(t) = T(t)y_0 + \int_0^t T(t-\theta)(M^{-1}u)(\theta)\,d\theta,$$

where $T(t) = \exp(-tM^{-1}L)$ is the semigroup. Since M^{-1} is a bounded operator, it is obvious that $T(t)$, $t \in R$, is a group of bounded linear operators in $H_0^1(\Omega)\cap H^2(\Omega)$. Notable properties of this class of systems are that the set of attainability can be described explicitly (unlike general parabolic problems) and that the semigroup $T(t)$ is, in fact, a group. It is easy to see that for any given point (target) $z\in H^1\cap H^2$, there exists a control u_z with values $u_z(t)=(1/\tau)MT(t-\tau)(z-T(\tau)y_0)$, $t\in[0,\tau]$, that transfers the system from the state y_0 to the state z. Thus, the attainable set is all of $H_0^1\cap H^2$. Since $H_0^1\cap H^2$ is dense in L_2, the pseudoparabolic problem is approximately controllable with respect to any target $z\in L_2$. For details the reader is referred to White [Wh.1].

Another interesting problem closely related to the problem of controllability is the question of observability or identification. Due to the limitation of space, we have not considered this problem. However, we mention that the work of Kobayashi [Ko.1, Ko.2] and Kitamura and Nakagiri [KN.1] is closely related to the quadratic optimization problems treated in Section 5.1. They have considered the problems of identification of initial state as well as parameters from known observations. For example, one may want to determine the parameter $g\in V^*$ appearing in the system equation $(dx/dt)+Ax=v(t)g$ from the knowledge of initial data and the measured output $y_e(t)=C(t)x(t)+e(t)$, $t\in I$, subject to error e. In conformity with the assumptions of Section 5.1, we may take $C\in L_\infty(I,\mathcal{L}(V,F))$, $e\in L_2(I,F)$, $v\in L_2(I,R)$, and $A\in\mathcal{L}(V,V^*)$ with $\langle A\xi,\xi\rangle\geq\alpha|\xi|_V^2$. The problem of identification of g may also be viewed as the problem of minimization of the measurement error $\|e(g)\|^2=\int_I|y_e(t)-C(t)x(t)|_F^2\,dt$ with respect to the parameter g. Under the given assumptions, the operator A is the generator of a c_0-semigroup $T(t)$, $t\geq 0$, in $H\supset V$ and, consequently, the output y, without the measurement error, is given by

$$y(t)=C(t)T(t)x_0+\int_0^t C(t)T(t-\theta)v(\theta)g\,d\theta,\qquad t\in I=[0,\tau].$$

Writing

$$y_e(t)-C(t)T(t)x_0\equiv y_0(t)\quad\text{and}\quad\int_0^t C(t)T(t-\theta)v(\theta)g\,d\theta\equiv(Wg)(t),$$

one can rewrite the error

$$\|e(g)\|^2=\|y_0\|^2-2\langle W^*y_0,g\rangle_{V-V^*}+\langle W^*Wg,g\rangle_{V-V^*},$$

where F is assumed to be self-adjoint. The parameter g $(\in V^*)$ that minimizes the error is given by the relation $W^*Wg=W^*y_0$. Note that $y_0\in L_2(I,F)$ and $W\in\mathcal{L}(V^*,L_2(I,F))$ and $W^*W\in\mathcal{L}(V^*,V)$. Clearly, if W^*W is positive or equivalently $\ker W=\{0\}$, then the system is observable at time τ in the sense that, from the measured data y_e, the parameter g is uniquely determined from the equation $g=(W^*W)^{-1}W^*y_0$. If W^*W is positive definite, its inverse is a continuous linear operator; otherwise a small measurement error can cause large identification error. Therefore, the given expression for g is not convenient for computation. To overcome this, Kobayashi suggests an approximation technique that amounts to adding to $\|e\|^2$ a positive definitive term $\delta\langle Ng, g\rangle$, $\delta>0$, with $N\in\mathcal{L}(V^*, V)$ and $\langle Ng, g\rangle_{V-V^*}\geq\nu|g|^2_{V^*}$, $\nu>0$. This leads to the approximate solution $g_\delta=(\delta N+W^*W)^{-1}_{W^*}y_0$, which is shown to converge weakly to g as $\delta\to 0$. For details see [Ko.1, Ko.2, KN.1]. A more interesting problem in application is the identification of the operator $B\in\mathcal{L}(H)$, appearing in the perturbed system $(dx/dt)+(A+B)x=f$. Using the theory of vector measures, Knowles [Kn.1, Kn.2] has given a model of infinite-dimensional systems as vector-valued measures and has studied time optimal control problems that include the questions of existence of optimal controls, necessary conditions of optimality, conditions for normality, and the bang-bang principle. For details the reader is referred to Knowles [Kn.1, Kn.2].

In Section 5.3, we have briefly considered a penalty function approach that, unlike other techniques, bypasses the problem of solving optimization problems involving constraints, differential or otherwise. The method essentially consists of converting the original problem with differential and other constraints into a suitable unconstrained problem by including in the cost function a penalty term for each of the constraints. The modified cost functional is then minimized over the product space of controls and states. Here we have given several results on the existence of optimal controls (Theorems 5.3.1–5.3.4). One of the merits of the approach is that even if the original problem (P_0) does not have a solution, the penalized problem (P_ε) gives a suboptimal solution for each $\varepsilon>0$, and if the original problem has a solution, the solution of the penalized problem (P_ε) converges to that of the original problem. This method was originally developed with a view to numerical applications by many authors (Yavon [Y.1], Balakrishnan [B.2], Bensoussan and Kenneth [BK.1], Julio [Ju.1]) and was reported in [Li.2]. There are other approximation techniques known as parabolic and elliptic regularizations [Li.2]. It is often convenient to approximate a hyperbolic system by a parabolic system or a parabolic system by an elliptic system and use known techniques for solving the approximate problems and show that the solutions of the approximate problems converge to those of the original problems.

In the final section, Section 5.4, we have considered a class of abstract nonlinear evolution equations and their optimal control. These results were originally developed by Ahmed [Ah.2, Ah.4] and Ahmed and Teo [AT.7]. The existence of solutions of the evolution equation as given in Theorem 5.4.1 follows from Theorem 2.5.3, which is originally due to Browder [Br.1]. For control problems considered here we have used the concept of strong solutions in the sense that $\dot{x}(t)\in R(t, x(t))\subset E^*$ for almost all $t\in I$, where $R(t, \xi)\in\mathcal{P}(E^*)$ is the velocity field [Definition 5.4.2, Equation (5.4.25)] at $(t, \xi)\in\Delta$. It is not clear whether mild (or weak) solutions, as considered in [Ah.2], can be used to prove the existence of optimal controls as done for time optimal control of linear systems (see Section 5.2.8). The selection theorem (Theorem 5.4.3) is due to Ahmed and Teo [AT.6]. This result, which is an extension of a result due to Cole [Co.1], removes the convexity assumption used by Cole. To our knowledge, Lemma 5.4.7 appears to have been reported here for the first time. Similar results for finite-dimensional problems are well known. Existence theorems for terminal control problems, time optimal control problems, and Bolza problems are given in Theorem 5.4.6–5.4.8. Theorems 5.4.6 and 5.4.8 are proved using compactness (weak compactness) and semicontinuity (weak lower semicontinuity) arguments. Here we have assumed that f is weakly continuous in u, which probably is not essential. The Lagrange problem (Theorem 5.4.9) is proved using the Filippov–Cesari approach, and this is reported here for the first time. The results on relaxed systems (Theorem 5.4.10) also appear to have been reported here for the first time. We wish to mention, however, that relaxed controls for finite-dimensional systems are well known in the literature [HL.1, War.1; see also the references listed in War.1]. In the Filippov–Cesari approach (Theorem 5.4.9), we have assumed that Q satisfies the weak Cesari property. It would be interesting to find the most general conditions under which this is true. Another interesting unresolved problem is the question of approximation of the relaxed trajectories by those of the original system. More precisely, it would be interesting to know whether the set of original trajectories X is dense in the set of relaxed trajectories X_0. This fact is well known for the finite-dimensional case [HL.1, Theorem 20.2]. There is no known result on the necessary conditions of optimality for the problems considered in this section. For the necessary conditions of optimality for strongly nonlinear parabolic problems the reader is referred to Ahmed [Ah.2, Ah.4]. For linear evolution equations in Hilbert space with state constraints see Barbu [Ba.1], and for nonlinear elliptic problems with state constraints see Michel [M.1]. For many interesting results on the existence of optimal controls and geometric and analytic views thereof the reader is referred to Berkovitz [Be.2, Be.3] and Cesari [Ce.1–Ce.6]. For limitation of space we have not considered stochastic systems. We mention only that optimal control of stochastic evolution equations with deterministic operator valued coefficients have been considered by Bensoussan and Viot [BV.1] and Balakrishnan [B.3] and with random operator valued coefficients by Ahmed [Ah.7]. For finite dimensional problems in this area interested reader is referred to Fleming and Rishel [FLR.1] and Ahmed [Ah.3] and the extensive bibliography therein.

BIBLIOGRAPHY

Adams, R. A.

[A.1] *Sobolev Spaces*, Academic Press, New York, 1975.

Ahmed, N. U.

[Ah.1] Optimal Control of Generating Policies in a Power System Governed by a Second Order Hyperbolic Partial Differential Equation, *SIAM Journal on Control and Optimization* 15:1016–1033 (1977).

[Ah.2] Optimal Control of a Class of Strongly Nonlinear Parabolic Systems, *Journal of Mathematical Analysis and Applications* 16:188–207 (1977).

[Ah.3] Optimal Control of Stochastic Systems, in *Probabilistic Analysis and Related Topics* (edited by A. T. Bharucha-Reid), Vol. 2, Academic Press, New York, 1979, pp. 1–68.

[Ah.4] Some Comments on Optimal Control of a Class of Strongly Nonlinear Parabolic Systems, *Journal of Mathematical Analysis and Applications* 68:595–598 (1979).

[Ah.5] Nonlinear Integral Equations on Reflexive Banach Spaces with Applications to Stochastic Integral Equations and Abstract Evolution Equations, *Journal of Integral Equations* 1:1–15 (1979).

[Ah.6] Stability Analysis of a Transmission Line with Variable Parameters, *IEEE Transaction on Circuit and Systems* CAS 25:878–885 (1978).

[Ah.7] Stochastic Control on Hilbert Space for Linear Evolution Equations with Random Operator Valued Coefficients, *SIAM Journal on Control and Optimization*, (to be published, 1980–81).

Ahmed, N. U., and Teo, K. L.

[AT.1] An Existence Theorem on Optimal Control of Partially Observable Diffusion, *SIAM Journal on Control* 12:351–355 (1974).

[AT.2] Stochastic Bang–Bang Control, *IEEE Transactions on Automatic Control* AC-19:73–75 (1974).

[AT.3] On the Optimal Parameter Selection for Stochastic Ito Differential Equation, *International Journal of Systems Science* 6:749–754 (1975).

[AT.4] Optimal Control of Stochastic Ito Differential Systems with Fixed Terminal Time, *Advances in Applied Probability* 17:154–178 (1975).

[AT.5] Necessary Conditions for Optimality of a Cauchy Problem for Parabolic Partial Differential Systems, *SIAM Journal on Control* 13:981–993 (1975).

[AT.6] Comments on a Selector Theorem in Banach Spaces, *Journal of Optimization Theory and Applications* 19:117–118 (1976).

[AT.7] Optimal Control of Systems Governed by a class of Nonlinear Evolution Equations in a Reflexive Banach Space, *Journal of Optimization Theory and Applications* 25:57–81 (1978).

Ahmed, N. U., and Wong, H. W.

[AW.1] A minimum Principle for Systems Governed by Ito Differential Equations with Markov Jump Parameters, in *Differential Games and Control Theory II* (edited by E. O. Roxin, P. T. Liu, and R. L. Sternberg), Lecture Notes in Pure and Applied Mathematics, No. 30, Marcel Dekker, New York, 1973.

Alvardo, F. L., and Mukundan, R.

[AM.1] An Optimization Problem in Distributed Parameter Systems, *International Journal of Control* 9:665–677 (1969).

Ames, W. F.

[Am.1] *Numerical Methods for Partial Differential Equations*, Nelson, London, 1969.

Aronson, D. G.

[Ar.1] Non-Negative Solutions of Linear Parabolic Equations, *Annali della Scuola Normale Superiore di Pisa* 22:607–694 (1968).

Aronson, D. G., and Serrin, J.

[AS.1] Local Behaviour of Solutions of Quasilinear Parabolic Equations, *Archive for Rational Mechanics and Analysis* 25:81–122 (1967).

Aziz, A. K.

[Az.1] *Control Theory of Systems Governed by Partial Differential Equations*, Academic Press, New York, 1977.

Balakrishnan, A. V.

[B.1] Optimal Control Problems in Banach Spaces, *SIAM Journal on Control* 3:152–180 (1965).

[B.2] A New Computing Technique in System Identification, *Journal of Computer and System Sciences* 2:102–116 (1968).

[B.3] *Applied Functional Analysis*, Springer-Verlag, Berlin, Heidelberg, New York, 1976.

Barbu, V.

[Ba.1] Convex Control Problems of Bolza in Hilbert Spaces, *SIAM Journal on Control and Optimization* 13:754–771 (1976).

Barnes, E. R.

[Bar.1] An Extension of Gilbert's Algorithm for Computing Optimal Controls, *Journal of Optimization Theory and Applications* 7:420–443 (1971).

Barron, E. N.

[Barr.1] Control Problems for Parabolic Equations on Controlled Domains, *Journal of Mathematical Analysis and Applications* 66:632–650 (1978).

Barros-Neto, J.

[Bn.1] *An Introduction to The Theory of Distributions*, Marcel Dekker, New York, 1973.

Bensoussan, A., Kenneth P.

[BK.1] *Cahiers, IRIA* 2 (1970).

Bensoussan, A., Viot, M.

[BV.1] Optimal Control of Stochastic Linear Distributed Parameter Systems, *SIAM Journal on Control* 13:904–926 (1975).

Berkovitz, L. D.

[Be.1] Existence Theorems in Problems of Optimal Controls, *Studia Mathematica* 65:275–285 (1972).

[Be.2] Existence and Lower Closure Theorems for Abstract Control Problems, *SIAM Journal on Control*,12:27–42 (1974).

[Be.3] A Lower Closure Theorem for Abstract Control Problems with L_p-bounded Control, *Journal of Optimization Theory and Applications* 14:521–528 (1974).

[Be.4] *Optimal Control Theory*, Springer-Verlag, New York, Heidelberg, Berlin, 1974.

Bourbaki, N.

[Bo.1] *General Topology*, Part 2, Addison-Wesley, Reading, 1966.

Boyd, I. E.

[Boy.1] Optimal Control of a Production Firm Model, M.Sc. Thesis, University of New South Wales, 1977.

Browder, F. E.

[Br.1] Nonlinear Equations of Evolution, *Annals of Mathematics* 80:485–523 (1964).

[Br.2] Nonlinear Initial Value Problems, *Annals of Mathematics* 81:51–87 (1965).

Butkovskiy, A. G.

[Bu.1] *Distributed Control Systems*, Elsevier, New York, 1969.

Butzer, P. L., and Berens, H.

[BB.1] *Semi-Groups of Operators and Approximation*, Springer-Verlag, Berlin, Heidelberg, New York, 1967.

Caligaris, O., and Oliva, P.

[CO.1] An Existence Theorem for Lagrange Problems in Banach Spaces, *Journal of Optimization Theory and Applications* 25:83–91 (1978).

[CO.2] Non-Convex Control Problems in Banach Spaces, *Applied Mathematics and Optimization* 5:315–329 (1979).

Carrol, R. W.

[C.1] *Abstract Methods in Partial Differential Equations*, Harper and Row, New York, 1969.

Casti, J., and Ljung, L.

[CL.1] Some New Analytic and Computational Results for Operator Riccati Equations, *SIAM Journal on Control* 13:817–826 (1975).

Cavin, R. K., Tandon, S. C., and Howze, J. W.

[Ca.1] Necessary Conditions for a Class of Smooth Second Order Distributed Systems, *IEEE Transactions on Automatic Control* AC-21:424–426 (1967).

Cesari, L.

[Ce.1] Existence Theorems for Weak and Usual Optimal Solutions in Lagrange Problems with Unilateral Constraints, I & II, *Transactions of the American Mathematical Society* 124:369–412 (1966).

[Ce.2] Existence Theorems for Multidimensional Lagrange Problems, *Journal of Optimization Theory and Applications* 1:87–112 (1967).

[Ce.3] Optimization with Partial Differential Equations in Dieudonńe-Rashevsky Form and Conjugate Problems, *Archive for Rational Mechanics and Analysis* 33:339–357 (1969).

[Ce.4] Geometric and Analytic Views in Existence Theorems for Optimal Control in Banach Spaces, I, Distributed Controls, *Journal of Optimization Theory and Applications* 14:505–520 (1974).

[Ce.5] Geometric and Analytic Views in Existence Theorems for Optimal Control in Banach Spaces, II, Distributed and Boundary Controls, *Journal of Optimization Theory and Applications* 15:467–497 (1975).

[Ce.6] Geometric and Analytic Views in Existence Theorems for Optimal Control in Banach Spaces, III, Weak Solution, *Journal of Optimization Theory and Applications* 19:185–214 (1976).

[Ce.7] Convexity of the Range of Certain Integrals, *SIAM Journal on Control* 13:666–676 (1975).

Chen, G.

[Ch.1] Energy Decay Estimates and Control Theory for the Wave Equation in a Bounded Domain, Ph.D. Dissertation, University of Wisconsin, 1977.

[Ch.2] Control and Stabilization for the Wave Equation in a Bounded Domain, *SIAM Journal on Control* 17:66–81 (1979).

Chewning, W. C.

[Che.1] Controllability of the Nonlinear Wave Equation in Several Space Variables, *SIAM Journal on Control* 14:19–25 (1976).

Cirina, M. A.

[Ci.1] Boundary Controllability of Nonlinear Hyperbolic Systems, *SIAM Journal on Control* 7:198–212 (1969).

Clarke, B. M. N.

[Cl.1] Boundary Controllability of Linear Symmetric Hyperbolic Systems, *Journal of the Institute of Mathematics and its Applications* 20:283–298 (1977).

Clarke, B. M. N., and Williamson, D.
[CW.1] Control Canonical Forms and Eigenvalue Assignment by Feedback for a Class of Linear Hyperbolic Systems, *SIAM Journal on Control and Optimization* (to be published, 1981).

Cole, J. K.
[Co.1] A Selector Theorem in Banach Spaces, *Journal of Optimization Theory and Applications* 7:170–172 (1971).

Conti, R.
[Con.1] On Some Aspects of Linear Control Theory, in *Mathematical Theory of Control* (edited by A. V. Balakrishnan and L. W. Neustadt), Academic Press, New York, 1967.
[Con.2] *Linear Differential Equations and Control*, Academic Press, New York, 1976.

Courant, R., and Hilbert, D.
[CH.1] *Methods of Mathematical Physics*, Vol. 1, John Wiley, New York, 1953.
[CH.2] *Methods of Mathematical Physics*, Vol. II: *Partial Differential Equations*, Interscience, New York, 1962.

Curtain, R. F., and Pritchard, A. J.
[CP.1] An Abstract Theory for Unbounded Control Action for Distributed Parameter Systems, *SIAM Journal on Control and Optimization* 15:566–611 (1977).

Datko, R.
[D.1] Measurability Properties of Set-Valued Mappings in a Banach Space, *SIAM Journal on Control* 8:226–238 (1970).
[D.2] A Linear Control Problem in an Abstract Hilbert Space, *Journal of Differential Equations* 9:346–359 (1971).

Da Prato, G.
[Da.1] Equations d'évolution dans des algèbres d'operateurs et applications, *Journal de mathématiques pures et appliquées* 48:59–107 (1969).

De Julio, S.
[De.1] On the Optimization of Infinite Dimensional Linear Systems, *Computing Methods in Optimization Problems—2* (edited by L. A. Zadeh, L. W. Neustadt, and A. V. Balakrishnan), Academic Press, New York, 1969.

Delfour, M. C., and Mitter, S. K.
[DM.1] Controllability Observability for Infinite Dimensional Systems, *SIAM Journal on Control* 10:329–333 (1972).

Di Pillo, G., and Grippo, L.
[DG.1] The Multiplier Method for Optimal Control Problems of Parabolic System, *Applied Mathematics and Optimization* 5:253–269 (1979).

Dunford, N., and Schwartz, J. T.
[DS.1] *Linear Operators*, Part 1, John Wiley, New York, 1958.

Egorov, A. I.
[E.1] Optimal Processes in Distributed Parameter Systems and Certain Problems in Invariance Theory, *SIAM Journal on Control* 4:601–661 (1966).

Egorov, Yu. V.

[Eg.1] Some Problems in the Theory of Optimal Control, *Soviet Mathematics* 3:1080–1084 (1962).

Falb, P.

[F.1] Infinite Dimensional Control Problems, I, On the Closure of the Set of Attainable States for Linear Systems, *Journal of Mathematical Analysis and Applications* 9:12–22 (1964).

Fattorini, H. O.

[Fa.1] Some Remarks on Complete Controllability, *SIAM Journal on Control* 4:686–694 (1966).

[Fa.2] Control in Finite Time of Differential Equations in Banach Space, *Communication on Pure and Applied Mathematics* 19:17–34 (1966).

[Fa.3] On Complete Controllability of Linear Systems, *Journal of Differential Equations* 3:391–402 (1967).

[Fa.4] A Remark on the Bang-Bang Principle for Linear Control Systems in Infinite-Dimensional Space, *SIAM Journal on Control* 6:109–113 (1968).

[Fa.5] Boundary Control Systems, *SIAM Journal on Control* 6:349–385 (1968).

[Fa.6] The Time Optimal Control Problem in Banach Spaces, *Applied Mathematics and Optimization* 1:163–188 (1974).

[Fa.7] Local Controllability of a Nonlinear Wave Equation, *Mathematical Systems Theory* 9:30–45 (1975).

[Fa.8] *The Time-Optimal Control for Boundary Control of the Heat Equation*, *Calculus of Variations and Control Theory*, Academic Press, New York, San Francisco, London, 1976.

[Fa.9] The Time Optimal Problem for Distributed Control of Systems Described by the Wave Equation, *Proceedings of Conference on Control Theory of Systems Governed by Partial Differential Equations*, *Naval Surface Weapons Centre*, *Baltimore*, Academic Press, New York, 1976, pp. 151–176.

Fattorini, H. O., and Russell, D. L.

[FR.1] Exact Controllability Theorems for Linear Parabolic Equations in One Space Dimension, *Archive for Rational Mechanics and Analysis* 43:272–292 (1971).

Fleming, W. H.

[Fl.1] Some Markovian Optimization Problems, *Journal of Mathematical Mechanics* 12:131–140 (1963).

[Fl.2] Optimal Control of Partially Observable Diffusions, *SIAM Journal on Control* 6:194–213 (1978).

[Fl.3] Optimal Continuous-Parameter Stochastic Control, *SIAM Review* 11:470–509 (1969).

Fleming, W. H., and Rishel, R. W.

[FlR.1] *Deterministic and Stochastic Optimal Control*, Springer-Verlag, Berlin, Heidelberg, New York, 1975.

Friedman, A.

[Fr.1] *Partial Differential Equations of Parabolic Type*, Prentice-Hall, Englewood Cliffs, New Jersey, 1964.

[Fr.2] Optimal Control for Parabolic Equations, *Journal of Mathematical Analysis and Applications* 18:479–491 (1967).

[Fr.3] *Partial Differential Equations*, Holt, Rinehart and Winston, New York, 1969.

Gal'chuk, L. I.

[G.1] Optimal Control of Systems Described by Parabolic Equations, *SIAM Journal on Control* 7:546–558 (1969).

Glashoff, K., and Gustafson, S. A.

[GG.1] Numerical Treatment of a Parabolic Boundary-Value Control Problem, *Journal of Optimization Theory and Applications* 19:645–663 (1976).

Goldwyn, R. M., Sriram, K. P., and Graham, M.

[GSG.1] The Optimal Control of a Linear Diffusion Process, *SIAM Journal on Control* 5:295–308 (1976).

Graham, K. D., and Russell, D. L.

[GR.1] Boundary Value Control of the Wave Equation in a Spherical Region, *SIAM Journal on Control* 13:174–196 (1975).

Henry, J.

[H.1] Étude de la controllabilité de certaines equations paraboliques non-lineaires, *Proceedings of IFIP Working Conference on Distributed Systems, Rome, June 1976.*

Hermes, H., and LaSalle, J. P.

[HL.1] *Functional Analysis and Time Optimal Control*, Academic Press, New York, 1969.

Hille, E., and Phillips, R. S.

[HP.1] *Functional Analysis and Semigroups*, American Mathematical Society Colloquim Publications, Vol. XXX1, American Mathematical Society, Providence, 1957.

Himmelberg, C. J., Jacobs, M. Q., and Van Vleck, F. S.

[HJV.1] Measurable Multifunctions, Selectors, and Filippov's Implicit Functions Lemma, *Journal of Mathematical Analysis and Applications* 25:276–284 (1969).

Holmes, R. B.

[Ho.1] *Geometric Functional Analysis and its Applications*, Springer-Verlag, Berlin, Heidelberg, New York, 1975.

Hullett, W.

[Hu.1] Optimal Estuary Aeration: An Application of Distributed Parameter Control Theory, *Applied Mathematics and Optimization* 1:20–63 (1974).

Il'in, A. M. Kalashnikov, A. S., and Oleinik, O. A.

[IKO.1] Linear Equations of the Second Order of Parabolic Type, *Russian Mathematical Surveys* 17:1–143 (1962).

Ioffe, A. D.

[Io.1] Survey of Measurable Selection Theorems: Russian Literature Supplement, *SIAM Journal on Control and Optimization* 16:728–732 (1978).

Jacobs, M. Q.

[J.1] Remarks on Some recent Extensions of Filippov's Implicit function Lemma, *SIAM Journal on Control* 5:622–627 (1967).

[J.2] Measurable Multivalued Mappings and Lusin's Theorem, *Transactions of American Mathematical Society* 134:471–481 (1968).

Jeffrey, A.

[Je.1] *Quasilinear Hyperbolic Systems and Waves*, Research Notes in Mathematics, Pitman Publishing, London, 1976.

Julio, S. D.

[Ju.1] On the Optimization of Infinite Dimensional Linear Systems, *Computing Methods in Optimization Problems—2* (edited by L. Zadeh, L. W. Neustadt, and A. V. Balakrishnan), Academic Press, New York, 1969, pp. 77–93.

Kato, T.

[K.1] Nonlinear Evolution Equations in Banach spaces, *Proceedings of Symposium on Applied Mathematics*, Vol. 17 American Mathematical Society, Providence, 1965, pp. 50–67.

[K.2] *Perturbation Theory for Linear Operators*, Springer-Verlag, Berlin, Heidelberg, New York, 1976.

Kenneth, P., Sibony, M., and Yavon, J. P.

[Ke.1] Penalization Techniques for Optimal Control Problems Governed by Partial Differential Equations, *Computing Methods in Optimization Problems—2* (edited by L. A. Zadeh, L. W. Neustadt, and A. V. Balakrishnan) Academic Press, New York, 1969, pp. 177–186.

Kim, M., and Erzberger, H.

[KE.1] On the Design of Optimum Distributed Parameter Systems with Boundary Functions, *IEEE Transactions on Automatic Control* AC-12:22–28 (1967).

Kim, M., and Gajwani, S. H.

[KG.1] A Variational Approach to Optimum Distributed Parameter Systems, *IEEE Transactions on Automatic Control* AC-13:191–193 (1968).

Kitamura, S., and Nakagiri, S.

[KN.1] Identifiability of Spatially Varying and Constant Parameters in Distributed Systems of Parabolic Type, *SIAM Journal on Control and Optimization* 15:785–802 (1977).

Knowles, G.

[Kn.1] Time Optimal Control of Infinite Dimensional Systems, *SIAM Journal on Control and Optimization* 14:919–933 (1976).

[Kn.2] Time-Optimal Control of Parabolic Systems with Boundary Conditions Involving Time Delays, (Journal of Optimization Theory and Applications) 25:563–574 (1978).

[Kn.3] Some Problems in the Control of Distributed Systems and Their Numerical Solution, *SIAM Journal on Control and Optimization* 17:5–22 (1979).

Kobayashi, T.

[Ko.1] Initial State Determination for Distributed Parameter Systems, *SIAM Journal on Control and Optimization* 14:934–944 (1976).

[Ko.2] A Well Posed Approximate Method for Initial State Determination of Discrete-Time Distributed Parameter Systems, *SIAM Journal on Control And Optimization* 15:947–958 (1977).

[Ko.3] Determination of Unknown Functions for a Class of Distributed Parameter Systems, *SIAM Journal on Control and Optimization* 17:469–476 (1979).

Komkov, V.

[Kom.1] The Optimal Control of a Transverse Vibration of a Beam, *SIAM Journal on Control* 6:401–421 (1968).

[Kom.2] Optimal Control of Vibrating Thin Plates, *SIAM Journal on Control* 8:273–304 (1970).

Krasnoselskii, M. A.

[Kr.1] *Topological Methods in the Theory of Nonlinear Integral Equations, English Translation, Macmillan, New York*, 1964 (*Gittl, Moscow*, 1956).

Kuratowski, K., and Ryll-Nardzewski, C.

[KR.1] A General Theorem on Selectors, *Bulletin de l'Académie polonaise des Sciences, série des sciences mathématiques, astronomiques et physiques*, 13:397–403 (1969).

Kushner, H. J.

[Kus.1] *Probability Methods for Approximations in Stochastic Control and for Elliptic Equations*, Academic Press, New York, 1977.

Ladyžhenskaja, O. A., Solonikov, V. A., and Ural'ceva, N.N.

[LSU.1] *Linear and Quasilinear Equations of Parabolic Type*, English Translation, American Mathematical Society, Providence, 1968.

Ladyžhenskaya, O. A., Ural'tseva, N. N.

[LU.1] Linear and Quasilinear Elliptic Equations, *Academic Press, New York*, 1968.

Lagnese, J.

[La.1] Boundary Value Control of a Class of Hyperbolic Equations, *SIAM Journal on Control* 15:973–983 (1977).

[La.2] Exact Boundary Value Controllability of a Class of Hyperbolic Equations, *SIAM Journal on Control* 16:1000–1017, (1978).

Larsen, R.

[Lar.1] *Functional Analysis*, Marcel Dekker, New York, 1973.

Lattés, R., and Lions, J. L.

[LL.1] *Méthode de quasi rèversibilitè et applications*, Dunod, Paris, 1967 (English Translation, Elsevier, New York, 1969).

Lions, J. L.

[Li.1] Optimisation pour certaines classes d'équations d'évolution non linéaires, *Annali di matematica pura ed applicata* 72:275–293 (1966).

[Li.2] *Optimal Control of Systems Governed by Partial Differential Equations*, Springer-Verlag, Berlin, Heidelberg, New York, 1971.

[Li.3] *Some Aspects of the Optimal Control of Distributed Parameter Systems*, Regional Conference Series in Applied Mathematics, SIAM, *1972*. J. W. Arrowsmith ltd, England.

[Li.4] Various Topics in the Theory of Optimal Control of Distributed Systems, Springer Lecture Notes in Economics and Mathematical Systems No. 15, Springer-Verlag, New York, 1974 pp. 166–308.

[Li.5] Various Topics in the Theory of Optimal Control of Distributed Systems, in *Economics and Mathematical Systems* (edited by B. J. Kirby), Springer Lecture Notes, No. 105, Springer-Verlag, New York, 1976, pp. 166–303.

[Li.6] Asymptotic Methods in the Optimal Control of Distributed Systems, *Automatica* 14:199–211 (1978).

Lions, J. L., and Magenes, E.

[LM.1] *Non-Homogeneous Boundary Value Problems and Applications*, I, Springer-Verlag, Berlin, Heidelberg, New York, 1972.

[LM.2] *Non-Homogeneous Boundary Value Problems and Applications*, II, Springer-Verlag, Berlin, Heidelberg, New York, 1972.

[LM.3] *Non-Homogeneous Boundary Value Problems and Applications*, III, Springer-Verlag, Berlin, Heidelberg, New York, 1972.

Lukes, D. L., and Russell, D. L.

[LR.1] The Quadratic Criterion for Distributed Systems, *SIAM Journal on Control* 7:101–121 (1969).

McCamy, R. C., Mizel, V. J., and Seidman, T. I.

[MMS.1] Approximate Boundary Controllability of the Heat Equation, *Journal of Mathematical Analysis and Applications* 23:699–703 (1968).

[MMS.2] Approximate Boundary Controllability of the Heat Equation, II, *Journal of Mathematical Analysis and Applications* 28:482–492 (1969).

McKnight, R. S., and Bosarge, W. E., Jr.

[MB.1] The Ritz-Galerkin Procedure for Parabolic Control Problems, *SIAM Journal on Control* 11:510–524 (1973).

Michel, P.

[M.1] Necessary Conditions for Optimality of Elliptic Systems with Positivity Constraints on the State, *SIAM Journal on Control and Optimization* 18:91–97 (1980).

Murat, F.

[Mu.1] Contre-examples pour divers problèmes où le controle intervient dan les coefficients, *Annali di matematica pura ed applicata* 112:49–68 (1977).

Nababan, S.

[N.1] On the Optimal control of Delay Differential Systems, Ph.D. Thesis, School of Mathematics, University of New South Wales, 1978.

Nababan, S., and Teo, K. L.

[NT.1] Necessary Conditions for Optimality of Cauchy Problems for Parabolic Partial Delay—Differential Equations, *Journal of Optimization Theory and Applications* 34(1) (to be published, 1981).

[NT.2] On the System Governed by Parabolic Partial Delay—Differential Equations with First Boundary Conditions, *Annali di matematica pura ed applicata* 119:39–57 (1979).

[NT.3] On the Existence of Optimal controls of the First Boundary Value Problems for Parabolic Delay—Differential Equations in Divergence Form, *Journal of the Mathematical Society of Japan* 32:343–362, (1980).

[NT.4] Necessary Conditions for Optimal Controls for Systems Governed by Parabolic Partial Delay—Differential Equations in Divergence Form with First Boundary Conditions, *Journal of Optimization Theory and Applications* 35(3) (to be published, 1981).

Neustadt, L. W.

[Ne.1] *Optimization: A Theory of Necessary Conditions*, Princeton University Press, Princeton, 1976.

Noussair, E. S., Nababan, S., and Teo, K. L.

[NNT.1] On the Existence of Optimal Controls for Quasi Linear Parabolic Partial Differential Equations, *Journal of Optimization Theory and Applications* 34(1) (to be published, 1981).

Oğuztöreli, M. N.

[O.1] *Time Lag Control Systems*, Academic Press, New York, 1966.

[O.2] Optimization in Distributed Parameter Control Systems: A Dynamic Programming Approach, University of Alberta, Department of Mathematics, Series A, Vol. 3, No. 28, 1967.

[O.3] On Sufficient Conditions for the Existence of Optimal Policies in Distributed Parameter Control Systems, *Accademia Nazionale dei Lincei, Rendiconti della classe di scienze fisiche, matematiche e naturali* 46:693–697 (1969).

[O.4] On the Optimal Controls in the Distributed Parameter Systems, *Rendiconti di matematica* 3:171–180 (1970).

[O.5] Optimal Controls in Distributed Parameter Control Systems, *Rendiconti di matematica* 4:55–74 (1971).

Parathasarathy, K. R.

[P.1] *Probability Measures on Metric Spaces*, Academic Press, New York, 1967.

Pierce, A.

[Pi.1] Unique Identification of Eigenvalues and Coefficients in a Parabolic Problem, *SIAM Journal on Control and Optimization* 17:494–499 (1979).

Pliś, A.

[Pl.1] Remark on Measurable Set-Valued Functions, *Bulletin de l'Acàdemie, Polonaise des Sciences, série des sciences mathématiques, astronomiques et physiques* 9:857–859 (1961).

Pontryagin, L. S., Boltyanskii, V. G. Gamkrelidze, R. V., and Mishchenko, E. F.

[PBGM.1] *The Mathematical Theory of Optimal Processes*, Interscience, New York, 1962.

Pulvirenti, G.

[Pu.1] Existence Theorems for an Optimal Control Problem Relative to a Linear, Hyperbolic Partial Differential Equation, *Journal of Optimization Theory and Applications* 7:109–117 (1971).

Pulvirenti, G., and Santagati, G.

[PS.1] On the Uniqueness of the Solution to a Minimum Problem for a Control Process with Distributed Parameters, *Journal of Optimization Theory and Applications* 28:109–124 (1979).

Quinn, J. P.

[Q.1] Time Optimal Control of Linear Distributed Parameter Systems, Ph.D. Thesis, Department of Mathematics, University of Wisconsin, Madison, 1969.

Reid, D. W

[R.1] On the Conputational Methods of Optimal Control Problems of Distributed Parameter Systems, Ph.D thesis, School of Mathematics, University of New South Wales, Australia, 1980.

Reid, D. W., and Teo, K. L.

[RT.1] Optimal Parameter Selection of Parabolic Systems, *Mathematics of Operations Research* (to appear).

[RT.2] On the Existence of Optimal Control for Systems governed by Parabolic Partial Differential Equations with Cauchy Boundary Conditions, *Annali di matematica pura ed applicata* (to appear).

[RT.3] First-Order Strong Variation Algorithm for a Class of Distributed Optimal Control Problems, *Applied Mathematics and Optimization* (to appear).

Robinson, A. C.

[Ro.1] A Survey of Optimal Control of Distributed Parameter Systems, *Automatica* 7:371–388 (1971).

Russell, D. L.

[Ru.1] Optimal Regulation of Linear Symmetric Hyperbolic Systems with Finite Dimensional Controls, *SIAM Journal on Control* 4:475–509 (1966).

[Ru.2] Nonharmonic Fourier Series in Control Theory of Distributed Parameter Systems, *Journal of Mathematical Analysis and Applications* 18:542–560 (1967).

[Ru.3] *On Boundary-Value Controllability of Linear Symmetric Hyperbolic Systems*, *Mathematical Theory of Control*, Academic Press, New York, 1967.

[Ru.4] Boundary Value Control Theory of the Higher Dimensional Wave Equation, *SIAM Journal on Control* 9:29–42 (1971).

[Ru.5] Boundary Value Control Theory of the Higher Dimensional Wave Equation, Part II, *SIAM Journal on Control* 9:401–419 (1971).

[Ru.6] Control Theory of Hyperbolic Equations Related to Certain Questions in Harmonic Analysis and Spectral Theory, *Journal of Mathematical Analysis and Applications* 40:336–368 (1972).

[Ru.7] A Unified Boundary Controllability Theory for Hyperbolic and Parabolic Partial Differential Equations, *Studies in Applied Mathematics* 52:189–211 (1973).

[Ru.8] Quadratic Performance Criteria in Boundary Control of Linear Symmetric Hyperbolic Systems, *SIAM Journal on Control* 11:475–509 (1973).

[Ru.9] Exact Boundary Value Controllability Theorems for Wave and Heat Processes in star-Complemented Regions, in *Differential Games and Control Theory* (edited by Roxin, Liu, and Sternberg), Marcel Dekker, New York, 1974.

[Ru.10] Controllability and Stabilizability Theory for Linear Partial Differential Equations: Recent Progress and Open Questions, *SIAM Review* 20:639–739 (1978).

Sakawa, Y.

[S.1] Solution of an Optimal Control Problem in a Distributed Parameter System, *IEEE Transactions on Automatic Control* AC-9:420–426 (1964).

[S.2] Optimal Control of a Certain Type of Linear Distributed Parameter Systems, *IEEE Transactions on Automatic Control* AC-11:35–41 (1966).

[S.3] Controllability for Partial Differential Equations of Parabolic Type, *SIAM Journal on Control* 12:389–400 (1974).

[S.4] Observability and Related Problems for Partial Differential Equations of Parabolic Type, *SIAM Journal on Control* 13:14–27 (1975).

Santagati, G.

[Sa.1] On an Optimal Control Problem for a Hyperbolic Partial Differential Equation, *Journal of Optimization Theory and Applications* 7:1–9 (1971).

Saul'yev, V. K.

[Sau.1] Integration of Equations of Parabolic Type by the Method of Nets, English Translation, Pergamon Press, Oxford, 1964 (Fizmatgiz, Moscow, 1960).

Schittkowski, K.

[Sc.1] Numerical Solution of a Time-Optimal Parabolic Boundary-Value Control Problem, *Journal of Optimization Theory and Applications* 27:271–290 (1979).

Schmaedeke, W.

[Sch.1] Mathematical Theory of Optimal Control for Semilinear Hyperbolic Systems in Two Independent Variables, *SIAM Journal on Control* 5:138–152 (1967).

Seidman, T. I.

[Se.1] *Boundary Observation and Control for the Heat Equation*, *Calculus of Variations and Control Theory*, Academic Press, New York, 1976.

Shin, Y. P., Chu, C. K.

[SC.1] On Optimal Feedback Control of a Class of Linear Distributed Parameter Systems, *Journal of Optimization Theory and Applications* 16: 327–338 (1975).

Slemrod, M.

[Sl.1] A Note on Complete Controllability and Stabilization for Linear Control Systems in Hilbert Space, *SIAM Journal on Control* 12:500–508 (1974).

Sobolev, S. L.

[So.1] *Applications of Functional Analysis in Mathematical Physics*, American Mathematical Society, Providence, 1963.

Stroock, D. W., and Varadan, S. R. S.

[SV.1] *Multidimensional Diffusion Processes*, Springer-Verlag, Berlin, Heidelberg, New York, 1979.

Suryanarayana, M. B.

[Su.1] Optimization Problems with Hyperbolic Partial Differential Equations, Ph.D. Thesis, Department of Mathematics, The University of Michigan, 1969.

[Su.2] Necessary Conditions for Optimization Problems with Hyperbolic Partial Differential Equations, *SIAM Journal on Control* 11:130–147 (1973).

[Su.3] Existence Theorems for Optimization Problems Concerning Hyperbolic Partial Differential Equations, *Journal of Optimization Theory and Applications* 15:361–392 (1975).

[Su.4] Existence Theorems for Optimization Problems Concerning Linear, Hyperbolic Partial Differential Equations without Convexity Conditions, *Journal of Optimization Theory and Applications* 19:47–61 (1976).

Temam, R.

[T.1] Sur l'equation de Riccati associée a des opérateurs nonbornés, *Journal of Functional Analysis* 7:85–115 (1971).

Teo, K. L.

[Te.1] Optimal Control of Systems Governed by Ito Stochastic Differential Equations, Ph.D. Thesis, Department of Electrical Engineering, University of Ottawa, 1974.

[Te.2] Optimal Control of Systems Governed by Time Delayed Second Order Linear Parabolic Partial Differential Equations with a First Boundary Condition, *Journal of Optimization Theory and Applications* 29:437–481 (1979).

[Te.3] Existence of Optimal Controls for Systems Governed by Second Order Linear Parabolic Partial Delay-Differential Equations with First Boundary Conditions, *Journal of Australian Mathematical Society*, *Series B* 21:21–36 (1979).

Teo, K. L., and Ahmed, N. U.

[TA.1] Optimal Feedback Control for a Class of Stochastic Systems, *International Journal of Systems Science* 5:357–365 (1974).

[TA.2] On the Optimal Control of Systems Governed by Quasilinear Integro-Partial Differential Equation of Parabolic Type, *Journal of Mathematical Analysis and Applications* 59:33–59 (1977).

[TA.3] On the Optimal Controls of a Class of Systems Governed by Second Order Parabolic Partial Delay-Differential Equations with First Boundary Conditions, *Annali di matematica pura ed applicata* 122:61–82 (1979).

Teo, K. L., Ahmed, N. U., and Wong, H. W.

[TAW.1] On Optimal Parameter Selection for Parabolic Differential Systems, *IEEE Transactions on Automatic Control* AC-19:286–287 (1974).

Teo, K. L., Reid, D. W., and Boyd, I. E.

[TRB.1] Stochastic Optimal Control Theory and Its Computational Methods, *International Journal of System Sciences* 11:77–95 (1980).

Tricomi, F. G.

[Tr.1] *Integral Equations*, Interscience, New York, 1959.

Tricario, M.

[Tri.1] Problemi al contorno per le equazioni ellittico et paraboliche di ordine 2m, *Annali di matematica pura ed applicata* 117:225–270 (1977).

Triggiani, R.

[Trig.1] A Note on the Lack of Exact Controllability and Optimization, *SIAM Journal on Control and Optimization* 15:407–411 (1977).

[Trig.2] On the Relationship Between First and Second Order Controllable Systems in Banach Spaces. *SIAM Journal on Control and Optimization* 16:847–859 (1978).

Vainberg, M. M.

[V.1] *Variational Methods for the Study of Nonlinear Operators*, English Translation, Holden-Day, San Francisco, 1964 (Moscow, 1956).

Vinter, R.B., and Johnson, T. L.

[VJ.1] Optimal Control of Nonsymmetric Hyperbolic Systems in N Variables on the Half-Space, *SIAM Journal on Control and Optimization* 15:129–143 (1977).

Wagner, D.

[W.1] Survey of Measurability of Set Valued Maps, *SIAM Journal on Control and Optimization* 15:857–903 (1977).

Wang, P. K. C.

[Wa.1] Optimal Control of Distributed Parameter Systems with Time Delays, *IEEE Transactions on Automatic Control* AC-9:13–22 (1964).

[Wa.2] Control of Distributed Parameter Systems, in *Advances in Control Systems, Theory and Applications* (edited by C. T. Leondes), Vol. 1, Academic Press, New York, 1964, pp. 75–172.

[Wa.3] On the Feedback Control of Distributed Parameter Systems, *Internations Journal on Control* 3:255–272 (1966).

[Wa.4] Optimal Control of a Class of Linear Symmetric Hyperbolic Systems with Applications to Plasma Confinement, *Journal of Mathematical Analysis and Applications* 28:594–608 (1969).

[Wa.5] Optimal Control of Parabolic Systems with Boundary Conditions Involving Time Delays, *SIAM Journal on Control* 13:274–293 (1975).

[Wa.6] Time Optimal Control of Time-Lag Systems with Time-Lag Controls, *Journal of Mathematical Analysis and Applications* 52:366–378 (1975).

Wang, P. K. C., and Tung, F.

[WT.1] Optimal control of Distributed Systems, *Transactions of ASME, Journal of Basic Engineering, Series D* 86:67–79 (1964).

Warga, J.

[War.1] *Optimal Control of Differential and Functional Equations*, Academic Press, New York, 1972.

Washburn, D.

[Was.1] A Bound on the Boundary Input Map for Parabolic Equations with Applications to Time Optimal Control, *SIAM Journal on Control and Optimization* 17:652–671 (1979).

White, L. W.

[Wh.1] Control of a Pseudo-Parabolic Initial-Value Problem to a Target Function, *SIAM Journal on Control and Optimization* 17:587–595 (1979).

Willard, S.

[Wi.1] *General Topology*, Addison-Wesley, Reading, 1970.

Wong, H. W.

[Wo.1] Optimal Control of Stochastic Differential Systems with Applications, Ph.D. Thesis, Department of Electrical Engineering, University of Ottawa, 1980.

Wong, H. W., and Ahmed, N. U.

[WA.1] Gradient Method for Computing Optimal Control for Stochastic Ito Differential Equations (submitted for publication).

Yavon, J. P.

[Y.1] *Cahiers, IRIA* 2 (1969).

Yeh, H. H., and Tou, J. T.

[YT.1] Optimal Control of a Class of Distributed-Parameter Systems, *IEEE Transaction on Automatic Control* AC-12:29–37 (1967).

Yosida, K.

[Yo.1] *Functional Analysis*, 2nd ed., Springer-Verlag, Berlin, Heidelberg, New York, 1968.

Zolezzi, T.

[Z.1] Teoremi d'existenza per problemi di controllo optimo retti da equazioni ellittiche o paraboliche, *Reddiconti del seminario matematico dell'Universita di Padova* 44:155–195 (1970).

[Z.2] Necessary Conditions for Optimal Control of Elliptic or Parabolic Problems, *SIAM Journal on Control* 10:594–607 (1972).

INDEX

T

U

V

W